T0178390

Lecture Notes in Computer Science 14428

Founding Editors

Gerhard Goos
Juris Hartmanis

Editorial Board Members

The series Lecture Notes in Computer Science (LNCS), including its subseries Lecture Notes in Artificial Intelligence (LNAI) and Lecture Notes in Bioinformatics (LNBI), has established itself as a medium for the publication of new developments in computer science and information technology research, teaching, and education.

LNCS enjoys close cooperation with the computer science R & D community, the series counts many renowned academics among its volume editors and paper authors, and collaborates with prestigious societies. Its mission is to serve this international community by providing an invaluable service, mainly focused on the publication of conference and workshop proceedings and postproceedings. LNCS commenced publication in 1973.

Qingshan Liu · Hanzi Wang · Zhanyu Ma ·
Weishi Zheng · Hongbin Zha · Xilin Chen ·
Liang Wang · Rongrong Ji
Editors

Pattern Recognition and Computer Vision

6th Chinese Conference, PRCV 2023
Xiamen, China, October 13–15, 2023
Proceedings, Part IV

Springer

Editors
Qingshan Liu (iD)
Nanjing University of Information Science
and Technology
Nanjing, China

Zhanyu Ma (iD)
Beijing University of Posts
and Telecommunications
Beijing, China

Hongbin Zha (iD)
Peking University
Beijing, China

Liang Wang
Chinese Academy of Sciences
Beijing, China

Hanzi Wang (iD)
Xiamen University
Xiamen, China

Weishi Zheng (iD)
Sun Yat-sen University
Guangzhou, China

Xilin Chen (iD)
Chinese Academy of Sciences
Beijing, China

Rongrong Ji (iD)
Xiamen University
Xiamen, China

ISSN 0302-9743   ISSN 1611-3349 (electronic)
Lecture Notes in Computer Science
ISBN 978-981-99-8461-9   ISBN 978-981-99-8462-6 (eBook)
https://doi.org/10.1007/978-981-99-8462-6

Preface

Welcome to the proceedings of the Sixth Chinese Conference on Pattern Recognition and Computer Vision (PRCV 2023), held in Xiamen, China.

PRCV is formed from the combination of two distinguished conferences: CCPR (Chinese Conference on Pattern Recognition) and CCCV (Chinese Conference on Computer Vision). Both have consistently been the top-tier conference in the fields of pattern recognition and computer vision within China's academic field. Recognizing the intertwined nature of these disciplines and their overlapping communities, the union into PRCV aims to reinforce the prominence of the Chinese academic sector in these foundational areas of artificial intelligence and enhance academic exchanges. Accordingly, PRCV is jointly sponsored by China's leading academic institutions: the Chinese Association for Artificial Intelligence (CAAI), the China Computer Federation (CCF), the Chinese Association of Automation (CAA), and the China Society of Image and Graphics (CSIG).

PRCV's mission is to serve as a comprehensive platform for dialogues among researchers from both academia and industry. While its primary focus is to encourage academic exchange, it also places emphasis on fostering ties between academia and industry. With the objective of keeping abreast of leading academic innovations and showcasing the most recent research breakthroughs, pioneering thoughts, and advanced techniques in pattern recognition and computer vision, esteemed international and domestic experts have been invited to present keynote speeches, introducing the most recent developments in these fields.

PRCV 2023 was hosted by Xiamen University. From our call for papers, we received 1420 full submissions. Each paper underwent rigorous reviews by at least three experts, either from our dedicated Program Committee or from other qualified researchers in the field. After thorough evaluations, 522 papers were selected for the conference, comprising 32 oral presentations and 490 posters, giving an acceptance rate of 37.46%. The proceedings of PRCV 2023 are proudly published by Springer.

Our heartfelt gratitude goes out to our keynote speakers: Zongben Xu from Xi'an Jiaotong University, Yanning Zhang of Northwestern Polytechnical University, Shutao Li of Hunan University, Shi-Min Hu of Tsinghua University, and Tiejun Huang from Peking University.

We give sincere appreciation to all the authors of submitted papers, the members of the Program Committee, the reviewers, and the Organizing Committee. Their combined efforts have been instrumental in the success of this conference. A special acknowledgment goes to our sponsors and the organizers of various special forums; their support made the conference a success. We also express our thanks to Springer for taking on the publication and to the staff of Springer Asia for their meticulous coordination efforts.

We hope these proceedings will be both enlightening and enjoyable for all readers.

October 2023

Qingshan Liu
Hanzi Wang
Zhanyu Ma
Weishi Zheng
Hongbin Zha
Xilin Chen
Liang Wang
Rongrong Ji

Organization

General Chairs

Hongbin Zha — Peking University, China
Xilin Chen — Institute of Computing Technology, Chinese Academy of Sciences, China
Liang Wang — Institute of Automation, Chinese Academy of Sciences, China
Rongrong Ji — Xiamen University, China

Program Chairs

Qingshan Liu — Nanjing University of Information Science and Technology, China
Hanzi Wang — Xiamen University, China
Zhanyu Ma — Beijing University of Posts and Telecommunications, China
Weishi Zheng — Sun Yat-sen University, China

Organizing Committee Chairs

Mingming Cheng — Nankai University, China
Cheng Wang — Xiamen University, China
Yue Gao — Tsinghua University, China
Mingliang Xu — Zhengzhou University, China
Liujuan Cao — Xiamen University, China

Publicity Chairs

Yanyun Qu — Xiamen University, China
Wei Jia — Hefei University of Technology, China

Local Arrangement Chairs

Xiaoshuai Sun	Xiamen University, China
Yan Yan	Xiamen University, China
Longbiao Chen	Xiamen University, China

International Liaison Chairs

Jingyi Yu	ShanghaiTech University, China
Jiwen Lu	Tsinghua University, China

Tutorial Chairs

Xi Li	Zhejiang University, China
Wangmeng Zuo	Harbin Institute of Technology, China
Jie Chen	Peking University, China

Thematic Forum Chairs

Xiaopeng Hong	Harbin Institute of Technology, China
Zhaoxiang Zhang	Institute of Automation, Chinese Academy of Sciences, China
Xinghao Ding	Xiamen University, China

Doctoral Forum Chairs

Shengping Zhang	Harbin Institute of Technology, China
Zhou Zhao	Zhejiang University, China

Publication Chair

Chenglu Wen	Xiamen University, China

Sponsorship Chair

Yiyi Zhou	Xiamen University, China

Exhibition Chairs

Bineng Zhong   Guangxi Normal University, China
Rushi Lan   Guilin University of Electronic Technology, China
Zhiming Luo   Xiamen University, China

Program Committee

Baiying Lei   Shenzhen University, China
Changxin Gao   Huazhong University of Science and Technology,
                          China
Chen Gong   Nanjing University of Science and Technology,
                          China
Chuanxian Ren   Sun Yat-Sen University, China
Dong Liu   University of Science and Technology of China,
                          China
Dong Wang   Dalian University of Technology, China
Haimiao Hu   Beihang University, China
Hang Su   Tsinghua University, China
Hui Yuan   School of Control Science and Engineering,
                          Shandong University, China
Jie Qin   Nanjing University of Aeronautics and
                          Astronautics, China
Jufeng Yang   Nankai University, China
Lifang Wu   Beijing University of Technology, China
Linlin Shen   Shenzhen University, China
Nannan Wang   Xidian University, China
Qianqian Xu   Key Laboratory of Intelligent Information
                          Processing, Institute of Computing
                          Technology, Chinese Academy of Sciences,
                          China
Quan Zhou   Nanjing University of Posts and
                          Telecommunications, China
Si Liu   Beihang University, China
Xi Li   Zhejiang University, China
Xiaojun Wu   Jiangnan University, China
Zhenyu He   Harbin Institute of Technology (Shenzhen), China
Zhonghong Ou   Beijing University of Posts and
                          Telecommunications, China

Contents – Part IV

Performance Evaluation and Benchmarks

Remote Sensing Image Interpretation

Pattern Classification and Cluster Analysis

Shared Nearest Neighbor Calibration for Few-Shot Classification

Rundong Qi, Sa Ning, Yong Jiang$^{(\boxtimes)}$, Yuwei Zhang, and Wenyu Yang

School of Computer Science and Technology, Southwest University of Science and Technology, Mianyang 621010, China
jiang_yong@swust.edu.cn

Abstract. Few-shot classification aims to classify query samples using very few labeled examples. Most existing methods follow the Prototypical Network to classify query samples by matching them to the nearest centroid. However, scarce labeled examples tend to bias the centroids, which leads to query samples matching the wrong centroids. In this paper, we address the mismatching problem of centroids and query samples by optimizing the matching strategy. The idea is to combine the Shared Nearest Neighbor similarity with cosine similarity proportionally to calibrate the matching results of the latter. Furthermore, we also improve a bias-diminishing approach to increase the number of shared nearest neighbors between query samples and the centroid of their class. We validate the effectiveness of our method with extensive experiments on three few-shot classification datasets: *mini*ImageNet, *tiered*ImageNet, and CUB-200-2011 (CUB). Moreover, our method has achieved competitive performance across different settings and datasets.

Keywords: Few-shot classification · metric learning · transductive

1 Introduction

Few-shot classification [1–9] [?] aims to classify unseen samples with few labeled examples. This is a challenging problem because learning a neural network with few examples is prone to overfitting. A mainstream framework for few-shot classification is based on meta-learning or learning-to-learn. Among the studies that have achieved remarkable results, metric-based methods adopting the meta-learning framework have attracted much research attention due to their malleability and effectiveness.

Metric-based few-shot classification methods [3–6] [?,?] mostly consist of a feature encoder E and a metric function M. Given a few-shot task consisting of labeled examples (support set X_s) and unlabeled samples (query set X_q), E extracts their features. M calculates the centroid of each class and predicts the categories for unlabeled query samples by matching the nearest centroid

This study is supported by the Sichuan Science and Technology Program (NO. 2021YFG0031).

(nearest-centroid matching strategy). It supposes that each query sample and the centroid closest to it belong to the same class. However, since there are very few support examples, the obtained centroids are inevitably biased, thus preventing some query samples from correctly classifying.

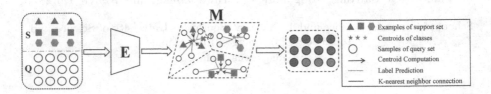

Fig. 1. Problem formulation.

As illustrated in Fig. 1, there are only three labeled examples in each class, and the centroids derived from them can hardly satisfy the assumption of the nearest-centroid matching strategy. As a result, the query sample "?" located at the boundary between the blue and green classes is assigned a blue label because it is closer to the blue centroid, while the true label of query sample "?" is green. Hence, the matching strategy between query samples and centroids needs to be optimized.

Many transductive inference methods have been proposed to optimize the matching strategy with task-level information [10–14]. BD-CSPN [10] proposes and verifies the existence of intra-class bias and cross-class bias and diminishes these two biases using label propagation and feature shifting, respectively. Proto-Com [12] utilizes the extra primitive knowledge of categories and visual attribute features to learn to complement the original centroids. LaplacianShot [13] minimizes a Laplacian-regularization objective to encourage neighboring unlabeled samples to have potentially identical label assignments. RDC-FT [14] uses the ranking of query samples in a task to calibrate the pairwise distances. However, none utilize the Shared Nearest Neighbors (SNN) to improve the matching strategy.

Our key observation is that the neighborhood overlap of two points in the same category tends to be larger than two points from different categories. As illustrated in Fig. 1, if we find $k-$nearest neighbors among all centroids and query samples for query sample "?", blue centroid and green centroid, respectively, we will find that the centroid sharing the greatest number of nearest neighbors with query sample "?" is green. And SNN measures similarity by overlapping the neighborhoods of two different samples. Thus, we proportionally combine the SNN similarity with cosine similarity to calibrate the biased results of the nearest-centroid matching strategy. Nevertheless, due to the diversity of image feature distributions, the weight between SNN and cosine similarity may require careful and tedious hand-tuning. Because of this, we develop a learning-to-learn algorithm to search for the weight hyperparameters. However, the centroids are still biased and decrease the number of nearest neighbors shared between centroids and query samples of the same category. To further diminish the bias, we also improve a centroid rectification method proposed in [10].

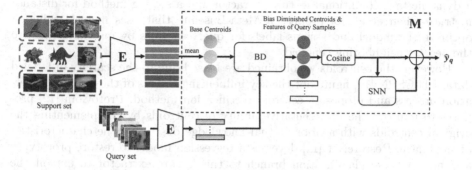

Fig. 2. Method overview. $\oplus$ denotes a weighted summation.

Our three main contributions are summarized below:

- To the best of our knowledge, we introduce SNN into the few-shot classification for the first time. Our work shows that SNN can help to improve classification performance.
- We combine SNN and cosine similarity proportionally to calibrate the matching results of the latter. Moreover, we adopt a learning-to-learn approach to search for the weight hyperparameters. We also develop a bias-diminishing method to increase the shared nearest neighbors between query samples and the centroid in the same class.
- We conduct extensive experiments and ablation studies on the four few-shot classification benchmarks (i.e., *mini*ImageNet, *tiered*ImageNet, CUB and *mini* → CUB). The experimental results show that our method can achieve competitive performance across different settings and datasets.

2   Related Work

Few-Shot Classification. The key idea of few-shot classification is to identify unlabeled samples with few labeled examples. Early attempt [2] introduces the concept of one-shot learning and proposes a variational Bayesian framework to solve this issue. Recently, researchers have made many attempts to address the problem of few-shot classification. There are two main streams of the existing few-shot classification methods: optimization-based methods and metric-based methods. Optimization-based methods [15–19] aim to improve the model's generalization ability, enabling it to adapt to unseen few-shot classification tasks efficiently. Metric-based methods predict categories for unlabeled query samples by calculating the similarity between unlabeled samples and labeled examples.

Metric-based methods are more malleable and effective than optimization-based approaches. Prototypical Network [4] proposes to use the mean vectors (centroids) of each class in the support set to represent the classes and classify query samples by matching them to the nearest centroid. TADAM [5] proposes a dynamic task-conditional feature extractor and a scaling method for distance metrics. Chen et al. [6] propose a Meta-Baseline that uses meta-learning to optimize the model and assigns labels for query samples by matching them to the nearest neighboring centroid.

However, the centroids we obtained are biased due to the scarcity of labeled data. BD-CSPN [10] figure out the key influencing factors of the centroid generation process and propose a centroid rectification method. ProtoCom [12] proposes to improve the representative ability of centroids by complementing the original centroids with a prior textual knowledge describing the characteristics of the image. RestoreNet [11] develops a regression model to restore prototypes and adds a label classification branch to the feature extractor to exploit the query samples by self-training. LaplacianShot [13] encourages neighboring samples in the feature space to have potentially identical label assignments using the Laplacian regularization term. Very recently, RDC-FT [14] uses the ranking of distance in a task to calibrate the pairwise distances.

Shared Nearest Neighbors. Sheared Nearest Neighbors (SNN) was first proposed in [20], which defines the number of nearest neighbors common to two points as the similarity, and the two points are similar only when they are one of the k nearest-neighbors of each other. [21] summarizes that when the points of different clusters are pairwise and stable, and the size of the nearest-neighbor query is equal to the size of the cluster, the nearest neighbors of the query point tend to be in the same cluster (class) as it. [22] finds that shared nearest neighbor similarity performs more steadily in high-dimensional data spaces than their related primary distance measures.

In this work, we try to tackle the problem of the biased centroid matching results in few-shot classification using SNN similarity. And we mention that rarely studies have used SNN in computer vision, and the SNN-based approaches we have found so far are solving problems in data mining.

3   Method

3.1   Preliminaries

Metric-based Few-Shot Classification Method. In the standard few-shot setup, we are given two datasets: the base dataset D_{base} and the novel dataset D_{novel} that do not intersect. The base dataset $D_{base} = \{x_i, y_i\}_{i=1}^O$, where x_i denotes the i-th sample in the D_{base} and y_i denotes the label of x_i. The goal of few-shot learning is to utilize the experience learned on D_{base} to perform few-shot tasks on D_{novel}. Each few-shot task T contains $N \times K + N \times Q$ objects (N-way K-shot and Q query), where N represents the number of categories. K and Q denote the number of labeled support examples and unlabeled query

---

Algorithm 1. Intra-domain Bias Diminution Algorithm

Input: threshold t, centroids V, query samples' features $\mathbb{F}_q$
Output: Rectified centroids V'

1: Initialize $i = 1$
2: **repeat**
3:     Calculate $W(f_i, v)$ using (6), where $f_i \in \mathbb{F}_q$ and $v \in V$;
4:     Find the top two scores: S_{1st} and S_{2nd};
5:     **if** $S_{1st} - S_{2nd} > t$:
        Assign the label of v in S_{1st} to f_i as a pseudo-label;
6:     $i = i + 1$
7: **until** $i = |\mathbb{F}_q|$
8: Initialize $l = 1$
9: **repeat**
10:    Calculate v'_l using (7);
11:    $l = l + 1$
12: **until** $l = N$
13: $V' = \left\{ v'_1, ..., v'_N \right\}$

---

samples in a class. The goal of the few-shot task is to use $N \times K$ labeled examples $x_s \in X_s$ to classify $N \times Q$ unlabeled samples $x_q \in X_q$.

Metric-based methods usually adopt the meta-learning framework. And such metric-based methods consist of two training stages: pre-training and meta-learning. In the pre-training stage, we train a convolutional neural network with D_{base} and remove the last full-connected layer to obtain the feature extractor E. In the meta-learning stage, we compute the centroids of the few-shot task and classify query samples in this task using the metric function M, which consists of a nearest neighbors classifier and a cosine similarity function. The meta-learning stage can be viewed as adapting E to the validation stage using the few-shot tasks sampled from the D_{base}. Specifically, E extracts the features, M then takes the extracted features as input and computes the centroid v_i:

$$v_i = \frac{1}{K} \times \sum_{x_s \in \mathbb{X}_s^i} E(x_s), \tag{1}$$

where $\mathbb{X}_s^i$ represents the set of support examples in class i. Subsequently, M predicts the probability for each query sample:

$$p(y = i \mid x_q) = \frac{exp(sim < E(x_q), v_i >)}{\sum_{j=1}^{N} exp(sim < E(x_q), v_j >)}, \tag{2}$$

where $sim < \cdot, \cdot >$ denotes the similarity of two vectors. The loss function is the cross-entropy loss. We evaluate the model's performance using the few-shot tasks sampled from C_{novel}.

Shared Nearest Neighbor. The basic form of SNN is "overlap". For a few-shot task, we first integrate query samples' feature vectors $\mathbb{F}_q = E(X_q)$ and centroids $V = \{v_1, ..., v_N\}$ into a set O and find k-nearest neighbors for each object $o_i \in O$ according to the cosine similarity. The overlap of objects o_i and o_j is computed as follows:

$$SNN_k(o_i, o_j) = \left| NN_k(o_i) \bigcap NN_k(o_j) \right|, \tag{3}$$

where $NN_k(o) \subseteq O$ represents the k-nearest neighbors to o, $k \in N^+$. As long as o_i and o_j are not both in each other's k-nearest neighbors, the $SNN_k(o_i, o_j)$ will be set to 0. Moreover, we set $k = Q + 1$ (1 centroid and Q query features per class).

3.2 Shared Nearest Neighbor Calibration (SNNC)

Mostly, M matches the nearest centroid for each query feature. Although this matching strategy is suitable for most cases, it performs poorly when processing query features locate at the inter-class junctions. To solve this problem, we propose SNNC. As shown in Fig. 2, SNNC consists of a bias diminishment (BD) module and a similarity calculation function that uses a combination of cosine similarity and SNN similarity to match centroids for query features:

$$sim < f_q, v_i >= \alpha SNN(f_q, v_i) + \beta cosine(f_q, v_i), \tag{4}$$

where $cosine(\cdot, \cdot)$ represents the cosine similarity of the two vectors, $f_q \in \mathbb{F}_q$, and $\alpha, \beta \in R^+$. To better balance the two metrics and avoid the exhaustive hand-tuning process, we set α and β as learnable parameters and search for suitable values for them in the meta-learning stage.

Since SNN requires both objects to be in each other's k-nearest neighbors, leading to a sparse SNN similarity matrix. To fully use the neighborhood overlap information to calibrate the matching results, we remove this restriction from SNN.

Bias Diminishing. Although the matching strategy is improved, the centroids are still biased. Intuitively, the process of bias diminishing can help query samples to share more nearest neighbors with the centroid of their category. We follow recent works [10, 13] and divide the bias-diminishing process into two steps: cross-domain bias reduction and intra-domain bias reduction.

The cross-domain bias b_c is the difference between the mean vector of the centroids (visible domain) and the mean feature vector of the query samples (invisible domain):

$$b_c = \frac{1}{|V|} \sum_{v_i \in V} v_i - \frac{1}{|\mathbb{F}_q|} \sum_{f_q \in \mathbb{F}_q} f_q. \tag{5}$$

We diminish the cross-domain bias for each query sample: $f_q = f_q + b_c$.

The intra-domain bias is the difference between the expected and computed centroid. To reduce the intra-domain bias, we rectify the centroids with the cross-domain bias-reduced query features. Previous works [10, 13] assign weights according to the prediction score. Unlike them, we only select query samples with a significant matching score difference between the most similar centroid and the second most similar centroid. Specifically, we assign pseudo-labels to query features when the score difference exceeds threshold t. The matching score W of query feature x_q and centroid v_i is defined as Equation (6):

$$W(f_q, v_i) = \frac{exp(cosine(f_q, v_i))}{\sum_{v_j \in V} exp(cosine(f_q, v_j))}, \tag{6}$$

where $v_i \in V$.

Table 1. Comparison to prior works on *mini*ImageNet and *tiered*ImageNet. The best results are reported in **bold font**.

Method	mini		tiered	
	5-way 1-shot	5-way 5-shot	5-way 1-shot	5-way 5-shot
Meta-Baseline [6]	63.17 ± 0.23	79.26 ± 0.17	68.62 ± 0.27	83.29 ± 0.18
BD-CSPN [10]	70.31 ± 0.93	81.89 ± 0.60	78.74 ± 0.95	86.92 ± 0.63
ProtoComNet [12]	73.13 ± 0.85	82.06 ± 0.54	81.04 ± 0.89	87.42 ± 0.57
LaplacianShot [13]	72.11 ± 0.19	82.31 ± 0.14	78.98 ± 0.21	86.39 ± 0.16
EPNet [23]	66.50 ± 0.89	81.06 ± 0.60	76.53 ± 0.87	87.32 ± 0.64
RGTransformer [24]	71.41 ± 0.91	**83.46 ± 0.57**	78.91 ± 0.99	86.99 ± 0.64
ALFA+MeTAL [33]	66.61 ± 0.28	81.43 ± 0.25	70.29 ± 0.40	86.17 ± 0.35
BaseTransformers [34]	70.88 ± 0.17	82.37 ± 0.19	72.46	84.96
MCGN [35]	67.32 ± 0.43	83.03 ± 0.54	71.21 ± 0.85	85.98 ± 0.98
SNNC(ours)	**73.75 ± 0.91**	83.39 ± 0.59	**82.13 ± 0.94**	**87.65 ± 0.74**

Then we rectify the centroids using the mean feature of the selected query samples:

$$v_i' = \frac{1}{2}(v_i + \frac{1}{|\mathbb{F}_{q,i}'|}\sum_{x_j \in \mathbb{F}_{q,i}'} E(x_j)), \tag{7}$$

where $\mathbb{F}_{q,i}'$ denotes the features of query samples with a i-class pseudo-label. We detail the process of intra-domain bias diminution in Algorithm 1.

4 Experiments

4.1 Experimental Setup

Dataset. We conduct few-shot classification experiments on four benchmarks: *mini*ImageNet [3], *tiered*ImageNet [25], CUB-200-2011(CUB) [26] and *mini*ImageNet → CUB.

Implementation Details. We take an 18-layer ResNet [31] as the backbone network of our method SNNC. In the pre-training phase, we train 100 epochs on the D_{base} and use an SGD optimizer with a momentum of 0.9. We set the batch size as 128 and the learning rate as 0.001. Moreover, the weight decay is 0.0005, and standard data augmentation methods are also applied. In the meta-learning phase, we train the SNNC with 100 epochs and choose SGD with a momentum of 0.9 as the optimizer. In particular, the learning rate is 0.001, which will decay at epochs 30 and 50, and the decay factor is 0.1. The learnable parameters α and β are initialized to 0.1 and 10, while the fixed threshold parameter t is initialized to 0.005.

Evaluation Protocol. To evaluate the performance of our proposed method, we take 10,000 N-way K-shot classification tasks from C_{novel}. We focus on the standard 5-way 1-shot and 5-way 5-shot task settings for each task. These few-shot tasks' average accuracy (in %) is reported with a 95% confidence interval.

Table 2. Comparison to prior works on CUB and $mini \rightarrow$ CUB benchmarks. The best results are reported in **bold font**.

Method	CUB		$mini$ImageNet$\rightarrow$ CUB
	5-way 1-shot	5-way 5-shot	5-way 5-shot
RestoreNet [11]	76.85 ± 0.95	–	–
LaplacianShot [13]	80.96	88.68	66.33
RDC-FT [14]	–	–	67.77 ± 0.4
EPNet [23]	82.85 ± 0.81	91.32 ± 0.41	–
NSAE [27]	–	–	68.51 ± 0.8
Fine-tuning [28]	–	–	63.76±0.4
CPDE [29]	80.11 ± 0.34	89.28 ± 0.33	–
ECKPN [30]	77.43 ± 0.54	**92.21 ± 0.41**	–
SNNC(ours)	**84.26 ± 0.87**	90.12 ± 0.45	**68.77 ± 0.61**

4.2   Results

General Few-Shot Classification. Following the standard setup, we conduct few-shot experiments on $mini$ImageNet and $tiered$ImageNet, and the results are shown in Table 1. We find that (i) SNNC outperforms ProtoComNet, BD-CSPN, EPNet, and LaplacianShot, which also employ transductive settings for few-shot classification. This demonstrates the effectiveness and advantage of SNNC. (ii) SNNC is 0.07% lower than RGTransformer on the 5-way 5-shot tasks of $mini$ImageNet, and outperforms RGTransformer by 0.6% to 3.2% in other settings. This shows that SNNC achieves competitive performance across different datasets and settings.

Fine-Grained Few-Shot Classification. We further evaluate the performance of SNNC on CUB [32]. Observing the experimental results in Table 2, we can find that our method achieves the best performance for 5-way 1-shot tasks on CUB.

Table 3. Experiments of SNN on *mini*ImageNet. $BD^\ddagger$ denotes the bias-diminishing method we proposed, S denotes the raw SNN. $S^\dagger$ denotes the unrestricted SNN, and S^* denotes the metric we proposed.

$BD^\ddagger$	S	$S^\dagger$	S^*	*mini*ImageNet	
				5-way 1-shot	5-way 5-shot
✓	✗	✗	✗	70.21 ± 0.97	81.93 ± 0.21
✓	✓	✗	✗	62.45 ± 0.99	72.03 ± 0.87
✓	✗	✓	✗	72.14 ± 0.93	82.27 ± 0.62
✓	✗	✗	✓	73.75 ± 0.91	83.39 ± 0.59

Table 4. Experiments of the centroid rectification method on *mini*ImageNet. S^* denotes the metric we proposed, BD denotes the original bias-diminishing method, and $BD^\ddagger$ denotes the bias-diminishing method we improved.

S^*	BD	$BD^\ddagger$	*mini*ImageNet	
			5-way 1-shot	5-way 5-shot
✓	✗	✗	67.23 ± 0.85	81.43 ± 0.57
✓	✓	✗	72.92 ± 0.97	82.64 ± 0.62
✓	✗	✓	73.75 ± 0.91	83.39 ± 0.59

For 5-way 5-shot tasks on CUB, SNNC outperforms LaplacianShot by 0.6%–1.2% and is on par with the second-best model (EPNet). In general, SNNC performs competitively in fine-grained few-shot classification scenarios.

Cross-Domain Few-Shot Classification. We evaluate the performance of SNNC by 5-way 5-shot tasks in cross-domain scenarios (train on *mini*ImageNet and test on CUB). The results are shown in Table 2, and SNNC achieves the best performance. Specifically, SNNC outperforms LaplacianShot by about 2.4%. This demonstrates that SNNC has competitive domain transfer capability.

4.3 Ablation Study

Are We Using SNN in the Right Way? In Table 3, we try to disassemble the SNNC and verify the effect of each modification we proposed on the metric function. The first row shows the performance of a model with the improved bias-diminishing method, and the second row shows the impact of replacing the cosine similarity with the raw SNN as the metric. We can find that the model's performance is significantly suppressed in the second row. And when we remove the restriction of mutual k-nearest neighbors from the SNN, the model's performance is enhanced. In particular, the model's performance is maximized when we proportionally combine the de-restricted SNN with the cosine similarity. Specifically, the fourth-row gains at least 1.4% improvement compared to the first row for both the 1-shot and 5-shot cases.

Can We Obtain a more Accurate Centroid? To demonstrate that our method can obtain a more accurate centroid, we change the bias-diminishing method in SNNC to the original method proposed in BD-CSPN [10]. The first row shows the performance of SNNC without bias-diminishing, and the second row shows the effect of changing the bias-diminishing method to the original method. Our improved bias-diminishing approach outperforms the original one by about 0.7% in both the 1-shot and the 5-shot cases.

Can SNN Work with Other Distance-based Metrics? To verify the applicability of SNNC to the other distance-based metrics, we reproduce the code of Meta-Basline using ResNet18 as the backbone network and combine SNN with Euclidean distance proportionally and search for weight hyperparameters during meta-learning. The experimental results are shown in Table 5. Our proposed SNNC (Euclid) is much improved compared to Meta-Baseline (Euclid), and this improvement comes from each component in the SNNC.

Table 5. Experiments of the distance-based metric on mini ImageNet.

Method	$mini$ImageNet	
	5-way 1-shot	5-way 5-shot
Meta-Baseline (Euclid)	62.21	80.47
Meta-Baseline (Cosine)	65.34	81.33
SNNC(Euclid)	69.63	82.05
SNNC(Cosine)	73.75	83.39

5   Conclusion

In this paper, we propose an SNNC method for few-shot classification. The core idea of SNNC is to use the Shared Nearest Neighbor similarity to calibrate the matching results of cosine similarity. SNNC can effectively weaken the impact of the data-scarce in two aspects: matching results calibration and bias diminishment. Extensive experiments show the effectiveness of our method, and either the metric or the bias diminishing module or the combination of both can help to improve classification performance.

References

1. Lake, B.M., Salakhutdinov, R., Tenenbaum, J.B.: Human-level concept learning through probabilistic program induction. Science **350**(6266), 1332–1338 (2015)
2. Fei-Fei, L., Fergus, R., Perona, P.: One-shot learning of object categories. IEEE Trans. Pattern Anal. Mach. Intell. **28**(4), 594–611 (2006)

3. Vinyals, O., Blundell, C., Lillicrap, T., Wierstra, D., et al.: Matching networks for one shot learning. In: Advances in Neural Information Processing Systems, vol. 29 (2016)
4. Snell, J., Swersky, K., Zemel, R.: Prototypical networks for few-shot learning. In: Advances in Neural Information Processing Systems, vol. 30 (2017)
5. Oreshkin, B., López, P.R., Lacoste, A.: TADAM: task dependent adaptive metric for improved few-shot learning. In: Advances in Neural Information Processing Systems, vol. 31 (2018)
6. Chen, Y., Liu, Z., Xu, H., Darrell, T., Wang, X.: Meta-baseline: exploring simple meta-learning for few-shot learning In: Proceedings of the IEEE/CVF International Conference on Computer Vision, pp. 9062–9071 (2021)
7. Gao, Z., Wu, Y., Jia, Y., Harandi, M.: Curvature generation in curved spaces for few-shot learning. In: Proceedings of the IEEE/CVF International Conference on Computer Vision, pp. 8691–8700 (2021)
8. Wu, Y., et al.: Object-aware long-short-range spatial alignment for few-shot fine-grained image classification. arXiv preprint arXiv:2108.13098 (2021)
9. Wu, Z., Li, Y., Guo, L., Jia, K.: PARN: Position-aware relation networks for few-shot learning. In: Proceedings of the IEEE/CVF International Conference on Computer Vision, pp. 6659–6667 (2019)
10. Liu, J., Song, L., Qin, Y.: Prototype rectification for few-shot learning. In: Vedaldi, A., Bischof, H., Brox, T., Frahm, J.-M. (eds.) ECCV 2020. LNCS, vol. 12346, pp. 741–756. Springer, Cham (2020). https://doi.org/10.1007/978-3-030-58452-8_43
11. Xue, W., Wang, W.: One-shot image classification by learning to restore proto-types. Proc. AAAI Conf. Artif. Intell. **34**, 6558–6565 (2020)
12. Zhang, B., Li, X., Ye, Y., Huang, Z., Zhang, L.: Prototype completion with primi-tive knowledge for few-shot learning. In: Proceedings of the IEEE/CVF Conference on Computer Vision and Pattern Recognition, pp. 3754–3762 (2021)
13. Ziko, I., Dolz, J., Granger, E., Ayed, I.B.: Laplacian regularized few-shot learning. In: International Conference on Machine Learning, pp. 11660–11670. PMLR (2020)
14. Li, P., Gong, S., Wang, C., Fu, Y.: Ranking distance calibration for cross-domain few-shot learning. In: Proceedings of the IEEE/CVF Conference on Computer Vision and Pattern Recognition, pp. 9099–9108 (2022)
15. Finn, C., Abbeel, P., Levine, S.: Model-agnostic meta-learning for fast adaptation of deep networks. In: International Conference on Machine Learning, pp. 1126–1135. PMLR (2017)
16. Lee, K., Maji, S., Ravichandran, A., Soatto, S.: Meta-learning with differentiable convex optimization. In: Proceedings of the IEEE/CVF Conference on Computer Vision and Pattern Recognition, pp. 10-657–10-665 (2019)
17. Tian, Y., Wang, Y., Krishnan, D., Tenenbaum, J.B., Isola, P.: Rethinking few-shot image classification: a good embedding is all you need? In: Vedaldi, A., Bischof, H., Brox, T., Frahm, J.-M. (eds.) ECCV 2020. LNCS, vol. 12359, pp. 266–282. Springer, Cham (2020). https://doi.org/10.1007/978-3-030-58568-6_16
18. Nichol, A., Schulman, J.: Reptile: a scalable metalearning algorithm. arXiv preprint arXiv:1803.02999, vol. 2, no. 3, p. 4 (2018)
19. Lee, K., Maji, S., Ravichandran, A., Soatto, S.: Meta-learning with differentiable convex optimization. In: Proceedings of the IEEE/CVF Conference on Computer Vision and Pattern Recognition, pp. 10-657–10-665 (2019)
20. Jarvis, R.A., Patrick, E.A.: Clustering using a similarity measure based on shared near neighbors. IEEE Trans. Comput. **100**(11), 1025–1034 (1973)

21. Bennett, K.P., Fayyad, U., Geiger, D.: Density-based indexing for approximate nearest-neighbor queries. In: Proceedings of the Fifth ACM SIGKDD International Conference on Knowledge Discovery and Data Mining, pp. 233–243 (1999)

22. Houle, M.E., Kriegel, H.-P., Kröger, P., Schubert, E., Zimek, A.: Can shared-neighbor distances defeat the curse of dimensionality? In: Gertz, M., Ludäscher, B. (eds.) SSDBM 2010. LNCS, vol. 6187, pp. 482–500. Springer, Heidelberg (2010). https://doi.org/10.1007/978-3-642-13818-8_34

23. Rodríguez, P., Laradji, I., Drouin, A., Lacoste, A.: Embedding propagation: smoother manifold for few-shot classification. In: Vedaldi, A., Bischof, H., Brox, T., Frahm, J.-M. (eds.) ECCV 2020. LNCS, vol. 12371, pp. 121–138. Springer, Cham (2020). https://doi.org/10.1007/978-3-030-58574-7_8

24. Jiang, B., Zhao, K., Tang, J.: Rgtransformer: region-graph transformer for image representation and few-shot classification. IEEE Signal Process. Lett. **29**, 792–796 (2022)

25. Ren, M., et al.: Meta-learning for semi-supervised few-shot classification. arXiv preprint arXiv:1803.00676 (2018)

26. Wah, C., Branson, S., Welinder, P., Perona, P., Belongie, S.: The Caltech-UCSD birds-200-2011 dataset (2011)

27. Liang, H., Zhang, Q., Dai, P., Lu, J.: Boosting the generalization capability in cross-domain few-shot learning via noise-enhanced supervised autoencoder. In: Proceedings of the IEEE/CVF International Conference on Computer Vision, pp. 9424–9434 (2021)

28. Guo, Y., et al.: A broader study of cross-domain few-shot learning. In: Vedaldi, A., Bischof, H., Brox, T., Frahm, J.-M. (eds.) ECCV 2020. LNCS, vol. 12372, pp. 124–141. Springer, Cham (2020). https://doi.org/10.1007/978-3-030-58583-9_8

29. Zou, Y., Zhang, S., Chen, K., Tian, Y., Wang, Y., Moura, J.M.F.: Compositional few-shot recognition with primitive discovery and enhancing. In: Proceedings of the 28th ACM International Conference on Multimedia, pp. 156–164 (2020)

30. Chen, C., Yang, X., Xu, C., Huang, X., Ma, Z.: ECKPN: explicit class knowledge propagation network for transductive few-shot learning. In: Proceedings of the IEEE/CVF Conference on Computer Vision and Pattern Recognition, pp. 6596–6605 (2021)

31. He, K., Zhang, X., Ren, S., Sun, J.: Deep residual learning for image recognition. In: Proceedings of the IEEE Conference on Computer Vision and Pattern Recognition, pp. 770–778 (2016)

32. Hilliard, N., Phillips, L., Howland, S., Yankov, A., Corley, C.D., Hodas, N.O.: Few-shot learning with metric-agnostic conditional embeddings. arXiv preprint arXiv:1802.04376 (2018)

33. Baik, S., Choi, J., Kim, H., Cho, D., Min, J., Lee, K.M.: Meta-learning with task-adaptive loss function for few-shot learning. In: Proceedings of the IEEE/CVF International Conference on Computer Vision, pp. 9465–9474 (2021)

34. Maniparambil, M., McGuinness, K., O'Connor, N.: Basetransformers: attention over base data-points for one shot learning. arXiv preprint arXiv:2210.02476 (2022)

35. Tang, S., Chen, D., Bai, L., Liu, K., Ge, Y., Ouyang, W.: Mutual CRF-GNN for few-shot learning. In: Proceedings of the IEEE/CVF Conference on Computer Vision and Pattern Recognition, pp. 2329–2339 (2021)

Prototype Rectification with Region-Wise Foreground Enhancement for Few-Shot Classification

Rundong Qi, Sa Ning, and Yong Jiang[✉]

School of Computer Science and Technology, Southwest University of Science and Technology, Mianyang 621010, China
jiang_yong@swust.edu.cn

Abstract. Few-Shot Classification is a challenging problem as it uses only very few labeled examples to assign labels for query samples. The Prototypical Network effectively addresses the issue by matching the nearest mean-based prototype for each query sample. However, the scarce labeled examples limit the ability of a model to represent the data distribution, which in turn biases the computed prototypes. In this paper, we propose a transductive bias diminishing method based on spatial similarity, which consists of a region-wise foreground enhancement (RFE) module and a prototype rectification (PR) module. RFE reconstructs the query sample's features to emphasize the discriminative parts while preserving local region consistency. PR uses the difference in prediction scores between the nearest and second nearest prototypes of the enhanced query samples to rectify the prototypes by label propagation. We validate the effectiveness of our method with extensive experiments on four Few-Shot Classification datasets: *mini*ImageNet, *tiered*ImageNet, Stanford Dogs, and CUB. Our method achieves competitive performance across different settings and datasets.

Keywords: Few-Shot Classification · metric learning · transductive

1 Introduction

Few-Shot Classification (FSC) has recently gained widespread attention thanks to its imitation of the human ability to learn new things [1,2]. FSC aims to improve the generalization ability of a model, allowing it to generalize efficiently to new classes when only very few labeled examples are provided. In FSC, we aim to utilize the prior knowledge learned on base classes with abundant labeled examples to classify samples in novel classes with scarce labeled examples. The classification process is performed on N-way K-shot few-shot tasks, where each task contains N classes with K labeled examples (support set) and some unlabeled samples (query set) to be classified.

This study is supported by the Sichuan Science and Technology Program (NO. 2021YFG0031).

Researchers have proposed many methods [3–7] to perform few-shot tasks, among which metric-based methods attracted considerable attention due to their simplicity and effectiveness. Metric-based methods aim to learn a transferable embedding feature space and predict labels by measuring the correlation between samples. For example, Prototypical Network [4] matches the nearest prototype for each query sample and assumes that the prototype shares the same label with the query sample. However, with sparse data, prototypes are always biased. As a result, the representation ability of prototypes is also weakened.

Many existing studies have attempted to improve the representation ability of prototypes. TAFSSL [8] proposes to remove noisy attributes utilizing feature space projection and rectifies the prototypes in a label propagation manner. BD-CSPN [9] proposes to rectify prototypes by diminishing cross-class bias and intra-class bias in two steps, where cross-class bias is reduced by eliminating the average feature difference between the support and query sets, and intra-class bias is diminished by updating prototypes by query samples with pseudo labels and weights. Nevertheless, these two methods do not fully exploit the correlation among the features' spatial locations. Some fine-grained classification studies [10,17,19] use spatial similarity to reduce the difference in discriminative object locations between the labeled and unlabeled instances. For example, BSFA [10] uses the spatial similarity between the query and the support features to reconstruct the feature maps of the support examples for aligning the discriminative objects' locations. However, the calculation of spatial similarity is highly susceptible to noisy deep local descriptors, impacting the consistency of regions within the features.

In this paper, we propose a method to improve the representation ability of prototypes using spatial similarity, which consists of a region-wise foreground enhancement (RFE) module and a prototype rectification (PR) module. RFE reconstructs the features of query samples, which divides the feature of each query sample into several connected regions according to the spatial similarity between deep local descriptors and assigns equal weight to all deep local descriptors within the corresponding area. PR uses the difference in predicted scores between the nearest and second nearest prototypes of the enhanced query samples to calculate the weights in prototype updates for improving the prototypes' representation ability. Moreover, to reduce the noise attributes in the feature space, we use the Principal Component Analysis (PCA) to search for a task-adaptive subspace during the test [8].

Our three main contributions are summarized below:

- We develop an FSC framework that can be optimized end-to-end, with only image-level labels available.
- Two well-designed modules are instantiated in our framework to implement region-wise foreground enhancement and prototype rectification, i.e., region-wise foreground enhancement module, and prototype rectification module.
- We conduct extensive experiments and ablation studies on the four FSC benchmarks. The experimental results show that our method can achieve competitive performance across different settings and datasets.

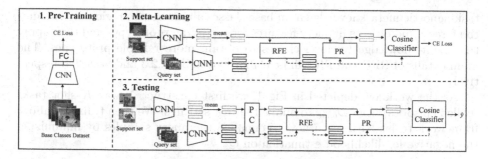

Fig. 1. Framework overview.

2   Method

2.1   Preliminaries

For the standard setting of FSC, two datasets: the base classes dataset (denoted as D_{base}) and the novel classes dataset (marked as D_{novel}), are provided. $D_{base} = \{(x_j, y_j)\}_{j=1}^W$ consists of W labeled images, where each image x_j is labeled with a class $y_j \in C_{base}$ (C_{base} represents the set of base classes). FSC aims to transfer the knowledge learned on D_{base} to perform few-shot tasks on D_{novel}. A few-shot task contains N classes, each with K labeled examples (support set S) and some unlabeled samples (query set Q). The goal of a few-shot task is to classify the unlabeled samples in Q using the labeled examples in S. Denote the set of novel classes as C_{novel}, and there is no intersection between C_{novel} and C_{base}.

For transductive FSC, we mark the set of query images $x \in Q$ as I. We aim to learn a task-specific classifier with the help of query images I, support set S, and the prior knowledge learned on D_{base}. In contrast, inductive FSC can only learn classifiers using S and D_{base}. We focus on transductive FSC in the following subsections to present our approach.

2.2   Overall Framework

As shown in Fig. 1, our framework consists of three stages: pre-training, meta-learning, and testing.

Pre-training. In the stage, we train a convolutional neural network f_θ with parameters θ by minimizing the classification loss on D_{base}. We use Cross-Entropy Loss (CE Loss) as the loss function. Then, we remove the last fully connected (FC) layer of f_θ and take the remainder as the feature encoder E. Following Meta-Baseline [6], we do not freeze E or introduce any additional parameters.

Meta-learning. We propose a region-wise foreground enhancement (RFE) module and a prototype rectification (PR) module. RFE learns to enhance the query samples in an episodic training paradigm, and PR rectifies prototypes using the reconstructed query samples. The purpose of such a design is to learn task-agnostic meta-knowledge from base classes about how to regionally enhance the foreground objects in query features for rectifying prototypes and then apply this meta-knowledge to novel classes to obtain more reliable prototypes. The main details of RFE and PR will be elaborated in Sect. 2.3 and Sect. 2.4, respectively.

As the workflow depicted in Fig. 1, we first construct N-way K-shot tasks (called episodes) from D_{base} to mimic the test settings. We then train the above framework to let RFE learn how to enhance the query samples by minimizing the negative log-likelihood estimation on Q:

$$\min_{\theta,\alpha} \mathbb{E}_{(S,Q)\in\mathbb{T}} \sum_{x,y\in Q} -log(P(y \mid x,S,I,\theta,\alpha)), \tag{1}$$

where $\mathbb{T}$ denotes the set of constructed N-way K-shot tasks, and α denotes the learnable parameter of RFE. Next we present the procedure for calculating the category probability $P(y \mid x,S,I,\theta,\alpha)$:

For each task, we first calculate the prototypes:

$$p_i = \frac{1}{K} \sum_{x\in S_i} f_\theta(x), \tag{2}$$

where S_i denotes the labeled examples with class i. Subsequently, we extract the query features: $F = f_\theta(I)$. Then, RFE is applied to enhance query features:

$$\hat{F} = RFE(F,p,\alpha), \tag{3}$$

where $p = \{p_1,...,p_N\}$. Further, to obtain more reliable prototypes, we rectify prototypes using the PR module:

$$\hat{p} = PR(\hat{F},p), \tag{4}$$

where $\hat{p} = \{\hat{p_1},...,\hat{p_N}\}$. The probability of each query sample $x \in Q$ belonging to class z is estimated according to the similarity between its enhanced feature and rectified prototypes $\hat{p}$. That is,

$$P(y = z \mid x,S,I,\theta,\alpha) = \frac{exp(cos \langle RFE(f_\theta(x),p,\alpha),\hat{p_z}\rangle \tau)}{\sum_o exp(cos \langle RFE(f_\theta(x),p,\alpha),\hat{p_o}\rangle \tau)}, \tag{5}$$

where $cos \langle \cdot,\cdot \rangle$ denotes the cosine similarity of two vectors, and τ is a learnable scale parameter [5,6].

Test. Following Eqs. (2) $\sim$ (5), we construct and perform N-way K-shot tasks for novel classes. Differently, to reduce the noisy attributes in the feature space, we map the features of the query samples and prototypes to an "optimal" (for the given task) subspace before Eq. (3), as TAFSSL [8] does.

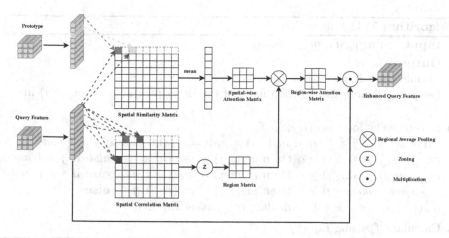

Fig. 2. Region-wise Foreground Enhancement (RFE) Module.

2.3 Region-Wise Foreground Enhencement

Previous methods reconstruct prototypes mainly employing label propagation [8, 9] or feature fusion [11]. There are scarce methods to rectify prototypes using the spatial similarity between features and prototypes. Inspired by the fine-grained FSC approaches [10,17,19], highlighting deep local descriptors in query features with high average similarity to each deep local descriptor within the prototype helps to reduce the differences in foreground objects between the query and support sets. However, spatial-wise enhancement of features can destroy the method's resistance to noisy deep local descriptors. In light of this, our proposed module goes two steps to enhance the features of query samples:

Step 1. As the Fig. 2, we first find a prototype for each query feature that is most similar to it according to cosine similarity. Subsequently, for a query feature $f_q \in \mathbb{R}^{c \times h \times w}$ and its most similar prototype $p_b \in \mathbb{R}^{c \times h \times w}$, we first reshape them to $\mathbb{R}^{hw \times c}$, i.e., $f'_q = \{q_i\}_{i=1}^{h \times w}$ and $p'_b = \{b_i\}_{i=1}^{h \times w}$, where q_i and b_i represent the i_{th} deep local descriptor in f'_q and p'_b, respectively. Then, we calculate the spatial similarity matrix $m_{p|q}$ between the reshaped p'_b and f'_q and the spatial correlation matrix $m_{q|q}$ within f'_q. Here, the resulting spatial similarity matrix $m_{p|q} \in \mathbb{R}^{hw \times hw}$ and spatial correlation matrix $m_{q|q} \in \mathbb{R}^{hw \times hw}$ are calculated by cosine similarity as $m_{p|q}(i,j) = \frac{q_i^T b_j}{\|q_i\| \times \|b_j\|}$ and $m_{q|q}(i,j) = \frac{q_i^T q_j}{\|q_i\| \times \|q_j\|}$. Further, a row-by-row mean operation is also applied to $m_{p|q}$ to reduce the impact of noisy deep local descriptors on the similarity calculation. Then we get the spatial-wise attention matrix $\hat{m}_{p|q}$:

$$\hat{m}_{p|q} = mean(m_{p|q}) = \frac{\sum_{j=1}^{h \times w} m_{p|q}(i,j)}{h \times w}. \tag{6}$$

Algorithm 1: Details of Step 2.

Input: α, $m_{q|q}$, f_q, $\hat{m}_{p|q}$, h, w

Output: $\hat{f}_q$

1 Initialize $i = 1$, $j = 1$, $v = 1$, $m_r = zeros(h, w)$;

2 Let $pos_{i,j}$ denote the position in f'_q of the descriptor at location (i, j) in f_q;

3 Iterate every location (i, j) in f_q:

4   **if** $i = 1$ *and* $j = 1$ **then** $m_r(i,j) = v$, $v = v + 1$ **else if** $j \neq 1$ *and* $m_{q|q}(pos_{i,j}, pos_{i,j-1}) \geq \alpha$ **then** $m_r(i,j) = m_r(i, j-1)$ **else if** $j = 1$ *and* $m_{q|q}(pos_{i,j}, pos_{i-1,j}) \geq \alpha$ **then** $m_r(i,j) = m_r(i-1, j)$ **else if** $i \neq 1$ *and* $m_{q|q}(pos_{i,j}, pos_{i-1,j}) \geq \alpha$ **then** $m_r(i,j) = m_r(i-1, j)$ **else** $m_r(i,j) = v$, $v = v + 1$ Calculate $\hat{m}'_{p|q}$ using Eq. (7);

5 Calculate $\hat{f}_q$ using Eq. (8);

Step 2. To keep the region consistency of the query feature f_q, we divide f_q into several connected regions according to the $m_{q|q}$. Specifically, suppose the similarity of two adjacent deep local descriptors exceeds a threshold α. In that case, we consider them to belong to the same connected region and assign the same value to the corresponding spatial locations in the area matrix m_r. Notably, α is a learnable parameter. For spatial locations that have the same tag in m_r, we correspondingly average their value in $\hat{m}_{p|q}$:

$$\hat{m}'_{p|q}(x_i, y_i) = \frac{\sum_{(x_j, y_j) \in V_z} \hat{m}_{p|q}(x_j, y_j)}{\|V_z\|}, \tag{7}$$

where V_z represents the set of all spatial locations in m_r with value z, and $(x_i, y_i) \in V_z$. As shown in Fig. 2, we employ the RFE module to calculate the region-wise attention matrix $\hat{m}'_{p|q}$, which is used to enhance the query feature f_q with its most similar prototype p_b. Formulaically, the enhanced feature $\hat{f}_q$ can be calculated by a spatial-wise multiplication as:

$$\hat{f}_q = \hat{m}'_{p|q} \cdot f_q. \tag{8}$$

We detail Step 2 in the Algorithm 1.

2.4 Prototype Rectification

To reduce the intra-class bias, BD-CSPN uses a pseudo-labeling strategy to rectify prototypes, which assigns labels and computes relation weight based on the prediction scores [12]. However, the prediction score is not enough to show the contribution of pseudo-labeled query features in prototype rectification compared to the difference between the prediction scores, e.g., a query feature that is in the class boundary region is not very useful for prototype rectification, but may have high similarity to several prototypes, and the differences between similarities will be slight.

In this paper, we develop a prototype rectification (PR) module to reduce intra-class bias, which calculates the pseudo-prototypes using the similarity difference between the closest and second closest prototypes to query features and takes the mean of the pseudo-prototypes and the base prototype as the new prototype. Specifically, we assign each query feature a pseudo-label that is identical to its nearest prototype based on cosine similarity, and we obtain a pseudo support set: S'. Since some pseudo-labeled features located at class boundaries may be misclassified, we use the difference between the maximum and sub-maximum prediction scores of each query feature as the basis for the weight computation. The pseudo-prototype of class n is computed by:

$$p'_n = \sum_{\hat{f}_q \in S'_n} t_{n,q} \times \hat{f}_q, \qquad (9)$$

where $\hat{f}_q \in S'_n$, S'_n denotes the set of query features pseudo-labeled with class n, and $t_{n,q}$ is the weight indicating the relation of the pseudo-labeled query feature $\hat{f}_q$ and the prototype of class n. Let $COS_{\hat{f}_q}$ denote the set of cosine similarities between each prototype and the query feature $\hat{f}_q$, then we compute the weight as follows:

$$t_{n,q} = \frac{e^{\varepsilon(max(COS_{\hat{f}_q})-submax(COS_{\hat{f}_q}))}}{\sum_{\hat{f}_i \in S'_n} e^{\varepsilon(max(COS_{\hat{f}_i})-submax(COS_{\hat{f}_i}))}} \qquad (10)$$

where ε is a scalar parameter. And we calculate the new prototypes by an average operation. That is,

$$\hat{p}_i = \frac{p'_n + p_n}{2}. \qquad (11)$$

In the pseudo-prototype computation, the higher the score difference of the query feature, the higher the proportion. In the new prototype computation, the proportion of labeled support features ($\frac{1}{2K}$) is theoretically larger than that of query features with pseudo-labels ($\approx \frac{1}{2(K+\|Q\|/N)}$). The rectified prototypes will be closer to the expectation than the basic prototypes.

3 Experiments

3.1 Experimental Setup

Datasets. We conduct FSC experiments on four datasets, including: *mini*ImageNet [3], *tiered*ImageNet [21], CUB [22], and Stanford Dogs [23]. The input images of all datasets are resized to 84×84 [3, 25].

Table 1. Comparison to prior works on *mini*ImageNet and *tiered*ImageNet.
Average accuracy (in %) is reported. The best results are reported in **bold font**.
∗ denotes the results we reproduced.

Method	Backbone	*mini*ImageNet		*tiered*ImageNet	
		5-way 1-shot	5-way 5-shot	5-way 1-shot	5-way 5-shot
ProtoNet [4]	ConvNet-64	60.37 ± 0.83	78.02 ± 0.57	65.65 ± 0.92	83.40 ± 0.65
Meta-Baseline [6]	ResNet-12	63.17 ± 0.23	79.26 ± 0.17	68.62 ± 0.27	83.29 ± 0.18
BD-CSPN [9]	ResNet-12	65.94 ± 0.93	81.89 ± 0.60	78.74 ± 0.95	86.92 ± 0.63
ProtoComNet [11]	ResNet-12	73.13 ± 0.85	82.06 ± 0.54	81.04 ± 0.89	87.42 ± 0.57
EPNet [14]	ResNet-12	66.50 ± 0.89	81.06 ± 0.60	76.53 ± 0.87	87.32 ± 0.64
RGTransformer [15]	ResNet-12	71.41 ± 0.91	83.46 ± 0.57	78.91 ± 0.99	86.99 ± 0.64
CGC [16]	ResNet-12	66.73 ± 0.22	80.57 ± 0.14	77.19 ± 0.24	86.18 ± 0.15
TAFSSL∗ [8]	ResNet-18	73.17 ± 0.27	82.14 ± 0.24	**81.87 ± 0.26**	87.43 ± 0.23
LaplacianShot [13]	ResNet-18	72.11 ± 0.19	82.31 ± 0.14	78.98 ± 0.21	86.39 ± 0.16
ours	ResNet-12	**73.21 ± 0.84**	**83.53 ± 0.87**	81.09 ± 0.81	**87.86 ± 0.92**

3.2   Implementation Details

We take the 12-layer ResNet [24] as the backbone network of our method. In
the pre-training phase, we train 200 epochs on D_{base} using an SGD optimizer
with a momentum of 0.9. We set the batch size as 128 and the learning rate as
0.001. Moreover, the weight decay is 0.0005, and standard data augmentation
methods like random resized crops are also applied. In the meta-learning phase,
we train 20 epochs and choose SGD with a momentum of 0.9 as the optimizer.
The learning rate is 0.001, which will decay at epochs 10 and 15, and the decay
factor is 0.1. The learnable parameter α and τ are initialized to 0.25 and 10,
respectively. The fixed scalar parameter ε is set to 10. In addition, for the 1-shot
and 5-shot settings, we set the number of principal components retained in the
PCA to 4 and 10, respectively.

Evaluation Protocol. To evaluate the performance of our proposed method,
we take 10,000 N-way K-shot classification tasks from D_{novel}. We focus on the
standard 5-way 1-shot and 5-way 5-shot task settings for each task. The average
accuracy of these few-shot tasks is reported with a 95% confidence interval.

Table 2. Comparison to prior on CUB and Stanford Dogs. Average accuracy
(in %) is reported. The best results are reported in **bold font**.

Method	Backbone	CUB		Stanford Dogs	
		5-way 1-shot	5-way 5-shot	5-way 1-shot	5-way 5-shot
OLSA [17]	Resnet-18	77.77 ± 0.44	89.87 ± 0.24	64.15 ± 0.49	78.28 ± 0.32
BSNet [18]	Resnet-18	73.48 ± 0.92	83.84 ± 0.59	61.95 ± 0.97	79.62 ± 0.63
TOAN [19]	Resnet-12	66.10 ± 0.86	82.27 ± 0.60	49.77 ± 0.86	69.29 ± 0.70
CAN [20]	Resnet-12	76.98 ± 0.48	87.77 ± 0.30	64.73 ± 0.52	77.93 ± 0.35
BSFA [10]	Resnet-12	82.27 ± 0.46	**90.76 ± 0.26**	69.58 ± 0.50	82.59 ± 0.33
ours	Resnet-12	**84.62 ± 0.89**	90.08 ± 0.97	**71.02 ± 0.84**	**83.44 ± 0.63**

3.3 Results

Results on General Few-Shot Classification. Following the standard setup [3,6], we conduct general FSC experiments on the *mini*ImageNet and the *tiered*ImageNet datasets. And the experimental results are shown in Table 1. On the *mini*ImageNet, the performance of our method is higher than Meta-Baseline [6] by 10.04% and 4.27% for the 5-way 1-shot and 5-way 5-shot settings, respectively. Our method achieves the best performance compared to other methods, with 73.21% and 83.53% accuracy in the 5-way 1-shot and 5-way 5-shot settings. On the *tiered*ImageNet, the performance of our method is higher than Meta-Baseline [6] by 12.27% and 4.57% for the 5-way 1-shot and 5-way 5-shot settings, respectively. Specifically, for the 1-shot setting, our method is second only to TAFSSL, achieving 81.09% accuracy. For the 5-shot setting, our method outperforms TAFSSL by 0.43%. The performance improvement of our method on the 5-shot setting is lower than that on the 1-shot setting. We attribute this phenomenon to the increase in the number of labeled examples in the 5-shot setting, where the computed prototypes are more in line with expectations, leading to a decrease in the performance improvement of our proposed method.

Results on Fine-Grained Datasets. We also conduct fine-grained FSC experiments on the CUB and the Stanford Dogs datasets. And we show the performance of our proposed method in Table 2. On the CUB, our proposed method achieve competitive or the best performance for the 5-way 1-shot and 5-way 5-shot settings. Specifically, our approach achieves 84.62% accuracy in the 1-shot setting, 2.35% higher than the performance of the BSFA method. For the 5-shot setting, our method achieves 90.08% accuracy, which is only second to BSFA. On the Stanford Dogs, our method outperforms BSFA by 3.3% and 0.45% for 5-way 1-shot and 5-way 5-shot tasks, respectively. Specifically, our method achieves 71.02% and 83.44% accuracy in the 5-way 1-shot and 5-way 5-shot settings.

3.4 Ablation Studies

Evaluating the Effect of Each Component. In Table 3, we disassemble our method into four parts: the backbone network (B), the region-wise foreground enhancement module(M_s), the prototype rectification module (M_r) and the PCA method(M_p). The first row shows the performance of B, and the second row shows the benefit of M_s to B. The third and fourth rows add M_p and M_r to the model in the second row, respectively. The fifth row adds M_p and M_r to the second row's model. From the third and fourth rows, we find that using RFE or PR alone can improve the model's performance. And as the fifth row shows, the model performance can be further enhanced when we use both modules.

Is the Region-Wise Foreground Enhancement Module more Effective? We conduct a comparison experiment to demonstrate the superiority of the

Table 3. Method disassembly. Average accuracy (%) is reported. The backbone network (B) is ResNet-12. The best results are reported in **bold font**.

B	M_p	M_s	M_r	miniImageNet	
				5-way 1-shot	5-way 5-shot
√	×	×	×	63.17 ± 0.23	79.26 ± 0.17
√	√	×	×	65.41 ± 1.67	78.85 ± 1.05
√	√	√	×	67.19 ± 0.76	80.52 ± 0.79
√	√	×	√	72.33 ± 0.92	82.94 ± 0.88
√	√	√	√	**73.21 ± 0.84**	**83.53 ± 0.87**

region-wise foreground enhancement (RFE) module. Specifically, we replace our RFE module with the foreground object alignment (FOA) module proposed in BSFA [10] while keeping other conditions constant. We show the experimental results in the Table 4, where $FOA + PR$ denotes our method of replacing RFE with FOA and $RFE + PR$ denotes our method. The results demonstrate that our RFE module performs better than the FOA module under the same conditions.

Is Our Prototype Rectification Module More effective? We conduct a comparison experiment to demonstrate the effectiveness of our proposed prototype rectification (PR) module. Specifically, we replace our PR module with the intra-class bias diminishing method (BD_{intra}) proposed in BD-CSPN [9] while keeping other conditions similar. The experimental results are shown in the Table 5, where $RFE + BD_{intra}$ denotes our method of replacing PR with the BD_{intra} and $RFE + PR$ denotes our method. We can easily find that our PR module outperforms the BD_{intra} method under the same conditions.

Table 4. Comparison of regional enhancement methods. Average accuracy (in %) is reported. The best results are reported in **bold font**.

Method	miniImageNet	
	5-way 1-shot	5-way 5-shot
$FOA + PR$	72.03 ± 1.37	82.86 ± 1.13
$RFE + PR$	**73.21 ± 0.84**	**83.53 ± 0.87**

Table 5. Comparison of prototype rectification methods. Average accuracy (in %) is reported. The best results are reported in **bold font**.

Method	miniImageNet	
	5-way 1-shot	5-way 5-shot
$RFE + BD_{intra}$	71.95 ± 0.99	82.71 ± 0.83
$RFE + PR$	$\mathbf{73.21 \pm 0.84}$	$\mathbf{83.53 \pm 0.87}$

4   Conclusion

The critical challenge of FSC is improving the prototypes' representation ability. While many approaches explore using prior knowledge to complete prototypes, we propose a prototype bias reduction method based on spatial similarity, which consists of a region-wise foreground enhancement (RFE) module for reconstructing query features using spatial similarity and a prototype rectification (PR) module for rectifying prototypes using the enhanced query features. Experimental results show the superior performance of our method, and the ablation studies demonstrate the proposed modules' effectiveness.

References

1. Lake, B.M., Salakhutdinov, R., Tenenbaum, J.B.: Human-level concept learning through probabilistic program induction. Science **350**(6266), 1332–1338 (2015)
2. Fei-Fei, L., Fergus, R., Perona, P.: One-shot learning of object categories. IEEE Trans. Pattern Anal. Mach. Intell. **28**(4), 594–611 (2006)
3. Vinyals, O., Blundell, C., Lillicrap, T., Wierstra, D., et al.: Matching networks for one shot learning. In: Advances in Neural Information Processing Systems, vol. 29 (2016)
4. Snell, J., Swersky, K., Zemel, R.: Prototypical networks for few-shot learning. In: Advances in Neural Information Processing Systems, vol. 30 (2017)
5. Oreshkin, B., Rodríguez López, P., Lacoste, A.: TADAM: task dependent adaptive metric for improved few-shot learning. In: Advances in Neural Information Processing Systems, vol. 31 (2018)
6. Chen, Y., Liu, Z., Xu, H., Darrell, T., Wang, X.: Meta-baseline: exploring simple meta-learning for few-shot learning. In: Proceedings of the IEEE/CVF International Conference on Computer Vision, pp. 9062–9071 (2021)
7. Sung, F., Yang, Y., Zhang, L., Xiang, T., Torr, P.H., Hospedales, T.M.: Learning to compare: relation network for few-shot learning. In: Proceedings of the IEEE Conference on Computer Vision and Pattern Recognition, pp. 1199–1208 (2018)
8. Lichtenstein, M., Sattigeri, P., Feris, R., Giryes, R., Karlinsky, L.: TAFSSL: task-adaptive feature sub-space learning for few-shot classification. In: Vedaldi, A., Bischof, H., Brox, T., Frahm, J.-M. (eds.) ECCV 2020. LNCS, vol. 12352, pp. 522–539. Springer, Cham (2020). https://doi.org/10.1007/978-3-030-58571-6_31
9. Liu, J., Song, L., Qin, Y.: Prototype rectification for few-shot learning. In: Vedaldi, A., Bischof, H., Brox, T., Frahm, J.-M. (eds.) ECCV 2020. LNCS, vol. 12346, pp. 741–756. Springer, Cham (2020). https://doi.org/10.1007/978-3-030-58452-8_43

10. Zha, Z., Tang, H., Sun, Y., Tang, J.: Boosting few-shot fine-grained recognition with background suppression and foreground alignment. IEEE Trans. Circ. Syst. Video Technol. (2023)

11. Zhang, B., Li, X., Ye, Y., Huang, Z., Zhang, L.: Prototype completion with primitive knowledge for few-shot learning. In: Proceedings of the IEEE/CVF Conference on Computer Vision and Pattern Recognition, pp. 3754–3762 (2021)

12. Li, X., et al.: Learning to self-train for semi-supervised few-shot classification. In: Advances in Neural Information Processing Systems, vol. 32 (2019)

13. Ziko, I., Dolz, J., Granger, E., Ayed, I.B.: Laplacian regularized few-shot learning. In: International Conference on Machine Learning. PMLR, pp. 11660–11670 (2020)

14. Rodríguez, P., Laradji, I., Drouin, A., Lacoste, A.: Embedding propagation: smoother manifold for few-shot classification. In: Vedaldi, A., Bischof, H., Brox, T., Frahm, J.-M. (eds.) ECCV 2020. LNCS, vol. 12371, pp. 121–138. Springer, Cham (2020). https://doi.org/10.1007/978-3-030-58574-7_8

15. Jiang, B., Zhao, K., Tang, J.: Rgtransformer: region-graph transformer for image representation and few-shot classification. IEEE Signal Process. Lett. **29**, 792–796 (2022)

16. Gao, Z., Wu, Y., Jia, Y., Harandi, M.: Curvature generation in curved spaces for few-shot learning. In: Proceedings of the IEEE/CVF International Conference on Computer Vision, pp. 8691–8700 (2021)

17. Wu, Y., Zhang, B., Yu, G., Zhang, W., Wang, B., Chen, T., Fan, J.: Object-aware long-short-range spatial alignment for few-shot fine-grained image classification. arXiv preprint arXiv:2108.13098 (2021)

18. Li, X., Wu, J., Sun, Z., Ma, Z., Cao, J., Xue, J.-H.: BSNet: bi-similarity network for few-shot fine-grained image classification. IEEE Trans. Image Process. **30**, 1318–1331 (2020)

19. Huang, H., Zhang, J., Yu, L., Zhang, J., Wu, Q., Xu, C.: TOAN: target-oriented alignment network for fine-grained image categorization with few labeled samples. IEEE Trans. Circuits Syst. Video Technol. **32**(2), 853–866 (2021)

20. Hou, R., Chang, H., Ma, B., Shan, S., Chen, X.: Cross attention network for few-shot classification. In: Advances in Neural Information Processing Systems, vol. 32 (2019)

21. Ren, M., et al.: Meta-learning for semi-supervised few-shot classification. arXiv preprint arXiv:1803.00676 (2018)

22. Wah, C., Branson, S., Welinder, P., Perona, P., Belongie, S.: The caltech-UCSD birds-200-2011 dataset (2011)

23. Khosla, A., Jayadevaprakash, N., Yao, B., Li, F.-F.: Novel dataset for fine-grained image categorization: Stanford dogs. In: Proceedings of the CVPR Workshop on Fine-Grained Visual Categorization (FGVC), vol. 2, no. 1. Citeseer (2011)

24. He, K., Zhang, X., Ren, S., Sun, J.: Deep residual learning for image recognition. In: Proceedings of the IEEE Conference on Computer Vision and Pattern Recognition, pp. 770–778 (2016)

25. Hilliard, N., et al.: Few-Shot learning with metric-agnostic conditional embeddings. arXiv preprint arXiv:1802.04376 (2018)

Rotation Augmented Distillation for Exemplar-Free Class Incremental Learning with Detailed Analysis

Xiuwei Chen[1] and Xiaobin Chang[1,2,3]([✉])

[1] School of Artificial Intelligence, Sun Yat-sen University, Guangzhou, China
chenxw83@mail2.sysu.edu.cn, changxb3@mail.sysu.edu.cn
[2] Guangdong Key Laboratory of Big Data Analysis and Processing,
Guangzhou 510006, People's Republic of China
[3] Key Laboratory of Machine Intelligence and Advanced Computing,
Ministry of Education, Guangzhou, China

Abstract. Class incremental learning (CIL) aims to recognize both the old and new classes along the increment tasks. Deep neural networks in CIL suffer from catastrophic forgetting and some approaches rely on saving exemplars from previous tasks, known as the exemplar-based setting, to alleviate this problem. On the contrary, this paper focuses on the Exemplar-Free setting with no old class sample preserved. Balancing the plasticity and stability in deep feature learning with only supervision from new classes is more challenging. Most existing Exemplar-Free CIL methods report the overall performance only and lack further analysis. In this work, different methods are examined with complementary metrics in greater detail. Moreover, we propose a simple CIL method, Rotation Augmented Distillation (RAD), which achieves one of the top-tier performances under the Exemplar-Free setting. Detailed analysis shows our RAD benefits from the superior balance between plasticity and stability. Finally, more challenging exemplar-free settings with fewer initial classes are undertaken for further demonstrations and comparisons among the state-of-the-art methods.

Keywords: Class Incremental Learning · Catastrophic Forgetting · Exemplar Free

1 Introduction

AI agents, e.g., deep neural networks (DNNs), deployed in the real world face an ever-changing environment, i.e., with new concepts and categories continually emerging [3–5]. Therefore, class incremental learning (CIL) [6–8] has attracted much attention as it aims to equip deep models with the capacity to continuously handle new categories. However, learning to discriminate sets of disjoint classes sequentially with deep models is challenging, as they are prone to entirely forgetting the previous knowledge, thus resulting in severe performance degradation of old tasks. It is known as catastrophic forgetting [9]. To maintain the old

Q. Liu et al. (Eds.): PRCV 2023, LNCS 14428, pp. 27–38, 2024.
https://doi.org/10.1007/978-981-99-8462-6_3

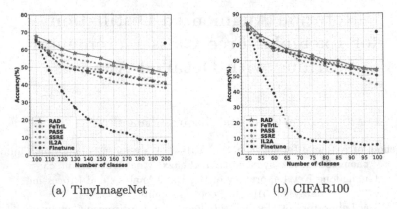

(a) TinyImageNet   (b) CIFAR100

Fig. 1. Incremental Accuracy of TinyImageNet and CIFAR100 with 10 incremental steps. The top-1 accuracy (%) after learning each task is shown. Existing SOTA methods achieve similar performance. The proposed Rotation Augmented Distillation (RAD) achieves the SOTA performance as well. The black dot in the upper right corner indicates the upper bound that the model trained on all the data. The black dotted line indicates the lower bound, a simple finetune method.

knowledge, a small number of exemplars from the previous tasks can be stored in the memory buffer, known as the exemplar-based methods [6,8,10]. However, exemplars from previous tasks can be unavailable due to constraints, e.g., user privacy or device limitations.

To this end, this paper focuses on exemplar-free class incremental learning (EFCIL) setting [11–13,15,17]. Model distillation [16] plays an important role in preserving past knowledge by existing methods [11–14]. Both the feature extractor and new classifier are learned with the classification loss on the new data and knowledge distillation from the previous model. The variations of deep features during the new task training are penalized by distillation loss. Other methods [15,18] learn a feature extractor in the initial task only and then fixed it for increment tasks. New classifiers are learned on the extracted features of new tasks while the old class statistics, e.g., prototype vectors and covariance matrices, are also computed and preserved. The state-of-the-art (SOTA) methods [11–13,15] follow either the two paradigms mentioned above. As shown in Fig. 1, these SOTA methods report similar overall performance, i.e., average incremental accuracy. However, further detailed analysis is not conducted within these methods. In this work, we further incorporate the measures of Forgetting and Intransigence as in [34]. The performance of SOTA EFCIL methods [11,12,15] under various settings is reproduced for more detailed comparisons and analysis.

As EFCIL is a challenging task, the existing experimental setting is with half of the dataset in the initial task. Therefore, a model trained on such an initial task may be strong enough for the following incremental tasks. A baseline method, named Feat*, is proposed for such demonstration and comparison. Specifically, a deep feature extractor is trained with the initial data and frozen. NME classifier is then used to discriminate different classes across incremental steps. However, Feat* achieves the SOTA level performance under the existing setting. In this work, we propose a simple yet effective method, Rotation Augmented Distillation (RAD), to enable the continuous training of the entire deep model along the increment tasks. On the one hand, the data augmentation used provides the plasticity by introducing the varied training samples during training. On the other hand, the knowledge distillation can achieve the stability by alleviating the forgetting of past knowledge. To alleviate the bias of the strong initial model, a more challenging setting with much fewer data in the initial task is introduced. Nevertheless, our RAD still achieves one of the best results among the SOTA methods due to its superior balance of stability and plasticity, as revealed by the detailed analysis.

The main contributions of this work are summarized as follows:

1. We provide a more detailed analysis of the EFCIL methods, rather than the overall results only in existing works. Specifically, the complementary metrics, forgetting and intransigence, are also used to evaluate SOTA methods. Detailed comparisons and analyses are thus enabled.
2. A simple and intuitive method, Rotation Augmented Distillation (RAD), is designed to better alleviate the plasticity-stability dilemma. Its effectiveness is demonstrated by its superior performance under various EFCIL settings.
3. A new challenging setting is provided to alleviate the bias brought by the strong initial model. Detailed comparisons and analyses are also conducted.

2   Related Work

Class incremental learning aims for a well-performing deep learner that can sequentially learn from the streaming data. Its main challenge is Catastrophic Forgetting [9] which depicts the deep models prone to entirely forgetting the knowledge learned from previous tasks. Different strategies have been proposed to handle this issue. Regularization strategies such as elastic weight consolidation (EWC) [19] use different metrics to identify and penalize the changes of important parameters of the original network when learning a new task. Rehearsal strategies [1,8,26,27] are widely adopted as well. The model is permitted to access data from previous tasks partially by maintaining a relatively small memory, enabling it to directly recall the knowledge from the previous data and mitigate forgetting. iCaRL [8] stores a subset of samples per class by selecting the good approximations to class means in the feature space. However, access to the exemplars of old tasks is not guaranteed and could be limited to data security and privacy constraints [25].

The exemplar-free class incremental learning (EFCIL) [2,11–13,15,17,18] is a challenging setting, where no data sample of old tasks can be directly stored. Existing EFCIL methods either use regularization to update the deep model for each incremental step [19,20] or adapt distillation to preserve past knowledge by penalizing variations for past classes during model updates [11–13]. Moreover, the prototypes of old tasks can be exploited in conjunction with distillation to improve overall performance, as shown in [11–13]. Specifically, PASS [12] proposes prototype augmentation in the deep feature space to improve the discrimination of classes learned in different increment tasks. A prototype selection mechanism is proposed in SSRE [11] to better enhance the discrimination between old and new classes. Feature generation for past classes is introduced in IL2A [13] by leveraging information about the class distribution. However, IL2A is inefficient to scale up for a large number of classes since a covariance matrix needs to be stored for each class. Inspired by the transfer learning scheme [28], independent shallow classifiers can be learned based on a fixed feature extractor, as in DeeSIL [18]. A pseudo-features generator is further exploited in FeTrIL [15] to create representations of past classes to improve classifier learning across tasks. The proposed Rotation Augmented Distillation (RAD) method is end-to-end trainable on the full model and simply based on a distillation strategy [16] and rotation data augmentation [21]. Detailed comparisons show that RAD benefits from the superior balance between stability and plasticity in EFCIL model training.

3   Methodology

3.1   Preliminary

In class incremental learning, a model is learned from an initial task (0) a sequence of T incremental tasks, where each task t has a set of n_t different classes $C_t = \{c_{t,1}, ..., c_{t,n_t}\}$. The classes in different tasks are disjoint, $C_i \cap C_j = \emptyset, i \neq j, i, j \in \{0, ..., T\}$. The training data of task t is denoted as D_t. D_t consists of data tuples (x, y) where x is an image and y is its corresponding ground-truth class label. A deep classification model Φ consists of two modules, the feature extractor F and the classifier head h. F is parameterized with θ and the feature representation of image x is obtained via $F(x) \in \mathbb{R}^d$. The classifier head h is parameterized with ω.

Under the exemplar-free class incremental learning (EFCIL) setting, a model can only access D_t when training on task t. At the initial task, the classification model Φ_0 is trained under the full supervision of D_0 and resulting in F_0 and h_0. At the incremental task $t, t \in \{1, ..., T\}$, Φ_t is partially initialized with θ_{t-1} and trained with D_t. The corresponding feature extractor F_t and classifier head h_t are learned. The overall classifier H_t is an aggregation of a set of task-specific classifiers $h_i, i = \{1, ..., t\}$, $H_t = \{h_0, h_2, \cdots, h_t\}$. During testing, data samples are from all observed classes so far with balanced distributions.

3.2   A Baseline Method: Feat*

The feature extractor learned at the initial task is frozen and denoted as F_0^*. The corresponding classifier h_0 is abandoned. F_0^* is used as the feature extractor across all tasks and no further training on the deep classifier is needed. This method is denoted as Feat*. F_0^* forwards all training samples in $D_t, t \in \{0, ..., T\}$

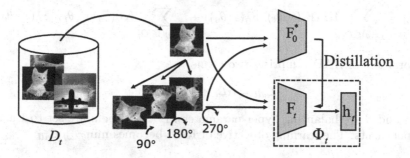

Fig. 2. Illustrations of the proposed Rotation Augmented Distillation (RAD) method for exemplar-free class incremental learning at task t. * indicates the corresponding module is frozen at training.

once for the class-specific mean feature vectors and such prototypes are preserved. NME classifier can then be applied to the seen prototypes so far during testing. Different from Feat*, a holistic classifier H_t for the seen classes till task t is trained in FeTrIL based on the pseudo-feature generation. During testing of FeTrIL, H_t rather than NME classifier is exploited.

3.3 Rotation Augmented Distillation

The deep classification model Φ_t can be end-to-end learned with two simple techniques, rotation data augmentation and knowledge distillation, as illustrated in Fig. 2. Specifically, for each class, we rotate each training sample x with 90, 180 and 270° [21] and obtained the augmented sample x':

$$\mathrm{x}' = \mathrm{rotate}(x, \delta), \delta \in \{90, 180, 270\}. \tag{1}$$

Each augmented sample x' subject to a rotated degree is assigned a new label y', extending the original K-class problem to a new 4K-class problem. The augmented dataset of task t is denoted as D'_t.

Both the feature extractor F_t and the classifier H_t of Φ_t are jointly optimized in RAD. Based on dataset D_t and the augmented one D'_t, the cross-entropy loss is computed,

$$\mathcal{L}_c = \sum_{(x,y) \in D_t} \mathrm{CE}(\Phi_t(x), y) + \sum_{(x',y') \in D'_t} \mathrm{CE}(\Phi_t(x'), y'). \tag{2}$$

To alleviate the mismatch between the saved old prototypes and the feature extractor, the knowledge distillation [16] is employed to regularize learning of the feature extractor. Specifically, we restrain the feature extractor by matching the features of new data extracted by the current model with that of the initial model F_0^*:

$$\mathcal{L}_{distil} = \sum_{(x,y)\in D_t} \mathrm{KL}(\mathrm{F}_t(x;\theta_t)||\mathrm{F}_0^*(x;\theta_0)) + \sum_{(x',y')\in D_t'} \mathrm{KL}(\mathrm{F}_t(x';\theta_t)||\mathrm{F}_0^*(x';\theta_0)).$$

$$(3)$$

The total loss of RAD comprises two terms,

$$\mathcal{L}_{all} = \alpha\mathcal{L}_c + \beta\mathcal{L}_{distil}, \tag{4}$$

with α and β are balancing hyper-parameters. Both of them are set to 1 across all experiments. The learning objective of RAD becomes $\min_{\theta_t,\omega_t} \mathcal{L}_{all}$.

4 Experiments

Datasets. The exemplar-free class incremental learning (EFCIL) is conducted on three datasets. Two large-scale datasets are TinyImageNet [29] and ImageNet100 [30]. The medium-size dataset used is CIFAR100 [31].

Protocols. Two EFCIL settings are followed in our experiments. (1) Conventional EFCIL setting. The conventional setting usually with half of the data as initial tasks. TinyImageNet contains images from 200 classes in total and the initial task (0) includes half of the dataset, i.e., images from 100 classes, denoted as **B100**. The data of the remaining classes are split into T tasks. For example, $T = 5$ corresponds to 5 incremental steps and each step contains the data of 20 classes. Therefore, this EFCIL setting of TinyImageNet is denoted as **B100 5 steps**. **B100 10 steps** and **B100 20 steps** are the other two EFCIL protocols of TinyImageNet. Both CIFAR100 and ImageNet100 are with 100 classes. They have the following three EFCIL protocols: **B50 5 steps**, **B50 10 steps** and **B40 20 steps**. (2) Challenging EFCIL setting. Training a method with half of the data can result in a strong initial model, which can introduce a non-negligible bias towards the following incremental learning performance. Therefore, a more challenging EFCIL setting is proposed by reducing the data in the initial task to half or even less. Specifically, the challenging EFCIL settings of TinyImageNet are **B50 5 steps**, **B50 10 steps** and **B50 25 steps**.

Implementation Details. The proposed Rotation Augmented Distillation (RAD) and the reproduced SOTA EFCIL methods [11,12,15] are implemented with PyTorch [32] and trained and tested on NVIDIA GTX 3080Ti GPU. Other results are from [15]. The initial model is trained for 200 epochs, and the learning rate is 0.1 and gradually reduces to zero with a cosine annealing scheduler. For incremental tasks, we train models for 30 epochs, and the initial learning rate is 0.001 with a cosine annealing scheduler. For a fair comparison, we adopt the default ConvNet backbone: an 18-layer ResNet [33]. The reproduced results of SOTA methods are with the same augmentation and assigned to incremental tasks using the same random seed as ours.

Evaluation Metrics. Three complementary metrics are used throughout the experiments. **Overall performance** is typically evaluated by average incremental accuracy [8]. After each batch of classes, the evaluation result is the classification accuracy curve. If a single number is preferable, the average of these accuracies is reported as average incremental accuracy. **Forgetting** [34] is

Table 1. Results of different EFCIL methods under the conventional setting. The overall performance, average incremental accuracy (%), is reported. Best results in red, second best in blue.

Methods	TinyImageNet			ImageNet100			CIFAR100		
	5 steps	10 steps	20 steps	5 steps	10 steps	20 steps	5 steps	10 steps	20 steps
Finetune	28.8	24.0	21.6	31.5	25.7	20.2	27.8	22.4	19.5
EWC [19]	18.8	15.8	12.4	-	20.4	-	24.5	21.2	15.9
LwF-MC [8]	29.1	23.1	17.4	-	31.2	-	45.9	27.4	20.1
DeeSIL [18]	49.8	43.9	34.1	67.9	60.1	50.5	60.0	50.6	38.1
LUCIR [24]	41.7	28.1	18.9	56.8	41.4	28.5	51.2	41.1	25.2
MUC [22]	32.6	26.6	21.9	-	35.1	-	49.4	30.2	21.3
SDC [17]	-	-	-	-	61.2	-	56.8	57.0	58.9
ABD [23]	-	-	-	-	-	-	63.8	62.5	57.4
IL2A [13]	47.3	44.7	40.0	-	-	-	66.0	60.3	57.9
PASS [12]	49.6	47.3	42.1	64.4	61.8	51.3	63.5	61.8	58.1
SSRE [11]	50.4	48.9	48.2	65.4	62.1	58.8	65.9	65.0	**61.7**
FeTrIL [15]	**54.8**	**53.1**	52.2	**72.2**	**71.2**	**67.1**	66.3	**65.2**	61.5
Feat*	53.3	53.0	**52.8**	70.1	69.9	65.1	63.9	63.6	59.8
RAD	55.9	55.6	55.2	72.4	71.8	67.4	66.5	65.3	61.9

defined to estimate the forgetting of previous tasks. The forgetting measure for the t-th task after the model has been incrementally trained up to task k as:

$$f_t^k = \max_{l \in \{0, \cdots, k-1\}} (a_{l,t} - a_{k,t}), \forall t < k. \qquad (5)$$

Note, $a_{m,n}$ is the accuracy of task n after training task m. The average forgetting at k-th task is then defined as $\mathbb{F}_k = \frac{1}{k} \sum_{i=0}^{k-1} f_i^k$. Lower $\mathbb{F}_k$ implies less forgetting on previous tasks. **Intransigence** [34] measures the inability of a model to learn new tasks. We train a reference model with dataset $\cup_{t=0}^k D_t$ and measure its accuracy on the held-out set of the k-th task, denoted as a_k^-. The intransigence for the k-th task as:

$$\mathbb{I}_k = a_k^- - a_{k,k}, \qquad (6)$$

where $a_{k,k}$ denotes the accuracy on the k-th task when trained up to task k in an incremental manner. Note, the lower the $\mathbb{I}_k$ the better the model.

4.1 Detailed Analysis Under Conventional EFCIL Setting

The overall results of different methods under conventional EFCIL settings of TinyImageNet, ImageNet100 and CIFAR100 are reported in Table 1. The simple baseline method, Feat*, achieves similar results compared to the SOTA methods. For example, the margins between FeTrIL and Feat* are no more than 2%. Feat* is observed to be slightly better (0.6% improvements) than FeTrIL under the TinyImageNet B100 20 steps setting. Therefore, it demonstrates that

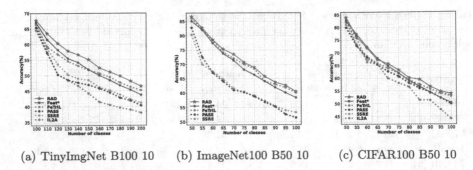

(a) TinyImgNet B100 10   (b) ImageNet100 B50 10   (c) CIFAR100 B50 10

Fig. 3. Incremental Accuracy Curves of the SOTA methods. Each point represents the incremental classification accuracy (%) after model learning on each task.

Table 2. Results on TinyImageNet and ImageNet100 with intransigence and forgetting metrics. $\mathbb{I}$ represents Intransigence (%), $\mathbb{F}$ represents Average Forgetting (%). Best results in red, second best in blue.

Methods	TinyImageNet						ImageNet100					
	5 steps		10 steps		20 steps		5 steps		10 steps		20 steps	
	$\mathbb{I}(\downarrow)$	$\mathbb{F}(\downarrow)$	$\mathbb{I}(\downarrow)$	$\mathbb{F}(\downarrow)$	$\mathbb{I}(\downarrow)$	$\mathbb{F}(\downarrow)$	$\mathbb{I}(\downarrow)$	$\mathbb{F}(\downarrow)$	$\mathbb{I}(\downarrow)$	$\mathbb{F}(\downarrow)$	$\mathbb{I}(\downarrow)$	$\mathbb{F}(\downarrow)$
PASS [12]	22.5	15.4	24.3	20.6	29.6	25.2	27.0	19.3	32.0	25.7	42.4	31.6
SSRE [11]	21.3	16.1	23.2	21.1	25.4	24.3	25.7	24.9	30.2	30.5	34.0	35.0
FeTrIL [15]	**18.9**	12.6	**19.4**	11.8	21.0	12.9	**21.7**	14.7	**23.4**	15.4	**27.7**	18.1
Feat*	20.9	**8.9**	20.9	9.0	**20.9**	8.9	25.1	8.9	25.1	9.8	31.9	10.4
RAD	18.0	8.7	18.1	**9.6**	18.4	**10.2**	21.5	**9.7**	23.0	**11.4**	27.6	**11.0**

a model trained with half of the dataset at the initial task may already equip sufficient classification capability for the subsequent incremental tasks even without further learning. This conclusion is frequently discussed in the previous sections of this paper and inspires the introduction of the challenging EFCIL setting. The proposed RAD consistently achieves the best results under conventional EFCIL settings of all datasets, as shown in Table 1. For example, RAD is better than FeTrIL on the largest dataset, TinyImageNet. The more incremental steps the larger improvements. Such improvement increased from 1.1% under 5 steps to 2.5% under 10 steps and reached 3.0% under 20 incremental steps. Detailed comparisons among different methods along the incremental learning procedure are also illustrated in Fig. 3. The curve of a superior method is usually above the inferior ones. For example, the superiority of RAD over FeTrIL can be demonstrated by their curves in Fig. 3(a).

To provide more detailed analyses on the SOTA EFCIL methods, their performance on another two metrics, forgetting and intransigence, are reported in Table 2. The proposed baseline Feat* achieves the lowest $\mathbb{F}$ under most cases. It shows that Feat* suffers the least from forgetting since the deep model is frozen at the initial state and does not fine-tuned on the following tasks. Moreover, the $\mathbb{I}$

Table 3. Ablative Study of the proposed RAD. Results of TinyImageNet B100 10 steps and ImageNet100 B50 10 steps are reported.

Components		TinyImageNet			ImageNet100		
Rotation	Distillation	Avg (↑)	$\mathbb{I}$ (↓)	$\mathbb{F}$ (↓)	Avg (↑)	$\mathbb{I}$ (↓)	$\mathbb{F}$ (↓)
		24.0	57.0	76.1	25.7	77.3	90.2
✓		11.2	60.1	74.3	13.7	78.7	87.7
	✓	52.9	21.1	11.6	70.0	24.2	26.7
✓	✓	55.6	18.1	9.6	71.8	23.0	11.4

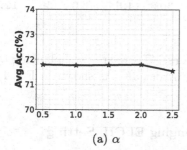 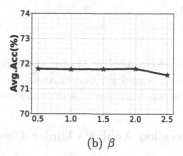

(a) α (b) β

Fig. 4. Impacts of varying α and β on overall results of ImgNet100 B50 10 steps.

of Feat* still achieves mid-level performance. It further demonstrates the that the initial model trained on half of the dataset can provide sufficient discriminative power for the rest classes, as described above. FeTrIL achieves the best overall performance among the exisitng SOTA methods, as shown in Table 1. Therefore, both $\mathbb{I}$ and $\mathbb{F}$ of FeTrIL are lower than those of its counterpart SOTA methods, as shown in Table 2. Similarly, the proposed RAD achieves the best overall performance on various datasets. This is consistent with the results in Table 2. The $\mathbb{I}$ and $\mathbb{F}$ of RAD are either the best or second best results. Specifically, RAD is only slightly worse than Feat* on the $\mathbb{F}$ metric. This is compensated by the best $\mathbb{I}$ performance of RAD. Moreover, RAD is consistently better than FeTrIL on both metrics. Therefore, the superiority of our RAD relies on the more balanced performance between plasticity and stability.

Ablation Study. Contributions of different components in the proposed RAD are illustrated in Table 3. First, applying data rotation in model finetuning harms the overall performance, as the augmented images seem to alleviate the forgetting issue (lower $\mathbb{F}$) but make less discriminative new knowledge learned (higher $\mathbb{I}$). Second, model distillation clearly boosts the overall performance. Distillation helps the model learning not only to defy forgetting the old knowledge but also to better capture the discriminative information from new tasks, as its $\mathbb{F}$ and $\mathbb{I}$ are relatively low. Finally, RAD combines these two techniques. Detailed analysis shows that model distillation benefits from rotation data augmentation and then achieves a superior balance between plasticity and stability.

Table 4. Results of TinyImageNet under challenging EFCIL setting with average incremental accuracy (Avg.), intransigence ($\mathbb{I}$) and forgetting metrics ($\mathbb{F}$) reported. Best results in red, second best in blue.

Methods	5 steps			10 steps			25 steps		
	Avg. ($\uparrow$)	$\mathbb{I}$ ($\downarrow$)	$\mathbb{F}$ ($\downarrow$)	Avg. ($\uparrow$)	$\mathbb{I}$ ($\downarrow$)	$\mathbb{F}$ ($\downarrow$)	Avg. ($\uparrow$)	$\mathbb{I}$ ($\downarrow$)	$\mathbb{F}$ ($\downarrow$)
PASS [12]	45.1	30.3	15.5	42.5	32.0	20.2	39.7	34.6	26.5
SSRE [11]	43.6	32.2	12.2	41.1	33.7	14.8	40.1	33.9	21.5
FeTrIL [15]	**47.9**	29.7	13.5	**46.5**	**30.4**	13.4	**45.1**	**30.8**	13.9
Feat*	44.1	34.6	10.1	43.3	34.6	9.9	42.7	34.6	10.1
RAD	48.6	**30.2**	**11.6**	47.7	30.3	**11.5**	47.0	30.2	**12.3**

Impacts of Balancing Hyper-parameters. There are two balancing hyper-parameters, α and β. They are set to 1.0 by default. As shown in Fig. 4, Our method is not sensitive to such changes.

4.2   Detailed Analysis Under Challenging EFCIL Setting

The results of challenging EFCIL setting are reported in Table 4. Comparing the overall performance, the average incremental accuracy, of the methods in Table 4 with those in Table 1, all of them experience significant drops, usually more than 6%, due to much less data being provided in the initial task under the new setting. Specifically, the baseline Feat* suffers from the worst performance degradation, from 9.2% to 10.1%, under different incremental steps. This is because Feat* is a simple CIL method that relies on the initial task training only. With less initial data provided, the corresponding model can be much less generalizable on the new data from unseen classes. Under the challenging EFCIL setting, the proposed RAD still achieves the best overall performance against the existing SOTA methods. With the $\mathbb{I}$ and $\mathbb{F}$ of different methods compared in Table 4, RAD consistently achieve either the best or second best performance on these metrics. It demonstrates that RAD can better alleviate the plasticity and stability dilemma in EFCIL than existing methods.

5   Conclusion

We provide a more detailed analysis and comparison of different exemplar-free class incremental learning (EFCIL) methods than many existing works. Besides the overall performance, i.e., the average incremental accuracy, two complementary metrics, the Forgetting and the Intransigence, are included to measure the EFCIL methods from the perspectives of stability and plasticity. A simple yet effective EFCIL method, Rotation Augmented Distillation (RAD), has been proposed. RAD consistently achieves one of the state-of-the-art overall performances under various EFCIL settings. Based on the detailed analysis, we find that the superiority of RAD comes from the more balanced performance between plasticity and stability. Moreover, a challenging EFCIL new setting with much fewer data in initial tasks is proposed. It aims to alleviate the bias of a strong initial pre-trained model and stand out the incremental learning performance.

Acknowledgement. This research is supported by the National Science Foundation for Young Scientists of China (No. 62106289).

References

1. Zhong, C., et al.: Discriminative distillation to reduce class confusion in continual learning. In: Proceedings of the Pattern Recognition and Computer Vision (PRCV) (2022)
2. Huang, T., Qu, W., Zhang, J.: Continual representation learning via auto-weighted latent embeddings on person ReIDD. In: Ma, H., et al. (eds.) PRCV 2021. LNCS, vol. 13021, pp. 593–605. Springer, Cham (2021). https://doi.org/10.1007/978-3-030-88010-1_50
3. Shaheen, K., Hanif, M.A., Hasan, O., et al.: Continual learning for real-world autonomous systems: algorithms, challenges and frameworks. J. Intell. Robot. Syst. (2022)
4. Geiger, A., Lenz, P., Urtasun, R.: Are we ready for autonomous driving? the KITTI vision benchmark suite. In: Proceedings of the IEEE Conference on Computer Vision and Pattern Recognition(CVPR) (2012)
5. Li, K., Chen, K., Wang, H., et al.: CODA: a real-world road corner case dataset for object detection in autonomous driving. In: Proceedings of the European Conference on Computer Vision (ECCV) (2022)
6. Rolnick, D., Ahuja, A., Schwarz, J., et al.: Experience replay for continual learning. In: Advances in Neural Information Processing Systems (NeurIPS) (2019)
7. Masana, M., Liu, X., Twardowski, B., et al.: Class-incremental learning: survey and performance evaluation on image classification. IEEE Trans. Pattern Anal. Mach. Intell. (TPAMI) (2022)
8. Rebuffi, S.A., Kolesnikov, A., Sperl, G., et al.: iCaRL: incremental classifier and representation learning. In: Proceedings of the IEEE Conference on Computer Vision and Pattern Recognition (CVPR) (2017)
9. French, R.M.: Catastrophic forgetting in connectionist networks. Trends Cogn. Sci. (1999)
10. Zhou, D.W., Yang, Y., Zhan, D.C.: Learning to classify with incremental new class. IEEE Trans. Neural Netw. Learn. Syst. (2021)
11. Zhu, K., et al.: Self-sustaining representation expansion for non-exemplar class-incremental learning. In: Proceedings of the IEEE Conference on Computer Vision and Pattern Recognition (CVPR) (2022)
12. Zhu, F., et al.: Prototype augmentation and self-supervision for incremental learning. In: Proceedings of the IEEE Conference on Computer Vision and Pattern Recognition (CVPR) (2021)
13. Zhu, F., et al.: Class-incremental learning via dual augmentation. In: Advances in Neural Information Processing Systems (NeurIPS) (2021)
14. Xu, Q., et al.: Constructing deep spiking neural networks from artificial neural networks with knowledge distillation. In: Proceedings of the IEEE/CVF Conference on Computer Vision and Pattern Recognition (CVPR) (2023)

15. Petit, G., et al.: FetrIL: feature translation for exemplar-free class-incremental learning. In: Proceedings of the IEEE/CVF Winter Conference on Applications of Computer Vision (WACV) (2023)
16. Hinton, G., Vinyals, O., Dean, J.: Distilling the knowledge in a neural network. arXiv preprint arXiv:1503.02531 (2015)
17. Yu, L., et al.: Semantic drift compensation for class-incremental learning. In: Proceedings of the IEEE Conference on Computer Vision and Pattern Recognition (CVPR) (2020)
18. Belouadah, E., Popescu, A.: DeeSIL: deep-shallow incremental learning. In: Proceedings of the European Conference on Computer Vision (ECCV) Workshop (2018)
19. Kirkpatrick, J., et al.: Overcoming catastrophic forgetting in neural networks. In: Proceedings of the National Academy of Sciences(PNAS) (2017)
20. Dhar, P., et al.: Learning without memorizing. In: Proceedings of the IEEE Conference on Computer Vision and Pattern Recognition (CVPR) (2019)
21. Lee, H., Ju Hwang, S., Shin, J.: Self-supervised label augmentation via input transformations. In: International Conference on Machine Learning (ICML) (2020)
22. Liu, Y., et al.: More classifiers, less forgetting: a generic multi-classifier paradigm for incremental learning. In: Proceedings of the European Conference on Computer Vision (ECCV) (2020)
23. Smith, J., et al.: Always be dreaming: a new approach for data-free class-incremental learning. In: Proceedings of the IEEE International Conference on Computer Vision (ICCV) (2021)
24. Hou, S., et al.: Learning a unified classifier incrementally via rebalancing. In: Proceedings of the IEEE Conference on Computer Vision and Pattern Recognition (CVPR) (2019)
25. Wang, Y., Huang, Z., Hong. X.: S-prompts learning with pre-trained transformers: an Occam's Razor for domain incremental learning. In: Advances in Neural Information Processing Systems (NeurIPS) (2022)
26. Wu, Y., et al.: Large scale incremental learning. In: Proceedings of the IEEE Conference on Computer Vision and Pattern Recognition (CVPR) (2019)
27. Zhao, B., et al.: Maintaining discrimination and fairness in class incremental learning. In: Proceedings of the IEEE Conference on Computer Vision and Pattern Recognition (CVPR) (2020)
28. Ginsca, A.L., Popescu, A., Le Borgne, H., Ballas, N., Vo, P., Kanellos, I.: Large-scale image mining with flickr groups. In: He, X., Luo, S., Tao, D., Xu, C., Yang, J., Hasan, M.A. (eds.) MMM 2015. LNCS, vol. 8935, pp. 318–334. Springer, Cham (2015). https://doi.org/10.1007/978-3-319-14445-0_28
29. Le, Y., Xuan Y.: Tiny imageNet visual recognition challenge. CS 231N 7.7, 3 (2015)
30. Russakovsky, O., et al.: ImageNet large scale visual recognition challenge. Int. J. Comput. Vis. (IJCV) (2015)
31. Krizhevsky, A., Hinton, G.: Learning multiple layers of features from tiny images. Technical Report TR 2009, University of Toronto, Toronto (2009)
32. Grossberg, S.T.: Studies of mind and brain: neural principles of learning, perception, development, cognition, and motor control. Springer Science & Business Media (2012). https://doi.org/10.1007/978-94-009-7758-7
33. He, K., et al.: Deep residual learning for image recognition. In: Proceedings of the IEEE Conference on Computer Vision and Pattern Recognition (CVPR) (2016)
34. Chaudhry, A., et al.: Riemannian walk for incremental learning: understanding forgetting and intransigence. In: Proceedings of the European Conference on Computer Vision (ECCV) (2018)

Nonconvex Tensor Hypergraph Learning for Multi-view Subspace Clustering

Xue Yao and Min Li[✉]

College of Mathematics and Statistics, Shenzhen University, Shenzhen 518060, China
`limin800@szu.edu.cn`

Abstract. Low-rank representation has been widely used in multi-view clustering. But the existing methods are matrix-based, which cannot well capture high-order low-rank correlation embedded in multiple views and fail to retain the local geometric structure of features resided in multiple nonlinear subspaces simultaneously. To handle this problem, we propose a nonconvex tensor hypergraph learning for multi-view subspace clustering. In this model, the hyper-Laplacian regularization is used to capture high-order global and local geometric information of all views. The nonconvex weighted tensor Schatten-p norm can better characterize the high-order correlations of multi-view data. In addition, we design an effective alternating direction algorithm to optimize this nonconvex model. Extensive experiments on five datasets prove the robustness and superiority of the proposed method.

Keywords: Hypergarph learning · Tensor Schatten-p norm · Low-rank representation

1 Introduction

Multi-view clustering has been a hot topic in artificial intelligence and machine learning [1,12,15,21]. Multi-view clustering aims to use heterogeneous features of data samples to partition data into several clustering. Extensive methods have been proposed for multi-view clustering, among which multi-view graph clustering and multi-view subspace clustering (MVSC) [3,7] are the most popular.

The graph-based methods usually represent each data point as a vertex and the pairwise similarity by edges [14]. For instance, Nie et al. [10] proposed a framework via the reformulation of the standard spectral learning model, and learnt an optimal weight for each graph automatically. Tang et al. [11] first proposed an effective and parameter-free method to learn a unified graph via cross-view graph diffusion. These methods have achieved a good results due

---

This work is supported in part by the National Nature Science Foundation of China (62072312, 61972264), in part by Shenzhen Basis Research Project (JCYJ20210324094009026, JCYJ20200109105832261).

Q. Liu et al. (Eds.): PRCV 2023, LNCS 14428, pp. 39–51, 2024.
https://doi.org/10.1007/978-981-99-8462-6_4

to well exploring the complementary information and global geometric structure. However, these graph methods ignore high-order information embedding in multi-view data.

The existing methods for multi-view subspace clustering always use self-representation to construct the similarity matrix or coefficient tensor [5]. Among multiple methods, low-rank representation (LRR) [16,25] is the most representative technology. There are two categories for LRR-based multi-view subspace clustering: 1)the matrix-based optimization methods [6,19], and 2)the tensor-based optimization methods [22,27]. Although these methods have impressive results, they still have shortcomings. For example, the matrix-based approaches could not well capture the high-order information among all views of data. The tensor-based methods often use tensor nuclear norm to regularize the coefficient tensor, which overlooks the difference of each singular value. In addition, the tensor-based methods usually rely solely on low-rank representation, which fails to characterize the multi-linear local geometric structure [16], which is important to cluster complex structured datasets, such as textual data and gene data.

To solve the aforementioned deficiencies, we propose the nonconvex tensor hypergraph learning for multi-view subspace clustering (THMSC). The main contributions of our method are summarized as follows:

- We propose to use the hyper-Laplacian regularization to capture global and local geometric structure in multiple self-representation matrices. We advocate adopting the weighted tensor Schatten p-norm to characterize the high-order low-rank information embedding in coefficient tensor. We integrates the tensor Schatten p-norm regularization and hyper-Laplacian regularization into a unified framework, which considers the prior information of singular values.
- We design an effective algorithm to solve this nonconvex model. Extensive experiments in several datasets show the robustness and effectiveness of our proposed method.

2   Related Work

In this section, we briefly review the multi-view clustering methods and the hyper-Laplacian learning. The notation and preliminaries of tensor used in this paper are introduced in works [4,20,28].

2.1   Multi-view Subspace Clustering

Multi-view subspace clustering has attracted extensive attention in machine learning and artificial intelligence [9], which aims to learn a new affinity matrix shared by different views. Based on multi-view subspace clustering, many methods have emerged in the past few years. For instance, Cao et al. [1] utilized the Hilbert Schmidt Independence Criterion to explore the complementarity information of multi-view data. Xia et al. [19] proposed a Markov chain method for

Robust Multi-view Spectral Clustering (RMSC) to construct a shared transition probability matrix. Wang et al. [18] proposed a multi-graph Laplacian regularized LRR to characterize local manifold structure. Zhang et al. [26] sought the underlying latent representation and explored the complementary information of multi-view. Zhang et al. [27] proposed a low-rank tensor constrained multi-view subspace clustering (LTMSC) method, which performed the Tucker tensor decomposition on the coefficient tensor. Xie et al. [22] proposed t-SVD based on multi-view subspace clustering (t-SVD-MSC).

2.2 Hypergraph Learning

Given a hypergraph $\mathbf{G} = (\mathbf{V}, \mathbf{E}, \mathbf{W})$, $\mathbf{V}$ represents the set of vertices, $\mathbf{E}$ is the set of hypergraph, and $\mathbf{W}$ is the weight matrix, whose elements are positive. We define the degrees of vertex $\mathbf{v}$ and edge $\mathbf{e}$ as $d(\mathbf{v}) = \sum_{\mathbf{e} \in \mathbf{E}} \mathbf{w}(\mathbf{e}) h(\mathbf{v}, \mathbf{e})$, $d(\mathbf{e}) = \sum_{\mathbf{v} \in \mathbf{V}} h(\mathbf{v}, \mathbf{e})$, where $h(\mathbf{v}, \mathbf{e})$ is the entry of the incidence matrix $\mathbf{H} \in \mathbb{R}^{|\mathbf{V}| \times |\mathbf{E}|}$, and $\mathbf{w}(\mathbf{e})$ is the entry associated with each hyperedge in weight matrix $\mathbf{W}$. $\mathbf{H}$ can represent the relationship between the vertices and the hyperedges, and is defined as follows:

$$h(\mathbf{v}, \mathbf{e}) = \begin{cases} 1, & \text{if } \mathbf{v} \in \mathbf{e} \\ 0, & otherwise. \end{cases} \tag{1}$$

Let $\mathbf{D_V}$ and $\mathbf{D_E}$ be diagonal matrices and correspond to the vertex degrees $d(\mathbf{v})$ and the hyperedge degrees $d(\mathbf{e})$, respectively. Then, the normalized hypergraph Laplacian matrix $\mathbf{L}_h = \mathbf{D_V} - \mathbf{H}\mathbf{W}\mathbf{D_E^{-1}}\mathbf{H}^T$.

For hyper-Laplacian regularizer, there is a assumption that if two data points x_i and x_j are close in the geometric structure of the original data distribution. Then, they also keep close in the new representation space where the neighboring points are linearly related [24]. So, the hyper-Laplacian regularizer on self-representation coefficient matrices can preserve the local geometric structure embedding in the high-dimensional representation space [23].

According to [24], the hyper-Laplacian regularizer on self-representation coefficients can be written as

$$\min_{\mathbf{Z}^{(v)}} \text{tr}(\mathbf{Z}^{(v)} \mathbf{L}_h^{(v)} \mathbf{Z}^{(v)^T}). \tag{2}$$

where $\mathbf{Z}^{(v)}$ and $\mathbf{L}_h^{(v)}$ are the representation coefficient matrix and hyper-graph Laplacian matrix of the v-th view, respectively.

3 The Proposed Approach

3.1 The Proposed Model

Tensor-based multi-view subspace clustering methods have got some promising results. However, most methods couldn't consider the prior information of matrix singular values based on tensor nuclear norm. In addition, the tensor nuclear norm regularizer can well capture the global low-dimensional structure, but can not obtain the local geometric structure of data features embedding in multiple nonlinear subspaces. To handle the limitations, we proposed a new hyper-Laplacian regularized tensor low-rank representation for multi-view subspace clustering.

Let $\mathbf{X}^{(v)} = [x_1^{(v)}, x_2^{(v)}, \ldots, x_N^{(v)}] \in \mathbb{R}^{d^{(v)} \times N}$ denotes the data matrix of the vth view, where $d^{(v)}$ is the dimension of data points, and N is the number of data points. $\mathbf{Z}^{(v)} = [z_1^{(v)}, z_2^{(v)}, \ldots, z_N^{(v)}] \in \mathbb{R}^{N \times N}$ is coefficient self-representation matrix of the vth view, and $\mathbf{E}^{(v)}$ denotes the noise and outliers. The proposed model is formulated as:

$$\min_{\mathcal{Z}, \mathbf{E}} \|\mathcal{Z}\|_{\omega, S_p}^p + \lambda_1 \|\mathbf{E}\|_{2,1} + \lambda_2 \sum_{v=1}^{M} \mathrm{tr}\left(\mathbf{Z}^{(v)} \mathbf{L}_h^{(v)} \mathbf{Z}^{(v)^T}\right)$$
$$\text{s.t. } \mathbf{X}^{(v)} = \mathbf{X}^{(v)} \mathbf{Z}^{(v)} + \mathbf{E}^{(v)}, v = 1, \ldots, M$$
$$\mathcal{Z} = \Phi\left(\mathbf{Z}^{(1)}, \mathbf{Z}^{(2)}, \ldots, \mathbf{Z}^{(M)}\right)$$
$$\mathbf{E} = \left[\mathbf{E}^{(1)}; \mathbf{E}^{(2)}; \cdots ; \mathbf{E}^{(M)}\right]. \tag{3}$$

where Φ denotes the construction of a tensor $\mathcal{Z} \in \mathbb{R}^{N \times M \times N}$ depending on $\mathbf{Z}^{(v)}$. For details, see [20]. $\mathcal{Z}(:, v, :) = \mathbf{Z}^{(v)}$ is the lateral slices of tensor $\mathcal{Z}$. $\mathbf{E}$ assembles those error matrices of each view by column alignment, $\mathbf{L}_h^{(v)}$ is the hyper-Laplacian matrix of the v-th view, and M is the number of multiple view. λ_1 and λ_2 are both non-negative parameters. The weighted tensor Schatten-p norm(WTSNM) $\|\mathcal{Z}\|_{\omega, S_p}^p$ in model (3) is defined as follows:

$$\|\mathcal{Z}\|_{\omega, S_p}^p = (\sum_{i=1}^{N} \sum_{j=1}^{h} \omega_j \cdot \sigma_j(\overline{\mathcal{Z}}^{(i)})^p)^{\frac{1}{p}}. \tag{4}$$

where p is the parameter of power, ω_j denotes the weighted term of the j-th singular value, and $\sigma_j(\overline{\mathcal{Z}}^{(i)})$ denotes the j-th sigular value of frontal slices $\overline{\mathcal{Z}}^{(i)}$, $h = min(N, M)$. $\overline{\mathcal{Z}}$ is the result of the fast Fourier transform of $\mathcal{Z}$. Intuitively, the WTSNM can employ certain weight and power to shrink the larger singular values less to better capture the high-order low-rank structure of the tensor. Then, the hyper-Laplacian regularizer can characterize the local geometric information. To solve model (3), we use the Augmented Lagrange Multiplier (ALM) and adopt alternating direction minimizing strategy. We introduce the M auxiliary matrices variable $\mathbf{Q}^{(v)}$ $(v = 1, \ldots, M)$ and one auxiliary tensor $\mathcal{J}$ to replace

$\mathbf{Z}^{(v)}$ and $\mathcal{Z}$, respectively. Then, the original problem can be rewritten as follows:

$$
\mathcal{L}\left(\mathbf{Z}^{(1)}, \ldots, \mathbf{Z}^{(M)}, \mathcal{J}, \mathbf{E}^{(1)}, \ldots, \mathbf{E}^{(M)}, \mathbf{Q}^{(1)}, \ldots, \mathbf{Q}^{(M)}\right)
$$

$$
= \|\mathcal{J}\|_{\omega, S_p}^p + \lambda_1 \|\mathbf{E}\|_{2,1} + \lambda_2 \sum_{v=1}^{M} \mathrm{tr}\left(\mathbf{Q}^{(v)} \mathbf{L}_h^v \mathbf{Q}^{(v)T}\right)
$$

$$
+ \sum_{v=1}^{M} \left(\left\langle \mathbf{Y}_1^{(v)}, \mathbf{X}^{(v)} - \mathbf{X}^{(v)} \mathbf{Z}^{(v)} - \mathbf{E}^{(v)} \right\rangle\right.
$$

$$
\left. + \frac{\mu_1}{2} \left\|\mathbf{X}^{(v)} - \mathbf{X}^{(v)} \mathbf{Z}^{(v)} - \mathbf{E}^{(v)}\right\|_F^2\right)
$$

$$
+ \sum_{v=1}^{M} \left(\left\langle \mathbf{Y}_2^{(v)}, \mathbf{Z}^{(v)} - \mathbf{Q}^{(v)} \right\rangle + \frac{\mu_2}{2} \left\|\mathbf{Z}^{(v)} - \mathbf{Q}^{(v)}\right\|_F^2\right)
$$

$$
+ \langle \mathcal{G}, \mathcal{Z} - \mathcal{J} \rangle + \frac{\rho}{2} \|\mathcal{Z} - \mathcal{J}\|_F^2. \tag{5}
$$

where $\mathcal{G}$, $\mathbf{Y}_1^{(v)}$, $\mathbf{Y}_1^{(v)}$ are the Lagrange multipliers, and μ_1, μ_2 and ρ are the penalty parameters.

3.2   Optimization

The optimization scheme could divided into four steps.

$\mathbf{Z}^{(v)}$-Subproblem: When $\mathbf{E}$, $\mathbf{Q}^{(v)}$, $\mathcal{J}$ are fixed, and $\Phi_{(v)}^{-1}(\mathcal{J}) = \mathbf{J}^{(v)}$ and $\Phi_{(v)}^{-1}(\mathcal{G}) = \mathbf{G}^{(v)}$, $\mathbf{Z}^{(v)}$ can be updated by solving the following subproblem:

$$
\underset{\mathbf{Z}^{(v)}}{\arg\min} \left(\left\langle \mathbf{Y}_1^{(v)}, \mathbf{X}^{(v)} - \mathbf{X}^{(v)} \mathbf{Z}^{(v)} - \mathbf{E}^{(v)} \right\rangle\right.
$$

$$
\left. + \frac{\mu_1}{2} \left\|\mathbf{X}^{(v)} - \mathbf{X}^{(v)} \mathbf{Z}^{(v)} - \mathbf{E}^{(v)}\right\|_F^2\right)
$$

$$
+ \left(\left\langle \mathbf{Y}_2^{(v)}, \mathbf{Z}^{(v)} - \mathbf{Q}^{(v)} \right\rangle + \frac{\mu_2}{2} \left\|\mathbf{Z}^{(v)} - \mathbf{Q}^{(v)}\right\|_F^2\right)
$$

$$
+ \langle \mathbf{G}^{(v)}, \mathbf{Z}^{(v)} - \mathbf{J}^{(v)} \rangle + \frac{\rho}{2} \|\mathbf{Z}^{(v)} - \mathbf{J}^{(v)}\|_F^2. \tag{6}
$$

By setting the derivative of Eq. (6) to zero, we obtain

$$
\mathbf{Z}_t^{(v)*} = (\mu_1 \mathbf{X}^{(v)T} \mathbf{X}^{(v)} + \mu_2 \mathbf{I} + \rho \mathbf{I})^{-1}
$$

$$
(\mu_1 \mathbf{X}^{(v)T} \mathbf{X}^{(v)} + \mathbf{X}^{(v)T} \mathbf{Y}_1^{(v)} + \rho \mathbf{J}^{(v)}
$$

$$
- \mu_1 \mathbf{X}^{(v)T} \mathbf{E}^{(v)} - \mathbf{G}^{(v)} - \mathbf{Y}_2^{(v)} + \mu_2 \mathbf{Q}^{(v)}). \tag{7}
$$

$\mathbf{Q}^{(v)}$-Subproblem: When $\mathbf{E}, \mathbf{Z}^{(v)}, \mathcal{J}$ are fixed, $\mathbf{Q}^{(v)}$ can be updated by solving the following subproblem:

$$\mathbf{Q}^* = \arg\min_{\mathbf{Q}^{(v)}} \lambda_2 \text{tr}\left(\mathbf{Q}^{(v)}\mathbf{L}_h^{(v)}\mathbf{Q}^{(v)^T}\right) + \langle \mathbf{Y}_2^{(v)}, \mathbf{Z}^{(v)} - \mathbf{Q}^{(v)}\rangle$$

$$+ \frac{\mu_2}{2}\left\|\mathbf{Z}^{(v)} - \mathbf{Q}^{(v)}\right\|_F^2$$

$$= (\mu_2\mathbf{Z}^{(v)} + \mathbf{Y}_2^{(v)})(2\lambda_2\mathbf{L}_h^{(v)} + \mu_2\mathbf{I})^{-1}. \tag{8}$$

$\mathbf{E}^{(v)}$-Subproblem: When $\mathbf{Z}^{(v)}$ are fixed, we have:

$$\mathbf{E}^* = \arg\min_{\mathbf{E}} \frac{\lambda_1}{\mu_1}\|\mathbf{E}\|_{2,1} + \frac{1}{2}\|\mathbf{E} - \mathbf{B}\|_F^2. \tag{9}$$

the solution of (9) is

$$\mathbf{E}^*_{:,i} = \begin{cases} \frac{\|\mathbf{B}_{:,i}\|_2 - \frac{\lambda_1}{\mu_1}}{\|\mathbf{B}_{:,i}\|_2}\mathbf{B}_{:,i} & \|\mathbf{B}_{:,i}\|_2 > \frac{\lambda_1}{\mu_1} \\ 0 & \text{otherwise} \end{cases}. \tag{10}$$

$\mathbf{B}_{:,i}$ is the ith column of $\mathbf{B} = [\mathbf{B}^{(1)}; \ldots; \mathbf{B}^{(M)}]$, $\mathbf{B}^{(v)} = \mathbf{X}^{(v)} - \mathbf{X}^{(v)}\mathbf{Z}^{(v)} + \frac{1}{\mu_1}\mathbf{Y}_1^{(v)}, v = 1, \ldots, M$.

$\mathcal{J}$-subproblem: Fixing variable $\mathbf{Z}^v$, then we have.

$$\mathcal{J}^* = \arg\min_{\mathcal{J}} \|\mathcal{J}\|_{\omega,S_p}^p + \frac{\rho}{2}\left\|\mathcal{Z} - \mathcal{J} + \frac{\mathcal{G}}{\rho}\right\|_F^2. \tag{11}$$

According to the algorithm introduced in [20], the solution of model (11) is

$$\mathcal{J}^* = \Gamma_{\frac{1}{\rho}\cdot n_3\cdot\omega}(\mathcal{Z} + \frac{1}{\rho}\mathcal{G}). \tag{12}$$

Updating Lagrange multiplier $\mathcal{G}, \mathbf{Y}_1^{(v)}, \mathbf{Y}_2^{(v)}$, they are solved by

$$\mathcal{G} = \mathcal{G} + \rho(\mathcal{Z} - \mathcal{J}). \tag{13}$$

$$\mathbf{Y}_1^{(v)} = \mathbf{Y}_1^{(v)} + \mu_1(\mathbf{X}^{(v)} - \mathbf{X}^{(v)}\mathbf{Z}^{(v)} - \mathbf{E}^{(v)}). \tag{14}$$

$$\mathbf{Y}_2^{(v)} = \mathbf{Y}_2^{(v)} + \mu_2(\mathbf{Z}^{(v)} - \mathbf{Q}^{(v)}). \tag{15}$$

In summary, the above optimization is dentoed aa Algorithm 1.

3.3 Complexity Analysis

The computational complexity of our proposed method mainly involves three variables $\mathcal{J}$, $\mathbf{E}$, and $\mathbf{L}_h^{(v)}$. The complexities for solving these variables iteratively are $\mathcal{O}(2N^2M\log(N) + N^2M^2)$, $\mathcal{O}(2N^2M)$ and $\mathcal{O}(N^2M\log(N))$. In addition, adding the computational complexity of the spectral clusteing is $\mathcal{O}(N^3)$, the total complexity of our method is $\mathcal{O}(T(2N^2M\log(N)+N^2M^2))+\mathcal{O}(N^3)$, where T is the number of iterations.

Algorithm 1. THMSC

Input: multi-view data $(\mathbf{X}^{(v)}, v = 1, 2, \ldots M)$; $\lambda_1, \lambda_2, \omega$.

Initialize: $\mathbf{Z}^{(v)}$, $\mathbf{E}^{(v)}$, $\mathbf{Y}_1^{(v)}, \mathbf{Y}_2^{(v)}$, $\mathcal{J}, \mathcal{G}$ initailized to $\mathbf{0}$; $\rho = \mu_1 = \mu_2 = 10^{-5}, \gamma = 2, \rho_{max} = \mu_{max} = 10^{10}, \epsilon = 10^{-7}$.

1: **while** not converged **do**
2: Update $\mathbf{Z}^{(v)}$ by (7);
3: Update $\mathbf{E}$ by (10);
4: Update $\mathbf{Q}^{(v)}$ by (8);
5: Update $\mathcal{J}$ by (12);
6: Updating Lagrange multiplier $\mathcal{G}, \mathbf{Y}_1^{(v)}, \mathbf{Y}_2^{(v)}$ according to (13), (14), (15);
7: Update parameters μ_1, μ_2 and ρ by $\mu_i = min(\gamma\mu_i, \mu_{max})i = 1, 2$ and $\rho = min(\gamma\rho, \rho_{max})$;
8: Check the convergence conditions:
$$max \left\{ \begin{matrix} \left\| \mathbf{X}^{(v)} - \mathbf{X}^{(v)}\mathbf{Z}^{(v)} - \mathbf{E}^{(v)} \right\|_\infty \\ \left\| \mathbf{Z}^{(v)} - \mathbf{J}^{(v)} \right\|_\infty \end{matrix} \right\} < \epsilon;$$
9: **end while**
10: Obtain affinity matrix $\mathbf{S} = \frac{1}{M} \sum_{v=1}^M \left| \mathbf{Z}^{(v)} \right| + \left| \mathbf{Z}^{(v)T} \right|$;
11: Apply spectral clustering on matrix $\mathbf{S}$;
Output: Clustering result $\mathcal{C}$.

---

4   Experimental Results and Analysis

4.1   Database and Competitors

We conduct the experiments on **ORL**[1], **Prokaryotic**, **BBC4view**[2], **BBC-Sport**[3] and **Scene-15** [22][4]. The experimental results of all compared methods are derived from their open source code, and the parameter settings follow the original paper. We choose the nine state-of-the-art clustering methods as competitors, which includes RMSC [19],DiMSC [1], LT-MSC [27], MLAN [8], ECMSC [17], t-SVD-MSC [22], GMC [13], LMSC [26] and LRTG [2].

Parameter Setting. In our experiments, we tune the parameter λ_1 and λ_2 in the range of [0.001, 0.01, 0.1, 0.2, 0.3, 0.4] and [1e-6, 1e-4, 1e-2, 0.1, 0.2, 0.4, 0.8, 1, 1.2, 1.4, 1.6, 1.8, 2, 5], respectively, the power $p \in (0, 1]$, and the weight $\omega_i \in (0, 100]$ to obtain the best results. Specifically, λ_1 and λ_2 are set to 0.2 and 0.4, p is set to 0.5, and the weight vector ω is set to (1;10;100) on ORL database; λ_1 and λ_2 are set to 0.2 and 2, p is set to 0.9, and the weight vector ω is set to (5;10) on Prokaryotic database; λ_1 and λ_2 are set to 0.01 and 1.2, p is set to 0.9, and the weight vector ω is set to (1;10) on BBCSport database;

---

[1] http://www.uk.research.att.com/facedatabase.html.
[2] http://mlg.ucd.ie/datasets/segment.html.
[3] http://mlg.ucd.ie/datasets/segment.html.
[4] http://www-cvr.ai.uiuc.edu/ponce_grp/data/.

Table 1. Clustering Results on BBC4view, BBCSport()

Dataset	Method	ACC	NMI	AR	F-score
BBC4view	RMSC	0.775 ± 0.003	0.616 ± 0.004	0.560 ± 0.002	0.656 ± 0.002
	DiMSC	0.892 ± 0.001	0.728 ± 0.002	0.752 ± 0.002	0.810 ± 0.002
	LT-MSC	0.591 ± 0.000	0.442 ± 0.005	0.400 ± 0.001	0.546 ± 0.000
	ECMSC	0.308 ± 0.028	0.047 ± 0.009	0.008 ± 0.018	0.322 ± 0.017
	MLAN	0.853 ± 0.007	0.698 ± 0.010	0.716 ± 0.005	0.783 ± 0.004
	t-SVD-MSC	0.858 ± 0.001	0.685 ± 0.002	0.725 ± 0.002	0.789 ± 0.001
	GMC	0.693 ± 0.000	0.563 ± 0.000	0.479 ± 0.000	0.633 ± 0.000
	LMSC	0.883 ± 0.000	0.699 ± 0.000	0.746 ± 0.000	0.806 ± 0.000
	LRTG	0.894 ± 0.000	0.769 ± 0.000	0.791 ± 0.000	0.839 ± 0.000
	THMSC	**0.999** ± 0.000	**0.995** ± 0.000	**0.975** ± 0.007	**0.981** ± 0.005
BBCSport	RMSC	0.826 ± 0.001	0.666 ± 0.001	0.637 ± 0.001	0.719 ± 0.001
	DiMSC	0.922 ± 0.000	0.785 ± 0.000	0.813 ± 0.000	0.858 ± 0.000
	LT-MSC	0.460 ± 0.046	0.222 ± 0.028	0.167 ± 0.043	0.428 ± 0.014
	ECMSC	0.285 ± 0.014	0.027 ± 0.013	0.009 ± 0.011	0.267 ± 0.020
	MLAN	0.721 ± 0.000	0.779 ± 0.010	0.591 ± 0.000	0.714 ± 0.000
	t-SVD-MSC	0.879 ± 0.000	0.765 ± 0.000	0.784 ± 0.000	0.834 ± 0.000
	GMC	0.807 ± 0.000	0.760 ± 0.000	0.722 ± 0.000	0.794 ± 0.000
	LMSC	0.847 ± 0.003	0.739 ± 0.001	0.749 ± 0.001	0.810 ± 0.001
	LRTG	0.943 ± 0.005	0.869 ± 0.009	0.840 ± 0.012	0.879 ± 0.010
	THMSC	**1.000** ± 0.000	**1.000** ± 0.000	**1.000** ± 0.000	**1.000** ± 0.000

Table 2. Clustering Results on Scene-15, ORL

Dataset	Method	ACC	NMI	AR	F-score
Scene-15	RMSC	0.503 ± 0.000	0.495 ± 0.000	0.325 ± 0.000	0.371 ± 0.000
	DiMSC	0.300 ± 0.010	0.269 ± 0.009	0.117 ± 0.012	0.181 ± 0.010
	LT-MSC	0.574 ± 0.009	0.571 ± 0.011	0.424 ± 0.010	0.465 ± 0.007
	ECMSC	0.457 ± 0.001	0.463 ± 0.002	0.303 ± 0.001	0.357 ± 0.001
	MLAN	0.331 ± 0.000	0.475 ± 0.000	0.151 ± 0.000	0.248 ± 0.000
	t-SVD-MSC	0.812 ± 0.007	0.858 ± 0.007	0.771 ± 0.003	0.788 ± 0.001
	GMC	0.381 ± 0.000	0.519 ± 0.000	0.191 ± 0.000	0.281 ± 0.000
	LMSC	0.563 ± 0.000	0.525 ± 0.000	0.397 ± 0.000	0.440 ± 0.000
	LRTG	0.615 ± 0.016	0.657 ± 0.005	0.486 ± 0.016	0.525 ± 0.014
	THMSC	**0.978** ± 0.000	**0.958** ± 0.000	**0.955** ± 0.000	**0.958** ± 0.000
ORL	RMSC	0.723 ± 0.007	0.872 ± 0.012	0.645 ± 0.003	0.654 ± 0.007
	DiMSC	0.838 ± 0.001	0.940 ± 0.003	0.802 ± 0.000	0.807 ± 0.003
	LT-MSC	0.795 ± 0.007	0.930 ± 0.003	0.750 ± 0.003	0.768 ± 0.004
	ECMSC	0.854 ± 0.011	0.947 ± 0.009	0.810 ± 0.012	0.821 ± 0.015
	MLAN	0.705 ± 0.022	0.854 ± 0.018	0.384 ± 0.010	0.376 ± 0.015
	t-SVD-MSC	0.970 ± 0.003	0.993 ± 0.002	0.967 ± 0.002	0.968 ± 0.003
	GMC	0.633 ± 0.000	0.857 ± 0.000	0.337 ± 0.000	0.360 ± 0.000
	LMSC	0.877 ± 0.024	0.949 ± 0.006	0.839 ± 0.022	0.843 ± 0.021
	LRTG	0.933 ± 0.003	0.970 ± 0.002	0.905 ± 0.005	0.908 ± 0.005
	THMSC	**1.000** ± 0.000	**1.000** ± 0.000	**1.000** ± 0.000	**1.000** ± 0.000

Table 3. Clustering Results on Prokaryotic

Dataset	Method	ACC	NMI	AR	F-score
Prokaryotic	RMSC	0.461 ± 0.049	0.315 ± 0.041	0.198 ± 0.044	0.447 ± 0.027
	DiMSC	0.395 ± 0.001	0.070 ± 0.000	0.053 ± 0.000	0.346 ± 0.000
	LT-MSC	0.431 ± 0.007	0.156 ± 0.020	0.051 ± 0.016	0.401 ± 0.006
	ECMSC	0.432 ± 0.001	0.193 ± 0.001	0.078 ± 0.001	0.383 ± 0.002
	MLAN	0.712 ± 0.002	0.387 ± 0.003	0.425 ± 0.003	0.618 ± 0.002
	t-SVD-MSC	0.523 ± 0.000	0.197 ± 0.000	0.137 ± 0.000	0.486 ± 0.000
	GMC	0.496 ± 0.000	0.193 ± 0.000	0.091 ± 0.000	0.461 ± 0.000
	LMSC	0.686 ± 0.002	0.306 ± 0.001	0.262 ± 0.001	0.603 ± 0.001
	LRTG	0.788 ± 0.000	0.484 ± 0.000	0.492 ± 0.000	0.671 ± 0.000
	THMSC	$\mathbf{0.799 \pm 0.000}$	$\mathbf{0.501 \pm 0.000}$	$\mathbf{0.561 \pm 0.000}$	$\mathbf{0.736 \pm 0.000}$

λ_1 and λ_2 are set to 0.01 and 1.2, p is set to 0.3, and the weight vector ω is set to (1;10;10;1) on BBC4view database; λ_1 and λ_2 are set to 0.01 and 1e-6, p is set to 0.5, and the weight vector ω is set to (1;10;100) on Scene-15 database. For each experiment, we run 10 times and report the average and standard deviation(average$\pm$ standard deviation). All experiments are accomplished in Matlab 2021a on a workstation with 3.2 GHz CPU and 16 GB RAM.

4.2 Experimental Results and Analysis

To evaluate the performance of our method, we use four common metrics including accuracy(ACC), normalized mutual information(NMI), adjusted rank index(AR), F-score. Tables 1, 2, 3 lists all the results of five databases. Through Tables 1, 2, 3, we have the following observation.

– Our method gets the best results on all the five databases. For example, on BBC4view database, our method indicates a significant increase of 10.4%, 22.6%, 18.4%, 14.2%, w.r.t. ACC, NMI, AR, F-score, respectively, over the second best baseline. On Scene-15 database, our method also has a significant increase of 16.6%, 10.0%, 18.3%, 17.0% w.r.t. ACC, NMI, AR, F-score, respectively, over the second best baseline. Especially on BBCSport and ORL, our method can accurately cluster all instances.
– On Scene-15, ORL databases, tensor-based method (t-SVD-MSC) performs better than graph-based method (LRTG). But, on BBCSport, BBC4view and Prokaryotic datasets, which have more complicated local geometric strcture. On the contrary, LRTG was superior to t-SVD-MSC.
– Our method combines hyper-graph learning and tensor constraint multi-view subspace clustering into a joint framework. It not only captures the global and local geometry structure, but also characterize the high-order correlation among multi-view data. Thus, our method have a superior performance compared with the other state-of-the-art methods. Figure 1 visualizes the confusion matrices of t-SVD-MSC and our method. We can see that our method wins in all categories.

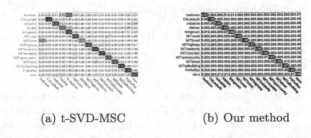

(a) t-SVD-MSC   (b) Our method

Fig. 1. Confusion matrices on Scene-15 dataset

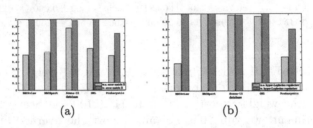

(a)   (b)

Fig. 2. (a) ACC on all datasets w.o. or w. error matrix $\mathbf{E}$. (b) ACC on all datasets w.o. or w. hyper-Laplacian regularizer

4.3   Parameter Analysis

We verify the effectiveness of the different components by related ablation studies, including hyper-Laplacian regularizer and error matrix $\mathbf{E}$. Figure 2 shows ACC on all datasets with(w.) or without(w.o.) the hyper-Laplacian regularizer and error matrix $\mathbf{E}$, respectively. It can be seen that error matrix $\mathbf{E}$ has an great influence on the clustering performance of all datasets. BBC4view database (textual dataset) and Prokaryotic database (gene dataset) are very sensitive to the hyper-Laplacian regularization. This phenomenon just shows that the hyper-Laplacian regularization can well retain the non-linear geometric structure and is helpful to cluster more complex stuctured datasets.

4.4   Convergence Analysis

Existing works have demonstrated that, when the number of variables is more than 2, it is an open problem to prove the convergence of inexact ALM. In Algorithm 1, we have four variables. Thus, we could not prove the convergence of our proposed THMSC. Therefore, we show the convergence of reconstruction error $\left\|\mathbf{X}^{(v)} - \mathbf{X}^{(v)}\mathbf{Z}^{(v)} - \mathbf{E}^{(v)}\right\|_{\infty}$ and variable error $\left\|\mathbf{Z}^{(v)} - \mathbf{J}^{(v)}\right\|_{\infty}$ of our algorithm on Scene-15 dataset in fig. 3. It can be seen that our method have a good convergence.

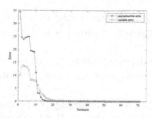

Fig. 3. Convergence curves on the Scene-15 dataset

5 Conclusion

We develop a nonconvex tensor hypergraph learning for multi-view subspace clustering. We propose to use the hyper-Laplacian regularization to capture high-order global and local geometric information of all views. We advocate adopting the nonconvex weighted tensor Schatten-p norm to characterize the high-order correlations of multi-view data. In addition, we also give an effective alternating direction algorithm to optimize this nonconvex model. Extensive experiments on five datasets illustrate the robustness and superiority of the proposed method.

References

1. Cao, X., Zhang, C., Fu, H., Liu, S., Zhang, H.: Diversity-induced multi-view subspace clustering. In: Proceedings of the IEEE Conference on Computer Vision and Pattern Recognition, pp. 586–594 (2015)
2. Chen, Y., Xiao, X., Peng, C., Lu, G., Zhou, Y.: Low-rank tensor graph learning for multi-view subspace clustering. IEEE Trans. Circuits Syst. Video Technol. **32**(1), 92–104 (2022). https://doi.org/10.1109/TCSVT.2021.3055625
3. Chen, Y., Xiao, X., Zhou, Y.: Multi-view clustering via simultaneously learning graph regularized low-rank tensor representation and affinity matrix. In: 2019 IEEE International Conference on Multimedia and Expo (ICME), pp. 1348–1353. IEEE (2019)
4. Kilmer, M.E., Braman, K., Hao, N., Hoover, R.C.: Third-order tensors as operators on matrices: a theoretical and computational framework with applications in imaging. SIAM J. Matrix Anal. Appl. **34**(1), 148–172 (2013)
5. Lu, C., Min, H., Zhao, Z., Zhu, L., Shuang Huang, D., Yan, S.: Robust and efficient subspace segmentation via least squares regression. In: ECCV (2012)
6. Luo, S., Zhang, C., Zhang, W., Cao, X.: Consistent and specific multi-view subspace clustering. In: Thirty-second AAAI Conference on Artificial Intelligence, pp. 3730–3737 (2018)
7. Najafi, M., He, L., Philip, S.Y.: Error-robust multi-view clustering. In: 2017 IEEE International Conference on Big Data (Big Data), pp. 736–745. IEEE (2017)
8. Nie, F., Cai, G., Li, J., Li, X.: Auto-weighted multi-view learning for image clustering and semi-supervised classification. IEEE Trans. Image Process. **27**(3), 1501–1511 (2018). https://doi.org/10.1109/TIP.2017.2754939
9. Nie, F., Cai, G., Li, X.: Multi-view clustering and semi-supervised classification with adaptive neighbours. In: AAAI (2017)

10. Nie, F., Li, J., Li, X.: Parameter-free auto-weighted multiple graph learning: a framework for multiview clustering and semi-supervised classification. In: IJCAI (2016)
11. Tang, C., et al.: CGD: multi-view clustering via cross-view graph diffusion. In: Proceedings of the AAAI Conference on Artificial Intelligence, vol. 34, pp. 5924–5931 (2020)
12. Tang, Y., Xie, Y., Yang, X., Niu, J., Zhang, W.: Tensor multi-elastic kernel self-paced learning for time series clustering. IEEE Trans. Knowl. Data Eng. 33(3), 1223–1237 (2021). https://doi.org/10.1109/TKDE.2019.2937027
13. Wang, H., Yang, Y., Liu, B.: GMC: graph-based multi-view clustering. IEEE Trans. Knowl. Data Eng. 32(6), 1116–1129 (2020). https://doi.org/10.1109/TKDE.2019.2903810
14. Wang, H., Cen, Y., He, Z., Zhao, R., Cen, Y., Zhang, F.: Robust generalized low-rank decomposition of multimatrices for image recovery. IEEE Trans. Multimedia 19(5), 969–983 (2016)
15. Wang, S., Chen, Y., Jin, Y., Cen, Y., Li, Y., Zhang, L.: Error-robust low-rank tensor approximation for multi-view clustering. Knowl.-Based Syst. 215, 106745 (2021). https://doi.org/10.1016/j.knosys.2021.106745
16. Wang, S., Chen, Y., Zhang, L., Cen, Y., Voronin, V.: Hyper-Laplacian regularized nonconvex low-rank representation for multi-view subspace clusteringd. IEEE Trans. Signal Inf. Process. Over Netw. 8, 376–388 (2022)
17. Wang, X., Guo, X., Lei, Z., Zhang, C., Li, S.Z.: Exclusivity-consistency regularized multi-view subspace clustering. In: 2017 IEEE Conference on Computer Vision and Pattern Recognition (CVPR), pp. 1–9 (2017). https://doi.org/10.1109/CVPR.2017.8
18. Wang, Y., Zhang, W., Wu, L., Lin, X., Fang, M., Pan, S.: Iterative views agreement: an iterative low-rank based structured optimization method to multi-view spectral clustering. arXiv preprint arXiv:1608.05560 (2016)
19. Xia, R., Pan, Y., Du, L., Yin, J.: Robust multi-view spectral clustering via low-rank and sparse decomposition. In: Proceedings of the AAAI Conference on Artificial Intelligence, vol. 28 (2014)
20. Xia, W., Zhang, X., Gao, Q., Shu, X., Han, J., Gao, X.: Multiview subspace clustering by an enhanced tensor nuclear norm. IEEE Trans. Cybern., 1–14 (2021). https://doi.org/10.1109/TCYB.2021.3052352
21. Xie, Y., et al.: Robust kernelized multiview self-representation for subspace clustering. IEEE Trans. Neural Netw. Learn. Syst. 32(2), 868–881 (2021). https://doi.org/10.1109/TNNLS.2020.2979685
22. Xie, Y., Tao, D., Zhang, W., Liu, Y., Zhang, L., Qu, Y.: On unifying multi-view self-representations for clustering by tensor multi-rank minimization. Int. J. Comput. Vis. 126(11), 1157–1179 (2018)
23. Xie, Y., Zhang, W., Qu, Y., Dai, L., Tao, D.: Hyper-laplacian regularized multilinear multiview self-representations for clustering and semisupervised learning. IEEE Trans. Cybern. 50(2), 572–586 (2020)
24. Yin, M., Gao, J., Lin, Z.: Laplacian regularized low-rank representation and its applications. IEEE Trans. Pattern Anal. Mach. Intell. 38(3), 504–517 (2016)
25. Zha, Z., Wen, B., Yuan, X., Zhou, J., Zhu, C.: Image restoration via reconciliation of group sparsity and low-rank models. IEEE Trans. Image Process. 30, 5223–5238 (2021)
26. Zhang, C., et al.: Generalized latent multi-view subspace clustering. IEEE Trans. Pattern Anal. Mach. Intell. 42(1), 86–99 (2020)

27. Zhang, C., Fu, H., Liu, S., Liu, G., Cao, X.: Low-rank tensor constrained multi-view subspace clustering. In: Proceedings of the IEEE International Conference on Computer Vision, pp. 1582–1590 (2015)
28. Zhang, Z., Ely, G., Aeron, S., Hao, N., Kilmer, M.: Novel methods for multilinear data completion and de-noising based on tensor-SVD. In: Proceedings of the IEEE Conference on Computer Vision and Pattern Recognition, pp. 3842–3849 (2014)

A Novel Method for Identifying Bipolar Disorder Based on Diagnostic Texts

Hua Gao, Li Chen, Yi Zhou, Kaikai Chi(✉), and Sixian Chan

College of Computer Science and Technology, Zhejiang University of Technology,
Hangzhou 310001, China
kkchi@zjut.edu.cn

Abstract. Bipolar disorder is a complex condition characterized by episodes of mixed manic and depressive states, exhibiting complexity and diversity in symptoms, with high rates of misdiagnosis and mortality. To improve the accuracy of automated diagnosis for bipolar disorder, this study proposes a text-based identification method. Our approach focuses on two extreme emotional features, utilizing two temporal networks in the emotion feature module to extract depressive phase features and manic phase features from the text. Simultaneously, mixed-dilated convolutions are introduced in TextCNN to extract local features with a larger receptive field. By integrating feature information captured from different perspectives, we construct a multi-scale feature model that emphasizes both state features. We utilized a self-collected dataset comprising symptom descriptions of bipolar disorder patients from hospitals, achieving an accuracy of 92.5%. This work provides an accurate assessment of bipolar disorder, facilitating individuals to gain a rapid understanding of their condition and holds significant social implications.

Keywords: Bipolar disorder · Text recognitnion · TextCNN · Deep learning

1 Introduction

Bipolar disorder, also known as manic-depressive disorder, is a severe mental health disease [13]. Classified as a mood disorder, it is characterized by intense fluctuations in mood, encompassing two extreme emotional states: manic episodes and depressive episodes. These two states exhibit distinct features and are sometimes even contradictory. It has been proven that such disorders impact social and psychological functioning in domains such as work, cognition, and interpersonal relationships. Moreover, they are associated with high rates of disability and mortality [9].

Currently, the diagnosis of bipolar disorder mainly relies on assessments conducted by experts in hospitals or related institutions. The assessments by experts are based on various sources, including psychological assessment questionnaires completed by patients, responses from patients and their family members, and

Q. Liu et al. (Eds.): PRCV 2023, LNCS 14428, pp. 52–63, 2024.
https://doi.org/10.1007/978-981-99-8462-6_5

some medical examinations. Automated diagnosis of bipolar disorder can provide quantitative indicators and assist doctors in their diagnosis, thereby reducing the workload burden caused by the significant shortage of psychiatrists.

In recent years, there have been significant advancements and refinements in the automated detection methods for bipolar disorder. Zain et al. [19] developed a three-classification method for recognizing the states of bipolar disorder based on clinical text analysis of patients' medical records. They first extracted features using the Term Frequency-Inverse Document Frequency(TF-IDF) to focus on important words. They trained the models by performing grid search optimization on decision trees, random forests, and Adaboost. Finally, they utilized a voting-based fusion approach to enhance accuracy and robustness. Aich et al. [2] extracted various features from patient dialogues, including time-related features such as the maximum duration of conversations and average duration of each conversation, classic Linguistic Inquiry and Word Count(LIWC) features, emotion features, and lexical diversity features. They achieved an F1 score of 93%-96% on their dataset by performing feature classification using models such as Logistic Regression(LR), Support Vector Machine(SVM), and Random Forest(RF).

Although existing research has achieved automated recognition of bipolar affective disorder, these methods still have limitations. For example, the feature extraction stage lacks deep feature extraction, and machine learning models are used for classification. As mentioned in the paper [6], there is very little research on bipolar affective disorder analysis using natural language processing. The majority of models rely on traditional machine learning methods such as SVM, random forest, and logistic regression, or deep learning methods such as CNN. These approaches have not effectively utilized deep learning for bipolar affective disorder research.

Moreover, due to the scarcity of publicly available datasets for bipolar disorder, particularly in Chinese, it is challenging to obtain relevant data. Therefore, through communication with medical experts and analyzing their diagnostic methods, we propose a bipolar disorder recognition method using a Chinese dataset collected from hospitals. The main contributions of this paper are as follows:

1) Our approach emphasizes the separate extraction of emotional features corresponding to the two extreme states of bipolar affective disorder, enabling the model to capture a more comprehensive range of emotional characteristics.
2) Introducing mixed dilated convolutions in TextCNN. we enhanced its feature-capturing performance by leveraging a larger receptive field.
3) We proposed a hybrid model that emphasized both state features. By integrating the two extreme emotion characteristics and multi-scale features, we further enhanced the performance of the model.

2   Related Work

In earlier research on the automated detection of bipolar disorder, statistical feature-based methods were commonly employed. These methods included the use of LIWC features to classify the vocabulary appearing in the text, thereby extracting information related to emotions, attitudes, and psychological states. Additionally, statistical information based on word frequency in the text was utilized to calculate TF-IDF features, which represent the importance of vocabulary. Kadkhoda et al. [7] created a dataset by tracking tweets from bipolar disorder patients and non-patients on the social network Twitter. Then they proposed a novel method for detecting bipolar disorder in Twitter users by extracting both static and dynamic features from user tweets. Abaeikoup et al. [1] extracted statistical features from the text, such as word count, type count, and average letter count per word. These features were then combined with simple acoustic features, and classification was performed using an ensemble classifier consisting of CNN and MLP stacked together, achieving an unweighted average recall(UAR) of 59.3%. Baki et al. [3] used the Google Speech Recognition tool to convert Turkish audio into English text. Then they extracted features from the LIWC text analysis tool, term frequency-inverse document frequency (TF-IDF), and three sentiment analysis tools (NLTK Vader, TextBlob, and Flair) to represent textual features. By combining audio and video features and using a Kernel ELM classifier, they achieved a maximum unweighted average recall of 64.8%. Although statistical features are easy to interpret and offer good explainability, they overlook the complex relationships and semantic information between words, limiting their ability to handle more sophisticated natural language processing tasks.

To better extract contextual information, textual semantics, and emotions from bipolar disorder text, deep learning models have been increasingly applied in research. Zhang et al. [20] proposed a fixed-length session-level paragraph vector representation for the textual modality in multi-modal feature extraction. They utilized paragraph vectors (PV) to embed interview records into doc2vec document representations, aiming to capture disorder-related features. Furthermore, they employed early fusion of multi-modal features for classification. Murarka et al. [12] employed the RoBERTa model to perform a five-class classification on mental health-related posts collected from social networks.

However, in current research on bipolar disorder, deep-learning techniques have not been fully utilized. In natural language processing, Cheng et al. [5] proposed a text classification model based on hierarchical self-attention capsule networks. The model first vectorizes the text and then directs it to the self-attention network for feature extraction at both word and sentence levels. The extracted features are then passed to the capsule network to refine the relationship between different parts of the text and further capture richer semantic information from the text. Performance improvement was achieved compared to baselines on public datasets. Liu et al. [11] combined contextualized features with a multi-stage attention model based on Time Convolutional Network (TCN) and CNN. By utilizing attention weights and incorporating multi-scale TCN, the model improved its parallel processing capability and ability to capture contextual features. The effectiveness of this approach in short texts was demonstrated on six public datasets. In addition to recurrent neural networks (RNN), CNN, and their variants, graph neural networks (GCN) have also been widely applied and developed in text classification [15,17,18]. GCN is used to model relationships between vocabulary in the text, such as co-occurrence and dependency, to improve model performance. The development of deep learning has provided support for the accurate diagnosis of bipolar disorder.

3   Proposed Method

We proposed a hybrid model that focused on two extreme emotional features. In the feature extraction stage, we extracted the emotional features and multi-scale features from the text. Specifically, the emotional features included manic features and depressive features. We trained two neural networks separately on preprocessed manic-phase text and depressive-phase text to extract the two types of emotional features. In the process of multi-scale feature extraction, we introduced mixed dilated convolutions to capture features with a larger receptive field in the text. Finally, we fused and classified the features from different perspectives, enabling a comprehensive analysis of bipolar disorder.

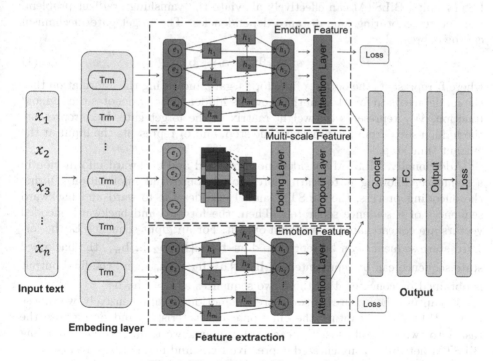

Fig. 1. The structure of our model.

Our model architecture is illustrated in Fig. 1. The left part represents the preprocessing process, where the processed text serves as the input. The right part represents the overall structure of the model, with the middle branch indicating the extraction of multi-scale features, and the upper and lower branches representing the extraction of emotional features.

3.1 Text Embedding

The BERT model consists of 12 layers of Transformer encoders. By fine-tuning BERT with our dataset, specifically retraining the 11th and 12th layers along with the fully connected (FC) layer, we aimed to enhance the model's applicability to our task.

We input the text sequences, represented as $\mathbf{x} = [x_1, x_2, ..., x_n]$,into BERT. BERT represents the entire text by mapping each word to a vector representation. When we input $\mathbf{x}$ to BERT, we obtain an output matrix $\mathbf{E} \in \mathbf{R}^{n \times d}$ where n denotes the number of words in the text and d represents the vector dimension.

3.2 Emotional Feature Extraction

As a variant of recurrent neural networks (RNNs), BiLSTM is well-suited for handling data with sequential structures. Due to the forget gate mechanism in LSTM units, BiLSTM can effectively alleviate the vanishing gradient problem, thus better capturing long-distance dependencies. The forget gate mechanism can be expressed as:

$$\mathbf{f}_t = \sigma(\mathbf{W}_f[\mathbf{h}_{t-1}, \mathbf{e}_t] + \mathbf{b}_f) \tag{1}$$

where $\mathbf{f}_t$ represents the output of the forget gate, indicating the information that the model needs to forget from the previous hidden state. σ denotes the sigmoid function, $\mathbf{W}_f$ represents the weight matrix of the forget gate, $\mathbf{h}_{t-1}$ represents the hidden state from the previous time step, and $\mathbf{e}_t$ represents the input at the current time step.

Additionally, BiLSTM considers both forward and backward information in a sequence, allowing it to learn more global contextual information. During the encoding process, the BiLSTM model encodes the forward and backward sequences of a sentence separately. Then, the forward and backward encoded vectors are concatenated as the final output. For an input sequence $\mathbf{E}$, the forward state sequence can be represented as $\overrightarrow{\mathbf{h}} = \{\overrightarrow{\mathbf{h}_1}, \overrightarrow{\mathbf{h}_2}, ..., \overrightarrow{\mathbf{h}_n}\}$, the backward state sequence can be represented as $\overleftarrow{\mathbf{h}} = \{\overleftarrow{\mathbf{h}_1}, \overleftarrow{\mathbf{h}_2}, ..., \overleftarrow{\mathbf{h}_n}\}$, and the final output is obtained by concatenating these two sequences as $\mathbf{h} = [\overrightarrow{\mathbf{h}}, \overleftarrow{\mathbf{h}}]$.

To capture the features of two extreme emotions comprehensively, we choose to use BiLSTM to capture the emotional characteristics and decompose the task into two subproblems. For each subproblem, we train two corresponding BiLSTM networks using curated depressive texts and manic texts, respectively. Formulas 2 and 3 represent the loss functions used for training the depressive network and manic network, respectively.

$$L_d = -\frac{1}{N_d} \sum_{i=1}^{N_d} d_i \cdot \log(p(d_i)) + (1 - d_i) \cdot \log(1 - p(d_i)) \tag{2}$$

where L_d represents the loss during the training process of the depressive network, N_d is the number of depressive samples, d_i is the binary label indicating depression or non-depression, $\cdot$ represents the multiplication operation between two scalars, and $p(d_i)$ is the probability of belonging to the label d_i.

$$L_m = -\frac{1}{N_m} \sum_{i=1}^{N_m} m_i \cdot \log(p(m_i)) + (1 - m_i) \cdot \log(1 - p(m_i)) \tag{3}$$

where L_m represents the loss during the training process of the manic network, N_m is the number of manic samples, m_i is the binary label indicating manic or non-manic, $\cdot$ represents the multiplication operation between two scalars, and $p(m_i)$ is the probability of belonging to the label m_i.

The two BiLSTM networks aim to learn the semantic information of different emotional features, comprehensively understanding the emotional characteristics from multiple levels and perspectives.

In the embedding layer, we obtained the vectorized representation $\mathbf{E} \in \mathbf{R}^{n \times d}$ of the text. By separately inputting the vectorized representations into two BiL-STM models dedicated to different emotional subtasks, we obtained distinct feature vectors $\mathbf{h_m} = [\overrightarrow{\mathbf{h_m}}, \overleftarrow{\mathbf{h_m}}]$ for manic feature extraction and $\mathbf{h_d} = [\overrightarrow{\mathbf{h_d}}, \overleftarrow{\mathbf{h_d}}]$ for depressive feature extraction. This approach ensures that each model focuses on its respective target emotional feature, thereby enhancing the overall diagnostic performance.

3.3   Multiscale Feature Extraction

In addition to emotional features, individuals with bipolar disorder also exhibit unique characteristics such as sleep disturbances, self-harm behaviors, attention deficits, and anxiety disorders. By extracting these features, further assistance can be provided in detecting bipolar disorder. TextCNN utilizes multiple convolutional kernels of different sizes for feature extraction. It performs a sliding window operation with fixed-sized kernels on the input and obtains features by computing the dot product between the kernel and the input. Standard convolution is the most commonly used convolution operation in TextCNN. Unlike standard convolution, dilated convolution can combine dilated convolutions with different dilation rates to increase the receptive field and capture features at different scales. It focuses on different granularities of information, allowing the model to better capture multiple features within the text.

We only applied dilated convolution operations with dilation rates of 1, 2, and 5 on the height of the feature matrix. By introducing mixed dilated convolutions in TextCNN, we extract features at different scales to obtain a larger receptive field. For the input matrix $\mathbf{E} \in \mathbf{R}^{n \times d}$, we perform mixed dilated convolution operations on matrix $\mathbf{E}$ to generate a new feature map $\mathbf{C} \in \mathbf{R}^{m \times e}$, where m represents the number of convolutional kernels and e represents the output channels. Specifically, for the convolutional kernel $W_{i,j}$ in the i-th group of convolution operations, each element $c_{i,j}$ in the feature map can be calculated as follows:

$$c_{i,j} = \mathrm{f}((\mathbf{W}_{i,j} * \mathbf{E}_{j:j+s-1}) + \mathbf{b}_{i,j}) \tag{4}$$

where $*$ denotes the convolution operation, $\mathbf{E}_{j:j+s-1}$ denotes the submatrix of matrix $\mathbf{E}$ ranging from column j to column $j + s - 1$, $\mathbf{b}_{i,j}$ denotes the bias term, and f represents the activation function. This operation is equivalent to applying mixed dilated convolution operations on each subsequence of length s in matrix $\mathbf{E}$, resulting in a new feature map.

3.4 Feature Fusion

We feed the feature matrix $\mathbf{E} \in \mathbf{R}^{n \times d}$ obtained from the embedding layer into both the emotion feature model and the multi-scale feature model, resulting in 256-dimensional manic feature $\mathbf{h}_m$, 256-dimensional depressive feature $\mathbf{h}_d$, and 288-dimensional feature vector $\mathbf{m}$ containing local information of the text. Then, we fuse multiple features to obtain a vector $\mathbf{o}$ and use fully connected layers and the softmax function for classification. The mathematical expression of vector $\mathbf{o}$ is represented by:

$$\mathbf{o} = [\mathbf{h}_m, \mathbf{h}_d, \mathbf{m}] \tag{5}$$

where $\mathbf{h}_m$ represents manic features, $\mathbf{h}_d$ represents depressive features, and $\mathbf{m}$ represents multi-scale features.

We use the cross-entropy loss function for forward propagation. In this case, the loss function can be expressed as follows:

$$L_b = -\frac{1}{N_b} \sum_{i=1}^{N_b} b_i \cdot \log(p(b_i)) + (1 - b_i) \cdot \log(1 - p(b_i)) \tag{6}$$

where L_b represents the loss of the model, N_b is the number of bipolar disorder samples, b_i is the binary label indicating bipolar or non-bipolar, $\cdot$ represents the multiplication operation between two scalars, and $p(b_i)$ is the probability of belonging to the label b_i.

4   Experiments and Results

4.1   Experimental Setup

Dataset. The dataset we used consists of responses from interviewees regarding symptoms related to bipolar affective disorder. It includes questions about emotions, sleep, health, appetite, and daily life, among others. The emotional questions encompass both manic and depressive symptoms. Our dataset comprises 67 samples from individuals with bipolar affective disorder and 72 samples from healthy individuals. We divided the dataset into trainset, validset, and testset in a ratio of 3:1:2.

Experiment Settings. The model was trained using Adam optimization with a batch size of 12 and a learning rate of 2e–5. The training process lasted for 100 epochs and included dropout regularization with a rate of 0.2. The hidden layer size of each BiLSTM layer is 128. For the convolutional layers, we used kernel sizes of [3, 4, 5], with 32 kernels and dilation factors of 1, 2, and 5.

4.2   Comparison with Recent Works

In classification tasks, recall, precision, accuracy, and F1 score are important evaluation metrics. We use these metrics to compare the performance of our model with recent research models.

Table 1. Comparison Results with Related Works.

Model	Recall	Precision	Accuracy	F1-score
SVM [4]	0.789	0.714	0.750	0.750
LR [14]	0.789	0.750	0.775	0.769
BERT-TCN [16]	0.842	0.889	0.875	0.865
BERTGCN [10]	0.842	0.842	0.850	0.842
Bi-GRU [8]	0.895	0.850	0.875	0.872
Ours	**0.947**	**0.900**	**0.925**	**0.923**

According to the results in Table 1, our model outperforms other models overall, and deep learning-based methods [8,10,16] show significant advantages over machine learning methods [4,14]. This is because machine learning methods struggle to understand contextual semantics and extract deeper-level features, which often leads to inferior performance in complex tasks.

In the deep learning models, relatively speaking, BERTGCN [10] falls slightly short in terms of performance. BERT-TCN [16] represents sequence features by using BERT's output as the input for TCN. However, the effectiveness of TCN, which captures local features of the input sequence through convolutional layers for classification, is inferior to that of Bi-GRU. Bi-GRU [8], a variant of BiLSTM, effectively captures long-term dependencies in sequences. Experimental evidence shows that using an RNN network like Bi-GRU to capture features yields better results when dealing with features related to bipolar disorder.

4.3   Ablation Study

Our model extracts emotion features and multi-scale features. In the process of emotion feature extraction, we utilize L_d and L_m loss functions to ensure that the model captures features from different emotional states.

To validate the effectiveness of our approach, we evaluate the impact of the two features and the L_d and L_m losses on the model in Experiment 1.

In Experiment 2, we conduct ablation comparisons for each feature extraction method, demonstrating the effectiveness of both feature extraction methods.

Experiment 1. We defined four scenarios: Multi-scale Feature, Emotion Feature, Multi-scale Feature + Emotion Feature (without $L_d + L_m$), and Multi-scale Feature + Emotion Feature (with $L_d + L_m$). We evaluated the performance differences of the model under these different scenarios. The experimental results are shown in Table 2.

Table 2. Experimental Results of Different Feature Extraction Methods.

Model	Recall	Precision	Accuracy	F1-score
Multi-scale Feature	0.947	0.857	0.900	0.900
Emotion Feature(with $L_d + L_m$)	0.895	0.895	0.900	0.895
Multi-scale + Emotion Feature (without $L_d + L_m$)	0.947	0.818	0.875	0.878
Multi-scale + Emotion Feature (with $L_d + L_m$)	**0.947**	**0.900**	**0.925**	**0.923**

From the experimental results in the table, it can be seen that when we solely use the emotion feature, the recall rate of the model decreases to 89.5%. We believe that this is because our emotion feature focuses on two emotional states. While this helps to avoid misclassifying samples with partially unrelated features, it also leads to misclassification of samples with less prominent emotions.

When using the multi-scale feature alone, the model's precision dropped to 85.7%. This is because, compared to emotion features, multi-scale features are more prone to learning some unrelated features while learning local features, resulting in the misclassification of samples. Therefore, models using multi-scale features have lower precision compared to models using emotion features.

In addition, we also removed the L_d and L_m losses from the model in the experiment. The results showed a significant decrease in model precision, even dropping to 81.8%. This further confirms the effectiveness of extracting two types of emotion features. Extracting these two emotion features can improve the model's precision and enhance overall performance.

Experiment 2. In our emotion feature extraction process, we trained two different neural networks to extract features corresponding to two emotional states. In the multi-scale feature extraction process, we introduced mixed-dilated convolutions in TextCNN. To further demonstrate the effectiveness of the two feature extraction methods, we conducted ablation experiments for each extraction method separately. Ablation experiments of the two feature extraction methods are shown in Table 3.

Multi-scale Feature Extraction. When extracting multi-scale features, the TextCNN model with mixed dilated convolutions outperforms the TextCNN model with regular convolutions. This is because mixed dilated convolutions have a larger receptive field, allowing for more comprehensive feature extraction.

Table 3. Ablation Experiments of the Two Feature Extraction Methods.

	Model	Recall	Precision	Accuracy	F1-score
Multi-scale extraction	BERT-TextCNN	0.842	0.800	0.825	0.821
	Multi-scale Feature	**0.947**	**0.857**	**0.900**	**0.900**
Emotion extraction	BERT-BiLSTM	0.895	0.810	0.850	0.850
	Emotion Feature(without $L_d + L_m$)	0.947	0.818	0.875	0.878
	Emotion Feature(with $L_d + L_m$)	**0.895**	**0.895**	**0.900**	**0.895**

Emotion Feature Extraction. The performance of the model using one BiLSTM network or two BiLSTM networks to extract emotion features is lower than our model with two separate losses for extracting different state features in the emotion feature extraction. Although using two BiLSTM networks achieves a recall rate of 94.7%, the precision of the model is significantly lower, with only 81.8%. This is because the two networks may not effectively learn the two different state features during the training process.

Experiment 3. In order to test the robustness of the model, we trained our model with learning rates 1e–6, 1e–5, 3e–5, and 5e–5. The accuracy, recall, precision, and F1 scores are shown in Table 4.

Table 4. The impacts of learning rate on model.

Learning Rate	Recall	Precision	Accuracy	F1-score
1e–6	0.895	0.850	0.875	0.872
1e–5	0.947	0.900	0.925	0.923
3e–5	0.947	0.900	0.925	0.923
5e–5	0.895	0.895	0.895	0.900

We also tested the model on epoch 50, 300, and 800. The experimental results are shown in Table 5.

Table 5. The impacts of epoch on model.

Epoch	Recall	Precision	Accuracy	F1-score
50	0.842	0.842	0.842	0.850
300	0.947	0.900	0.925	0.923
800	0.947	0.900	0.925	0.923

It can be seen from the results that when the learning rates are 1e–6 and 5e–5, the performance of the model is not very good, because if the learning rate is too small, the model will fall into the optimal solution, while if the learning rate is too large, the model will not be able to learn features better. When epoch is 50, the model has not yet converged, so the performance of the model has not yet reached the optimal level.

5   Conclusion

This paper proposes a method for identifying bipolar disorder based on diagnostic texts. By extracting two types of emotion features and analyzing the multi-scale local features of the text, the model achieves higher accuracy in identifying bipolar disorders. Effectively recognizing and extracting the two types of extreme emotion features provide strong evidence for detecting bipolar disorder.

There is still room for further improvement in our method. For example, in this study, we only analyzed the two types of emotion features from a textual perspective. However, the manic phase of bipolar disorder is characterized by increased tone and accelerated speech rate. Therefore, when extracting emotion features, combining text and audio information could be considered. Moreover, the symptoms of bipolar disorder overlap with other mental disorders. Therefore, in future work, we hope to adopt multimodal approaches to differentiate various diseases more accurately.

References

1. AbaeiKoupaei, N., Al Osman, H.: A multi-modal stacked ensemble model for bipolar disorder classification. IEEE Trans. Affect. Comput. **14**(1), 236–244 (2023)
2. Aich, A., et al.: Towards intelligent clinically-informed language analyses of people with bipolar disorder and schizophrenia. In: Findings of the Association for Computational Linguistics: EMNLP 2022, pp. 2871–2887 (2022)
3. Baki, P., Kaya, H., Çiftçi, E., Güleç, H., Salah, A.A.: A multimodal approach for mania level prediction in bipolar disorder. IEEE Trans. Affect. Comput. **13**(4), 2119–2131 (2022)
4. Chang, C.C., Lin, C.J.: Libsvm: a library for support vector machines. ACM Trans. Intell. Syst. Technol. (TIST) **2**(3), 1–27 (2011)
5. Cheng, Y., et al.: Hsan-capsule: a novel text classification model. Neurocomputing **489**, 521–533 (2022)
6. Jan, Z., et al.: The role of machine learning in diagnosing bipolar disorder: scoping review. J. Med. Internet Res. **23**(11), e29749 (2021)
7. Kadkhoda, E., Khorasani, M., Pourgholamali, F., Kahani, M., Ardani, A.R.: Bipolar disorder detection over social media. Inf. Med. Unlocked **32**, 101042 (2022)
8. Khodeir, N.A.: BI-GRU urgent classification for MOOC discussion forums based on bert. IEEE Access **9**, 58243–58255 (2021)
9. Laksshman, S., Bhat, R.R., Viswanath, V., Li, X.: Deepbipolar: identifying genomic mutations for bipolar disorder via deep learning. Hum. Mutat. **38**(9), 1217–1224 (2017)
10. Lin, Y., et al.: Bertgcn: transductive text classification by combining gcn and bert. arXiv preprint arXiv:2105.05727 (2021)

11. Liu, Y., Li, P., Hu, X.: Combining context-relevant features with multi-stage attention network for short text classification. Comput. Speech Lang. **71**, 101268 (2022)
12. Murarka, A., Radhakrishnan, B., Ravichandran, S.: Classification of mental illnesses on social media using roberta. In: Proceedings of the 12th International Workshop on Health Text Mining and Information Analysis, pp. 59–68 (2021)
13. Rowland, T.A., Marwaha, S.: Epidemiology and risk factors for bipolar disorder. Therap. Adv. Psychopharmacol. **8**(9), 251–269 (2018)
14. Shah, K., Patel, H., Sanghvi, D., Shah, M.: A comparative analysis of logistic regression, random forest and KNN models for the text classification. Augment. Hum. Res. **5**, 1–16 (2020)
15. She, X., Chen, J., Chen, G.: Joint learning with BERT-GCN and multi-attention for event text classification and event assignment. IEEE Access **10**, 27031–27040 (2022)
16. Wang, F., Liu, G., Hu, Y., Wu, X.: Affective tendency of movie reviews based on bert and TCN. In: 2021 2nd International Seminar on Artificial Intelligence, Networking and Information Technology (AINIT), pp. 244–247. IEEE (2021)
17. Wang, K., Han, S.C., Poon, J.: Induct-GCN: inductive graph convolutional networks for text classification. In: 2022 26th International Conference on Pattern Recognition (ICPR), pp. 1243–1249. IEEE (2022)
18. Yao, L., Mao, C., Luo, Y.: Graph convolutional networks for text classification. In: Proceedings of the AAAI Conference on Artificial Intelligence, vol. 33, pp. 7370–7377 (2019)
19. Zain, S.M., Mumtaz, W.: Tri-model ensemble with grid search optimization for bipolar disorder diagnosis. In: 2022 International Conference on Frontiers of Information Technology (FIT), pp. 24–29. IEEE (2022)
20. Zhang, Z., Lin, W., Liu, M., Mahmoud, M.: Multimodal deep learning framework for mental disorder recognition. In: 2020 15th IEEE International Conference on Automatic Face and Gesture Recognition (FG 2020), pp. 344–350. IEEE (2020)

Deep Depression Detection Based on Feature Fusion and Result Fusion

Hua Gao⑩, Yi Zhou, Li Chen, and Kaikai Chi[(✉)]⑩

College of Computer Science and Technology, Zhejiang University of Technology,
Hangzhou 310023, China
kkchi@zjut.edu.cn

Abstract. Depression, as a severe mental disorder, has significant impacts on individuals, families, and society. Accurate depression detection is of great significance. To this end, we propose a deep depression detection based on feature fusion and result fusion. We have introduced a dataset consisting of answers that correspond to four questions. Using the answers to the first three questions provided by the patients, we propose a text feature extraction method based on a pre-trained BERT model and an improved TextCNN model to extract features and introduce an attention mechanism for feature fusion. The vector obtained by fusing the features of the first three questions is inputted into a classifier to obtain classification results. These results, along with the classification results obtained from the fourth question, are fused using result fusion techniques to obtain the final outcome. To evaluate our method, we conducted experiments on the dataset collected from the specialized hospital for mental illness. We compare our depression detection algorithm with mainstream ones. Our method achieves an accuracy of 94.2% on our dataset. The results show that our method achieves an accuracy rate of 7.7% higher than the highest accuracy rate among the compared mainstream models.

Keywords: Depression recognition · Feature fusion · Result fusion · BERT · TextCNN

1 Introduction

Depression is a prevalent and debilitating mental disorder characterized by persistent feelings of sadness, loss of interest, and a decrease in the ability to experience pleasure in daily life. Accurately identifying depression [25] and providing timely intervention is of utmost importance, benefiting not only individuals but also their families and society as a whole.

Text-based depression detection methods are a promising approach for detecting depression, supported by substantial evidence [4], and have demonstrated high performance. Early text-based methods for depression detection predominantly employed a bottom-up approach, incorporating deep learning and machine learning techniques. Traditional machine learning algorithms, such as

© The Author(s), under exclusive license to Springer Nature Singapore Pte Ltd. 2024
Q. Liu et al. (Eds.): PRCV 2023, LNCS 14428, pp. 64–74, 2024.
https://doi.org/10.1007/978-981-99-8462-6_6

Support Vector Machines(SVM), Random Forests(RF), were commonly applied for depression classification tasks. Deep learning methods have made significant advancements in depression detection. Facilitates automatic feature extraction from raw data and end-to-end model training using training samples. Most research on depression detection focuses on analyzing text generated by users on social media platforms. However, they [7,20] have noted that a simplistic analysis of users' posts may not accurately identify their depressive tendencies.

In order to improve depression detection, this study focuses on detecting depression based on text features, using a deep depression detection model based on feature fusion and result fusion The main contributions of this paper are as follows:

(1) We collected a dataset from a specialized psychiatric hospital to enhance the professionalism and persuasiveness of our results. We proposed a depression detection method that trains the model based on specific questions. The acquired text data was categorized into recent situations, mood, sleep, and tendencies and behaviors related to suicidal inclination. This method has improved the accuracy of depression detection.
(2) We improved the TextCNN model by incorporating both k-max pooling and max pooling, which enables the extraction of richer and more comprehensive local feature representations. This approach captures a wider range of contextual information and extracts multi-scale features.
(3) We have utilized the specificity of suicide-related issues, and we have constructed a dual-branch model. This model incorporated an attention mechanism to fuse the features of the first three questions and then merge the results from the two branches to obtain the final outcome.

2   Related Work

Traditional methods for detecting depression primarily rely on two aspects: the subjective description provided by patients and the assessment through questionnaires. However, depressive individuals often fail to seek timely medical attention when their condition goes unnoticed by themselves and their family members, leading to worsening symptoms and increased difficulty in achieving recovery. It is crucial to identify depression accurately in order to facilitate rehabilitation. To address this issue, numerous researchers have embarked on relevant studies in this field.

Early depression detection methods used machine learning technologies, where artificial intelligence (AI) and natural language processing (NLP) techniques aid in identifying depression from emotions or sentiments [22]. Researchers typically employ classical machine learning (ML) algorithms such as RF, SVM, Decision Trees (DT), etc. Priya et al. [16] proposed a five-level prediction of depression, stress, and anxiety using five ML algorithms. Mori et al. [14] and others utilized machine learning algorithms on four types of Twitter information, including network, statistical word, temporal, and bag-of-words. This study considered 24 personality traits and 239 features. However, the efficacy of traditional ML techniques is limited due to the significant increase in data volume and the number of correlations [18].

Currently, mainstream research on depression primarily relies on deep learning models. J. Kim et al. [9] collected posts from a mental health community and accurately identified whether users' posts belonged to specific mental disorders using CNN and XGBoost models. Lin et al. [12] proposed a deep visual-text multimodal learning approach to reveal the psychological state of users on social networks. Deep features were extracted from images and text posted by users using a Convolutional Neural Networks(CNN) classifier and Bidirectional Encoder Representations from Transformers(BERT). Respectively, These visual and textual features were then combined to reflect users' emotional expressions. Li et al. [11] proposed a Multimodal Hierarchical Attention (MHA) model for depression detection on social media. Orabi et al. [15] employed CNN and Recurrent Neural Networks (RNN) for diagnosing depression. He L et al. [8] reviewed the recognition of depression using both unimodal and multimodal audio-visual cues, mainly focusing on deep learning-based depression recognition tasks. Luna Ansari et al. [1] improved depression performance by examining and comparing the integration and fusion of these two methods. Kour H et al. [10] explored the possibility of predicting users' mental states by classifying depression and non-depression using Twitter data. Deep learning models were used to analyze the semantic context in textual narratives based on the textual content of users' tweets. However, the proposed models were hybrids of two deep learning architectures, Convolutional Neural Network (CNN), and Bidirectional Long Short-Term Memory (BiLSTM). Mao et al. [13] proposed a depression severity prediction model based on bidirectional LSTM and temporally distributed CNN, utilizing rhythm and semantic features.

3   Proposed Method

This study proposes a question-level text feature extraction model, where features are extracted separately for each question's corresponding answers. Two-step operations are performed based on the specific nature of the questions. In the first step, an approach is introduced using a pre-trained BERT model and an enhanced TextCNN model for extracting text features by incorporating improved local information extraction. The feature representation is enriched by incorporating k-max pooling, and feature fusion is achieved through the addition of an attention mechanism in a hybrid model. In the second step, leveraging the specificity of the "suicide trendies" question, features from answers related to suicide are extracted using the BERT model and the improved TextCNN model. The results obtained from the softmax layer of this step are then combined with the results obtained in the first step.

The proposed method demonstrates an improved accuracy rate compared to the results of the first-step model and outperforms existing models. Our model structure is shown in Fig. 1.

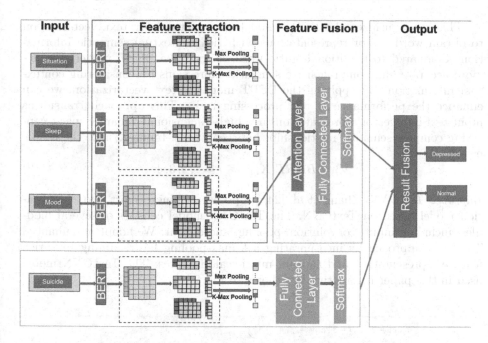

Fig. 1. The framework of our proposed method

3.1 Model Structure

Input Module. In this paper, the input is a text data $\mathbf{X}$, which includes answers to four questions about recent events, sleep, mood, and suicidal tendencies, denoted as $\mathbf{X_1}, \mathbf{X_2}, \mathbf{X_3}$, and $\mathbf{X_4}$. Corresponding texts are separately fed into corresponding models for training:

$$\mathbf{X} = (\{\mathbf{X}_1, \mathbf{X}_2, \mathbf{X}_3, \mathbf{X}_4\}, y) \tag{1}$$

where y is the label of the data.

Feature Extraction. The primary task of feature extraction is to select or extract an appropriate subset of features. Feature selection involves choosing the most relevant or representative features from the original feature set to reduce dimensionality and improve the generalization ability of the model. On the other hand, feature extraction creates new feature representations by transforming, converting, or computing derived features to capture useful information in the data.

BERT. It is a language representation model developed by Google [5]. It is a pretrained model based on the Transformer architecture and has been trained on a large-scale unlabeled corpus to learn deep contextual language representations.

This study utilizes the BERT model as a method for text vectorization to obtain word vector representations with rich contextual semantic information. Compared to traditional static word vectors, the BERT model provides more accurate and comprehensive semantic expressions by considering contextual information. By applying the BERT model to text vectorization, we can enhance the performance of text-processing tasks. This approach transforms input text into vector representations in a high-dimensional semantic space, capturing complex semantic relationships between words. Input $\mathbf{X}_i$ into the BERT model:

$$\mathbf{V}_i = BERT(\mathbf{X}_i) \qquad i = 1, 2, 3, 4 \qquad (2)$$

Improved TextCNN. Zhang et al. [24] proposed a Chinese short text classification model based on TextCNN. The TextCNN model consists of several modules, including input, convolution, pooling, and output. We adopt an enhanced TextCNN approach by incorporating a k-max-pooling layer, aiming to enrich feature representation and capture multi-scale features. The TextCNN model used in this paper is structured as shown in Fig. 2.

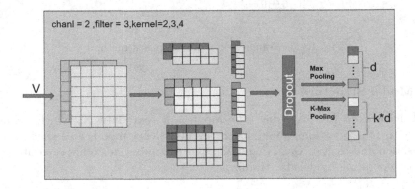

Fig. 2. The structure of improved TextCNN

The input feature vectors $\mathbf{V}_i$ are fed into the convolution module where multiple convolutional filters of different sizes are applied to capture local patterns and extract features from the input vectors. Specifically, with 2 channels, a filter size of 3, and kernel sizes of 2, 3, and 4, the following equation is used:

$$\mathbf{C}_i^{(j)} = \sigma(\mathbf{W}^{(j)} * \mathbf{V}_i) \qquad (3)$$

where $\mathbf{C}_i^{(j)}$ represents the j-th channel feature map obtained by convolving the j-th filter with the input vector $\mathbf{V}_i$, $\mathbf{W}^{(j)}$ represents the weights of the j-th filter, and $*$ denotes the convolution operation, σ represents the activation function applied element-wise to the convolution result.

After the convolution operation $\mathbf{C}_i^{(j)}$, the TextCNN model employs a pooling module to reduce the dimensionality of the feature maps and capture the most important information. In this study, two types of pooling are used: max pooling and k-max pooling.

Max pooling selects the maximum value from each feature map, capturing the most salient features within each map. On the other hand, k-max pooling [6] selects the k largest values from each feature map, capturing multiple important features.

$$\mathbf{M}_i = P_{max}(\mathbf{C}_i^{(j)}) \tag{4}$$

$$\mathbf{K}_i = P_{kmax}(\mathbf{C}_i^{(j)}) \tag{5}$$

where $P_{max}(\cdot)$ represents the max pooling operation, and $P_{kmax}(\cdot)$ represents the k-max pooling operation. Subsequently, the output module takes the pooled feature maps and combines $\mathbf{M}_i$ and $\mathbf{K}_i$ into a vector $\mathbf{Q}_i$.

Feature Fusion. Feature fusion is the combination of features from different layers or branches, widely used in modern network architectures. It is typically achieved through simple operations such as summation or concatenation, but this may not always be the optimal choice. To better fuse semantic features, we introduce attention mechanisms, which weigh the input features for fusion and utilize linear layers for feature transformation and prediction.

In the attention mechanism, we extract features from the answers provided for three specific questions. Next, we calculate attention weights. Attention weights measure the level of attention each answer receives in relation to the question and can be computed by comparing the similarity between the question and the answer. This can be achieved through the following formula:

$$\mathbf{A} = \frac{\exp(f(\mathbf{Q}_i))}{\sum_{j=1}^{N} \exp(f(\mathbf{Q}_i))} \tag{6}$$

where, N is the total number of features, and $f(\cdot) : R^{108} \to R$ is a learnable function that computes the weight for each feature. By applying attention weights, we can perform a weighted fusion of input features to better capture semantic information.

Next, we perform feature transformation and prediction on the fused features using linear layers. A linear layer can be represented by the following equation:

$$\mathbf{U} = \mathbf{AQ}_i + \mathbf{b} \tag{7}$$

Through linear transformation, we can map the fused features to the final output space for the prediction task.

Output Module. In depression detection, result fusion is a widely used approach to combine the output results of multiple models or features to improve overall accuracy and robustness.

Considering question four regarding suicidal tendencies, the feature vectors $\mathbf{Q_i}$ obtained from the aforementioned feature extraction are inputted into a fully connected layer for classification. Let's denote the classification output result as R_1. Simultaneously, we input the feature-fused result U into a classifier to obtain another classification result R_2. Next, we perform result ensemble on R_1 and R_2 to obtain a more accurate final classification result.

To perform result ensemble, various methods can be employed, such as voting, stacking, or weighted averaging. We use a weighted averaging method to fuse R_1 and R_2. Let the weight assigned to the prediction result R_1 be e_1, and the weight assigned to the prediction result R_2 be e_2, and $e_1 + e_2 = 1$. We can calculate the final result after weighted fusion.

$$R = e_1 R_1 + e_2 R_2 \tag{8}$$

After conducting experiments, we set e_1 and e_2 to 0.5 respectively. If R is greater than or equal to 0.5, we classify it as depression. By performing a result ensemble, we can leverage the strengths of both classification results and demonstrate through experiments the improvement in accuracy rate.

4   Experiment

4.1   Dataset

The dataset used in this study is derived from hospital outpatient data. It contains data from a total of 156 individuals, including 77 patients with depression and 79 individuals in the control group. During the data processing, we excluded data from patients unrelated to the disease condition and only prescribed medication. Subsequently, we organized the obtained valid text data and divided each patient's response into four samples, based on the information contained in the text and the issues of concern to the hospital. These samples include the patient's recent condition, recent sleep status, recent mood, and whether the patient has suicidal tendencies.

4.2   Experiment Settings

This study focuses on the classification of depression based on text. We utilize the dataset introduced in Sect. 4.1, where the samples undergo necessary preprocessing. The dataset is then split into training, validation, and testing sets with a ratio of 3:1:1. Subsequently, model training and parameter tuning are performed using the training set. Various techniques such as regularization, optimization algorithms, and hyperparameter tuning are applied to improve the model's generalization and performance. The model is iteratively updated based on a chosen loss function, such as cross-entropy, to minimize the discrepancy between predicted and actual labels. After model training and tuning, the trained models are evaluated on the testing set to obtain the final performance metrics, including accuracy, precision, recall, and F1 score.

Overall, this study follows a rigorous experimental setup involving dataset preparation, training-validation-testing split, model training, parameter tuning, and performance evaluation using standard metrics. These steps ensure a systematic and objective assessment of the proposed method's performance in text-based depression classification.

4.3 Comparative Experiments

In order to evaluate the performance of the proposed model in the task of classifying text depression, this study selected seven representative methods for experimental validation. In the experiments, we compare the proposed method with the existing methods in accuracy, precision, recall, and F1 score index.

The approach presented in this paper involves dividing the text into four segments and training them separately, while the compared methods train the four segments together to obtain the results. Through this comparison, we aim to demonstrate the effectiveness of our method.

Table 1. The Table of Comparative Experimental Results.

Model	Accuracy	Recall	Precision	F1 Score
LR [19]	0.769	0.679	0.814	0.760
SVM [3]	0.828	0.750	0.875	0.808
BERT [5]	0.808	0.714	0.909	0.800
TextCNN [23]	0.846	0.928	0.813	0.867
TCN [2]	0.808	0.750	0.875	0.808
BERT+BiLSTM [21]	0.865	0.892	0.862	0.877
BiLSTM+Attention [17]	0.827	0.786	0.880	0.830
Ours	**0.942**	**0.964**	**0.931**	**0.947**

According to the data in Table 1, apart from the method proposed in this paper, the BERT+BiLSTM [21] method achieved the highest accuracy among these comparison models, reaching an accuracy of 86.5%. The BERT [5] pre-training model achieved a precision value of 90.9%, and the TextCNN [23] model achieved a recall rate of 92.8%. In comparison, the method proposed in this paper showed superior performance in terms of accuracy, recall rate, precision, and F1 score. It achieved an accuracy of 94.2%, a recall rate of 96.4%, a precision of 93.1%, and an F1 score of 94.7%. Therefore, the method proposed in this paper outperforms existing models in terms of performance.

4.4 Ablation Study

In the subsequent experiments, we have conducted two ablation studies. Experiment 1 focused on the fusion module, while Experiment 2 targeted the k-max pooling module.

Experiment 1. We have compared four different model configurations, including single-branch without fusion, four-branch feature fusion, four-branch result fusion, and our final model (three-branch feature fusion combined with the specificity of the "suicide" problem). Table 2 presents the results of our comparative analysis.

Table 2. Ablation experiments of our method

Model	Accuracy	Recall	Precision	F1 Score
Baseline	0.865	0.857	0.889	0.873
Result Fusion	0.904	0.821	1.00	0.901
Feature Fusion	0.923	0.929	0.929	0.929
Feature Fusion + Result Fusion	**0.942**	**0.964**	**0.931**	**0.947**

As seen in Table 2, our model has achieved a significant 3.8% improvement in accuracy compared to the model that solely performed result fusion. Furthermore, it has outperformed the model that focused exclusively on feature fusion by 1.9%. Remarkably, when compared to the approach of inputting text into the model altogether, our model exhibited a substantial accuracy improvement of 7.7%.

Experiment 2. We have conducted an ablation study on the k-max pooling layer in our final model by removing it. We compared the results between the baseline TextCNN module and our final model. The comparative results are shown in Table 3.

Table 3. Ablation study with/without k-max pooling in our method

Model	Accuracy	Recall	Precision	F1 Score
Ours(without k-max pooling)	0.885	0.929	0.867	0.897
Ours(with k-max pooling)	**0.942**	**0.964**	**0.931**	**0.940**

According to Table 3, the addition of the k-max pooling layer resulted in a significant improvement in the performance metrics. Specifically, there have been an increase in accuracy by 5.7%, a rise in the recall by 3.5%, a boost in precision by 6.4%, and a substantial enhancement in the F1 score by 4.3%.

5   Conclusion

Traditional depression detection methods mostly involve directly inputting text into models for recognition. In this paper, a question-level approach is proposed, where the answer to the four questions are separately trained and inputted, and we utilize a deep depression detection based on feature fusion and result fusion we proposed, significantly increasing the accuracy of recognition. We use a large-scale pre-trained model, BERT, to encode text and obtain word vectors with rich semantic information. These vectors are then fed into an improved TextCNN model for text feature extraction. By adding a dual-channel K-max pooling layer, combined with max pooling, more comprehensive and localized features can be obtained. This enables the model to capture a wider range of contextual information and extract multi-scale features, enhancing the robustness of the model. An attention mechanism is applied to fuse the features obtained from this layer and perform dual-branch classification with specific questions. The final results are then fused to achieve an accuracy of 94.2%.

This work has great potential for further research in the future. For example, exploring biological markers such as genes, electroencephalograms (EEG), neuroimaging, etc. can be combined with training on textual, audio, and other information to improve the accuracy of the model. Since depression is a complex mental health issue involving multiple levels and factors interacting with each other, future research needs to strengthen interdisciplinary cooperation. Our future work aims to detect other mental disorders related to depression and capture prevalent mental health issues in daily life.

References

1. Amanat, A., et al.: Deep learning for depression detection from textual data. Electronics **11**(5), 676 (2022)
2. Bai, S., Kolter, J.Z., Koltun, V.: An empirical evaluation of generic convolutional and recurrent networks for sequence modeling. arXiv preprint arXiv:1803.01271 (2018)
3. Chang, C.C., Lin, C.J.: LibSVM: a library for support vector machines. ACM Trans. Intell. Syst. Technol. (TIST) **2**(3), 1–27 (2011)
4. Deng, T., Shu, X., Shu, J.: A depression tendency detection model fusing Weibo content and user behavior. In: 2022 5th International Conference on Artificial Intelligence and Big Data (ICAIBD), pp. 304–309. IEEE (2022)
5. Devlin, J., Chang, M.W., Lee, K., Toutanova, K.: BERT: pre-training of deep bidirectional transformers for language understanding. In: 2019 Conference of the North American Chapter of the Association for Computational Linguistics: Human Language Technologies (Naacl Hlt 2019), pp. 4171–4186 (2019)
6. Grefenstette, E., Blunsom, P., et al.: A convolutional neural network for modelling sentences. In: The 52nd Annual Meeting of the Association for Computational Linguistics, Baltimore, Maryland (2014)
7. Harris, J.R.: No Two Alike: Human Nature and Human Individuality. WW Norton & Company, New York (2010)
8. He, L., et al.: Deep learning for depression recognition with audiovisual cues: a review. Inf. Fusion **80**, 56–86 (2022)
9. Kim, J., Lee, J., Park, E., Han, J.: A deep learning model for detecting mental illness from user content on social media. Sci. Rep. **10**(1), 1–6 (2020)

10. Kour, H., Gupta, M.K.: An hybrid deep learning approach for depression prediction from user tweets using feature-rich CNN and bi-directional LSTM. Multimedia Tools Appl. **81**(17), 23649–23685 (2022)
11. Li, Z., An, Z., Cheng, W., Zhou, J., Zheng, F., Hu, B.: MHA: a multimodal hierarchical attention model for depression detection in social media. Health Inf. Sci. Syst. **11**(1), 6 (2023)
12. Lin, C., et al.: Sensemood: depression detection on social media. In: Proceedings of the 2020 International Conference on Multimedia Retrieval, pp. 407–411 (2020)
13. Mao, K., et al.: Prediction of depression severity based on the prosodic and semantic features with bidirectional LSTM and time distributed CNN. IEEE Trans. Affect. Comput. (2022)
14. Mori, K., Haruno, M.: Differential ability of network and natural language information on social media to predict interpersonal and mental health traits. J. Pers. **89**(2), 228–243 (2021)
15. Orabi, A.H., Buddhitha, P., Orabi, M.H., Inkpen, D.: Deep learning for depression detection of twitter users. In: Proceedings of the Fifth Workshop on Computational Linguistics and Clinical Psychology: From Keyboard to Clinic, pp. 88–97 (2018)
16. Priya, A., Garg, S., Tigga, N.P.: Predicting anxiety, depression and stress in modern life using machine learning algorithms. Procedia Comput. Sci. **167**, 1258–1267 (2020)
17. Rahali, A., Akhloufi, M.A., Therien-Daniel, A.M., Brassard-Gourdeau, E.: Automatic misogyny detection in social media platforms using attention-based bidirectional-LSTM. In: 2021 IEEE International Conference on Systems, Man, and Cybernetics (SMC), pp. 2706–2711. IEEE (2021)
18. Rao, G., Zhang, Y., Zhang, L., Cong, Q., Feng, Z.: MGL-CNN: a hierarchical posts representations model for identifying depressed individuals in online forums. IEEE Access **8**, 32395–32403 (2020)
19. Shah, K., Patel, H., Sanghvi, D., Shah, M.: A comparative analysis of logistic regression, random forest and KNN models for the text classification. Aug. Hum. Res. **5**, 1–16 (2020)
20. Sisask, M., Värnik, A., Kolves, K., Konstabel, K., Wasserman, D.: Subjective psychological well-being (who-5) in assessment of the severity of suicide attempt. Nord. J. Psychiatry **62**(6), 431–435 (2008)
21. Zeberga, K., Attique, M., Shah, B., Ali, F., Jembre, Y.Z., Chung, T.S.: A novel text mining approach for mental health prediction using bi-LSTM and BERT model. In: Computational Intelligence and Neuroscience 2022 (2022)
22. Zehra, W., Javed, A.R., Jalil, Z., Khan, H.U., Gadekallu, T.R.: Cross corpus multilingual speech emotion recognition using ensemble learning. Complex Intell. Syst. **7**(4), 1845–1854 (2021). https://doi.org/10.1007/s40747-020-00250-4
23. Zhang, Q., Zheng, R., Zhao, Z., Chai, B., Li, J.: A textCNN based approach for multi-label text classification of power fault data. In: 2020 IEEE 5th International Conference on Cloud Computing and Big Data Analytics (ICCCBDA), pp. 179–183. IEEE (2020)
24. Zhang, T., You, F.: Research on short text classification based on textCNN. J. Phys. Conf. Ser. **1757**, 012092. IOP Publishing (2021)
25. Zou, B., et al.: Semi-structural interview-based Chinese multimodal depression corpus towards automatic preliminary screening of depressive disorders. IEEE Trans. Affect. Comput. 1–16 (2022)

Adaptive Cluster Assignment
for Unsupervised Semantic Segmentation

Shengqi Li, Qing Liu, Chaojun Zhang, and Yixiong Liang[✉]

School of Computer Science and Engineering, Central South University,
Changsha, China
yxliang@csu.edu.cn

Abstract. Unsupervised semantic segmentation (USS) aims to identify semantically consistent regions and assign correct categories without annotations. Since the self-supervised pre-trained vision transformer (ViT) can provide pixel-level features containing rich class-aware information and object distinctions, it has recently been widely used as the backbone for unsupervised semantic segmentation. Although these methods achieve exceptional performance, they often rely on the parametric classifiers and therefore need the prior about the number of categories in advance. In this work, we investigate the process of clustering adaptively for the current mini-batch of images without having prior on the number of categories and propose Adaptive Cluster Assignment Module (ACAM) to replace parametric classifiers. Furthermore, we optimize ACAM to generate weights via the introduction of contrastive learning, which is used to re-weight features, thereby generating semantically consistent clusters. Additionally, we leverage image-text pre-trained models, CLIP, to assign specific labels to each mask obtained from clustering and pixel assignment. Our method achieves new state-of-the-art results in COCO-Stuff and Cityscapes datasets.

Keywords: Unsupervised semantic segmentation · Contrastive learning · Image-text retrieval

1 Introduction

In computer vision, semantic segmentation is a fundamental task that involves assigning every pixel in an image to a particular known category. The applications of semantic segmentation are numerous, ranging from medical image analysis to autonomous driving [11,37,40,43,47]. However, the process of pixel-level annotation required for semantic segmentation is labor-intensive and time-consuming. As such, researchers have focused on developing methods that can guide model training with few or no labels. A large number of weakly supervised semantic segmentation models have been explored that use image- [1,38], scribble- [26], box- [14,34], or point-level [4] supervision information to reduce the need for manual annotation. In contrast, the task of unsupervised semantic

© The Author(s), under exclusive license to Springer Nature Singapore Pte Ltd. 2024
Q. Liu et al. (Eds.): PRCV 2023, LNCS 14428, pp. 75–86, 2024.
https://doi.org/10.1007/978-981-99-8462-6_7

segmentation (USS) is even more challenging, as it involves capturing pixel-level semantic information from unlabeled datasets by mining the intrinsic characteristics of the data.

Early USS approaches usually utilized convolutional neural network (CNN) as the backbone and leveraged clustering as a proxy task to facilitate representation learning [12,20]. However, these methods required the training of both the backbone network and the clustering process from scratch, which often leads to optimization difficulties. Recently, the seminal ViT [15] brings attention-based transformer [36] to computer vision and has taken this community by storm [46]. As the self-supervised ViT (e.g. DINO [8,28]) can generate patch-level feature relationships with rich semantic information, recent USS approaches [17,27,31,33,42] often directly leverage them as the frozen backbone to extract generic features, which can effectively circumvent the challenge of optimization difficulties arising from training from scratch. Although significant advancements have been made, existing USS methods usually rely on parameterized classifiers to acquire class centers that represent the entire dataset. Consequently, they necessitate knowing the number of categories in the dataset beforehand. Furthermore, as discussed in [39,50], using a fixed parameterized classifier fails to consider the presence of intra-class variance, leading to suboptimal performance.

Fig. 1. Qualitative results of ACAM and DINO on the COCO-Stuff-27 dataset.

In this paper, we propose a simple but effective adaptive cluster assignment module (ACAM) to dynamically weigh the pixel-level features generated by pretrained ViT to produce adaptive cluster centers, while assigning pixels to the nearest cluster centers for unsupervised semantic segmentation. Adaptive pixel assignment is reflected in the fact that the representation and number of cluster centers change with the image features in the current mini-batch, so the cluster representation varies with different scene images, effectively alleviating the influence of intra-class variance. We use the text features of CLIP [29] as a classifier to replace a parameterized classifier so we do not need to know the number of categories in the dataset in advance. We assign a specific category to clusters by calculating the similarity between cluster features and text features. In our method, we incorporate contrastive loss and diversity loss to optimize the network. Figure 1 are the qualitative results of our method. To sum up, the main contributions of our method are as follows:

- We propose ACAM to generate adaptive clusters for unsupervised semantic segmentation.
- We leverage both the supervised contrastive and diversity loss to learn the proposed ACAM in an end-to-end manner.
- We perform extensive experiments on unsupervised semantic segmentation datasets, including COCO-Stuff and Cityscapes, and both achieve new state-of-the-art (SOTA) performance.

2   Related Work

2.1   Unsupervised Semantic Segmentation

We have witnessed unprecedented progress in semantic segmentation during the past decade, whereas massive methods have tackled segmentation in either full-supervised [11,37,39,40,43] or semi-/weakly-supervised setting [1,14,34,38]. Recently there are also a few emerging works focused on achieving semantic segmentation without labels [8,12,20,25,35,42]. These techniques often draw inspiration from self-supervised visual representation learning. Early techniques usually focused on pixel-level self-supervised training from scratch [12,20,21,35]. However, their training process highly relies on data augmentation and prior knowledge. Recently, using the self-supervised ViT (*i.e.* DINO [8]) as a backbone is a successful attempt for USS [17,31,42,45]. For instance, STEGO [17] and HP [31] add a task-related segmentation head after the backbone and further generate new features to transfer knowledge. TransFGU [42] activates semantically related regions based on the relationship between features from self-supervised ViT and then trains a segmentation model with generated pseudo-labels. While the aforementioned studies obtain discriminative features by fine-tuning or self-training on the dataset, these methods need to train parametric classifiers using pseudo-labels, requiring prior knowledge about the number of categories in the dataset.

In contrast to the utilization of parametric classifiers, AcSeg [25] takes another direction by directly optimizing a set of learnable prototypes which indicate semantic segments. The optimization process for prototypes interacts with image features and adaptively generates the underlying concepts at the dataset level. However, features within the same class of different scenes often exhibit significant variations [39] and therefore using prototypes for the whole dataset may not be flexible enough to match the distribution of pixels belonging to that class in the feature space. Instead, our method generates adaptive cluster centers in the mini-batch and each clustering feature only accounts for the scene-level semantic information in the batch, enabling a more accurate representation of the data.

2.2 Self-supervised Visual Feature Learning

Learning meaningful visual representations without human annotation is a challenging and long-standing goal in computer vision. Self-supervised learning (SSL) [2] aims to achieve this goal by using some proxy tasks that provide supervised information for unlabeled data, enabling the learning of general features in datasets without manual labeling. In recent years, SSL has evolved from using simple proxy tasks such as learning spatial context [9,23,24] to contrastive learning [7,10,16,19] and masked image modeling (MIM) [3,18,41,49]. SSL has been successfully used to pre-train both CNN and ViT [15]. Specifically, DINO [8] explores a self-distillation strategy on ViT, showing that ViT contains explicit information that can be used for the semantic segmentation of an image. DINOv2 [28] further explores the potential of SSL by enhancing data quality, refining models, and employing advanced training strategies. Many unsupervised semantic segmentation methods have benefited from the properties of DINO [8,28], exploiting the frozen features extracted from self-supervised ViT [8,28]. In our method, we also adopt the frozen ViT trained via DINO [8] as the backbone for a fair comparison.

2.3 Zero-Shot Semantic Segmentation with Pretrained Image-Language Model

The pretrained image-language model has exhibited the strong ability of zero-shot transfer on various downstream vision tasks [29,30]. For semantic segmentation, MaskCLIP [48] modifies the visual encoder of CLIP to align the text-based classifier at the pixel level. RECO [32] leverages text to retrieve images by CLIP [29], and uses the correspondence between image features to generate prototypes to co-segment entities. In our method, we adopt MaskCLIP [48] to assign class labels to the generated clusters.

3 Method

3.1 Overview

We approach USS via a two-stage frame as illustrated in Fig. 2. To begin with, we leverage a frozen self-supervised ViT, *i.e.* DINO [8], to produce the pixel-level features of images in the batch. These valuable features are then fed into the proposed ACAM for clustering and segmentation. In the inference stage, we simply input the image and corresponding segmented masks into a pretrained image-language model [29] to assign a specific category to each segmented mask. Specifically, in ACAM, we first learn K pointwise convolutions to directly predict the confidence of each pixel to K clusters and allocate pixels to their respective

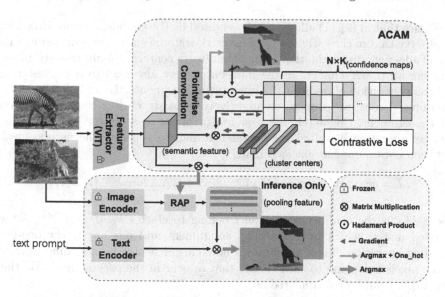

Fig. 2. The framework of our method. In the training stage, ACAM generates class-agnostic segmentations for a mini-batch of training images. In the inference stage, features extracted from the CLIP image encoder within each cluster are combined by region average pooling, and then assign the pre-defined semantic category to every cluster by calculating the similarity with text-encoded features, achieving semantic segmentation without ground-truth.

clusters. By doing so, we can obtain the clusters in the mini-batch and adaptively generate cluster centers by weighting the pixel-level features in the clusters with their respective confidences. Ultimately, we obtain class-agnostic segmentation by assigning each pixel to the nearest cluster center in the feature space. We will illustrate each module in detail next.

3.2 Adaptive Clustering Assignment Module

Given a set of training images $X = \{x_b\}_{b=1}^{B}$ with batch size of B, we first obtain a feature map $F \in \mathbb{R}^{N \times D}$ by feeding them into the frozen feature extractor $\mathcal{F}$, where $N = B \times H \times W$ is the number of pixel features in the mini-batch and D is the dimension of pixel feature, H and W are the height and width of feature maps respectively. We then generate the confidence scores of pixel features belonging to each cluster $M \in \mathbb{R}^{N \times K}$ via pointwise convolutions. By multiplying the pixel features F with the confidence map M through matrix multiplication, we adaptive obtain the class centers for mini-batch images, $i.e.$

$$C = M^T \otimes F, \tag{1}$$

where $\otimes$ is matrix multiplication and $C \in \mathbb{R}^{K \times D}$ is the matrix of generated cluster centers.

Notice that in Eq. (1) all pixels are weighted by the confidence map. However, we observe that in the early stages of network training such a scheme results in a lack of adequate discrimination among the cluster centers. Simultaneously, incorporating all the K clusters in the mini-batch may also lead to over-clustering and the generation of additional segmentation fragments. To address the aforementioned issues, we propose to adaptively generate cluster centers based on pixel assignment. Concretely, we first assign pixels to a cluster by performing arg max operation based on the confidence map to generate the pseudo mask and then obtain one-hot encoded matrix, $i.e.$

$$\tilde{M} = \texttt{one\_hot}(\ \arg\max_{k}(M_{ik})\ ),\qquad i \in [1, ..., N],\ k \in [1, ..., K], \qquad (2)$$

where M_{ik} is the confidence of pixel i belong to cluster k, $\tilde{M} \in \mathbb{R}^{N \times K}$ is the encoded matrix which indicates pixel assignment and $\texttt{one\_hot}$ is the one-hot encoding operation. Then we obtain a new cluster center by assigning weights to the pixels belonging to each cluster that appear in the pseudo mask. By this way, Eq. (1) can be reformulated as

$$C = (M \odot \tilde{M})^{T} \otimes F, \qquad (3)$$

where $\odot$ is the Hadamard product.

It should be noticed that in Eq. (2) we presumably assign pixels to the clusters based on the confidence M_{ik}. Some clusters may be not assigned any pixels and therefore the corresponding column of $\tilde{M}$ are zeros, resulting in some dummy zero-clusters in C. We simply remove them while the remaining non-zero cluster centers are denoted by $\tilde{C} \in \mathbb{R}^{K' \times D}(K' \leq K)$. Then the similarity matrix of each pixel to all valid K' clusters can be obtained by

$$S = F \otimes \tilde{C}^{T} \in \mathbb{R}^{N \times K'}. \qquad (4)$$

Again, by applying the arg max operation on each row of S, we can obtain the segmentation results of images.

3.3   Objective Function

In order to train our ACAM, we first consider the supervised contrastive loss (SCL) [22] by contrasting semantically similar (positive) and dissimilar (negative) pairs. Intuitively, the cluster centers obtained by Eq. (3) can be directly used as the anchors and the pixels assigned to each anchor can be treated as the positive samples. However, as the final segmentation is based on the arg max operation on the soft assignment of pixels in Eq. (4), we use them instead of Eq. (3) to determine the anchors and corresponding positive samples. Specifically, we re-assign each pixel according to the soft assignments in Eq. (4) and based on them we select the cluster centers of non-empty clusters as anchor features and all pixels belonging to the current cluster as positive samples of anchors while other the remaining as negatives, resulting the following contrastive loss

$$L^{con} = -\frac{1}{K'} \sum_k^{K'} \frac{1}{|\mathcal{P}_k|} \sum_{p \in \mathcal{P}_k} \log \frac{\exp(\cos(c_k, f_p)/\tau)}{\sum_{n=1}^{N} \exp(\cos(c_k, f_n)/\tau)}, \quad (5)$$

where $\mathcal{P}_k$ is the index set of pixels assigned to valid cluster c_k (*i.e.* anchor) and $|\mathcal{P}_k|$ is the cardinality of $\mathcal{P}_k$, $\cos(\cdot, \cdot)$ is the cosine similarity function and τ denotes the scalar temperature parameter.

In addition to the above contrastive loss, we also introduce the diversity loss [42] to enforce the diversity of cluster centers. Specifically, the diversity loss aims to maximize the distance between pairwise cluster centers by minimizing their cosine similarity

$$L^{div} = \frac{1}{K'(K'-1)} \sum_{k \neq j} \frac{\cos(c_k, c_j)}{\sqrt{D}}. \quad (6)$$

The overall loss function for end-to-end training of our ACAM is as follows

$$L = L^{div} + \beta L^{con}. \quad (7)$$

3.4   Inference

Inference is a two-stage process. Firstly, ACAM generates cluster centers dynamically for the min-batch images, and the pixels are assigned to the nearest cluster centers in the feature space, thereby generating class-agnostic object segmentation. In the second stage, following [48], we exploit the image encoders of CLIP [29] to obtain the pixel-level representations and then leverage the region average pooling to produce the representation of each cluster. We also generate text features via the CLIP text encoder [29] for predefined categories, and compute the similarity between cluster representations and text features for classification. By doing this, the final semantic segmentation is obtained.

4   Experiments

4.1   Datasets and Experimental Settings

Datasets. COCO-Stuff [5] is a large-scale scene-centric dataset that contains 80 things classes and 90 stuff classes. The total number of images in the training set and validation set are 117,266 and 5,000, respectively. We evaluate COCO-Stuff based on previous work [17,31,42] on the "curated" dataset (2,175 images total). Concretely, in order to demonstrate the flexibility of our method, we evaluate its performance within two distinct category granularity settings: 1) COCO-Stuff-27 which divides all classes into 27 superclasses (12 things categories and 15 stuff categories), and 2) COCO-Stuff-171 which is the original 171 things and stuff categories. Cityscapes [13] is a dataset of urban street scenes, containing

2,975 training images and 500 validation images. Following [12,17], we categorize this dataset into 27 classes, which is a challenging benchmark. We adopt mean intersection over union (mIoU) for qualitative evaluation, following previous works [17,25] and other detailed experimental settings are provided in the supplementary material.

Table 1. Comparative results on COCO-Stuff-27 validation set. † denotes our reproduction results.

method	backbone	CLIP	mIoU
IIC [20] (CVPR2019)	R18+FPN		6.7
PICIE [12] (CVPR2021)	R18+FPN		13.8
MaskCLIP [48] (ECCV2022)	ViT-B/16	√	19.0
RECO [32] (Neurips2022)	DeiT-SIN	√	26.3
DINO [8] (ICCV2021)	ViT-S/8		11.3
+TransFGU [42] (ECCV2022)	ViT-S/8		17.5
+STEGO [17] (ICLR2022)	ViT-S/8		24.5
+HP [31] (CVPR2023)	ViT-S/8		24.6
+AcSeg [25] (CVPR2023)	ViT-S/8	√	28.1
+ACAM (ours)	ViT-S/8	√	**29.6**
DINO† [8] (ICCV2021)	ViT-B/8		13.6
+STEGO [17] (ICLR2022)	ViT-B/8		28.2
+ACAM (ours)	ViT-B/8	√	**31.0**

4.2 Quantitative Comparative Results

We compare our proposed method with previous SOTA techniques and the results are listed in Table 1 and Table 2. Our method consistently outperforms previous SOTA methods on both the COCO-Stuff-27 and Cityscapes datasets. We defer to the supplementary material the comparative results on COCO-Stuff-171 dataset which also shows a similar conclusion and reflects the flexibility that our method is not limited by the granularity of categories.

4.3 Ablation Studies

This section is dedicated to a comprehensive analysis of ablation studies to analyze our method.

Ablation Study of ACAM. In order to assess the efficacy of ACAM, we conduct a comparative analysis with a newly introduced baseline approach called DINO+CLIP. We implemented the baseline method. Specifically, we employ the K-means algorithm to cluster the features generated by the DINO model,

Table 2. Comparative results on Cityscapes validation set.

method	backbone	CLIP	mIoU
MDC [6] (ECCV2018)	R18+FPN		7.1
IIC [20] (CVPR2019)	R18+FPN		6.4
PICIE [12] (CVPR2021)	R18+FPN		12.3
MaskCLIP [48] (ECCV2022)	ViT-B/16	√	14.1
RECO [32] (Neurips2022)	DeiT-SIN	√	19.3
DINO [8] (ICCV2021)	ViT-S/8		10.9
+TransFGU [42] (ECCV2022)	ViT-S/8		16.8
+HP [31] (CVPR2023)	ViT-S/8		18.4
+ACAM (ours)	ViT-S/8	√	**22.3**
DINO [8] (ICCV2021)	ViT-B/8		11.8
+STEGO [17] (ICLR2022)	ViT-B/8		21.0
+HP [31] (CVPR2023)	ViT-B/8		18.4
+ACAM (ours)	ViT-B/8	√	**22.5**

with the number of clusters aligned with the categories within the respective datasets. Subsequently, we employ the text encoder from the CLIP model to assign a specific label to each cluster. Our proposed ACAM method surpasses the traditional clustering procedure. Our experimental evaluations are conducted on the COCO-Stuff27 and Cityscapes datasets, and the results of our empirical evaluations are presented in Table 3.

Ablation Study of Backbones. We explore the performance of various backbones trained via different self-supervised ways for unsupervised semantic segmentation. The three backbones generate generic features by discriminative learning where DINO [8] is implemented at the image-level and SelfPatch [44] at the patch-level. Additionally, DINOv2 [28] is trained at both image- and patch-levels. We employ the K-means algorithm to cluster the features generated by the backbone networks for semantic segmentation as the baseline. Our method adaptively generates the cluster centers according to the images within the mini-batch in the inference stage. As shown in Table 4, our method consistently improves performance across three representative backbones.

Table 3. Ablation results for ACAM.

dataset	method	backbone	mIoU
COCO-Stuff-27	DINO+CLIP	ViT-B/8	26.5
COCO-Stuff-27	+ACAM	ViT-B/8	**30.53**
Cityscapes	DINO+CLIP	ViT-B/8	15.96
Cityscapes	+ACAM	ViT-B/8	**22.5**

Table 4. Ablation results with various backbones on the COCO-Stuff-27 validation set.

method	backbone	mIoU
DINO [8]	ViT-S/16	8.0
ours	ViT-S/16	**29.0**
SelfPatch [44]	ViT-S/16	12.3
ours	ViT-S/16	**28.0**
DINOv2 [28]	ViT-S/14	23.6
ours	ViT-S/14	**30.0**

5   Conclusion

In this paper, we proposed a novel adaptive cluster assignment module for unsupervised semantic segmentation, which dynamically adjusts the number of clusters within a mini-batch of images, effectively addressing the challenges of over-clustering and under-clustering and generating semantically consistent clusters which are assigned to specific category by image-language pre-training model. The experimental results demonstrate that our method can obtain new SOTA performance on both the COCO-Stuff and Cityscapes datasets.

Acknowledgments. The authors wish to acknowledge High Performance Computing Center of Central South University for computational resources.

References

1. Ahn, J., Kwak, S.: Learning pixel-level semantic affinity with image-level supervision for weakly supervised semantic segmentation. In: CVPR, pp. 4981–4990 (2018)
2. Balestriero, R., et al.: A cookbook of self-supervised learning. arXiv preprint arXiv:2304.12210 (2023)
3. Bao, H., Dong, L., Piao, S., Wei, F.: BEiT: BERT pre-training of image transformers. In: ICLR (2022)
4. Bearman, A., Russakovsky, O., Ferrari, V., Fei-Fei, L.: What's the point: semantic segmentation with point supervision. In: Leibe, B., Matas, J., Sebe, N., Welling, M. (eds.) ECCV 2016. LNCS, vol. 9911, pp. 549–565. Springer, Cham (2016). https://doi.org/10.1007/978-3-319-46478-7_34
5. Caesar, H., Uijlings, J., Ferrari, V.: COCO-stuff: thing and stuff classes in context. In: CVPR, pp. 1209–1218 (2018)
6. Caron, M., Bojanowski, P., Joulin, A., Douze, M.: Deep clustering for unsupervised learning of visual features. In: Ferrari, V., Hebert, M., Sminchisescu, C., Weiss, Y. (eds.) Computer Vision – ECCV 2018. LNCS, vol. 11218, pp. 139–156. Springer, Cham (2018). https://doi.org/10.1007/978-3-030-01264-9_9

7. Caron, M., Misra, I., Mairal, J., Goyal, P., Bojanowski, P., Joulin, A.: Unsupervised learning of visual features by contrasting cluster assignments. In: NeurIPS, vol. 33, pp. 9912–9924 (2020)

8. Caron, M., et al.: Emerging properties in self-supervised vision transformers. In: ICCV, pp. 9650–9660 (2021)

9. Chen, P., Liu, S., Jia, J.: Jigsaw clustering for unsupervised visual representation learning. In: CVPR, pp. 11526–11535 (2021)

10. Chen, T., Kornblith, S., Norouzi, M., Hinton, G.: A simple framework for contrastive learning of visual representations. In: ICML, pp. 1597–1607. PMLR (2020)

11. Cheng, B., Misra, I., Schwing, A.G., Kirillov, A., Girdhar, R.: Masked-attention mask transformer for universal image segmentation. In: CVPR, pp. 1290–1299 (2022)

12. Cho, J.H., Mall, U., Bala, K., Hariharan, B.: PiCIE: unsupervised semantic segmentation using invariance and equivariance in clustering. In: CVPR, pp. 16794–16804 (2021)

13. Cordts, M., et al.: The cityscapes dataset for semantic urban scene understanding. In: CVPR, pp. 3213–3223 (2016)

14. Dai, J., He, K., Sun, J.: Boxsup: exploiting bounding boxes to supervise convolutional networks for semantic segmentation. In: ICCV, pp. 1635–1643 (2015)

15. Dosovitskiy, A., et al.: An image is worth 16x16 words: transformers for image recognition at scale. In: ICLR (2021)

16. Grill, J.B., et al.: Bootstrap your own latent-a new approach to self-supervised learning. In: NeurIPS, vol. 33, pp. 21271–21284 (2020)

17. Hamilton, M., Zhang, Z., Hariharan, B., Snavely, N., Freeman, W.T.: Unsupervised semantic segmentation by distilling feature correspondences. In: ICLR (2022)

18. He, K., Chen, X., Xie, S., Li, Y., Dollár, P., Girshick, R.: Masked autoencoders are scalable vision learners. In: CVPR, pp. 16000–16009 (2022)

19. He, K., Fan, H., Wu, Y., Xie, S., Girshick, R.: Momentum contrast for unsupervised visual representation learning. In: CVPR, pp. 9729–9738 (2020)

20. Ji, X., Henriques, J.F., Vedaldi, A.: Invariant information clustering for unsupervised image classification and segmentation. In: ICCV, pp. 9865–9874 (2019)

21. Ke, T.W., Hwang, J.J., Guo, Y., Wang, X., Yu, S.X.: Unsupervised hierarchical semantic segmentation with multiview cosegmentation and clustering transformers. In: CVPR, pp. 2571–2581 (2022)

22. Khosla, P., et al.: Supervised contrastive learning. In: NeurIPS, vol. 33, pp. 18661–18673 (2020)

23. Kim, D., Cho, D., Yoo, D., Kweon, I.S.: Learning image representations by completing damaged jigsaw puzzles. In: WACV, pp. 793–802. IEEE (2018)

24. Komodakis, N., Gidaris, S.: Unsupervised representation learning by predicting image rotations. In: ICLR (2018)

25. Li, K., et al.: ACSeg: adaptive conceptualization for unsupervised semantic segmentation. In: CVPR (2023)

26. Lin, D., Dai, J., Jia, J., He, K., Sun, J.: Scribblesup: scribble-supervised convolutional networks for semantic segmentation. In: CVPR, pp. 3159–3167 (2016)

27. Melas-Kyriazi, L., Rupprecht, C., Laina, I., Vedaldi, A.: Deep spectral methods: a surprisingly strong baseline for unsupervised semantic segmentation and localization. In: CVPR, pp. 8364–8375 (2022)

28. Oquab, M., et al.: DINOv2: learning robust visual features without supervision. arXiv preprint arXiv:2304.07193 (2023)

29. Radford, A., et al.: Learning transferable visual models from natural language supervision. In: ICML, pp. 8748–8763. PMLR (2021)

30. Rao, Y., et al.: DenseCLIP: language-guided dense prediction with context-aware prompting. In: CVPR, pp. 18082–18091 (2022)
31. Seong, H.S., Moon, W., Lee, S., Heo, J.P.: Leveraging hidden positives for unsupervised semantic segmentation. In: CVPR (2023)
32. Shin, G., Xie, W., Albanie, S.: ReCo: retrieve and co-segment for zero-shot transfer. In: NeurIPS (2022)
33. Shin, G., Xie, W., Albanie, S.: Namedmask: distilling segmenters from complementary foundation models. In: CVPRW (2023)
34. Song, C., Huang, Y., Ouyang, W., Wang, L.: Box-driven class-wise region masking and filling rate guided loss for weakly supervised semantic segmentation. In: CVPR, pp. 3136–3145 (2019)
35. Van Gansbeke, W., Vandenhende, S., Georgoulis, S., Van Gool, L.: Unsupervised semantic segmentation by contrasting object mask proposals. In: ICCV, pp. 10052–10062 (2021)
36. Vaswani, A., et al.: Attention is all you need. In: NeurIPS, vol. 30 (2017)
37. Wang, J., et al.: Deep high-resolution representation learning for visual recognition. TPAMI **43**(10), 3349–3364 (2020)
38. Wang, Y., Zhang, J., Kan, M., Shan, S., Chen, X.: Self-supervised equivariant attention mechanism for weakly supervised semantic segmentation. In: CVPR, pp. 12275–12284 (2020)
39. Wu, D., Guo, Z., Li, A., Yu, C., Gao, C., Sang, N.: Semantic segmentation via pixel-to-center similarity calculation. arXiv preprint arXiv:2301.04870 (2023)
40. Xie, E., Wang, W., Yu, Z., Anandkumar, A., Alvarez, J.M., Luo, P.: SegFormer: simple and efficient design for semantic segmentation with transformers. In: NeurIPS, vol. 34, pp. 12077–12090 (2021)
41. Xie, Z., et al.: SimMIM: a simple framework for masked image modeling. In: CVPR, pp. 9653–9663 (2022)
42. Yin, Z., et al.: TransFGU: a top-down approach to fine-grained unsupervised semantic segmentation. In: Avidan, S., Brostow, G., Cissé, M., Farinella, G.M., Hassner, T. (eds.) ECCV 2022. LNCS, vol. 13689, pp. 73–89. Springer, Cham (2022). https://doi.org/10.1007/978-3-031-19818-2_5
43. Yuan, Y., Chen, X., Wang, J.: Object-contextual representations for semantic segmentation. In: Vedaldi, A., Bischof, H., Brox, T., Frahm, J.-M. (eds.) ECCV 2020. LNCS, vol. 12351, pp. 173–190. Springer, Cham (2020). https://doi.org/10.1007/978-3-030-58539-6_11
44. Yun, S., Lee, H., Kim, J., Shin, J.: Patch-level representation learning for self-supervised vision transformers. In: CVPR, pp. 8354–8363 (2022)
45. Zadaianchuk, A., Kleindessner, M., Zhu, Y., Locatello, F., Brox, T.: Unsupervised semantic segmentation with self-supervised object-centric representations. In: ICLR (2023)
46. Zhai, X., Kolesnikov, A., Houlsby, N., Beyer, L.: Scaling vision transformers. In: CVPR, pp. 12104–12113 (2022)
47. Zheng, S., et al.: Rethinking semantic segmentation from a sequence-to-sequence perspective with transformers. In: CVPR, pp. 6881–6890 (2021)
48. Zhou, C., Loy, C.C., Dai, B.: Extract free dense labels from CLIP. In: Avidan, S., Brostow, G., Cissé, M., Farinella, G.M., Hassner, T. (eds.) ECCV 2022. LNCS, vol. 13688, pp. 696–712. Springer, Cham (2022). https://doi.org/10.1007/978-3-031-19815-1_40
49. Zhou, J., et al.: Image BERT pre-training with online tokenizer. In: ICLR (2022)
50. Zhou, T., Wang, W., Konukoglu, E., Van Gool, L.: Rethinking semantic segmentation: a prototype view. In: CVPR, pp. 2582–2593 (2022)

Confidence-Guided Open-World Semi-supervised Learning

Jibang Li, Meng Yang$^{(\boxtimes)}$, and Mao Feng

School of Computer Science and Engineering, Sun Yat-sen University,
Guangzhou, China
lijb55@mail2.sysu.edu.cn, yangm6@mail.sysu.edu.cn

Abstract. Open-world semi-supervised learning differs from traditional semi-supervised learning by allowing labeled and unlabeled data from different data distributions, i.e., unlabeled data may contain novel classes. The goal is to identify samples of known classes while simultaneously detecting and clustering samples of novel classes found in the unlabeled data. To address this challenging task, we propose a novel open-world semi-supervised learning model, which integrates two new confidence-guided techniques to better leverage unlabeled data, especially those from novel classes. Specifically, we first approximate the confidence of model on unlabeled samples by inspecting the prediction consistency through the training process, which guides the temperature scaling for cross-entropy loss. Then, we introduce an adaptive class-specific reweighting contrastive learning method, which further synchronizes the learning speed of known and unknown classes, while reducing the influence of noisy labels. Extensive empirical results show our approach achieves significant performance improvement in both seen and unseen classes compared with previous studies, especially in challenging scenarios with limited labeled data.

Keywords: open-world semi-supervised learning · confidence estimation · contrastive learning

1 Introduction

Deep learning has achieved great success in various tasks by leveraging a sufficient amount of labeled training data. Recently, many robust semi-supervised learning (SSL) methods [2,5,20] have been used to supplement annotated training data with abundant unlabeled data, achieving outstanding performance while reducing the dependence on labeled data. Despite recent success, SSL approaches work under the closed-set assumption, in which the categories in labeled and unlabeled subsets share the same underlying class label space. This assumption is difficult to satisfy in many real-world scenarios, which limits their application in practical situations. To address this issue, [12,15] proposed the task of

This work is partially supported by National Natural Science Foundation of China (Grants no. 62176271), and Science and Technology Program of Guangzhou (Grant no. 202201011681).

Novel Category Discovery (NCD), which aims to learn novel classes from unlabeled data based on a set of known classes. Recent work ORCA [3] extends the NCD problem to a more challenging environment, termed by Open-World Semi-Supervised Learning (OWSSL), where the unlabeled data contains known and novel classes. It aims not only to classify known classes, but to identify novel classes samples and cluster them.

Generally, existing OWSSL methods can be roughly divided into two categories: two-stage methods and single-stage methods. Most of current methods are two-stage, such as ORCA [3], OPENLDN [18] and NACH [10]. Typically, two-stage methods rely on some sort of feature pretraining approach, and then fine-tune the network through semi-supervised learning paradigms. Two-stage methods provide a complicated solution and require more time for training. Recently, TRSSL [19] proposes the first single-stage method, which introduces an effective approach for generating pseudo-labels and does not require pre-training. However, we find that single-stage methods depend on the accuracy of pseudo-labels, and for novel class, the generated pseudo-labels are more unreliable. While some previous works have explored the use of pseudo-labels for OWSSL, few have paid attention to the quality and consistency of the generated pseudo-labels. Moreover, this kind of method rely heavily on the label space and overlook the importance of underlying feature space.

To solve OWSSL problem and avoid cumbersome solutions, we focus on the single-stage approach and introduce two confidence-guided techniques to facilitate the model's learning, which are confidence-guided temperature scaling technique and adaptive class-specific re-weighting contrastive module. We first propose confidence-guided temperature scaling technique, which uses the consistency of the predictions through the training process as a reference for estimating sample confidence. We treat the samples with a high frequency of getting the same prediction in sequential training iterations as reliable samples. Encouraging sharper predicted distributions for high confident data helps model mitigate the impact of unreliable pseudo-labels. Then we introduce an adaptive threshold adjusting scheme to guide contrastive learning. Specifically, we adaptively adjust the threshold based on the learning state of each class. For samples with a prediction probability higher than the threshold, we consider the sample from the same class as positive pairs. And when the probability is lower than the threshold, we only use their augmented versions as positive samples to prevent wrongly predicted label from misleading the model's learning. Through the threshold strategy, we select more reliable labels to guide the model's feature learning. And the re-weighted strategy further emphasizes the learning of samples with high confidence while reducing the influence of uncertain noisy samples.

The main contributions of our work in this paper can be summarized as follows:

- We proposes a new consistency-guided temperature scaling technique to alleviate the impact of uncertainty in pseudo-labeling.

- An adaptive class-specific re-weighting contrastive loss is introduced to reduce confirmation bias via simply combining two contrastive learning methods guided by an adaptive thresholding.
- We conduct extensive experiments on three popular datasets. The results show that our method offers competitive performance compared with other methods.

2 Related Work

ORCA [3] is the first work that addresses this challenging problem(OWSSL). This method is a two-stage approach, first obtaining better discriminative features by pretraining, then gradually reducing the model's overfitting degree and enhancing its discriminability in the second stage using an uncertainty-adaptive margin mechanism. GCD [22] considers a similar problem setting to OWSSL. Different from our setting, they primarily focused on leveraging large-scale pretraining and the performance on the Semantic Shift Benchmark. [10] discovers novel class samples through the pairwise similarity between examples, and then synchronizes the learning speed between known classes and novel classes through an adaptive threshold with distribution alignment. [21] introduces prototype-based contrastive learning algorithms for discovering novel classes and learning discriminative representations. OPENLDN [18] uses pairwise similarity to discover novel classes and introduces iterative pseudo-labeling technique to achieve additional performance. Two-stage methods provide a cumbersome solution. To the best of our knowledge, TRSSL [19] is the only prior work that solves this challenging problem using the single-stage method. TRSSL achieves very promising performance in comparison to other two-stage method. To alleviate the computational burden and improve the efficiency, TRSSL generates pseudo-labels for unlabeled data using prior target distribution and trains the network using common semi-supervised paradigm. But it does not take into account the information provided by the feature space (Fig. 1).

3 Methodology

Due to the mixture of known and novel classes in the unlabeled data, the reliability of pseudo-labels will be more important in OWSSL than in SSL. We propose a dynamic temperature technique that introduces prediction consistency as guidance for sample confidence, so as to make sharper predictions on more reliable samples. Unlike previous methods that mainly focus on label or feature information, we select reliable samples for contrastive learning based on their confidence scores to fully exploit feature space information, and use re-weighting to emphasizes learning on confident samples and prevent overfitting of confident but incorrect samples.

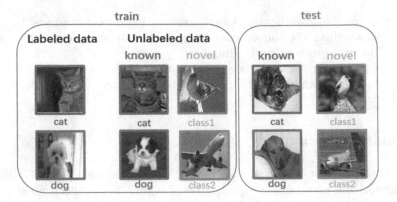

Fig. 1. In open-world semi-supervised learning, the unlabeled data may contain classes that are not present in the labeled data. During testing, we need to identify the known classes and partition the examples of novel classes into appropriate clusters.

3.1   Problem Formulation

In OWSSL, the training data consists of a labeled set D_L and an unlabeled set D_U. The labeled set D_L includes n labeled samples $D_L = \{(x_i^l, y_i)\}_{i=1}^n$, where x_i^l is the a labeled sample and y_i is its corresponding label belonging to one of the C_L known classes. On the other hand, the unlabeled set D_U consists of m (in practice, $m \gg n$) unlabeled samples $D_U = \{(x_i^u)\}_{i=1}^m$, where x_i^u is an unlabeled sample belonging to one of the C_U all classes. The main difference between Closed-world and Open-world semi-supervised learning is that Closed-world assumes $C_L = C_U$, whereas in OWSSL, $C_L \subseteq C_U$. We refer to $C_U \backslash C_L$ as novel classes or C_N.(Following [3,10,18,19], we assume that the number of C_N is known.) During testing, the goal is to assign samples from novel classes to the corresponding novel classes in C_N, and classify samples from known classes into one of the classes in C_L.

In this section, we first introduce the overall framework of our method and then introduce the consistency-guided temperature scaling method and the adaptive class-specific re-weighting contrastive loss.

3.2   Overall Framework

The overall framework of our paper is shown in Fig. 2, which consists of a classification branch and a contrastive learning branch. We construct the classification branch based on existing works TRSSL [19] (as shown in the upper half of Fig. 2) and additionally incorporate a contrastive learning branch to enhance the quality of representations(as shown in the lower half of Fig. 2). For the classification branch, referring to the self-labeling loss method [1], we first generate pseudo-labels for unlabeled data, and then use the generated pseudo-labels to perform self-training on the model. The cross-entropy loss is defined as follows:

$$L_{ce}(\hat{y}_j^{(i)}, y_j^{(i)}) = -\frac{1}{N} \sum_{i=0}^{N} \sum_{j=0}^{C} y_j^{(i)} log\hat{y}_j^{(i)} \tag{1}$$

where $\hat{y}_j^{(i)}$ is the j-th element of model predictive distribution $\hat{y}^{(i)}$, $y_j^{(i)}$ is the j-th element of pseudo-labels $y^{(i)}$. For labeled data, we can directly use ground truth labels. To generate pseudo-labels for unlabeled data, referring to [19], we model the problem as an optimal transport problem and use the fast Sinkhorn-Knopp algorithm [7] to find an approximate assignment. These pseudo-labels are soft

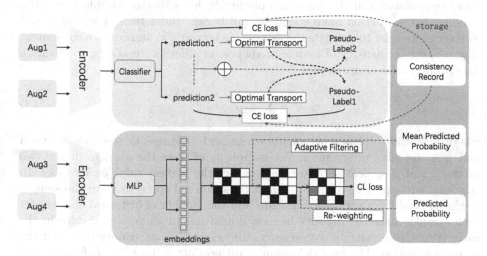

Fig. 2. Architecture of the proposed method. Aug1 and Aug2 are the two augmented views of the classification branch, while Aug3 and Aug4 are the two augmented views of the contrastive learning branch. "Consistency Record" stores the historical consistency information of sample prediction. "Mean Predicted Probability" stores the average predicted probabilities of all samples in each class. "Predicted Probability" represents the predicted probability for each sample. The red lines represent the two confidence-based guiding strategies proposed in our paper. (Color figure online)

probabilities instead of discrete one-hot codes, which are reported to get better performance [8]. Particularly, following [19], we convert soft labels to hard labels for unlabeled samples of novel classes with a prediction confidence exceeding 0.5, which encourages confident learning for novel classes. For classification branch, we generate pseudo-labels for two different views of each training images. To encourage the network to output consistent predictions for two augmented views of the same image, we adopt the prediction swapping task [4].

$$L_{CE} = L_{ce}(\hat{y}_2, sk(\hat{y}_1)) + L_{ce}(\hat{y}_1, sk(\hat{y}_2)) \tag{2}$$

where $sk(\hat{y}_1)$ and $sk(\hat{y}_2)$ are the soft pseudo-labels obtained through the Sinkhorn-Knopp algorithm.

To transfer knowledge between labeled and unlabeled data, we couple the learning of labeled and unlabeled data by sharing the encoder. Additionally, we adopt mixup technique [16] between labeled and unlabeled data to enhance the robustness of the network. Furthermore, we add an adaptive class-specific re-weighting contrastive loss as an auxiliary loss in the contrastive learning branch.

3.3 Consistency-Guided Temperature Scaling

Temperature scaling is a strategy for modifying the softness of output probability distributions. According to [14], samples with high confidence should be assigned lower temperatures. If the network's prediction for a specific sample is certain, making that prediction more confident will benefit the network's learning, and vice versa. In order to estimate the sample uncertainty, we introduce a convenient way, i.e., the consistency of sample prediction in different iterations, to estimate uncertainty c_i for each sample.

$$c_i = \frac{1}{T} \sum_{t=1}^{T} \mathbb{1}(\hat{y}_i^t = \hat{y}_i^{t+1}) \tag{3}$$

where T is the total number of iterations and t represents the t-th iteration and $\hat{y}$ is the label of sample i. For sample i, we take the average of the two prediction of classification branch and select the highest value as its label. In this way, we only need a small amount of additional space to store the sample prediction results of previous iterations and the number of historical changes for each sample. Since the network has low confidence in all unlabeled data in the early stages, to accelerate convergence, we weight the uncertainty based on the training iteration. The final definition of uncertainty in time t is as follows:

$$u^t(x^{(i)}) = G(c_i)F(\frac{t}{T}) \tag{4}$$

where $G(\cdot)$ and $F(\cdot)$ are mapping functions. The mapping function G should be monotonically decreasing, while F should be monotonically increasing. The former is because as the consistency of sample prediction increases, that is, the confidence level becomes higher, the temperature should be lower; the latter is because the low confidence level of early model predictions (especially for novel classes) leads to a high initial temperature, which makes it difficult for the model to learn information from unlabeled samples. Therefore, the temperature in the early stage should be lowered based on iterations. We have found that simply setting $G(x) = 1 - x$ and $F(x) = x^2$ yields promising results.

In the final step, we utilize the temperature to adjust the predicted distribution of the model $\hat{y}$. For convenience, we represent new softmax probability as $\tilde{y}$, which is calculated as follows:

$$\tilde{y}_j^{(i)} = \frac{exp(\hat{y}_j^{(i)}/u(x^{(i)}))}{\sum_c exp(\hat{y}_c^{(i)}/u(x^{(i)}))} \tag{5}$$

3.4   Adaptive Class-Specific Re-weighting Contrastive Module

In open-world setting, unsupervised contrastive learning cannot effectively assist sample discrimination in the label space. When using pseudo labels to guide supervised contrastive learning for unlabeled data, the noise of pseudo labels, especially on novel classes, can cause the model to bias towards known classes and result in a decrease in accuracy in discriminating novel classes. To address this issue, we solve it through simple adaptive thresholding and class-specific re-weighting.

We seamlessly integrate clustering and contrastive learning in the feature space and apply re-weighting to training. Recent semi-supervised learning methods have shown that thresholding on pseudo labels can further improve semi-supervised performance. In recent semi-supervised learning method [5], the mean confidence of all samples per class can serve as an indicator of the network's learning progress for that particular class. By selecting samples above the sample prediction mean, we can obtain high-quality samples for contrastive learning while mitigating the learning discrepancies across various classes. Specifically, we maintain a prediction mean table for each class. To guarantee the stability of the estimation, we update the confidence using exponential moving average (EMA) with a momentum λ over previous batches:

$$
\widetilde{Tp_t}(c) = \begin{cases} 0 & if\ t = 0 \\ \lambda \widetilde{Tp_{t-1}}(c) + (1-\lambda)\frac{1}{N(c)}\sum^{D_U} max(\hat{y})\mathbb{1}(max(\hat{y}) = c) & otherwise \end{cases}
$$
$$(6)$$

where $N(c) = \sum^{D_U} \mathbb{1}(max(\hat{y}) = c)$ is the number of samples per class in the unlabeled dataset, and $\widetilde{Tp_t} = [\widetilde{Tp_t}(1), \widetilde{Tp_t}(2), \cdots, \widetilde{Tp_t}(c)]$ contains the average probability per class predicted by the model in iteration t, which also serves as the dynamic threshold for each class. $\hat{y}$ is the mean of predicted probabilities from two augmented versions.

We define the set of samples with confidence scores higher than the threshold as Z^h with $(max(\hat{y}) > \widetilde{Tp_t})$, and the set of samples with confidence scores below the threshold as Z^l with $(max(\hat{y}) < \widetilde{Tp_t})$. For anchor sample z_i in Z^h, belonging to class c, we select its positive sample set as follows:

$$
Z_i^+ = \{z_j \in Z^h | z_j\ belong\ to\ c\ and\ q^j > \widetilde{Tp_t}(c)\} \tag{7}
$$

where $q^j = max(\hat{y}^{(j)})$ is the confidence score of sample j. For samples with high confidence scores, we define the loss as follows:

$$
L_i^h = \frac{1}{|Z_i^+|} \sum_{z_j \in Z_i^+} q_i \cdot q_j log \frac{exp(z_i \cdot z_j/\tau)}{\sum_{z_k \in Z^h} \mathbb{1}(j \neq i)exp(z_i \cdot z_k/\tau)} \tag{8}
$$

where z_i and z_j represent the embeddings of sample i and j, q_i and q_j are the confidence score. We use these probabilities to re-weight each sample to focus on learning from high-confidence clean data. For samples with low confidence scores $max(p) < \widetilde{Tp_t}(c)$, we follow the contrastive learning mechanism [6] and treat the augmented version z_i^* as the positive sample and the other instances as negative samples.

$$L_i^l = -q_i \cdot q_j \cdot log \frac{exp(z_i \cdot z_i^*/\tau)}{\sum_{j=1}^{2N} \mathbb{1}(j \neq i)exp(z_i \cdot z_j/\tau)} \qquad (9)$$

Finally, we have the following training loss:

$$L = L_{CE} + \lambda_{cl}(\sum^H L_i^h + \sum^L L_i^l) \qquad (10)$$

where λ_{cl} is a hyperparameter, which controls the weight of contrastive learning loss. H and L represent the sets of samples with confidence scores above and below the threshold, respectively.

4 Experiments

4.1 Datasets and Evaluation Metrics

To demonstrate the effectiveness of our method, we conduct experiments on three common benchmark datasets: CIFAR-10, CIFAR-100 and Tiny ImageNet. Following [19], we divide each of these datasets based on the percentage of known and novel classes. We first divide classes into 50% known classes and 50% novel classes, then select 50% and 10% of seen classes as the labeled data respectively, and the rest as unlabeled data.

We follow the evaluation strategy proposed by ORCA [3] and report the following metrics: (1)accuracy for known classes, (2)accuracy for novel classes, and (3)accuracy for all classes. When reporting accuracy for novel classes and all classes, we use the Hungarian algorithm [17] to align the predictions and ground-truth labels.

We compare our method with various existing methods. This includes modified methods derived from semi-supervised methods [20], as well as novel class discovery methods [9,11–13]. Additionally, we also compared two-stage methods [3,10,18,21] and single-stage methods [19].

4.2 Implementation Details

For a fair comparison with the existing methods, we use ResNet-18 as the backbone. The model is trained by using standard SGD with a momentum of 0.9 and a weight decay of 0.0001. We train the model for 200 epochs with a batch size of 256. Specifically, for the contrastive loss, we set the weight λ_{cl} to 0.2 for CIFAR10, and 0.5 for the other dataset. Besides, for unlabeled data, we clip the consistency-guided temperature scaling between 0.1 and 1.0. For labeled data, the temperature is set to 0.1 by default. Since the CIFAR-10 dataset is relatively simple, the network learns to classify it faster. Setting the temperature too low for the labeled dataset may cause the model to overfit on the known classes and

Table 1. The results are averaged over three independent runs. Average accuracy on the CIFAR-10, CIFAR-100, and Tiny ImageNet datasets with with 10% labeled data.

Mehtods	Cifar10			Cifar100			Tiny imagenet		
	seen	novel	all	seen	novel	all	seen	novel	all
FixMatch [20]	64.3	49.4	47.3	30.9	18.5	15.3	–	–	–
DS3L [11]	70.5	46.6	43.5	33.7	15.8	15.1	–	–	–
DTC [13]	42.7	31.8	32.4	22.1	10.5	13.7	13.5	12.7	11.5
RankStats [12]	71.4	63.9	66.7	20.4	16.7	17.8	9.6	8.9	6.4
UNO [9]	86.5	71.2	78.9	53.7	33.6	42.7	28.4	14.4	20.4
ORCA [3]	82.8	85.5	84.1	52.5	31.8	38.6	–	–	–
Opencon [21]	–	–	–	62.5	44.4	48.2	-	–	–
NACH [10]	91.8	89.4	90.6	65.8	37.5	49.2	-	–	–
OpenLDN [18]	92.4	**93.2**	92.8	55.0	40.0	47.7	–	–	–
TRSSL [19]	94.9	89.6	92.2	68.5	52.1	60.3	39.5	**20.5**	30.3
Ours	**95.0**	91.5	**93.2**	**71.1**	**53.1**	**62.1**	**45.8**	20.0	**33.3**

cause insufficient learning for novel classes. Therefore, we set the temperature to 0.15 with 10% labeled data and 0.5 with 50%. We primarily use UNO [9] augmentations for classification branch and Simclr [6] augmentations for contrastive branch. Finally, We inherit the hyperparameters for the Sinkhorn-Knopp algorithm [7] from the former work [19], i.e., $\epsilon = 0.05$ and the iteration number is 3.

Table 2. The results are averaged over three independent runs. Average accuracy on the CIFAR-10, CIFAR-100, and Tiny ImageNet datasets with 50% labeled data.

Mehtods	Cifar10			Cifar100			Tiny imagenet		
	seen	novel	all	seen	novel	all	seen	novel	all
FixMatch [19]	71.5	50.4	49.5	39.6	23.5	20.3	–	–	–
DS3L [11]	77.6	45.3	40.2	55.1	23.7	24.0	–	–	–
DTC [13]	53.9	39.5	38.3	31.3	22.9	18.3	28.8	16.3	19.9
RankStats [12]	86.6	81.0	82.9	36.4	28.4	23.1	5.7	5.4	3.4
UNO [9]	91.6	69.3	80.5	68.3	36.5	51.5	46.5	15.7	30.3
ORCA [3]	88.2	90.4	89.7	66.9	43.0	48.1	–	–	–
NACH [10]	89.5	92.2	91.3	68.7	47.0	52.1	–	–	–
Opencon [21]	89.3	91.1	90.4	69.1	47.8	52.7	–	–	–
OpenLDN [18]	95.7	**95.1**	**95.4**	73.5	46.8	60.1	58.3	25.5	41.9
TRSSL [19]	**96.8**	92.8	94.8	80.2	49.3	64.7	59.1	24.2	41.7
Ours	95.1	**95.1**	95.1	**80.3**	**51.5**	**65.9**	**60.0**	**26.7**	**43.4**

4.3 Experimental Results and Analysis

In Table 1, we report the results when only 10% of the labels are available for the seen categories. Across all three datasets, our method achieved overall classification accuracy that outperforms the state-of-the-art (SOTA) methods. On CIFAR-10 we observe that two-stage approach demonstrated superior performance compared to the single-stage method. Due to the relatively simple nature of the CIFAR-10 dataset, its performance is close to saturation. However, we still achieved a 1.0% improvement in overall performance. Our proposed method achieves significant improvements on all three evaluation metrics, with performance gains of 2.6%, 1.0%, and 1.8% over TRSSL on CIFAR-100 dataset, respectively. Notably, the proposed method achieves increasingly significant improvements as the difficulty of the dataset increases, with particularly impressive gains observed on the challenging Tiny ImageNet dataset, with 200 classes. We achieved improvements of 6.3% on seen classes, and 3.0% on overall performance over the current SOTA methods on the three metrics, respectively.

To further demonstrate the effectiveness of our proposal under different label sizes, we conducted additional experiments when 50% of labels were available for the seen categories. As shown in Table 2, we still achieved overall performance improvements on CIFAR-100 and Tiny ImageNet. On CIFAR-10, we achieved performance close to the current SOTA method.

4.4 Ablation and Analysis

Temperature Module and Contrastive Module. In Table 3, we investigate two main techniques of our method, including consistency-guided temperature scaling and self-adaptive thresholding class-aware contrastive module. The temperature module allows the model to distinguish confidence levels between different samples, which is beneficial for sample selection in contrastive learning. Simultaneously, contrastive learning utilizes the information in the sample feature space to reduce label noise. Our results show that both the temperature module of our approach and the contrastive module are beneficial for improving the accuracy, and combining these two components yields the highest performance.

Table 3. Ablation experiments on CIFAR-100 dataset with 10% labeled data. "uncer" denotes consistency-guided temperature scaling, and "CL" denotes self-adaptive thresholding class-aware contrastive module.

uncer	CL	seen	novel	all
		65.7	48.1	56.9
✓		68.4	51.8	60.2
	✓	69.4	49.6	59.5
✓	✓	**71.1**	**53.1**	**62.1**

Table 4. Ablation of different contrastive learning methods. Experiments on CIFAR-100 dataset with 10% labeled data. "unsupCon" refers to unsupervised contrastive loss, "supCon" refers to supervised contrastive loss.

Method	seen	novel	all
unsupCon	64.4	45.4	54.1
supCon	**72.1**	46.4	59.2
w/o threshold	**72.1**	48.2	60.1
w/o re-weighting	68.1	51.2	59.8
Ours	71.1	**53.1**	**62.1**

Threshold Selection and Re-weighting. Table 4 serves as a validation of the efficacy of threshold selection and re-weighting in our model. The experiments demonstrate that unsupervised contrastive loss significantly reduces the performance, whereas supervised contrastive loss biases the model towards the known classes. Our proposed threshold selection and re-weighting techniques effectively balance the learning rates between visible and novel classes, resulting in an overall optimal performance.

5   Conclusion

In this paper, we introduced consistency-guided temperature scaling and adaptive threshold-based class-specific contrastive learning strategies for open-world semi-supervised learning. The consistency-guided temperature scaling approach fully leverages the information of high-confidence pseudo-labels while avoiding overfitting on low-confidence ones with almost no additional computational overhead. Additionally, the class-specific contrastive learning method under the adaptive threshold strategy can effectively reduce the noise of the pseudo-labels. The performance improvements achieved on multiple benchmark image datasets validate the effectiveness of our proposed methods. Our approach also achieves more significant improvements in situations with fewer labeled samples and more classes.

References

1. Asano, Y.M., Rupprecht, C., Vedaldi, A.: Self-labelling via simultaneous clustering and representation learning. In: International Conference on Learning Representations (ICLR) (2020)
2. Berthelot, D., Carlini, N., Goodfellow, I., Papernot, N., Oliver, A., Raffel, C.A.: Mixmatch: a holistic approach to semi-supervised learning. In: Advances in Neural Information Processing Systems, vol. 32 (2019)
3. Cao, K., Brbić, M., Leskovec, J.: Open-world semi-supervised learning. In: International Conference on Learning Representations (ICLR) (2022)

4. Caron, M., Misra, I., Mairal, J., Goyal, P., Bojanowski, P., Joulin, A.: Unsupervised learning of visual features by contrasting cluster assignments. In: Advances in Neural Information Processing Systems, vol. 33, pp. 9912–9924 (2020)
5. Chen, H., et al.: Softmatch: addressing the quantity-quality trade-off in semi-supervised learning. arXiv preprint arXiv:2301.10921 (2023)
6. Chen, T., Kornblith, S., Norouzi, M., Hinton, G.: A simple framework for contrastive learning of visual representations. In: International Conference on Machine Learning, pp. 1597–1607. PMLR (2020)
7. Cuturi, M.: Sinkhorn distances: lightspeed computation of optimal transport. In: Advances in Neural Information Processing Systems, vol. 26 (2013)
8. Doersch, C., Gupta, A., Efros, A.A.: Unsupervised visual representation learning by context prediction. In: Proceedings of the IEEE International Conference on Computer Vision, pp. 1422–1430 (2015)
9. Fini, E., Sangineto, E., Lathuilière, S., Zhong, Z., Nabi, M., Ricci, E.: A unified objective for novel class discovery. In: Proceedings of the IEEE/CVF International Conference on Computer Vision, pp. 9284–9292 (2021)
10. Guo, L.Z., Zhang, Y.G., Wu, Z.F., Shao, J.J., Li, Y.F.: Robust semi-supervised learning when not all classes have labels. In: Advances in Neural Information Processing Systems, vol. 35, pp. 3305–3317 (2022)
11. Guo, L.Z., Zhang, Z.Y., Jiang, Y., Li, Y.F., Zhou, Z.H.: Safe deep semi-supervised learning for unseen-class unlabeled data. In: International Conference on Machine Learning, pp. 3897–3906. PMLR (2020)
12. Han, K., Rebuffi, S.A., Ehrhardt, S., Vedaldi, A., Zisserman, A.: Automatically discovering and learning new visual categories with ranking statistics. arXiv preprint arXiv:2002.05714 (2020)
13. Han, K., Vedaldi, A., Zisserman, A.: Learning to discover novel visual categories via deep transfer clustering. In: Proceedings of the IEEE/CVF International Conference on Computer Vision, pp. 8401–8409 (2019)
14. Hinton, G., Vinyals, O., Dean, J.: Distilling the knowledge in a neural network. In: NIPS'15 Proceedings of the 28th International Conference on Neural Information Processing Systems, vol. 2, pp. 287–295 (2015)
15. Hsu, Y.C., Lv, Z., Kira, Z.: Learning to cluster in order to transfer across domains and tasks. arXiv preprint arXiv:1711.10125 (2017)
16. Huang, L., Zhang, C., Zhang, H.: Self-adaptive training: beyond empirical risk minimization. In: Advances in Neural Information Processing Systems, vol. 33, pp. 19365–19376 (2020)
17. Kuhn, H.W.: The Hungarian method for the assignment problem. Naval Res. Logist. Q. **2**(1–2), 83–97 (1955)
18. Rizve, M.N., Kardan, N., Khan, S., Shahbaz Khan, F., Shah, M.: OpenLDN: learning to discover novel classes for open-world semi-supervised learning. In: Avidan, S., Brostow, G., Cissé, M., Farinella, G.M., Hassner, T. (eds.) ECCV 2022. LNCS, vol. 13691, pp. 382–401. Springer, Cham (2022). https://doi.org/10.1007/978-3-031-19821-2_22
19. Rizve, M.N., Kardan, N., Shah, M.: Towards realistic semi-supervised learning. In: Avidan, S., Brostow, G., Cissé, M., Farinella, G.M., Hassner, T. (eds.) ECCV 2022. Part XXXI. LNCS, vol. 13691, pp. 437–455. Springer, Cham (2022). https://doi.org/10.1007/978-3-031-19821-2_25
20. Sohn, K., et al.: Fixmatch: simplifying semi-supervised learning with consistency and confidence. In: Advances in Neural Information Processing Systems, vol. 33, pp. 596–608 (2020)

21. Sun, Y., Li, Y.: OpenCon: open-world contrastive learning. Trans. Mach. Learn. Res. (2023). https://openreview.net/forum?id=2wWJxtpFer
22. Vaze, S., Han, K., Vedaldi, A., Zisserman, A.: Generalized category discovery. In: Proceedings of the IEEE/CVF Conference on Computer Vision and Pattern Recognition, pp. 7492–7501 (2022)

SSCL: Semi-supervised Contrastive Learning for Industrial Anomaly Detection

Wei Cai[1(✉)] and Jiechao Gao[2]

[1] University of Science and Technology Beijing,
Haidian District, Beijing, Beijing, China
cw.caiwei1@gmail.com
[2] University of Virginia, Charlottesville, VA 22904, USA
jg5ycn@virginia.edu

Abstract. Anomaly detection is an important machine learning task that aims to identify data points that are inconsistent with normal data patterns. In real-world scenarios, it is common to have access to some labeled and unlabeled samples that are known to be either normal or anomalous. To make full use of both types of data, we propose a semi-supervised contrastive learning method that combines self-supervised contrastive learning and supervised contrastive learning, forming a new framework: SSCL. Our method can learn a data representation that can distinguish between normal and anomalous data patterns, based on limited labeled data and abundant unlabeled data. We evaluate our method on multiple benchmark datasets, including MNIST, CIFAR-10 and industrial anomaly detection MVtec, STC. The experimental results show that our method achieves superior performance on all datasets compared to existing state-of-the-art methods.

Keywords: Semi-supervised classification · Anomaly detection · Contrastive learning · Data representation

1 Introduction

Anomaly detection (AD) [1] has emerged as a critical task in the realm of data analysis, aiming to identify unusual samples within datasets. Traditional unsupervised AD methods [2] assume that the majority of samples are normal and endeavour to learn a compact representation of the data. For instance, in the field of one-class classification [3], the main objective is to identify data points that belong to the majority of the dataset, while considering any data points not belonging to this subset as anomalies. Traditional unsupervised AD methods, such as the One-Class SVM [4], Kernel Density Estimation [5] and Isolation Forest [6], often rely on manual feature engineering to perform well with complex, high-dimensional datasets. Additionally, these methods face limitations when applied to large-scale datasets due to scalability issues. Consequently, there has

Q. Liu et al. (Eds.): PRCV 2023, LNCS 14428, pp. 100–112, 2024.
https://doi.org/10.1007/978-981-99-8462-6_9

been considerable interest in developing new deep-learning approaches for unsupervised anomaly detection [7,14].

In practical situations, it is typical to have datasets that contain a combination of unlabeled and labeled data. In such cases, the problem becomes semi-supervised anomaly detection. The task is to learn a model that effectively characterizes the "normal class" using a combination of labeled and unlabeled data. The samples consist of unlabeled instances, which are mostly normal but may include some anomalous samples, along with labeled samples which include normal samples and anomalous samples (see Fig. 1 for detail). The goal is to create a model that accurately describes the typical or normal data in a concise way. This model should capture the key features and patterns of the normal class, enabling it to distinguish anomalies from the majority of normal samples effectively.

The majority of existing methods for semi-supervised anomaly detection, both shallow and deep, focus solely on labeled normal samples and do not take into account labeled anomalies. These methods are categorized as "Learning from Positive and Unlabeled Examples". In contrast, most research on deep semi-supervised learning has primarily concentrated on classification tasks, where the cluster assumption assumes that similar data points belong to the same class. However, this assumption does not hold true for the "anomaly class" because anomalies can vary significantly from one another. Thus, in semi-supervised anomaly detection, the main challenge lies in finding a concise representation of the normal class while accurately discriminating the labeled anomalies by utilizing all labeled and unlabeled data.

Contrastive learning [8] has extensive applications in domains such as computer vision [17], natural language processing [19], and recommender systems [18]. It offers several advantages for anomaly detection. Firstly, contrastive learning eliminates the need for labeled data, reducing the burden of data collection and annotation. Secondly, it enables the automatic learning of discriminative feature representations that effectively differentiate between normal and abnormal patterns. Lastly, contrastive learning facilitates efficient anomaly detection by clustering normal samples together, enabling effective detection based on distance measurements. Building upon these advantages, we introduce contrastive learning to the semi-supervised AD setting for improved anomaly detection.

In this paper, we propose Semi-supervised Contrastive Learning(SSCL) for industrial anomaly detection in this work. Our main contributions are summarized as followed:

- We consider a realistic industrial anomaly detection scenario - semi-supervised anomaly detection. Due to the scarcity of labeled data in reality, we think of combining labeled and ulabeled data to break through the limitations of the traditional anomaly detection paradigm.
- Combining self-supervised contrastive learning with supervised contrastive learning, we propose SSCL which encourage the model to learn more informative and discriminative representations by jointly considering labeled and unlabeled sample.

- We conduct extensive experiments on MNIST, CIFAR-10, MVtec and Shang-haiTech Campus (STC) [22] datasets. The proposed method outperforms state-of-the-art methods on multiple AD benchmarks, yielding appreciable performance improvements even when provided with only little labeled data.

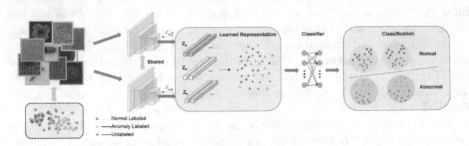

Fig. 1. The framework of semi-supervised contrastive learning for anomaly detection. We start with a training dataset that contains normal, anomaly and unlabeled samples. Next, we leverage specific loss functions (see details in Sect. 3) to train a model that effectively learns the underlying features of the data. Subsequently, we employ a classifier to further train the model, enabling it to effectively differentiate between normal and anomalous data instances. The classifier leverages the learned features to make accurate distinctions.

2   Related Works

Unsupervised Anomaly Detection. Unsupervised anomaly detection methods [9], such as OC-SVM, KDE and Isolation Forest (IF), aim to detect anomalies in unlabeled data without relying on explicit class labels. While this is an advantage in scenarios where labeled data is scarce or unavailable, it also means that these methods cannot make use of explicit class information to guide the learning process. Without labels, the model may struggle to differentiate between anomalies and normal instances accurately. Furthermore, unsupervised anomaly detection methods assume that normal data follow a certain distribution, such as being clustered or having a specific density. If the data distribution is complex or exhibits significant variations, these methods may struggle to accurately capture the normal behaviour, leading to high false-positive rates or missed anomalies. Therefore, we consider semi-supervised anomaly detection to solve this problem.

Semi-supervised Anomaly Detection. [10] involves detecting anomalies in data when only a limited amount of labeled data is available, along with a larger amount of unlabeled data. This approach combines the benefits of supervised and unsupervised methods, using labeled data to guide the anomaly detection process while leveraging unlabeled data for a broader understanding of the data distribution. This addresses the limitations of unsupervised anomaly detection methods. Deep Semi-Supervised Anomaly Detection (DeepSAD) [10] is a popular method in this category. DeepSAD combines deep learning techniques with a semi-supervised framework, using an autoencoder architecture to learn a compact representation of the data. It minimizes the reconstruction error of normal instances, allowing it to accurately reconstruct normal patterns. However, a drawback of semi-supervised anomaly detection methods is the limited generalization to unseen anomalies.

Contrastive Learning [8]. To address the limitations of insufficient generalization in semi-supervised anomaly detection, we have introduced contrastive learning. It effectively learns meaningful and discriminative representations of both normal and anomalous instances. By maximizing the similarity of positive pairs and minimizing the similarity of negative pairs, contrastive learning captures subtle differences between them, enhancing the model's discriminative capabilities in anomaly detection. Data augmentation techniques in contrastive learning enable the model to generalize better to unseen variations and anomalies. Self-supervision through contrastive learning provides additional guidance for differentiating between normal and anomalous patterns. These advantages effectively solve the problem of limited generalization.

3   Method

We now introduce our new method: SSCL combines the advantages of self-supervised contrastive learning [12] and supervised contrastive learning [8] to form a unique structure.

In our method, when given a batch of input data, we will apply data augmentation ($Aug(\cdot)$) twice, resulting in two identical copies of the batch. Both copies are then passed through an encoder network($Enc(\cdot)$). During training, this embedding is further processed by a projection network($Proj(\cdot)$), which is not used during inference. The self-supervised contrastive loss and supervised contrastive loss are calculated based on the outputs of the projection network. To utilize the trained model for classification purposes, we train a linear classifier on top of the fixed embeddings using a cross-entropy loss.

3.1   Self-supervised Contrastive Learning

For a randomly sampled set of N sample/label pairs, $\{x_k, y_k\}_{k=1...N}$, the corresponding training batch consists of $2N$ pairs, $\{\tilde{x}_\ell, \tilde{y}_\ell\}_{\ell=1...2N}$, where $\tilde{x}_{2k}$ and $\tilde{x}_{2k-1}$ are two random augmentations or "views" of $x_k(k = 1...N)$, and $\tilde{y}_{2k-1} = \tilde{y}_{2k} = y_k$. Throughout the remainder of this paper, we will refer to a set of N samples as a "batch" and the set of $2N$ augmented samples as a "multiviewed batch". In other words, for each original sample, we generate two different augmented views, and they are included in the training batch along with their corresponding original samples and labels. The term "batch" is used to refer to the set of original samples, while "multiviewed batch" refers to the set containing the augmented samples.

Within a multiviewed batch, we denote the index of an augmented sample as i, belonging to the set I, which encompasses values from 1 to $2N$. Additionally, we define $j(i)$ as the index of the other augmented sample originating from the same source sample. In self-supervised contrastive learning approaches [13], the loss function is formulated as follows:

$$\mathcal{L}_{\text{self}} = \sum_{i \in I} \mathcal{L}_{\text{self}}^i = -\sum_{i \in I} \log \frac{\exp\left(z_i \cdot z_{j(i)}/\tau\right)}{\sum_{a \in A(i)} \exp\left(z_i \cdot z_a/\tau\right)} \tag{1}$$

where $z_\ell = Proj\left(Enc\left(\tilde{x}_\ell\right)\right) \in \mathcal{R}^{D_P}$ represents the projection of the encoded representation (Enc) of the augmented sample $\tilde{x}_\ell$ in the $\mathcal{R}^{D_P}$ space. The D_P means output vector of size or just a single linear layer of size. The $\bullet$ denotes the inner product operation. The scalar temperature parameter $\tau \in \mathcal{R}^+$ is introduced to control the sharpness of the probability distribution. Each value of i serves as an anchor index, associated with one positive pair $j(i)$ and $2(N-1)$ negative pairs ($\{k \in A(i)\backslash\{j(i)\}$), where N is the total number of samples. It's important to note that for every anchor i, there are a total of $2N-2$ negative pairs and $2N-1$ terms (including both the positive and negative pairs) in the denominator of the loss function.

3.2 Supervised Contrastive Learning

In the case of supervised learning, the self-supervised contrastive loss is not suitable for handling scenarios where multiple samples belonging to the same class are known due to the availability of labels. To generalize the contrastive loss to handle an arbitrary number of positive pairs, different approaches can be considered. The supervised contrastive loss [8,15] present straightforward methods to extend the self-supervised contrastive loss and incorporate supervision:

$$\mathcal{L}_{\text{sup}} = \sum_{i \in I} \mathcal{L}_{\text{sup}}^i = \sum_{i \in I} \frac{-1}{|P(i)|} \sum_{p \in P(i)} \log \frac{\exp\left(z_i \cdot z_p/\tau\right)}{\sum_{a \in A(i)} \exp\left(z_i \cdot z_a/\tau\right)} \tag{2}$$

$P(i) \equiv \left\{p \in A(i) : \tilde{y}_p = \tilde{y}_i\right\}$ is defined as the set of indices of all positive pairs in the multiviewed batch, excluding i, and where the labels $\tilde{y}_p$ and $\tilde{y}_i$ are the same. The cardinality of this set, denoted as $|P(i)|$, represents the number of positive pairs. In supervised contrastive loss, the summation over the positive pairs ($P(i)$) is positioned outside the logarithm function, specifically in the expression $\mathcal{L}_{\text{sup}}$.

3.3   Semi-supervised Contrastive Learning

In our approach, we assume the availability of two types of samples: unlabeled samples $x_1, ... x_n \subseteq \chi$, and labeled samples $(x_1, y_1)...(x_n, y_n) \subseteq \chi \times \psi$. Here, χ represents a subset of D-dimensional real numbers, while ψ is a label set consisting of $\{-1, 1\}$. In the labeled samples, $y = +1$ denotes known normal samples, while $y = -1$ indicates known anomalies. Consequently, we have three types of samples: labeled normal samples, labeled anomaly samples, and unlabeled samples.

For unlabeled data, we employ contrastive learning loss (i.e., Eq. 1). While we introduce supervised contrastive learning loss (i.e., Eq. 2) for the labeled data in our SSCL objective.

Depending on self-supervised contrastive learning and Supervised Contrastive Learning. So, We can define our SSCL objective as follows:

$$\mathcal{L} = \mathcal{L}_{\text{self}} + \alpha \mathcal{L}_{\text{sup}} \tag{3}$$

$$\mathcal{L} = -(1 + \alpha \sum_{i \in I} \frac{-1}{|P(i)|}) \sum_{i \in I} \log \frac{\exp(z_i \cdot z_{j(i)}/\tau)}{\sum_{a \in A(i)} \exp(z_i \cdot z_a/\tau)} \tag{4}$$

where α represents the trade-off parameter.

4   Experiments

We conduct an extensive evaluation of our proposed SSCL method on multiple benchmark datasets, including MNIST, CIFAR-10 and industrial anomaly detection MVtec, STC. We compare the performance of SSCL against various existing approaches, including shallow unsupervised, semi-supervised, and deep semi-supervised methods.

4.1   Competition Methods

We consider several baselines for our experiments, including shallow unsupervised methods such as OC-SVM [4], IF [6] and KDE [5]. For deep unsupervised methods, while for more general semi-supervised AD approaches that make use of labeled anomalies, we employ the state-of-the-art shallow SSAD method with Gaussian kernel [11]. For deep semi-supervised methods, we will utilize Deep-SAD [10], PaDiM [20] and PatchCore [21].

4.2   Evaluation Metrics

We use AUROC as the evaluation metric, which quantifies the overall performance of the classification models. AUROC is widely used in binary classification tasks, indicating the quality of the predictor. A higher AUROC score signifies better performance and accuracy. By employing AUROC as our evaluation metric, we aim to objectively assess and compare the effectiveness of different approaches, providing a standardized measure of their predictive capabilities.

4.3   Scenario Setup

In our semi-supervised anomaly detection setup, we evaluate the performance of SSCL on the MNIST and CIFAR-10 datasets, commonly used for image classification tasks. These datasets contain images of handwritten digits and real-world objects, serving as benchmarks for model verification. Additionally, we apply SSCL to the MVtec and STC dataset, a real-world industrial defect detection dataset. This allows us to assess SSCL's performance in challenging environments and detect and localize various types of defects across multiple object categories.

Following previous works, AD setups on MNIST and CIFAR10. While for MVtec, we cteate respectively AD setup depending on the number of classes in the subset. In each setup, we designate one of all classes as the normal class, while the remaining classes represent anomalies.

To train our model, we use the original training data of the normal class as the unlabeled part of our training set. This ensures that we start with a clean AD setting where all unlabeled samples are assumed to be normal. The training data of the anomaly classes then form a data pool from which we randomly select anomalies for training, allowing us to create different scenarios.

Table 1. Complete results of experimental scenario (i), where we increase the ratio of labeled anomalies γ_l in the training set. We report the avg. AUC.

Data	γ_l	OC-SVM	IF	KDE	SSAD	DeepSAD	PaDiM	PatchCore	SSCL
MNIST	.00	96.0	85.4	95.0	96.0	92.8	96.8	95.8	**97.5**
	.01				96.6	96.4	97.2	97.6	**98.6**
	.05				93.3	96.1	98.8	98.5	**99.2**
	.10				90.7	96.9	99.3	99.1	**99.5**
	.20				87.2	96.9	99.6	99.4	**99.6**
CIFAR10	.00	62.0	60.0	59.9	62.0	60.9	**80.2**	79.3	78.5
	.01				73.0	72.6	82.3	83.2	**83.6**
	.05				71.5	77.9	86.5	87.1	**87.5**
	.10				70.1	79.8	87.2	88.0	**88.5**
	.20				67.4	81.9	88.9	89.1	**89.2**

In our study, we explore three distinct scenarios that are based on prior research and involve widely used datasets such as MNIST, CIFAR10, MVTec and STC. In short, we conduct simulation experiments on MNIST and CIFAR10 to verify the superiority of SSCL. We then experiment with the method on the real MVTec and STC datasets to evaluate its performance in more challenging environments and to detect and localize different classes of defects. These scenarios are carefully chosen to provide representative examples for our experimentation.

4.4 Experimental Scenarios on MNIST and CIFAR-10

(i) **Labeled Anomalies Experiment:** In this particular scenario, we aim to examine the impact of incorporating labeled anomalies during the training process on the detection performance and to determine the advantages of using a general semi-supervised anomaly detection approach compared to other methods. To conduct this experiment, we gradually increase the ratio of labeled training data (denoted as $\gamma_l = m/(n + m)$) by introducing more known anomalies (labeled as $x_1, ...x_n$) with corresponding labels ($y_j = -1$) into the training set. These labeled anomalies are sampled from all anomaly classes. For the unlabeled portion of the training set, we retain the training data belonging to the respective class, keeping it free from pollution ($\gamma_p = 0$) in this specific experimental setup. During testing, we consider the remaining nine classes as anomalies, simulating the unpredictable nature of anomalies. This means that there are eight novel classes at testing time. We repeat the process of generating the training set for each AD setup, considering all nine respective anomaly classes. We then calculate and report the average results over the ten AD setups multiplied by the nine anomaly classes, resulting in a total of 90 experiments for each labeled ratio γ_l.

(ii) **Polluted Training Data Experiment:** In this experiment, we focus on examining the robustness of different methods to an increasing pollution ratio γ_p in the training set caused by unlabeled anomalies. To accomplish this, we introduce anomalies from all nine respective anomaly classes into the unlabeled

Table 2. Complete results of experimental scenario (ii), where we pollute the unlabeled part γ_p of the training set with (unknown) anomalies. We report the avg. AUC.

Data	γ_p	OC-SVM	IF	KDE	SSAD	DeepSAD	PaDiM	PatchCore	SSCL
MNIST	.00	96.0	85.4	95.0	97.9	96.7	98.5	98.3	**98.5**
	.01	94.3	85.2	91.2	96.6	95.5	96.5	97.0	**97.2**
	.05	91.4	83.9	85.5	93.3	93.5	**95.1**	94.6	94.8
	.10	88.8	82.3	82.1	90.7	91.2	92.0	91.1	**92.2**
	.20	84.1	78.7	77.4	87.4	86.6	86.9	87.3	**87.9**
CIFAR10	.00	62.0	60.0	59.9	73.8	77.9	83.1	**83.8**	83.5
	.01	61.9	59.9	72.8	76.5	80.6	80.9		**81.7**
	.05	61.4	59.6	58.1	71.5	74.0	77.3	78.1	**79.6**
	.10	60.8	58.8	57.3	69.3	71.8	72.3	74.5	**76.3**
	.20	60.3	57.9	56.2	67.8	68.5	69.3	70.9	**72.1**

Table 3. Results of experiments on MVtec. We set $\gamma_k = 0.6$ for each dataset category. The method with the highest mean of these two values is highlighted in boldface for each texture and object category.

Category		SSCL	PaDiM	PatchCore	DeepSAD	SSAD	OC-SVM	KDE	IF
texture	carpet	0.72	**0.74**	0.68	0.47	0.46	0.50	0.36	0.37
	grid	0.69	0.75	0.71	0.8	0.79	0.78	**0.82**	0.67
	leather	0.79	0.73	0.76	0.75	0.79	**0.88**	0.44	0.57
	tile	**0.83**	0.81	0.81	0.80	0.80	0.77	0.76	0.78
	wood	0.87	0.83	0.80	0.86	0.82	0.82	0.87	**0.93**
	average	**0.78**	0.77	0.75	0.74	0.73	0.75	0.65	0.66
object	hazelnut	**0.87**	0.79	0.73	0.72	0.71	0.68	0.63	0.04
	bottle	0.80	0.85	0.82	**0.95**	0.91	0.88	0.88	0.73
	cable	0.82	0.80	**0.83**	0.82	0.73	0.79	0.79	0.65
	capsule	**0.81**	0.77	0.75	0.78	0.71	0.57	0.52	0.49
	metal nut	0.75	**0.78**	0.76	0.74	0.56	0.54	0.55	0.55
	pill	**0.82**	0.75	0.77	0.82	0.70	0.66	0.65	0.69
	screw	**0.72**	0.65	0.69	0.28	0.16	0.01	0.03	0.00
	toothbrush	0.75	0.83	0.81	**0.98**	0.90	0.78	0.94	0.94
	transistor	**0.82**	0.78	0.74	0.77	0.75	0.73	0.73	0.67
	zipper	0.75	**0.81**	0.77	0.80	0.80	0.75	0.66	0.72
	average	**0.79**	0.78	0.77	0.77	0.69	0.64	0.64	0.55
	average	**0.79**	0.78	0.76	0.76	0.71	0.68	0.64	0.59

portion of the training set for each AD setup. By doing so, we simulate the effect of increasing pollution in the training data. To maintain consistency, we set the ratio of labeled training samples (γ_l) at a fixed value of 0.05. We iterate the process of generating the training set for each AD setup, considering all nine respective anomaly classes. We then calculate and report the average results over the resulting 90 experiments for each pollution ratio γ_p. We hypothesize that employing a semi-supervised anomaly detection approach, which involves learning from labeled anomalies, mitigates the negative impact caused by pollution in the unlabeled data. This is because similar unknown anomalies present in the unlabeled data may be identified through the use of labeled anomalies.

Results on the MNIST and CIFAR10: The outcomes of scenarios (i) to (ii) are presented in Table 1 to 2, showcasing the key findings of our study. Table 1 shows that SSCL has a good effect on the MNIST and CIFAR10 datasets. With the ratio of labeled anomalies γ_l increasing, SSCL has a better effect. Table 2 demonstrates a decrease in the detection performance of all methods as the level of data pollution increases. Once again, SSCL exhibits the highest level of robustness, particularly on the CIFAR-10 dataset. Regardless of the detection effect or stability, SSCL has shown excellent performance and potential in the simulation experiments of MNIST and CIFAR10 datasets. Whether it is detection effect or stability, SSCL has shown excellent performance and potential in simulation experiments on MNIST and CIFAR10.

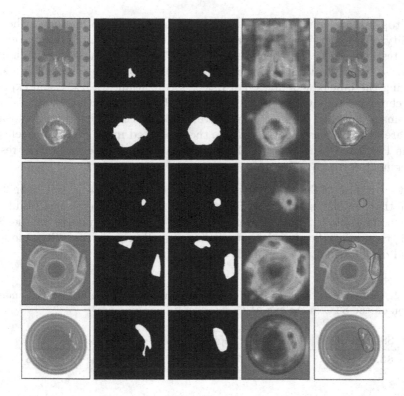

Fig. 2. Qualitative results of anomaly localization for SSCL on the MVTec dataset From left to right, input images, ground-truthmask, predicted mask, predicted heat map and segmentation result.

4.5   Experimental Scenarios on Industrial Anomaly Detection MVtec and STC

SSCL Experimental Setting on MVtec and STC: In our approach, we make the assumption that we have access to a subset of labeled normal samples (χ_n) and abnormal samples (χ_a) during the training phase. We consider one class as the normal set, while the remaining classes are treated as anomalies. The samples in both χ_n and χ_a are randomly selected. To quantify the extent of labeled data available, we define the labeled ratio (γ_k) for both χ_n and χ_a. This ratio indicates the proportion of labeled samples in relation to the total number of samples. Then we train our model with resnet50 about 1000 epochs to make sure that our model has learned good features. Next, train a classifier to classify defective images and non-defective images.

Results on the MVtec: We present the results in Table 3, which proves that SSCL also has a good effect on the MVtec dataset that is closer to the real world. Table 3 illustrates the advantages of our semi-supervised approach to anomaly detection (AD), particularly on the challenging MVtec dataset. SSCL emerges as the top performer in this scenario. Furthermore, Table 3 highlights the vulnerability of a supervised classification approach when facing novel anomalies during testing, especially when the amount of labeled training data is limited. In contrast, SSCL demonstrates the ability to generalize to novel anomalies while leveraging the labeled examples. For defect localization, we use visual explanation techniques, GradCAM [16], to highlight the area affecting the decision of the anomaly detector. We show qualitative results in the fourth column of Fig. 2, which are visually pleasing. We will see the predicted mask perform excellently and the Predicted heat map fits. The last column means Segmentation results that segment correctly.

Results on the STC: For further evaluation, we also test SSCL on STC dataset that simulates video surveillance from a static camera. The results are as shown in Table 4. SSCL also exhibited excellent performance on the STC dataset, thereby confirming the outstanding contribution of our SSCL model in the field of semi-supervised anomaly detection.

Table 4. Comparison of our SSCL model with the state-of-the-art for the anomaly detection on the STC in the avg.AUC.

SSCL	PaDiM	PatchCore	DeepSAD	SSAD	OC-SVM	KDE	IF
83.8	83.2	81.9	79.3	76.6	73.3	72.5	74.8

4.6 Ablation Study

In our ablation studies, we evaluated the effectiveness of the SCL and CL modules by removing each component individually. Results in Table 5 show significant performance improvements and a higher AUC score compared to the baseline when both modules are included. This confirms the effectiveness of both SCL and CL. The best performance is achieved when the SCL and CL modules are combined, indicating their highly complementary nature.

Table 5. Result of ablation studies, where $\gamma_l = 0.05$, $\gamma_p = 0.1$ and $\gamma_k = 0.6$. We report the avg. AUC.

SCL	CL	MNIST	CIFAR10	MVtec
×	×	87.5	73.6	71.2
×	✓	90.5	76.9	75.3
✓	×	91.2	78.1	78.2
✓	✓	**93.5**	**81.3**	**79.6**

5   Conclusion and Future Work

We proposed a semi-supervised contrastive learning method, SSCL, for industrial anomaly detection. By combining self-supervised contrastive learning with supervised contrastive learning, our approach effectively utilizes both labeled and unlabeled data to learn informative and discriminative representations. Experimental results on benchmark datasets, including MNIST, CIFAR-10, MVtec and STC, demonstrated the superior performance of SSCL compared to state-of-the-art methods, even with limited labeled data. For future work, it is important to evaluate SSCL on larger and more diverse datasets, explore alternative contrastive learning techniques, investigate interpretability of learned representations, and extend the method to other related tasks such as outlier and novelty detection.

References

1. Pang, G., Shen, C., Cao, L., Hengel, A.V.D.: Deep learning for anomaly detection: a review. ACM Comput. Surv. (CSUR) **54**(2), 1–38 (2021)
2. Goldstein, M., Uchida, S.: A comparative evaluation of unsupervised anomaly detection algorithms for multivariate data. PLoS ONE **11**(4), e0152173 (2016)
3. Seliya, N., Abdollah Zadeh, A., Khoshgoftaar, T.M.: A literature review on one-class classification and its potential applications in big data. J. Big Data **8**(1), 1–31 (2021). https://doi.org/10.1186/s40537-021-00514-x
4. Amer, M., Goldstein, M., Abdennadher, S.: Enhancing one-class support vector machines for unsupervised anomaly detection. In: Proceedings of the ACM SIGKDD Workshop on Outlier Detection and Description, pp. 8–15 (2013)
5. Chen, Y.C.: A tutorial on kernel density estimation and recent advances. Biostat. Epidemiol. **1**(1), 161–187 (2017)
6. Liu, F.T., Ting, K.M., Zhou, Z.H.: Isolation forest. In: 2008 Eighth IEEE International Conference on Data Mining, pp. 413–422. IEEE (2008)
7. Pang, G., Shen, C., van den Hengel, A.: Deep anomaly detection with deviation networks. In: Proceedings of the 25th ACM SIGKDD International Conference on Knowledge Discovery and Data Mining, pp. 353–362 (2019)
8. Khosla, P., et al.: Supervised contrastive learning. In: Advances in Neural Information Processing Systems, vol. 33, pp. 18661–18673 (2020)
9. Budiarto, E.H., Permanasari, A.E., Fauziati, S.: Unsupervised anomaly detection using k-means, local outlier factor and one class SVM. In: 2019 5th International Conference on Science and Technology (ICST), vol. 1, pp. 1–5. IEEE (2019)
10. Ruff, L., et al.: Deep semi-supervised anomaly detection. arXiv preprint arXiv:1906.02694 (2019)
11. Görnitz, N., Kloft, M., Rieck, K., Brefeld, U.: Toward supervised anomaly detection. J. Artif. Intell. Res. **46**, 235–262 (2013)
12. Jaiswal, A., Babu, A.R., Zadeh, M.Z., Banerjee, D., Makedon, F.: A survey on contrastive self-supervised learning. Technologies **9**(1), 2 (2020)
13. Tian, Y., Krishnan, D., Isola, P.: Contrastive multiview coding. In: Vedaldi, A., Bischof, H., Brox, T., Frahm, J.-M. (eds.) ECCV 2020. LNCS, vol. 12356, pp. 776–794. Springer, Cham (2020). https://doi.org/10.1007/978-3-030-58621-8_45

14. Hendrycks, D., Mazeika, M., Dietterich, T.: Deep anomaly detection with outlier exposure. arXiv preprint arXiv:1812.04606 (2018)
15. Zheng, M., et al.: Weakly supervised contrastive learning. In: Proceedings of the IEEE/CVF International Conference on Computer Vision, pp. 10042–10051 (2021)
16. Selvaraju, R.R., Cogswell, M., Das, A., Vedantam, R., Parikh, D., Batra, D.: Grad-CAM: visual explanations from deep networks via gradient-based localization. In: Proceedings of the IEEE International Conference on Computer Vision, pp. 618–626 (2017)
17. Chen, T., Kornblith, S., Norouzi, M., Hinton, G.: A simple framework for contrastive learning of visual representations. In: International Conference on Machine Learning, pp. 1597–1607. PMLR (2020)
18. Liu, Z., Ma, Y., Ouyang, Y., Xiong, Z.: Contrastive learning for recommender system. arXiv preprint arXiv:2101.01317 (2021)
19. Wu, Z., Wang, S., Gu, J., Khabsa, M., Sun, F., Ma, H.: Clear: contrastive learning for sentence representation. arXiv preprint arXiv:2012.15466 (2020)
20. Defard, T., Setkov, A., Loesch, A., Audigier, R.: PaDiM: a patch distribution modeling framework for anomaly detection and localization. In: Del Bimbo, A., et al. (eds.) ICPR 2021. LNCS, vol. 12664, pp. 475–489. Springer, Cham (2021). https://doi.org/10.1007/978-3-030-68799-1_35
21. Roth, K., Pemula, L., Zepeda, J., Schölkopf, B., Brox, T., Gehler, P.: Towards total recall in industrial anomaly detection. In: Proceedings of the IEEE/CVF Conference on Computer Vision and Pattern Recognition, pp. 14318–14328 (2022)
22. Liu, W., Luo, W., Lian, D., Gao, S.: Future frame prediction for anomaly detection-a new baseline. In: Proceedings of the IEEE Conference on Computer Vision and Pattern Recognition, pp. 6536–6545 (2018)

One Step Large-Scale Multi-view Subspace Clustering Based on Orthogonal Matrix Factorization with Consensus Graph Learning

Xinrui Zhang[1], Kai Li[1,2](✉), and Jinjia Peng[1,2](✉)

[1] School of Cyber Security and Computer, Hebei University, Baoding 071000, China
zhxr@stumail.hbu.edu.cn, {likai,pengjinjia}@hbu.edu.cn
[2] Hebei Machine Vision Engineering Research Center, Baoding, China

Abstract. Multi-view clustering has always been a widely concerned issue due to its wide range of applications. Since real-world datasets are usually very large, the clustering problem for large-scale multi-view datasets has always been a research hotspot. Most of the existing methods to solve the problem of large-scale multi-view data usually include several independent steps, namely anchor point generation, graph construction, and clustering result generation, which generate the inflexibility anchor points, and the process of obtaining the cluster indicating matrix and graph constructing are separating from each other, which leads to suboptimal results. Therefore, to address these issues, a one-step multi-view subspace clustering model based on orthogonal matrix factorization with consensus graph learning(CGLMVC) is proposed. Specifically, our method puts anchor point learning, graph construction, and clustering result generation into a unified learning framework, these three processes are learned adaptively to boost each other which can obtain flexible anchor representation and improve the clustering quality. In addition, there is no need for post-processing steps. This method also proposes an alternate optimization algorithm for convergence results, which is proved to have linear time complexity. Experiments on several real world large-scale multi-view datasets demonstrate its efficiency and scalability.

Keywords: Multi-view clustering · Self-expressive subspace clustering · Large-scale datasets · Orthogonal matrix factorization

1 Introduction

In the last decade, multi-view clustering of data containing heterogeneous information has received a lot of attention [15,16]. Data can be obtained from different

Supported by Central Government Guides Local Science and Technology Development Fund Projects (236Z0301G); Hebei Natural Science Foundation (F2022201009); Science and Technology Project of Hebei Education Department (QN2023186).

Q. Liu et al. (Eds.): PRCV 2023, LNCS 14428, pp. 113–125, 2024.
https://doi.org/10.1007/978-981-99-8462-6_10

feature extractors, for example, the website can be composed of Chinese, English and pictures, so how to utilize this heterogeneous information effectively to carry out multi-view clustering is a hotspot in research. A lot of research has been done in the field of multi-view clustering, mainly including classical clustering methods [2], non-negative matrix factorization methods [9], spectral clustering methods [7,8], and other typical methods [3,11,13]. However, with the explosive growth of visual data, how to cluster large-scale image data effectively has become a challenging problem. Existing single-view and multi-view clustering methods are often not effectively applied to the clustering of large-scale multi-view data due to the high computational complexity and high storage costs.

Aiming at the problem of high computational cost and high memory overhead of large-scale multi-view clustering, binary code learning provides a good solution for large-scale problems. Zhang et al. [21] learn collaborative discrete representations and binary clustering structure jointly. Zhao et al. [22] orthogonalized the mapping matrix of each view and embedded a binary graph junction into a unified binary multi-view clustering model. Jiang et al. [5] performed discrete hash encoding based on graph clustering to learn efficient binaries. Wang et al. [17] performed a discrete hash encoding method based on auto-encoder to learn efficient binaries.

In addition, there is an anchor point-based method, which is an improvement on multi-view subspace clustering. The anchor point-based method selects a group of anchor points as representatives. Kang et al. [6] construct a two-part plot between the original data points and the generated anchor points. Wang et al. [18] captured the correspondence between feature information and structural information to obtain aligned anchor points. Liu et al. [10] first utilized graph filtering technology to eliminate high-frequency noise, and designed a new strategy to select anchor points.

However, the anchor point, obtained by most of the above strategies, and the subsequent graph fusion process are split, a separate step is also required to obtain the clustering results, which makes the clustering results inflexible and leads to suboptimal results. Therefore, in order to solve these problems, the proposed framework integrates the three steps of the traditional anchor method to improve the quality of the obtained consensus graph through adaptive optimization of the generated anchor point. They act on orthogonal matrix factorization to improve cluster quality. The three steps reinforce each other and can be carried out adaptively throughout the optimization process.

The contribution of the proposed methodology is summarized below:

- This paper proposed an adaptively weighted large-scale multi-view subspace clustering method which integrates anchor learning, graph construction, and matrix factorization into a unified framework. All these three components can negotiate with each other to promote clustering quality. Moreover, there is an orthogonal constraint on the learned anchors, which makes the anchors more flexible and discriminating.
- The clustering results are obtained directly through orthogonal matrix factorization without any additional steps. An orthogonal constraint is put on the

centroid matrix to have strict cluster interpretation while the nuclear norm is introduced on the centroid matrix to obtain a more compact structure.

- An alternating optimization algorithm is proposed to solve the optimization problem, which makes the method have nearly linear complexity. A large number of experiments have been carried out on multiple real-world large-scale datasets to verify the effectiveness of our proposed method.

2   Proposed Method

Notations. This paper utilizes capital letters to denote matrices. $\|\cdot\|_F$ represents the Frobenius norm of the matrix, and $\|\cdot\|_*$ represents the nuclear norm of the matrix. We use A_v to denote the v-th view of A, I_d to denote the identity matrix of dimension d and $\mathbf{1}$ to denote the unit vector with all elements being 1.

2.1   Overview and Objective Function

To obtain high-quality bases anchor, this paper proposes an effective algorithm which combines anchor selection, graph construction, and cluster allocation into an overall framework, these three processes can promote each other and optimize together. The overall flowchart of this method is shown in Fig. 1.

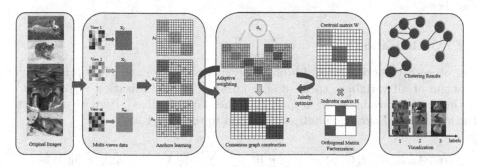

Fig. 1. The overview of the proposed CGLMVC. This method adaptively learns the consensus low-dimension anchors with multi-view information and simultaneously constructs the anchor graph, meanwhile, the orthogonal matrix factorization has been put on the self-expressive subspace graph Z to obtain clustering results directly. These three parts are jointly optimized and boosted.

We add an orthogonal constraint to the learned anchors. Since multi-view clustering is based on a common assumption that multiple views share a potentially consistent representation of the data distribution [20], we jointly learn a common self-representation graph for all views. Meanwhile, we introduce an orthogonal non-negative matrix factorization to the obtained self-expressive matrix Z, for more representative results, we remove the non-negative constraints imposed on the matrix. Then, the nuclear norm is introduced on the common centroid representation matrix to improve clustering performance, and our overall objective function is as follows

$$
\mathcal{J}(\alpha_v, A_v, Z, W, H) = \min_{\alpha_v, A_v, Z, W, H} \sum_{v=1}^{m} \alpha_v^2 \|X_v - A_v Z\|_F^2 + \beta \|Z\|_F^2
$$

$$
+ \lambda \|Z - WH\|_F^2 + \|W\|_*,
$$

$$
s.t. \quad \alpha^\top \mathbf{1} = 1, A_v^\top A_v = I_k, Z \geq 0, Z^\top \mathbf{1}_k = \mathbf{1}_n, W^\top W = I_d, \tag{1}
$$

$$
H \in \{0,1\}^{c \times n}, \sum_{i=1}^{d} H_{ij} = 1, \forall j = 1, 2, \ldots\ldots, n,
$$

since each view is weighted differently, α_v can adaptively learn the weight of every view. β is the non negative equilibrium coefficient, $X_v \in \mathbb{R}^{d_v \times n}$ represents the v-th view of the original data, while d_v is the dimension of the view and n is the number of samples. $A_v \in \mathbb{R}^{d_v \times k}$ is anchor matrix for v-th view, where k is the number of anchor points. $Z \in \mathbb{R}^{k \times n}$ is the consensus graph we learned as a common subspace representation matrix, it combines the anchor representation of all views and constructs them into a monolithic diagram. Here, $W \in \mathbb{R}^{k \times c}$ is the centroid matrix, where c is the number of clusters, and $H \in \mathbb{R}^{c \times n}$ for cluster allocation matrix. λ is a non-negative regularization parameter, the last term is the kernel norm applied on W to guarantee the quality of the learned centroid matrix.

2.2 Optimization

Our objective function in Eq. 1 is obviously non-convex for the joint optimization problem of all variables, so we design an alternate optimization algorithm to optimize each variable while keeping the other variables fixed. Here we introduce the augmented Lagrange multiplier method to turn Eq. 1 into

$$
\mathcal{J}(\alpha_v, A_v, Z, W, H, G, S) = \min_{\alpha_v, A_v, Z, W, H, G, S} \sum_{v=1}^{m} \alpha_v^2 \|X_v - A_v Z\|_F^2 + \beta \|Z\|_F^2
$$

$$
+ \lambda \|Z - GH\|_F^2 + \|W\|_* + Tr(S^\top(W - G)) + \frac{\mu}{2} \|W - G\|_F^2,
$$

$$
s.t. \quad \alpha^\top \mathbf{1} = 1, A_v^\top A_v = I_k, Z \geq 0, Z^\top \mathbf{1} = 1, W^\top W = I_d, \tag{2}
$$

$$
H \in \{0,1\}^{c \times n}, \sum_{i=1}^{d} H_{ij} = 1, \forall j = 1, 2, \ldots\ldots, n,
$$

$S \in \mathbb{R}^{k \times c}$ here stands for the Lagrange multiplier. $G \in \mathbb{R}^{k \times c}$ is used for substituting W to complete the optimization of the nuclear norm, which accomplishes the requirement of parameters for the Lagrange multiplier method. μ is a positive penalty parameter.

Update A_v. By fixing other variables, the objective function can be rewritten as

$$\min_{A_v} \sum_{v=1}^{m} \alpha_v^2 \|X_v - A_v Z\|_F^2,$$
$$s.t. \quad A_v^\top A_v = I_k, \tag{3}$$

expanding the Frobenius norm in Eq. 3, we can obtain the following form

$$\max_{A_v} Tr(A_v^\top B_v),$$
$$s.t. \quad A_v^\top A_v = I_k, \tag{4}$$

where $B_v = X_v Z^\top$, according to [19], if the result of B_v's singular value decomposition (SVD) is $U_{bv} \Sigma_{bv} V_{bv}^\top$, then we can obtain that the optimal solution for A_v is $U_{bv} V_{bv}^\top$.

Update Z. To obtain Z, we fix all variables except Z, so that the optimization function for updating Z can be rewritten as

$$\min_{Z} \sum_{v=1}^{m} \alpha_v^2 \|X_v - A_v Z\|_F^2 + \beta\|Z\|_F^2 + \lambda\|Z - GH\|_F^2,$$
$$s.t. \quad Z \geq 0, Z^\top 1 = 1, \tag{5}$$

the above optimization problem on Z can be divided into the following n subproblems, which can be rewritten as the following quadratic programming (QP) problem

$$\min \frac{1}{2} Z_{:,j}^\top C Z_{:,j} + f^\top Z_{:,j},$$
$$s.t. \quad Z_{:,j}^\top 1 = 1, Z \geq 0, \tag{6}$$

where $C = 2(\sum_{v=1}^{m} \alpha_v^2 + \beta + \lambda)I_k$, $f^\top = -2\sum_{v=1}^{m} \alpha_v^2 X_{v[:,j]}^\top A_v - 2\lambda H_{:,j}^\top G^\top$, it can be optimized by solving the QP problem for each column of Z.

Update W. To obtain W, we fix the rest of the extraneous variables so that we can get the following optimization problem about W

$$\min_{W} \|W\|_* + Tr(S^\top (W - G)) + \frac{\mu}{2}\|W - G\|_F^2,$$
$$s.t. \quad W^\top W = I_d, \tag{7}$$

by folding Eq. 7, we can obtain the following form

$$\min_{W} \frac{1}{\mu}\|W\|_* + \frac{1}{2}\|W - (G - \frac{1}{\mu}S)\|_F^2,$$
$$s.t. \quad W^\top W = I_d, \tag{8}$$

we can get the optimal solution of W with this form of the objective function by the singular value threshold (SVT) method. The obtained results are then orthogonalized.

Update H. The optimization problem of H can be obtained by treating other quantities that are not related to H as constants

$$\min_{H} \lambda \|Z - GH\|_F^2,$$

$$s.t. \quad H \in \{0,1\}^{c \times n}, \sum_{i=1}^{c} H_{ij} = 1, \forall j = 1, 2, \ldots \ldots, n, \tag{9}$$

since each column of H is a one-hot vector, optimizing H as a whole is difficult. Because each column of H represents the variety each sample belongs, and samples are independent of each other, it can be divided into the following subproblems

$$\min_{H_{:,j}} \|Z_{:,j} - GH_{\cdot,j}\|_F^2,$$

$$s.t. \quad H_{:,j} \in \{0,1\}^{c \times n}, \sum_{i=1}^{c} H_{ij} = 1, \forall j = 1, 2, \ldots \ldots, n, \tag{10}$$

for the problem in Eq. 10, we have $H_{i,j} = 1$, where i is the optimal row to be solved, except for i-th row, the value of the other rows of this column is 0. i can be obtained by the following equation

$$i = \arg\min_{i} \|Z_{:,j} - G_{:,i}\|_F^2, \tag{11}$$

Update α_v. Fixing other variables that are not related to α_v, we can get optimization problem for updating α_v

$$\min_{\alpha_v} \sum_{v=1}^{m} \alpha_v^2 \|X_v - A_v Z\|_F^2,$$

$$s.t. \quad \alpha^\top 1 = 1, \tag{12}$$

let $D_v = \|X_v - A_v Z\|_F$, according to Cauchy Schmidt's inequality, we can directly obtain the optimal solution of α_v, i.e. $\alpha_v = \frac{M}{D_v}$, $M = \frac{1}{\sum_{v=1}^{m} \frac{1}{D_v}}$.

Update G. By treating other extraneous variables as constants, the optimization function of G can be transformed into

$$\min_{G} \lambda \|Z - GH\|_F^2 + Tr(S^\top(W - G)) + \frac{\mu}{2}\|W - G\|_F^2, \tag{13}$$

as can be seen from the optimization objective function above, this is a convex function, so the final solution can be obtained directly by taking the derivative of the equation, so the calculated solution about G is of the form

$$G = \frac{1}{2\lambda + \mu}(S + \mu W + 2\lambda Z H^\top), \tag{14}$$

Update S. S is the Lagrange multiplier, so it can be optimized according to the following update rules

$$S = S + \mu(W - G), \tag{15}$$

2.3 Complexity Analysis

The computational burden of the proposed method contains the optimization of each variable. The time cost of updating A_v is $O(d_v k)$ for matrix multiplication and $O(d_v k^2)$ for SVD. To update Z, the QP problem requires $O(nk^3)$ for all columns. To update W, it needs $O(k^2 c)$ for the SVT process. For updating H, the time cost is $O(knc)$. To update G, it requires $O(nck)$. And the time cost for α_v and S are both $O(1)$. Since $n \gg k$, $n \gg c$, the complexity of the method proposed is nearly linear to $O(n)$.

As for the complexity of space, the major memory costs of this method are matrices $A_v \in \mathbb{R}^{d_v \times k}$, $Z \in \mathbb{R}^{k \times n}$, $W \in \mathbb{R}^{k \times c}$, $H \in \mathbb{R}^{c \times n}$, $S \in \mathbb{R}^{k \times c}$ and $G \in \mathbb{R}^{k \times c}$. Therefore, the time complexity is $O((k + c)n + (m + 3c)k)$, where $m = \sum_{i=1}^{v} d_i$. Therefore the space complexity is nearly $O(n)$.

3 Experiments

3.1 Datasets and Comparison Methods

Our proposed method uses eight widely used large-scale datasets: Caltech101-20, CCV, Caltech101-all, SUNRGBD, NUSWIDEOBJ, AwA, MNIST and YouTube-Face. Caltech101 is an object image dataset. Caltech101-20 is a subset of Caltech101. CCV is a database of YouTube videos containing 20 semantic categories. SUNRGBD is an indoor image dataset. NUS-WIDE is a multi-label network image dataset. The Animals With Attributes (AwA) dataset is a dataset of animal images. MNIST is a handwritten piece of data. YouTubeFace is a dataset of face videos obtained from YouTube.

In our experiments, we compare the methods of our method with the following state-of-the-art multi-view clustering methods. Multi-view Subspace Clustering (MVSC) [4], Parameter Free Auto-Weighted Multiple Graph Learning (AMGL) [12], Multi-view Low-rank Sparse Subspace Clustering (MLRSSC) [1], Large-scale Multi-view Subspace Clustering in Linear Time (LMVSC) [6], Scalable Multi-view Subspace Clustering with Unified Anchors (SMVSC) [14], and Fast Parameter-Free Multi-View Subspace Clustering With Consensus Anchor Guidance (FPMVS-CAG) [20].

3.2 Experimental Setup

In our experiment, we initialize A, Z, and W to zero, and H to the identity matrix, α_v is initialized to $\frac{1}{v}$, where v is the number of views, and we input the number of clusters as c, then we select the number of anchor points within

$\{c, 2c, 3c\}$ for our experiments, and set the value of the common dimension d to be equal to the number of clusters c. Our experiment consists of two hyperparameters β and λ, where the range of β we choose to vary between $[0.01, 0.1, 1, 10]$, and the range of λ we choose to be between $[0.1, 1, 10, 20, 30]$ to perform the experiment. For the methods used for comparison, we all use the best results they published. We use the widely used Normalized Interaction Information (NMI), Accuracy (ACC), Purity, and Fscore to evaluate clustering performance.

Our experiment was implemented on a standard Linux computer with Intel(R) Xeon(R) Silver 4214R CPU @2.40 GHz and 503 GB RAM, MATLAB 2021a (64-bit).

3.3 Experiments Results

Table 1 lists the clustering performance of our method and the performance of the other 6 comparison algorithms on 8 benchmark datasets. From the results of the comparison, the following can be observed

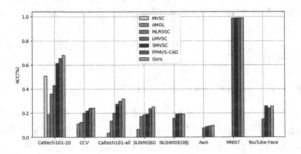

Fig. 2. Clustering performance about ACC on eight datasets.

Clustering Performance. The information on clustering results on eight datasets are shown in Table 1, which include scores of ACC, NMI, Purity, Fscore of six multi-view clustering methods and our method. The best clustering results are highlighted in bold and the second best with an underlining. In addition, '–' means the method is out-of-memory on the corresponding dataset or we have not run the best results about this. In Fig. 2, it shows the clustering performance of ACC on the three datasets to illustrate the comparison results more directly.

According to the contents of Table 1, the method proposed in this paper always achieves the best or the second result. For example, on the Caltech101-20, Caltech101-all, and SUNRGBD datasets, the proposed method achieves improvements of 2.67%, 1.68%, and 1.46% in ACC, respectively, and achieves optimal or suboptimal results on datasets with more than 10,000 samples.

At the same time, it can be seen from the table that anchor-based methods (LMVSC, FPMVS-CAG, SMVSC and the proposed method) are more suitable

Table 1. Clustering performance of compared methods. "-" means out-of-memory failure. The best results are highlighted in boldface and the underlines indicate the second-best competitors.

Datasets	Metrics	MVSC	AMGL	MLRSSC	LMVSC	SMVSC	FPMVS-CAG	Ours
Caltech101-20	ACC	0.5080	0.1876	0.3600	0.4304	0.6132	<u>0.6547</u>	**0.6814**
	NMI	0.5271	0.1101	0.2008	0.5553	0.5873	<u>0.6326</u>	**0.6439**
	Purity	0.7125	0.6313	0.4476	0.7125	0.6999	<u>0.7368</u>	**0.7841**
	Fscore	0.4329	0.4661	0.3069	0.3414	0.6699	<u>0.6905</u>	**0.7123**
CCV	ACC	–	0.1102	0.1259	0.2014	0.2182	<u>0.2399</u>	**0.2426**
	NMI	–	0.0758	0.0471	0.1657	0.1684	**0.1760**	<u>0.1753</u>
	Purity	–	0.2021	0.1307	0.2396	0.2439	<u>0.2605</u>	**0.2680**
	Fscore	–	0.1215	0.1215	0.1194	0.1307	<u>0.1419</u>	**0.1521**
Caltech101-all	ACC	–	0.0359	0.1365	0.2005	0.2750	<u>0.3015</u>	**0.3183**
	NMI	–	0.0187	0.1066	**0.4155**	0.3510	0.3549	<u>0.3676</u>
	Purity	–	**0.4311**	0.1371	0.3975	0.3395	0.3460	<u>0.4048</u>
	Fscore	–	**0.3617**	0.0815	0.1586	0.2224	<u>0.2326</u>	0.2296
SUNRGBD	ACC	–	0.0643	0.1741	0.1858	0.1930	<u>0.2392</u>	**0.2538**
	NMI	–	0.0371	0.1108	0.2607	0.2007	<u>0.2418</u>	**0.2644**
	Purity	–	0.2411	0.1741	0.3818	0.2971	<u>0.3400</u>	**0.3521**
	Fscore	–	0.1894	0.1453	0.1201	0.1279	<u>0.1597</u>	**0.1613**
NUSWIDEOBJ	ACC	–	–	–	0.1583	0.1916	**0.1946**	<u>0.1922</u>
	NMI	–	–	–	<u>0.1337</u>	0.1272	**0.1351**	0.1286
	Purity	–	–	–	**0.2488**	0.2331	<u>0.2382</u>	0.2310
	Fscore	–	–	–	0.0990	0.1365	<u>0.1372</u>	**0.1428**
AwA	ACC	–	–	–	0.0770	0.0878	<u>0.0919</u>	**0.0990**
	NMI	–	–	–	0.0879	<u>0.1061</u>	**0.1083**	0.1033
	Purity	–	–	–	0.0957	**0.0993**	<u>0.0961</u>	0.0939
	Fscore	–	–	–	0.0378	0.0636	<u>0.0640</u>	**0.0687**
MNIST	ACC	–	–	–	0.9852	0.9875	<u>0.9884</u>	**0.9891**
	NMI	–	–	–	0.9576	<u>0.9627</u>	**0.9651**	0.9625
	Purity	–	–	–	0.9852	0.9875	<u>0.9884</u>	**0.9896**
	Fscore	–	–	–	0.9704	0.9751	**0.9768**	<u>0.9753</u>
YouTubeFace	ACC	–	–	–	0.1479	**0.2587**	0.2414	<u>0.2562</u>
	NMI	–	–	–	0.1327	0.2292	<u>0.2433</u>	**0.2486**
	Purity	–	–	–	0.2816	**0.3321**	0.3279	<u>0.3313</u>
	Fscore	–	–	–	0.0849	0.0849	**0.1433**	<u>0.1328</u>

for large-scale datasets than traditional methods, and traditional methods usually encounter errors such as insufficient memory to calculate clustering results, so these anchor-based methods can be suitable for more situations. The good performance of the proposed method on large-scale datasets proves the superiority of combining anchor learning, graph construction and clustering processes into one step, so that better results can be obtained than other methods.

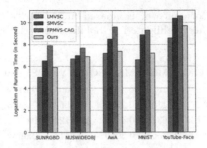

Fig. 3. Comparison results on the large-scale multi-view datasets in terms of running time in seconds.

Running Time. Figure 3 shows the time taken by the anchor-based method to run on five datasets with more than 10,000 samples, which have been processed with logarithmic cardinality for visualization. It can be seen that the proposed method achieves good results in calculation time and clustering performance. Compared with these methods, the proposed method has less time consumption and is also very effective in terms of performance. Therefore, in general, in terms of clustering results and time required, the proposed method adaptively generates orthogonal anchor bases for each view, while obtaining clustering results directly from the constructed graph, which shows better performance in the whole clustering process. In summary, the proposed method strikes a balance in time and performance, so it is more suitable for clustering of large-scale datasets.

Parameter Analysis. In this section, we illustrate the parameter sensitivity on two typical datasets, Caltech101-20 and Caltech101-all. In the proposed method, there are two parameters need to be concerned about, namely β and λ, and they are all related to the number of selected anchors. Therefore we show their sensitivity about the number of anchors in Fig. 4(1). We can see that the proposed

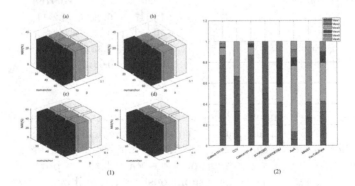

Fig. 4. (1) Parameter analysis results on Caltech101 datasets. a, b for Caltech101-20, and c, d for Caltech101-all.(2) The learned coefficients of view on eight datasets.

method shows a robust to these two parameters, the results won't flatter while changing the anchors. Thus, it proves the robustness of the proposed method.

Coefficients of Views. As is shown in Fig. 4(2), we demonstrate the learned coefficients of each view on eight datasets. In the proposed method, it has been proved that the algorithm learns the coefficients adaptively during the whole optimization process. We put a constraint on the view coefficients which make the sum of each view into one. With this constraint, it can learn the weight of each view according to its values, which makes the important view have a bigger effect in the clustering process. From Fig. 4(2), we can learn that the self-adaptive weights have been jointly optimized, which shows more practical and reasonable application.

Convergence Analysis. Since the objective function in Eq. 1 is non-convex, and in order to optimize the kernel norm, the alternating optimization minimization algorithm of the Lagrange multiplier method is used to solve the problem in the objective equation. Since the Lagrangian multiplier method is non-increasing with the increase of the number of iterations, the final convergence of the algorithm can be guaranteed. This section records the target values on some of the benchmark datasets, as shown in the Fig. 5. It can be observed that as the number of iterations increases, it generally remains stable after about 15 times, which means that the proposed method can finally converge after a limited number of operations.

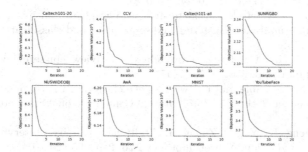

Fig. 5. The convergence analysis through iterations on 8 benchmark datasets.

4   Conclusion

In this paper, a new unified model has been proposed for solving large-scale multi-view clustering. It combines anchor learning, graph building, and clustering into a unified framework. We also propose an alternative minimum optimization strategy to optimize the objective function, which has an almost linear time complexity. Experiments are conducted on several large real-world datasets to demonstrate the superiority of the proposed method.

References

1. Brbić, M., Kopriva, I.: Multi-view low-rank sparse subspace clustering. Pattern Recogn. **73**, 247–258 (2018)
2. Cai, X., Nie, F., Huang, H.: Multi-view k-means clustering on big data. In: Twenty-Third International Joint Conference on Artificial Intelligence (2013)
3. Chen, M.S., Huang, L., Wang, C.D., Huang, D., Lai, J.H.: Relaxed multi-view clustering in latent embedding space. Inf. Fusion **68**, 8–21 (2021)
4. Gao, H., Nie, F., Li, X., Huang, H.: Multi-view subspace clustering. In: Proceedings of the IEEE International Conference on Computer Vision, pp. 4238–4246 (2015)
5. Jiang, G., Wang, H., Peng, J., Chen, D., Fu, X.: Graph-based multi-view binary learning for image clustering. Neurocomputing **427**, 225–237 (2021)
6. Kang, Z., Zhou, W., Zhao, Z., Shao, J., Han, M., Xu, Z.: Large-scale multi-view subspace clustering in linear time. In: Proceedings of the AAAI Conference on Artificial Intelligence, vol. 34, pp. 4412–4419 (2020)
7. Kumar, A., Rai, P., Daume, H.: Co-regularized multi-view spectral clustering. In: Advances in Neural Information Processing Systems, vol. 24 (2011)
8. Li, Y., Nie, F., Huang, H., Huang, J.: Large-scale multi-view spectral clustering via bipartite graph. In: Proceedings of the AAAI Conference on Artificial Intelligence, vol. 29 (2015)
9. Liu, J., Wang, C., Gao, J., Han, J.: Multi-view clustering via joint nonnegative matrix factorization. In: Proceedings of the 2013 SIAM International Conference on Data Mining, pp. 252–260. SIAM (2013)
10. Liu, L., Chen, P., Luo, G., Kang, Z., Luo, Y., Han, S.: Scalable multi-view clustering with graph filtering. Neural Comput. Appl. **34**(19), 16213–16221 (2022)
11. Liu, X., et al.: Late fusion incomplete multi-view clustering. IEEE Trans. Pattern Anal. Mach. Intell. **41**(10), 2410–2423 (2018)
12. Nie, F., Li, J., Li, X., et al.: Parameter-free auto-weighted multiple graph learning: a framework for multiview clustering and semi-supervised classification. In: IJCAI, pp. 1881–1887 (2016)
13. Nie, F., Li, J., Li, X., et al.: Self-weighted multiview clustering with multiple graphs. In: IJCAI, pp. 2564–2570 (2017)
14. Sun, M., et al.: Scalable multi-view subspace clustering with unified anchors. In: Proceedings of the 29th ACM International Conference on Multimedia, pp. 3528–3536 (2021)
15. Wang, H., Feng, L., Yu, L., Zhang, J.: Multi-view sparsity preserving projection for dimension reduction. Neurocomputing **216**, 286–295 (2016)
16. Wang, H., Jiang, G., Peng, J., Deng, R., Fu, X.: Towards adaptive consensus graph: multi-view clustering via graph collaboration. IEEE Trans. Multimedia (2022)
17. Wang, H., Yao, M., Jiang, G., Mi, Z., Fu, X.: Graph-collaborated auto-encoder hashing for multiview binary clustering. IEEE Trans. Neural Netw. Learn. Syst. (2023)
18. Wang, S., et al.: Align then fusion: generalized large-scale multi-view clustering with anchor matching correspondences. arXiv preprint arXiv:2205.15075 (2022)
19. Wang, S., et al.: Multi-view clustering via late fusion alignment maximization. In: IJCAI, pp. 3778–3784 (2019)
20. Wang, S., et al.: Fast parameter-free multi-view subspace clustering with consensus anchor guidance. IEEE Trans. Image Process. **31**, 556–568 (2021)

21. Zhang, Z., Liu, L., Shen, F., Shen, H.T., Shao, L.: Binary multi-view clustering. IEEE Trans. Pattern Anal. Mach. Intell. **41**(7), 1774–1782 (2018)
22. Zhao, J., Kang, F., Zou, Q., Wang, X.: Multi-view clustering with orthogonal mapping and binary graph. Expert Syst. Appl. **213**, 118911 (2023)

Deep Multi-task Image Clustering with Attention-Guided Patch Filtering and Correlation Mining

Zhongyao Tian[1], Kai Li[1,2($\boxtimes$)], and Jinjia Peng[1,2($\boxtimes$)]

[1] School of Cyber Security and Computer, Hebei University, Baoding 071000, China
[2] Hebei Machine Vision Engineering Research Center, Baoding 071000, Hebei, China
likai@hbu.edu.cn

Abstract. Deep Multi-task image clustering endeavors to leverage deep learning techniques for the simultaneous processing of multiple clustering tasks. Current multi-task deep image clustering approaches typically rely on conventional deep convolutional neural networks and transfer learning technologies. However, suboptimal clustering results are produced in the execution of each task including irrelevant redundant information. This paper proposes a novel end-to-end deep multi-task clustering framework named Deep Multi-Task Image Clustering with Attention-guided Patch Filtering and Correlation Mining (APFMTC) that eliminates redundant information between different tasks while extracting relevant information to achieve improved cluster division. Specifically, APFMTC partitions image samples into several patches, treating each patch as a word thus each image is regarded as an article, and the process of determining the cluster to which an image belongs is likened to categorizing articles. During the clustering process, several parts of each image sample generally carry more significance. Therefore, a weights estimation module is designed to evaluate the importance of different visual words extracted by the key patch filter for different categories. Ultimately, in each task, the final cluster division is determined by assigning weights to the words contained within the image samples. To evaluate the effectiveness of the proposed method, it is tested on multi-task datasets created from four datasets: NUS-Wide, Pascal VOC, Caltech-256, and Cifar-100. The experimental results substantiate the efficacy of the proposed method.

Keywords: Multi-task clustering · Deep clustering · Image clustering

1 Introduction

In the era of big data, the rapid development of information technology has led to the proliferation of a substantial amount of unlabeled image data, thus the utilization of unsupervised image clustering methods [7,14] have become prevalent in the field of machine vision. It is common for image data from different

Supported by Central Government Guides Local Science and Technology Development Fund Projects (236Z0301G); Hebei Natural Science Foundation (F2022201009); Science and Technology Project of Hebei Education Department (QN2023186).

Q. Liu et al. (Eds.): PRCV 2023, LNCS 14428, pp. 126–138, 2024.
https://doi.org/10.1007/978-981-99-8462-6_11

domains to be associated with distinct tasks. Besides that, the samples within these tasks often exhibit a high degree of correlation. For instance, consider a scenario where one task involves image samples related to bicycles, while another task comprises image samples associated with motorcycles. Although these two categories belong to different classes, they share similar local features such as wheels, handlebars, etc. Consequently, exploring and extracting relevant information from dissimilar but analogous samples across multiple tasks can greatly facilitate the attainment of enhanced cluster division within each individual task, which benefits from the discovery of shared patterns and features. However, each traditional image clustering algorithm is a single-task clustering algorithm, which cannot mine the correlation between different but related image data sets belonging to different tasks. Consequently, many methods of multi-task image clustering have been proposed.

Multi-task Clustering (MTC) aims to improve the performance of each task by exploiting related information between tasks. In recent years, many researchers applied the MTC method to different multi-task environments and achieved better results. For example, Cao et al. [2] proposed an MTC method for clustering by exploring inter-task, inter-cluster and feature-to-feature correlations, Yan et al. [20] and Zhang et al. [25] adopted EM distance and JS Divergence to measure the distance between paired clusters from different tasks and discover potential correlations. Zhang et al. [23] and Yang et al. [21] represent task dependencies by mapping multi-task data to subspaces [13] to achieve good results. Zhong and Pun [26] transfer knowledge between tasks by introducing joint projections of heterogeneous features while exploring discriminative information in low-dimensional subspaces to obtain clustering results for multiple tasks. Yan et al. [17] formulated the MTC problem as an information loss minimization function, and quantified multiple action clustering tasks through the distribution correlation of clusters in different tasks, sharing information among them to improve the performance of individual tasks. However, due to the problems of the existing MTC methods such as they cannot capture the nonlinear nature of complex multi-task data, researchers began to focus on combining deep image clustering methods to solve the MTC problem.

Motivated by the powerful nonlinear learning ability of Deep Neural Network (DNN), many deep image clustering methods have been proposed in recent years, such as some deep clustering methods based on semi-supervised learning [9] [10] [22] [11] and unsupervised methods [3,8,15] that use pseudo-labels as supervised information to reduce DNN loss. In addition, Xu et al. [16] combined contrastive learning with neighborhood relationship mining to propose nearest-neighbor contrastive clustering. Vilhagra et al. [12] used DNN to perform feature conversion on samples and then used traditional clustering loss for clustering. Asano et al. [1] transformed the image clustering problem into a self-labeling task. Ji et al. [6] directly train a randomly initialized neural network into a classification function. Promising performance have been obtained for all the above deep image clustering methods focusing on improving the clustering performance of a single task through the powerful feature extraction ability of DNN. Therefore, research on using DNN to solve MTC problems has emerged. For example, Yan et al. [18] performed clustering of multiple tasks by maximizing the relevant information and minimizing the irrelevant information between multiple tasks. Zhang et al. [24] employed graph neural networks to learn node embeddings that facilitate downstream clustering, and proposed a multi-task embedding learning method for attribute graph clustering. Yan et al. [19] explored multi-task correlation and clustered images simultaneously by transferring training parameters among multiple deep convolutional neural networks. However, this transfer learning method based on the parameter sharing mechanism shares relevant information between multiple tasks while also making irrelevant redundant information participate in the execution of each task, which may lead to poor clustering results.

The proposed APFMTC addresses the challenge of removing redundant information between different tasks while extracting relevant information to enhance cluster division. Specifically, the method partitions the image sample into multiple patches, considering each patch as a discrete "word." Consequently, the image is treated as an "article," and the clustering process is framed as categorizing the "article" into specific categories. Key patches are extracted and bag of visual words is constructed from multi-task image samples. Furthermore, a weight estimation module is devised to assess the significance of different visual words with respect to various categories. Ultimately, within each task, the final cluster division is determined by assigning appropriate weights to the words contained in the image samples. This approach enables the extraction of crucial information while minimizing the influence of redundant information, ultimately leading to improved cluster division within each individual task.

The main contributions of this study are summarized as follows:

(1) A novel key information extraction mechanism is proposed, patches with higher attention scores obtained by the Transformer training are extracted as key information to remove redundant information in image samples and form bag of visual words.
(2) A weight estimation module is devised within APFMTC, which aims to effectively capture the relevance and significance of each word in the clustering process, and allows for a more accurate representation of the visual characteristics and discriminative information contained within the image samples.

2   Multi-task Image Clustering with Attention-Guided Patch Filtering and Correlation Mining

2.1   Overview

The overall framework of the proposed APFMTC is depicted in Fig. 1. Initially, the original image data undergoes auxiliary clustering to generate pseudo-labels. These pseudo-labels are then utilized for training the Vision Transformer (ViT) network comprehensively. Subsequently, key patch filter is designed to extract patches from each image sample that contain more valuable information, and a distance-based clustering approach is employed to construct a bag of visual words (BoVW) with extracted patches.

To assign appropriate weights to each word in BoVW, a specifically designed weights estimation module is incorporated. This module evaluates the importance and relevance of each word for different categories. Finally, the weights obtained from the weights estimation module are utilized in subsequent multiple clustering tasks. These weights play a crucial role in influencing the clustering outcomes within each task.

The overall framework integrates the steps of auxiliary clustering, Vision Transformer training, key patch filter, BoVW construction, and weights estimation module. This method enables improved performance and accuracy in multi-task image clustering.

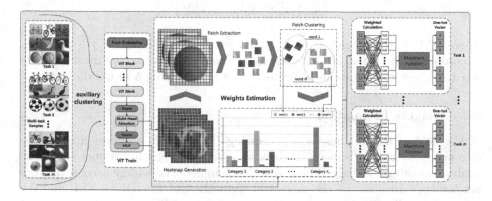

Fig. 1. The overall framework of the algorithm proposed in this paper.

2.2   Key Patch Filter

Building upon the recent success of ViT, which inspired us to partition images into patches and treat each patch as a "word," this paper has developed an end-to-end deep multitasking clustering network. To begin, the method proposed in this paper employs a distance-based clustering algorithm to facilitate the clustering process and generate pseudo labels. These pseudo labels are then used for comprehensive training of the ViT network, where the parameters are optimized using the pseudo labels obtained from the auxiliary clustering step. Subsequently, attention scores obtained from the trained attention layers are utilized to construct an attention heatmap. This heatmap can identify the regions of the image sample that significantly influence the classification results. By leveraging the attention heatmap, the more valuable patches regarded as key information can be extracted effectively.

The attention heatmap generation method employed in this study is derived from the work of Chefer et al. [4], which builds upon the deep Taylor decomposition approach. It facilitates the identification of key regions and aids in the subsequent extraction of more informative patches, thus contributing to the overall performance improvement of the proposed clustering algorithm. The heatmap generation formula is as follows:

$$s_i^j = S_\theta(x_i^j), \qquad 1 \le j \le m, 1 \le i \le N^j \tag{1}$$

where s_i^j is the attention score metric of the ith image in the jth data set generated by the attention heatmap generating function, N^j is the number of images in the ith data set, θ is the parameters of ViT. Then the matrix is divided into T patches, and the attention score for each patch is calculated. The n patches with the highest attention score can be obtained:

$$\max_n \{(s_p^1)_i^j, (s_p^2)_i^j, ..., (s_p^T)_i^j\} \tag{2}$$

where $(s_p^t)_i^j$ is the attention score of tth patch of ith image in jth data set. According to the n patches with the highest attention score, the most valuable n patches in each image can be extracted for constructing a visual dictionary.

2.3   Weights Estimation

In the field of natural language processing (NLP), it is a common means to give weight to keywords in articles according to TF-IDF, and conduct sentiment analysis or category analysis on articles, which has been widely used with good performance. Inspired by the successful application of TF-IDF in NLP and the successful application of image patching in the field of CV, the method proposed in this paper first regards each patch as a word and each image sample as an article composed of words. Unlike the application of TF-IDF, the words in the article are exactly the same, and identical words will always appear in different articles. The patches in the image will not be exactly the same, but objects belonging to the same category in the image will be represented as similar objects, with closer distances between them. For example, in different vehicle images, patches containing wheels will have closer distances, as these patches describe the same type of object, Therefore, we can represent them as the same category, which means treating these different patches as the same words.

Therefore, after obtaining valuable patches based on the attention heatmap, we need to cluster them and record their centroids, using the cluster number to which each patch belongs to represent them, the popular k-means clustering is then used to cluster these patches to construct BoVW.

After obtaining the visual dictionary, the designed weights estimation formula was used to assign weights to different words, with the specific formula 3:

$$w_k = pw_k/aw_k \times (1 - npw_k/naw_k) \tag{3}$$

where pw_k is the number of occurrences of a visual word in category k, and aw_k is the total number of words in category k, npw_k represents the number of occurrences of this visual word in all classes except class k, and naw_k represents the total number of words in all classes except class k.

2.4   Multi-task Clustering

In each individual task, after dividing the image into several patches with the same dimension as the vocabulary in the visual dictionary, the Euclidean distance between each patch and each vocabulary in the visual dictionary is calculated, so that each patch determines which visual vocabulary should belong to. The word representation function is as follows:

$$w^t = v((x_p^t)_i) \qquad (4)$$

where $(x_p^t)_i$ is the tth patch of the ith image sample, and $v(\cdot)$ is the word representation function. Thus the visual vocabulary w^t to which patch $(x_p^t)_i$ belongs can be computed. The ith image sample can be represented as a vector of visual words in the following form:

$$x_i = \{w^1, w^2, ..., w^t, ..., w^T\} \qquad (5)$$

where T is the number of patches of each image. Each element in the vector represents a vocabulary. These vocabularies should have different weights according to their importance to different categories. According to the weights obtained in the weights estimation module, these words are weighted, the formula is as follows:

$$S_k = \sum_{t=1}^{T} w_k^t \qquad (6)$$

where w_k^t is the weight of the tth patch to the kth cluster, and S_k is the score of the sample for the kth cluster. The number of S_k calculated by each sample should be K_{gt}, and the highest S_k is selected after comparison. At this time, k is considered to be the cluster that the sample should belong to. The final cluster division can be obtained by calculating S_k of all samples and selecting the largest S_k of each sample.

3   Experiments

3.1   Data Sets

In order to verify the effectiveness of APFMTC, we used four multi-task image datasets [19], the samples of which were from the Caltech-256 dataset, NUS-Wide dataset, Cifar-100 dataset and Pascal Voc data set, each multi-task image

Table 1. Detailed information of the constructed multi-task data sets, from the NUS-Wide dataset.

	Task 1	Task 2	Task 3	Task 4
Class 1	Activity.Dance(314)	Activity.Run(341)	Activity.Throw(470)	Activity.Walk(225)
Class 2	Animal.Horse(356)	Animal.Cow(412)	Animal.Tiger(361)	Animal.Zebra(412)
Class 3	Vehicle.Bicycle(471)	Vehicle.Car(435)	Vehicle.Motorbike(451)	Vehicle.Bus(472)
Class 4	Flower.Lily(439)	Flower.Tulip(392)	Flower.Rose(298)	Flower.Orchild(471)

Table 2. Detailed information of the constructed multi-task data sets, from the Caltech-256 dataset.

	Task 1	Task 2	Task 3
Class 1	Gun.Ak47(314)	Gun.Rifle(106)	Gun.Ak47(99)
Class 2	Aircraft.Jet(99)	Aircraft.Copter(88)	Aircraft.Plane(140)
Class 3	Hat.Tophat(80)	Hat.Cowboyhat(110)	Hat.Yarmulke(84)
Class 4	Bike.Tourbike(110)	Bike.Motorbike(98)	Bike.Hillbike(82)
Class 5	Bird.Cormorant(106)	Bird.Ibis(101)	Bird.Duck(87)
Class 6	Ball.Soccer(106)	Ball.Golf(101)	Ball.Tennis(87)

Table 3. Detailed information of the constructed multi-task data sets, from the Cifar-100 dataset.

	Task 1	Task 2	Task 3
Class 1	Rodents.hamster(400)	Rodents.mouse(400)	Rodents.squirrel(400)
Class 2	Tree.oak(400)	Tree.palm(400)	Tree.pine(400)
Class 3	Pantherinae.cat(400)	Pantherinae.lion(400)	Pantherinae.tiger(400)
Class 4	Fish.dolphin(400)	Fish.whale(400)	Fish.shark(400)
Class 5	Insect.bee(400)	Insect.beetle(400)	Insect.cock(400)

Table 4. Detailed information of the constructed multi-task data sets, from the Pascal VOC dataset.

	Task 1	Task 2
Class 1	Livestock.sheep(300)	Livestock.cow(300)
Class 2	Bike.bicycle(300)	Bike.motor(300)
Class 3	Vehicle.car(300)	Vehicle.bus(300)
Class 4	Furniture.chair(300)	Furniture.sofa(300)

data set contains several different tasks, but there are some similarities between the categories in different tasks to a certain extent. The details of the data set are shown in Tables 1, 2, 3 and 4.

3.2   Baselines

The method proposed in this paper is compared with several deep clustering algorithms named DC [3], DCCM [15], SL [1] respectively, and their full-task versions namely ALLDC, ALLDCCM, ALLSL, multi-task clustering algorithms named MTMVC [25] and MICCP [5], a multi-task deep clustering algorithm DMTC [19].

3.3   Experimental Setup

The experiments use the Vision Transformer as the backbone network for auxiliary clustering, during which several different values of auxiliary clustering number of K_{fake} are used as a guide. After obtaining a patch with a high attention score, several different n_w are used to cluster these patches to generate BoVW. The experiment also tried different n for patch extraction, different numbers of patches with the highest attention scores were extracted in each sample. All baselines are reproduced according to the settings in the experiments conducted by their authors for reference.

The experiment selects two commonly used evaluation indicators in image clustering as performance references, namely ACC and NMI, and their values obtained in each method are the average values obtained after 10 experiments. All experiments are conducted on hardware consisting of eight Intel(R) Xeon(R) Platinum 8352Y CPUs @ 2.20 GHz and two NVIDIA A6000 GPUs.

3.4   Compared with the Latest Method

The results of comparing APFMTC with the baselines used as a reference on constructed multi-task dataset from NUS-Wide are shown in Table 5. Except that the NMI of Task 3 and Task 4 are slightly lower than DMTC, the rest of the indicators are the best results among all baselines. On the multi-task dataset constructed from Caltech-256 as shown in Table 6, only the NMI of Task 3 is not ideal, and the remaining indicators are the best results among all baselines.

Table 5. Clustering results(%) on NUS-Wide. The best results are highlighted in bold.

Method	Task 1		Task 2		Task 3		Task 4	
	ACC	NMI	ACC	NMI	ACC	NMI	ACC	NMI
DC	20.45	17.08	25.62	19.73	38.01	25.10	32.40	31.41
DCCM	19.76	9.93	18.82	16.85	23.31	19.49	25.35	20.43
SL	28.36	14.99	37.09	16.26	41.46	15.60	34.91	21.77
ALLDC	22.48	22.69	21.99	16.44	34.09	20.34	27.24	22.38
ALLDCCM	24.71	11.95	22.90	12.62	20.31	11.45	22.24	19.44
ALLSL	33.30	9.55	36.66	14.85	36.19	15.17	33.35	14.39
MTMVC	35.68	5.62	38.03	7.10	40.56	6.09	37.48	12.67
MICCP	58.11	18.33	55.52	18.98	51.13	17.31	49.32	15.51
DMTC	58.69	28.12	62.94	29.81	68.00	**40.09**	72.66	**41.75**
APFMTC	**62.26**	**29.02**	**64.29**	**31.83**	**71.49**	39.66	**73.22**	39.36

Table 6. Clustering results(%) on Caltech-256. The best results are highlighted in bold.

Method	Task 1		Task 2		Task 3	
	ACC	NMI	ACC	NMI	ACC	NMI
DC	26.43	24.99	28.96	22.47	28.12	26.45
DCCM	27.68	17.67	28.58	26.83	33.39	24.78
SL	32.93	19.14	33.77	13.94	39.61	18.41
ALLDC	25.50	18.26	21.43	18.77	34.11	27.50
ALLDCCM	29.03	17.66	28.62	21.63	31.24	24.99
ALLSL	27.46	18.38	33.44	20.07	42.62	20.31
MTMVC	34.80	18.29	34.04	15.96	43.21	25.31
MICCP	56.92	27.97	56.26	32.49	47.43	26.07
DMTC	64.09	50.44	60.87	46.06	64.74	**57.65**
APFMTC	**65.29**	**53.39**	**63.76**	**57.93**	**65.59**	52.19

According to Table 7 and Table 8, the proposed APFMTC achieves the best results compared with the baselines on Cifar-100, the clustering results of Task 1 on Pascal VOC obtained by APFMTC are the best, and the ACC and NMI of Task 2 are lower than DMTC.

Table 7. Clustering results(%) on Cifar-100. The best results are highlighted in bold.

Method	Task 1		Task 2		Task 3	
	ACC	NMI	ACC	NMI	ACC	NMI
DC	31.15	30.74	23.28	11.09	28.01	28.75
DCCM	47.59	32.12	32.86	23.23	37.94	32.14
SL	43.89	17.90	37.88	17.49	49.83	26.17
ALLDC	31.26	24.07	24.39	9.82	27.65	24.06
ALLDCCM	50.47	34.04	35.11	21.66	37.33	32.46
ALLSL	54.62	26.84	35.65	12.33	45.30	22.53
MTMVC	32.17	17.22	30.98	17.42	40.64	23.71
MICCP	24.90	4.85	21.31	2.28	23.19	4.40
DMTC	66.55	41.28	55.96	31.33	53.94	29.11
APFMTC	**71.38**	**47.14**	**62.33**	**38.29**	**61.02**	**42.32**

3.5   Impact of the Number of Clusters for Auxiliary Clustering

In order to explore the influence of the number of clusters used in auxiliary clustering on the final clustering results, different K_{fake} are used in the process of auxiliary clustering, and the values are $\{K_{gt}-2, K_{gt}-1, K_{gt}, K_{gt}+1, K_{gt}+2\}$,

Table 8. Clustering results(%) on Pascal VOC. The best results are highlighted in bold.

Method	Task 1		Task 2	
	ACC	NMI	ACC	NMI
DC	30.88	7.20	35.36	19.49
DCCM	24.79	4.97	21.40	9.98
SL	35.75	16.32	35.79	15.04
ALLDC	32.72	11.23	32.70	16.39
ALLDCCM	28.99	7.91	20.35	8.10
ALLSL	34.10	11.89	35.76	14.07
MTMVC	32.14	6.82	34.24	9.91
MICCP	25.71	3.15	29.76	4.81
DMTC	66.55	30.74	**79.39**	**55.90**
APFMTC	**69.72**	**47.87**	78.26	51.28

where K_{gt} are the ground truth numbers of cluster in each task. As shown in Fig. 2, it can be found that the best results can be obtained when K_{fake} is set to the same value as K_{gt}, the reason may be the desined key patch filter relies on the statistics of the attention score of each patch, and the attention score is strongly related to the supervision information. When K_{fake} is set to be the same as the value of K_{gt}, most of the clusters formed by auxiliary clustering are samples with strong correlation and high similarity, so the best results can be obtained.

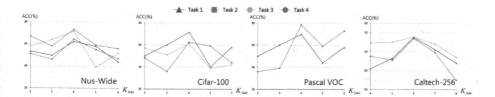

Fig. 2. The Impact of K_{fake}.

3.6   Impact of the Number of Extracted Patches

In the designed key patch filter, the n patches with the highest attention scores in each sample are extracted. To explore the influence of different values of n on the clustering results, we additionally set n to $\{5, 10, 13, 15, 18, 20\}$. As shown in Fig. 3, it can be found that when n is set to 13, the best results can be obtained. The reason may be that when the value of n is less than 13, extracting The local features obtained are few, which are not enough to represent the key information in the sample. When the value of n is greater than 13, the redundant

information that is weakly correlated with the category of each sample will be
extracted, resulting in the degradation of clustering performance.

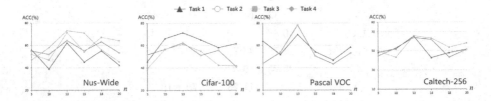

Fig. 3. The Impact of n.

3.7   Impact of Word Count in Visual Dictionaries

A BoVW is constructed in the proposed APFMTC for clustering. In this process,
n patches extracted from each sample are clustered into n_w clusters, and the
obtained n_w centroids are taken as separate representations for each word in
the visual dictionary. To observe the influence of different n_w, five experiments
on the four datasets were performed. As shown in Fig. 4, different clustering
results are associated with different n_w within $\{120, 150, 200, 250, 300\}$, the best
clustering results can be obtained when the value of n_w is 250. It is obvious that
different n_w leads to different degrees of changes in the clustering results.

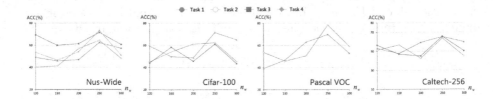

Fig. 4. The Impact of n_w.

4   Conclusions

Inspired by the successful application of many NLP methods and ideas in
machine vision, the proposed APFMTC divides the image into patches of the
same size, so that each patch is regarded as a word, and the image sample is
regarded as an article composed of several words, the process of determining the
cluster to which an image belongs is likened to categorizing articles. Ultimately,
the clustering performance of multiple tasks can be improved by sharing local
details and key information in different images.

In the future, more methods and ideas in NLP can be considered to be applied
to multi-task depth image clustering, or other computer vision work. Researchers
in the field of computer vision should not stick to the methods in this field. We
believe that the existing methods can be greatly improved by using the methods
in other fields for reference.

References

1. Asano, Y., Rupprecht, C., Vedaldi, A.: Self-labelling via simultaneous clustering and representation learning. In: 2020 International Conference on Learning Representations
2. Cao, W., Wu, S., Yu, Z., Wong, H.S.: Exploring correlations among tasks, clusters, and features for multitask clustering. IEEE Trans. Neural Netw. Learn. Syst. **30**(2), 355–368 (2019)
3. Caron, M., Bojanowski, P., Joulin, A., Douze, M.: Deep clustering for unsupervised learning of visual features. In: Proceedings of the European Conference on Computer Vision (ECCV), pp. 132–149 (2018)
4. Chefer, H., Gur, S., Wolf, L.: Transformer interpretability beyond attention visualization. In: Proceedings of the IEEE/CVF Conference on Computer Vision and Pattern Recognition, pp. 782–791 (2021)
5. Hu, S., Yan, X., Ye, Y.: Multi-task image clustering through correlation propagation. IEEE Trans. Knowl. Data Eng. **33**(03), 1113–1127 (2021)
6. Ji, X., Henriques, J.F., Vedaldi, A.: Invariant information clustering for unsupervised image classification and segmentation. In: Proceedings of the IEEE/CVF International Conference on Computer Vision,. pp. 9865–9874 (2019)
7. Jiang, G., Wang, H., Peng, J., Chen, D., Fu, X.: Graph-based multi-view binary learning for image clustering. Neurocomputing **427**, 225–237 (2021)
8. Park, S., et al.: Improving unsupervised image clustering with robust learning. In: Proceedings of the IEEE/CVF Conference on Computer Vision and Pattern Recognition, pp. 12278–12287 (2021)
9. Ren, Y., Hu, K., Dai, X., Pan, L., Hoi, S.C., Xu, Z.: Semi-supervised deep embedded clustering. Neurocomputing **325**, 121–130 (2019)
10. Shukla, A., Cheema, G.S., Anand, S.: Semi-supervised clustering with neural networks. In: 2020 IEEE Sixth International Conference on Multimedia Big Data (BigMM), pp. 152–161. IEEE (2020)
11. Sun, B., Zhou, P., Du, L., Li, X.: Active deep image clustering. Knowl.-Based Syst. **252**, 109346 (2022)
12. Vilhagra, L.A., Fernandes, E.R., Nogueira, B.M.: Textcsn: a semi-supervised approach for text clustering using pairwise constraints and convolutional siamese network. In: Proceedings of the 35th Annual ACM Symposium on Applied Computing, pp. 1135–1142 (2020)
13. Wang, H., Feng, L., Yu, L., Zhang, J.: Multi-view sparsity preserving projection for dimension reduction. Neurocomputing **216**, 286–295 (2016)
14. Wang, H., Yao, M., Jiang, G., Mi, Z., Fu, X.: Graph-collaborated auto-encoder hashing for multiview binary clustering. IEEE Transactions on Neural Networks and Learning Systems (2023)
15. Wu, J., et al.: Deep comprehensive correlation mining for image clustering. In: Proceedings of the IEEE/CVF International Conference on Computer Vision, pp. 8150–8159 (2019)
16. Xu, C., Lin, R., Cai, J., Wang, S.: Deep image clustering by fusing contrastive learning and neighbor relation mining. Knowl.-Based Syst. **238**, 107967 (2022)
17. Yan, X., Hu, S., Ye, Y.: Multi-task clustering of human actions by sharing information. In: Proceedings of the IEEE Conference on Computer Vision and Pattern Recognition, pp. 6401–6409 (2017)
18. Yan, X., Mao, Y., Li, M., Ye, Y., Yu, H.: Multitask image clustering via deep information bottleneck. IEEE Transactions on Cybernetics (2023)

19. Yan, X., Shi, K., Ye, Y., Yu, H.: Deep correlation mining for multi-task image clustering. Expert Syst. Appl. **187**, 115973 (2022)
20. Yan, Y., Ricci, E., Liu, G., Sebe, N.: Egocentric daily activity recognition via multitask clustering. IEEE Trans. Image Process. **24**(10), 2984–2995 (2015)
21. Yang, Y., Ma, Z., Yang, Y., Nie, F., Shen, H.T.: Multitask spectral clustering by exploring intertask correlation. IEEE Trans. Cybern. **45**(5), 1069–1080 (2015)
22. Zhang, H., Zhan, T., Basu, S., Davidson, I.: A framework for deep constrained clustering. Data Min. Knowl. Disc. **35**, 593–620 (2021)
23. Zhang, X.L.: Convex discriminative multitask clustering. IEEE Trans. Pattern Anal. Mach. Intell. **37**(01), 28–40 (2015)
24. Zhang, X., Liu, H., Zhang, X., Liu, X.: Attributed graph clustering with multi-task embedding learning. Neural Netw. **152**, 224–233 (2022)
25. Zhang, X., Zhang, X., Liu, H., Liu, X.: Multi-task multi-view clustering. IEEE Trans. Knowl. Data Eng. **28**(12), 3324–3338 (2016)
26. Zhong, G., Pun, C.M.: Local learning-based multi-task clustering. Knowl. Based Syst. **255**, 109798 (2022)

Deep Structure and Attention Aware Subspace Clustering

Wenhao Wu(ID), Weiwei Wang(✉)(ID), and Shengjiang Kong(ID)

School of Mathematics and Statistics, Xidian University, Xian 710071, China
wwwang@mail.xidian.edu.cn

Abstract. Clustering is a fundamental unsupervised representation learning task with wide application in computer vision and pattern recognition. Deep clustering utilizes deep neural networks to learn latent representation, which is suitable for clustering. However, previous deep clustering methods, especially image clustering, focus on the features of the data itself and ignore the relationship between the data, which is crucial for clustering. In this paper, we propose a novel Deep Structure and Attention aware Subspace Clustering (DSASC), which simultaneously considers data content and structure information. We use a vision transformer to extract features, and the extracted features are divided into two parts, structure features, and content features. The two features are used to learn a more efficient subspace structure for spectral clustering. Extensive experimental results demonstrate that our method significantly outperforms state-of-the-art methods. Our code will be available at https://github.com/cs-whh/DSASC.

Keywords: Deep clustering · Subspace clustering · Transformer

1 Introduction

Clustering, aiming to partition a collection of data into distinct groups according to similarities, is one fundamental task in machine learning and has wide applications in computer vision, pattern recognition, and data mining. In recent years, a significant amount of research has been dedicated to Self-Expressive-based (SE) subspace clustering due to its effectiveness in dealing with high-dimensional data. SE-based subspace clustering primarily assumes that each data point can be expressed as a linear combination of other data points within the same subspace. It generally consists of two phases. In the first phase, a self-representation matrix is computed to capture similarity between data points. Subsequently, spectral clustering is employed to obtain the data segmentation.

Learning an effective self-representation matrix is critical for clustering. Therefore, various regularization methods have been designed to pursue clustering-friendly properties of the self-representation matrix. For example, Sparse Subspace Clustering (SSC) [10] uses the ℓ_1 norm regularization to learn sparse representation. The Low-Rank Representation (LRR) [16] uses the nuclear

© The Author(s), under exclusive license to Springer Nature Singapore Pte Ltd. 2024
Q. Liu et al. (Eds.): PRCV 2023, LNCS 14428, pp. 139–150, 2024.
https://doi.org/10.1007/978-981-99-8462-6_12

norm regularization to induce low-rank representation. You et al. [25] introduced an elastic net regularization to balance subspace preservation and connectivity. Although these methods have demonstrated impressive clustering performance, they are ineffective in dealing with complicated real data, which have non-linearity [13] and contain redundant information and corruptions.

Thanks to the powerful capability of Deep Neural Networks (DNNs) in representation learning, deep SE-based subspace clustering employs DNNs to integrate feature learning and self-representation, hoping that the latent features can reduce the redundancy and corruptions meanwhile satisfy the self-expressive assumption. Convolutional AutoEncoder (CAE) has been widely used for image clustering. For instance, the Deep Subspace Clustering network (DSC) [13] adds a self-representation layer between the convolutional encoder and the decoder to learn the self-representation matrix in the latent subspaces. However, the convolutional operations focus on local spatial neighborhoods, having limited ability to capture long-range dependencies in images. To exploit the long-term dependence and extract more effective features, self-attention mechanisms and transformer architectures have emerged as leading techniques, delivering state-of-the-art performance in various computer vision tasks, including image classification [9], object recognition [4], and semantic segmentation [22].

The traditional DNNs learn the feature representation of each data independently, ignoring the relationship between data points. In contrast, the Graph Convolutional Network (GCN) [15] learns the feature representation of connected data points collaboratively, which is helpful in exploiting the inherent feature of data. Based on GCN, the Attributed Graph Clustering (AGC) [27] and the Structural Deep Clustering Network (SDCN) [1] have been proposed for clustering data. However, these works focus on data that naturally exist in graphs, such as citation and community networks, and few pay attention to image datasets.

The self-attention mechanisms and the transformer architectures are effective in capturing long-range dependencies in images. At the same time, the GCN is advantageous in exploring the inherent feature of similar data. That inspires us to propose the Deep Structure and Attention aware Subspace Clustering (DSASC) for image datasets. Specifically, the proposed network couples the Vision Transformer (ViT) [9] and the GCN. The ViT is pre-trained on a large-scale image dataset using the state-of-the-art representation learning framework DINO (self-DIstillation with NO labels) [6], which is trained with two ViT of the same architecture, one called the teacher network and the other called the student network. The teacher network receives input from a global perspective, and the student network receives input from a local perspective. DINO expects the output of the student network and the teacher network to be consistent, thus learning the semantic features of the data. Once the training is completed, the teacher network is used for feature extraction. However, DINO only considers the information within each image, ignoring the cluster structure within similar images. To capture the cluster structure within similar images, we construct a K-Nearest Neighbor (KNN) [8] graph for intermediate features of ViT and use the GCN to extract the features from the graph perspective. We refer to the output of GCN as structural features and the output of ViT as content features.

To explore the underlying subspace structure of images, we learn the self-representation matrix for the content and structure features separately. It should be noted that the self-representation matrices of the content features and the structure features are learned collaboratively to facilitate each other. Finally, we use the fused self-representation matrix for spectral clustering. Our main contribution can be summarized as follow:

- We propose a novel Deep Structure and Attention aware Subspace Clustering method that takes into account both data content and structural information, enhancing the performance of image clustering.
- We combining the strengths of a vision transformer and graph convolutional network to enhance feature learning efficiency and clustering performance.
- Extended experiments show that the proposed method outperforms some state-of-the-art baselines.

2 Proposed Method

ViT effectively captures long-range dependencies within images through self-attention mechanisms and the transformer architecture. We propose to use the pre-trained ViT to learn the content features of images. GCN can capture the structural features of data that exist in graphs form. To capture the cluster structure within similar images, we organize the image as a graph using the KNN and use GCN to learn the structural features. Note that the ViT and the GCN are coupled. The ViT, comprising 12 layers, extracts content features across different depths. Our approach employs features from the third, sixth, and ninth ViT layers to create KNN graphs and utilizes GCN to learn structural features. Finally, we fuse the features extracted from multiple layers into a unified structural feature. The self-representation matrices for the content features and structure features are learned to explore the underlying subspace structure of images. Fig. 1 illustrates our proposed network, which contains two coupled feature learning modules and two self-representation modules.

The Content Feature Learning Module includes a ViT with an added learnable class embedding (cls token) and image patches, which are processed by the Transformer Encoder. The cls token's output serves as features (yellow patches). The Structure Feature Learning Module utilizes ViT's intermediate features and employs GCN for feature extraction. The Self-Representation Module is a fully-connected layer without bias and activation function to simulate the self-expressive process.

To show the effectiveness of our method, we select 1000 sample images for 10 clusters (100 samples for each cluster) from the dataset STL-10 and obtain the affinity matrices using SSC [10] and our method, respectively. Fig. 2 illustrates the results. Note that SSC computes the self-representation of the raw images directly, while our method learns the self-representation of the features. The affinity matrix is obtained by adding the absolute value of the self-representation and its transpose. SSC's affinity matrix lacks favorable diagonal blocks for subspace clustering. Conversely, our method's matrix displays distinct diagonal block structure.

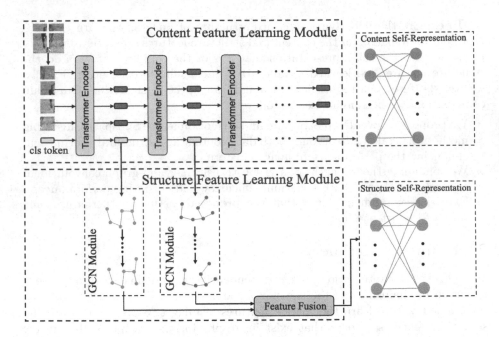

Fig. 1. The overall framework of our model.

2.1   Content Feature Learning

DINO uses two networks with the same architecture during training, the teacher network g_{θ_t} and the student network g_{θ_s} with parameters θ_t and θ_s, respectively. The output of both networks is followed by a softmax layer with temperature parameters to map the features into a probability distribution $P_t(x)$ and $P_s(x)$.

To train the two networks, DINO employs a multi-crop strategy [5] to segment each image x into multiple views, each with distinct resolutions. More precisely, starting from an input image x, a set V of diverse views is generated, comprising two global views denoted as x_1^g and x_2^g, along with several local views of lower resolutions. The student network mainly receives input from the local view, while the teacher network receives input from the global view.

At each iteration, the weights θ_s are updated through a gradient descent to minimize the cumulative cross-entropy loss over a minibatch of input images.

$$\min_{\theta_s} \sum_{x \in \{x_1^g, x_2^g\}} \sum_{\substack{x' \in V \\ x' \neq x}} -P_t(x) \log P_s(x') \tag{1}$$

The parameters θ_t of the teacher network are derived by using the exponential moving average (EMA) technique on the student parameters θ_s.

$$\theta_t \leftarrow \lambda \theta_t + (1 - \lambda)\theta_s \tag{2}$$

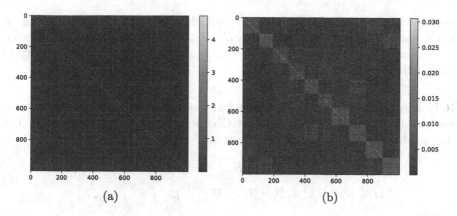

Fig. 2. (a) Affinity matrix obtained by SSC (b) Affinity matrix obtained by our method.

where the value of λ is selected based on a cosine schedule. To avoid training collapse, the centering operation is conducted to the teacher network.

$$g_t(x) \leftarrow g_t(x) + c \tag{3}$$

where c is updated every mini-batch.

$$c \leftarrow mc + (1 - m)\frac{1}{|B|} \sum_{i:x_i \in B} g_{\theta_t}(x_i) \tag{4}$$

where $m > 0$ is a rate parameter and B is the batch size. In the training process, the local view is required to match the global view, thus compelling the network to learn semantic features rather than low-level features. After DINO training, the g_{θ_t} is used for feature extraction, and the content features are obtained using the formula $F_A = g_{\theta_t}(x)$.

2.2   Structure Feature Learning

ViT Intermediate Feature Extraction. ViT consists of a sequence of Transformer Encoders that take a sequence as input and produce an output sequence of equal length. The output sequence has been feature extracted using the attention mechanism. Let the raw image be X, the image is divided into b blocks, and an additional cls token has been added. Consider these $b+1$ blocks as the input sequence, denoted as $M_0 \in \mathbb{R}^{(b+1) \times d}$, where d is the dimension of each block. Feeding data into a ViT can be formalized as follows.

$$M_{i+1} = T(M_i), i = 0, 1, ..., L_1 \tag{5}$$

where T is the transformer encoder, L_1 is the number of layers of ViT. We extracted some intermediate layers and let the indices of the selected layers be S. Then the extracted features can be denoted as $\{M_k[0, :], k \in S\}$, where $M_k[0, :]$

refers to the first row of the matrix M_k, which is the feature corresponding to the cls token. We denote the features as $H_1, H_2, ..., H_t$ with $t = |S|$ for simplicity, and construct KNN graphs measuring by cosine similarity for each feature and obtain adjacency matrix: $A_1, A_2, ..., A_t$.

GCN Module. To effectively extract features and learn node representations from graph structures, we utilize the GCN module for feature extraction. The GCN module can be expressed as follows.

$$H_i^{(0)} = H_i, \quad i = 1, 2..., t$$
$$H_i^{(l)} = \phi(D_i^{-\frac{1}{2}} \tilde{A}_i D_i^{-\frac{1}{2}} H_i^{(l-1)} W_i^{(l-1)}), \quad l = 1, 2, ..., L_2$$

(6)

where H_i is the feature obtained from the ViT intermediate layer. $W_i^{(l-1)}$ is the learnable parameter, D_i is the degree matrix, L_2 is the number of layers of graph convolution, $\tilde{A}_i = A_i + I$ takes into account the self-connectivity of nodes and ϕ is the activation function.

Features Fusion. To fuse the learned structure features, we use an adaptive method with learnable parameters W_i. This method avoids some deviations caused by artificial design, and allows the network to identify which feature is more important.

$$F_S = \sum_{i=1}^{t} H_i^{(L_2)} W_i$$

(7)

2.3   Training Strategy

To explore the internal subspace structure of the data, we use a fully-connected layer without bias and activation function to simulate the self-expressive process and learn the self-expressive matrix of content features. In particular, we minimize the following objective.

$$\min_{C_A} \|C_A\|_1 + \lambda_1 \|F_A - F_A C_A\|_F^2$$

(8)

where $C_A \in \mathbb{R}^{n \times n}$ is the self-expressive matrix of content. Similar to the content self-representation, we adopt the same way to learn the self-representation matrix C_S of structural features.

$$\min_{C_S} \|C_S\|_1 + \lambda_2 \|F_S - F_S C_S\|_F^2$$

(9)

After acquiring the C_A and C_S, we combine them to construct the final affinity graph C_F for spectral clustering. Instead of adopting the method proposed by previous work [18], we adopt a simple yet effective fusion strategy: we add two self-expressive matrix to obtain the fused self-expressive matrix.

$$C_F = C_A + C_S$$

(10)

This approach can enhance self-expressive consistency and alleviate the error caused by random noise. We also add a sparsity constraint to C_F, ensuring that a meaningful representation is learned.

$$\min_{C_F} \|C_F\|_1 \tag{11}$$

The total loss function is:

$$\mathcal{L} = \|C_A\|_1 + \lambda_1 \|F_A - F_A C_A\|_F^2 + \|C_S\|_1 \\ + \lambda_2 \|F_S - F_S C_S\|_F^2 + \|C_F\|_1 \tag{12}$$

3   Experiments

In this section, we first present the datasets used in our experiment, the implementation details, and the baselines. Then we present the experimental results and conduct an analysis to validate the efficacy of our proposed method.

3.1   Experiments Settings

Datasets. We evaluate our method on four image benchmark datasets including CIFAR10, STL-10, Fashion-MNIST and CIFAR100. The CIFAR-10 dataset comprises 60,000 RGB images depicting 10 distinct objects. The STL-10 dataset consists of 13,000 RGB images, depicting 10 distinct objects, with each image having a resolution of 96 × 96 pixels. The Fashion-MNIST dataset consists of 70,000 grayscale images of various fashion products, classified into 10 categories, with a resolution of 28 × 28 pixels. The CIFAR100 dataset is a more challenging dataset that consists of 60,000 images for 100 different objects. There are 600 images per class.

For the computation efficiency of the self-representation matrix, we randomly take 1000 images from CIFAR10, STL-10, and Fashion-MNIST for cluster analysis. For the CIFAR100 dataset, we randomly take 3000 images of the 100 objects (300 images for each object) for clustering.

Implementation Details. We use the ViT-S/8 architecture, which was trained on ImageNet-1k (without labels) for content feature extraction. We use the pretrained parameters provided by the authors. During our training, the parameters of the ViT are frozen. The GCN module consists of 2 layers of graph convolution, and the ReLU activation function is used. To construct the KNN graph, we set $K = 10$. The λ_1 and λ_2 in Eq. 12 are set at 1. The Adam optimizer is used to minimize the loss, and the learning rate is set to 10^{-5}. We train for 2000 epochs and report the final results.

Table 1. Clustering results of different methods on benchmark datasets. The top-performing results are highlighted in bold, while the second-best results are marked with an underline.

Method	CIFAR10		STL10		Fashion-MNIST		CIFAR100	
	ACC	NMI	ACC	NMI	ACC	NMI	ACC	NMI
k-means	0.2290	0.0840	0.1920	0.1250	0.4740	0.5120	0.1300	0.0840
LSSC	0.2114	0.1089	0.1875	0.1168	0.4740	0.5120	0.1460	0.0792
LPMF	0.1910	0.0810	0.1800	0.0960	0.4340	0.4250	0.1180	0.0790
DEC	0.3010	0.2570	0.3590	0.2760	0.5180	0.5463	0.1850	0.1360
IDEC	0.3699	0.3253	0.3253	0.1885	0.5290	0.5570	0.1916	0.1458
DCN	0.3047	0.2458	0.3384	0.2412	0.5122	0.5547	0.2017	0.1254
DKM	0.3526	0.2612	0.3261	0.2912	0.5131	0.5557	0.1814	0.1230
VaDE	0.2910	0.2450	0.2810	0.2000	0.5039	0.5963	0.1520	0.1008
DSL	0.8340	0.7132	0.9602	0.9190	0.6290	0.6358	0.5030	0.4980
EDESC	0.6270	0.4640	0.7450	0.6870	0.6310	0.6700	0.3850	0.3700
CDEC	0.5640	0.5778	0.7328	0.7183	0.5351	0.5404	-	-
DML	<u>0.8415</u>	<u>0.7170</u>	<u>0.9645</u>	<u>0.9211</u>	0.6320	0.6480	<u>0.5068</u>	<u>0.5019</u>
SENet	0.7650	0.6550	0.8232	0.7541	**0.6970**	<u>0.6630</u>	-	-
Our	**0.8740**	**0.8059**	**0.9740**	**0.9491**	<u>0.6680</u>	**0.6739**	**0.5200**	**0.6865**

Comparison Algorithms. We compare our method DSASC with three traditional methods: k-means [17], large-scale spectral clustering (LSSC) [7], locality preserving non-negative matrix factorization (LPMF) [2], and ten deep clustering methods: deep embedding clustering (DEC) [23], improved deep embedding clustering (IDEC) [12], deep clustering network (DCN) [24], deep k-means (DKM) [11], variational deep embedding (VaDE) [14], deep successive subspace learning (DSL) [19], efficient deep embedded subspace clustering (EDESC) [3], contrastive deep embedded clustering (CDEC) [21], deep multi-representation learning (DML) [20], self-expressive network (SENet) [26].

3.2 Results and Discussion

We use two commonly used metrics to evaluate clustering performance, including clustering accuracy (ACC) and normalized mutual information (NMI). The clustering results are presented in Table 1, where the best results are emphasized in bold the second best results are marked with an underline. As evident from the table, DSASC significantly improves the clustering performance. DSASC improves ACC and NMI of DML in CIFAR10, respectively, by 3.25% and 8.89%. The non-significant improvement in clustering performance of DSASC on Fashion-MNIST is due to the fact that Fashion-MNIST is gray-scale images, while our ViT is pre-trained on color images and suffers from domain shift. But our method still achieves competitive results on Fashion-MNIST. That suggests

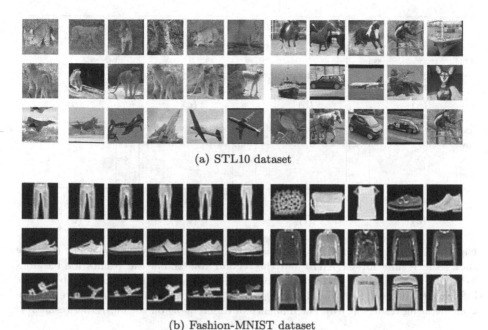

(a) STL10 dataset

(b) Fashion-MNIST dataset

Fig. 3. The first column is the anchor image. We sort the numerical values in the self-expressive matrix. Where columns 2 to 6 are images corresponding to top-5 values (most similar) and columns 7 to 11 are images corresponding to bottom-K values (least similar).

Table 2. The ablation experiments results. The top-performing results are highlighted in bold.

Datasets	Content SE		Our Method	
	ACC	NMI	ACC	NMI
CIFAR10	0.8370	0.7430	**0.8740**	**0.8059**
STL10	0.9380	0.8886	**0.9740**	**0.9491**
Fashion-MNIST	0.6560	0.6741	**0.6680**	**0.6739**
CIFAR100	0.3527	0.5220	**0.5200**	**0.6865**

that combining structure features and content features effectively improve the discrimination of the feature representation and clustering performance.

In order to confirm the efficacy of our proposed method, we conduct ablation experiments on structure feature learning module. We remove the structural feature learning module and learn the content self-representation matrix C_A using Eq. 8, then perform spectral clustering. We denote the method as Content SE. Table 2 presents the results. It can be seen that joint content features and structural features can significantly improve the clustering performance.

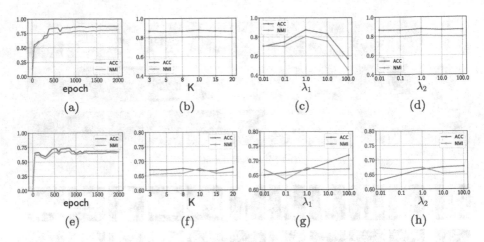

Fig. 4. Cluster convergence analysis and parameter sensitivity analysis. The first row shows the results for CIFAR10 and the second row shows the results for Fashion-MNIST. (a)(e) Changes of clustering ACC and NMI with training epochs. (b)(f) The influence of different K on the clustering results. (c)(g) The influence of different λ_1 on the clustering results. (d)(h) The influence of different λ_2 on the clustering results.

To further examine the quality of the self-expressive matrix obtained through our approach, we randomly selected three anchor images from the STL10 and Fashion-MNIST datasets, respectively. Subsequently, we have determined the five largest and five smallest representation coefficients of anchor images derived from the self-representation matrix C_F. These coefficients indicate the most similar and least similar images, respectively. This analysis provides insights into the effectiveness of the matrix and its ability to capture meaningful patterns. The result is shown in Fig. 3. Where the first column is the anchor images, the next five are the most similar images, and the last are the most dissimilar images. It can be seen that our method can effectively learn the subspace structure, leading to satisfactory clustering outcomes.

Figure 4(a) and Fig. 4(e) show the change of ACC and NMI during training, and it can be seen that the accuracy gradually improves as the training proceeds, which indicates that our method can effectively converge.

We additionally investigate the impact of hyperparameters on the experimental results. The hyperparameters include the value of K in the GCN Module, λ_1, and λ_2 in Eq. 12. In the following experiments, the default value of K is 10, the default value of λ_1 and λ_2 is 1, and when one parameter is investigated, the other parameters are kept as default. We select $K \in \{3, 5, 8, 10, 15, 20\}$ and $\lambda_1, \lambda_2 \in \{0.01, 0.1, 1, 10, 100\}$ to evaluate their influence. The corresponding ACC and NMI of the clustering results are presented in Figs. 4(b) to 4(d) for the dataset CIFAR10 and Figs. 4(f) to 4(h) for the dataset Fashion-MNIST, respectively. As we can see from these figures, our method is not sensitive to the number of neighbors selection, and λ_2 also has less impact on the clustering performance. The content self-representation coefficient λ_1 dominates the clustering results. When λ_1 increases, the clustering performance of CIFAR10 decreases, while the trend is the opposite in Fashion-MNIST. We chose $\lambda_1 = 1$ for all datasets in our experiments, which produced competitive results.

4 Conclusion

We propose a Deep Structure and Attention aware Subspace Clustering framework. Specifically, to discover the latent subspace structure of images, our network learns content features and structure features, as well as their sparse self-representation. Extended experiments show that our proposed method is highly effective in unsupervised representation learning and clustering.

Acknowledgements. The authors would like to thank the editors and the anonymous reviewers for their constructive comments and suggestions. This paper is supported by the National Natural Science Foundation of China (Grant Nos. 61972264, 62072312), Natural Science Foundation of Guangdong Province (Grant No. 2019A1515010894) and Natural Science Foundation of Shenzhen (Grant No. 20200807165235002).

References

1. Bo, D., Wang, X., Shi, C., Zhu, M., Lu, E., Cui, P.: Structural deep clustering network. In: Proceedings of the Web Conference 2020, pp. 1400–1410 (2020)
2. Cai, D., He, X., Wang, X., Bao, H., Han, J.: Locality preserving nonnegative matrix factorization. In: Twenty-first International Joint Conference on Artificial Intelligence, pp. 1010–1015 (2009)
3. Cai, J., Fan, J., Guo, W., Wang, S., Zhang, Y., Zhang, Z.: Efficient deep embedded subspace clustering. In: Proceedings of the IEEE/CVF Conference on Computer Vision and Pattern Recognition, pp. 1–10 (2022)
4. Carion, N., Massa, F., Synnaeve, G., Usunier, N., Kirillov, A., Zagoruyko, S.: End-to-end object detection with transformers. In: Vedaldi, A., Bischof, H., Brox, T., Frahm, J.-M. (eds.) Computer Vision – ECCV 2020: 16th European Conference, Glasgow, UK, August 23–28, 2020, Proceedings, Part I, pp. 213–229. Springer International Publishing, Cham (2020). https://doi.org/10.1007/978-3-030-58452-8_13
5. Caron, M., Misra, I., Mairal, J., Goyal, P., Bojanowski, P., Joulin, A.: Unsupervised learning of visual features by contrasting cluster assignments. Adv. Neural. Inf. Process. Syst. **33**, 9912–9924 (2020)
6. Caron, M., et al.: Emerging properties in self-supervised vision transformers. In: Proceedings of the IEEE/CVF International Conference on Computer Vision. pp. 9650–9660 (2021)
7. Chen, X., Cai, D.: Large scale spectral clustering with landmark-based representation. In: Twenty-fifth AAAI Conference on Artificial Intelligence (2011)
8. Cover, T., Hart, P.: Nearest neighbor pattern classification. IEEE Trans. Inf. Theory **13**(1), 21–27 (1967)
9. Dosovitskiy, A., et al.: An image is worth 16x16 words: Transformers for image recognition at scale. In: International Conference on Learning Representations (2020)

10. Elhamifar, E., Vidal, R.: Sparse subspace clustering: Algorithm, theory, and applications. IEEE Trans. Pattern Anal. Mach. Intell. **35**(11), 2765–2781 (2013)
11. Fard, M.M., Thonet, T., Gaussier, E.: Deep k-means: jointly clustering with k-means and learning representations. Pattern Recogn. Lett. **138**, 185–192 (2020)
12. Guo, X., Gao, L., Liu, X., Yin, J.: Improved deep embedded clustering with local structure preservation. In: Proceedings of the Twenty-Sixth International Joint Conference on Artificial Intelligence, pp. 1753–1759 (2017)
13. Ji, P., Zhang, T., Li, H., Salzmann, M., Reid, I.: Deep subspace clustering networks. In:Advances in Neural Information Processing Systems **30** (2017)
14. Jiang, Z., Zheng, Y., Tan, H., Tang, B., Zhou, H.: Variational deep embedding: an unsupervised and generative approach to clustering. arXiv preprint arXiv:1611.05148 (2016)
15. Kipf, T.N., Welling, M.: Semi-supervised classification with graph convolutional networks. arXiv preprint arXiv:1609.02907 (2016)
16. Liu, G., Lin, Z., Yan, S., Sun, J., Yu, Y., Ma, Y.: Robust recovery of subspace structures by low-rank representation. IEEE Trans. Pattern Anal. Mach. Intell. **35**(1), 171–184 (2012)
17. Lloyd, S.: Least squares quantization in pcm. IEEE Trans. Inf. Theory **28**(2), 129–137 (1982)
18. Peng, Z., Liu, H., Jia, Y., Hou, J.: Adaptive attribute and structure subspace clustering network. IEEE Trans. Image Process. **31**, 3430–3439 (2022)
19. Sadeghi, M., Armanfard, N.: Deep successive subspace learning for data clustering. In: 2021 International Joint Conference on Neural Networks (IJCNN), pp. 1–8. IEEE (2021)
20. Sadeghi, M., Armanfard, N.: Deep multirepresentation learning for data clustering. IEEE Transactions on Neural Networks and Learning Systems (2023)
21. Sheng, G., Wang, Q., Pei, C., Gao, Q.: Contrastive deep embedded clustering. Neurocomputing **514**, 13–20 (2022)
22. Strudel, R., Garcia, R., Laptev, I., Schmid, C.: Segmenter: transformer for semantic segmentation. In: Proceedings of the IEEE/CVF International Conference on Computer Vision, pp. 7262–7272 (2021)
23. Xie, J., Girshick, R., Farhadi, A.: Unsupervised deep embedding for clustering analysis. In: International Conference on Machine Learning, pp. 478–487. PMLR (2016)
24. Yang, B., Fu, X., Sidiropoulos, N.D., Hong, M.: Towards k-means-friendly spaces: Simultaneous deep learning and clustering. In: International Conference on Machine Learning, pp. 3861–3870. PMLR (2017)
25. You, C., Li, C.G., Robinson, D.P., Vidal, R.: Oracle based active set algorithm for scalable elastic net subspace clustering. In: Proceedings of the IEEE Conference on Computer Vision and Pattern Recognition, pp. 3928–3937 (2016)
26. Zhang, S., You, C., Vidal, R., Li, C.G.: Learning a self-expressive network for subspace clustering. In: Proceedings of the IEEE/CVF Conference on Computer Vision and Pattern Recognition, pp. 12393–12403 (2021)
27. Zhang, X., Liu, H., Li, Q., Wu, X.M.: Attributed graph clustering via adaptive graph convolution. arXiv preprint arXiv:1906.01210 (2019)

Broaden Your Positives: A General Rectification Approach for Novel Class Discovery

Yaqi Cai[1], Nan Pu[2], Qi Jia[1], Weimin Wang[1], and Yu Liu[1(✉)]

[1] International School of Information Science and Engineering,
Dalian University of Technology, Dalian, China
caiyaqi1998@mail.dlut.edu.cn, {jiaqi,wangweimin,liuyu8824}@dlut.edu.cn
[2] The Department of Information Engineering and Computer Science,
University of Trento, Trento, Italy
nan.pu@unitn.it

Abstract. Novel category discovery (NCD), which is a challenging and emerging task, aims to cluster unlabelled instances with knowledge information transferred from labelled ones. A majority of recent state-of-the-art methods leverage contrastive learning to model labelled and unlabelled data simultaneously. Nevertheless, they suffer from inaccurate and insufficient positive samples, which are detrimental to NCD and even its generalized class discovery (GCD) setting. To solve this problem, we propose positive-augmented contrastive learning (PACL), which can mine more positive samples and additional pseudo-positive samples, while augmenting the loss cost corresponding to these positive pairs. Consequently, PACL alleviates the imbalance between positive and negative pairs in contrastive learning, and facilitates the knowledge transfer for novel class discovery. In addition, we develop a general feature rectification approach based on PACL to rectify the representation learning achieved by existing NCD or GCD models. Extensive experiments on three datasets exhibit the necessity and effectiveness of our approach for both NCD and GCD tasks, without loss of generality.

Keywords: Novel Class Discovery · Contrastive Learning · Rectification

1 Introduction

Supervised deep learning models have demonstrated exceptional performance in both upstream and downstream tasks when trained on extensive sets of labelled data, even surpassing human capabilities [1,2,19,23]. However, such superior performance is always evaluated via recognizing or segmenting images from a set of pre-defined categories/classes, which have been fully annotated by human experts during training phase. Consequently, their effectiveness is heavily restricted when encountering unlabelled data or unknown classes in real-world scenarios. To explore and address this problem, Novel Class Discovery (NCD) [5,7,8] has been proposed to leverage knowledge information from

Q. Liu et al. (Eds.): PRCV 2023, LNCS 14428, pp. 151–162, 2024.
https://doi.org/10.1007/978-981-99-8462-6_13

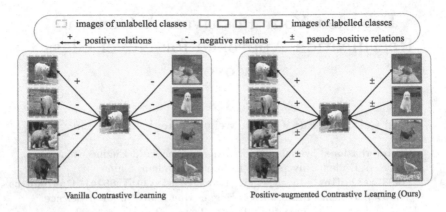

Fig. 1. Conceptual comparison between vanilla contrastive learning and our positive-augmented contrastive learning (PACL). Our PACL can rescue positive samples from false negative relations, and further explore more pseudo-positive relations.

labelled instances of known classes so as to group unlabelled instances into new and unknown classes, which are related yet disjoint with the known classes. However, it is unrealistic to assume the unlabelled data are all from unknown classes, but are never from known classes at all. To this end, NCD has recently been extended to Generalized Category Discovery (GCD) [21], enabling the model to cluster arbitrary unlabelled data from both known and unknown classes.

The key challenge for NCD is how to adapt the model pre-trained on labelled classes to discover unlabelled ones. To achieve the adaptability, existing methods are dedicated to exploiting unsupervised learning objectives for training the unlabelled data. Specifically, a line of early works [5,7,10,25–27] employ various clustering algorithms (*e.g.*, *k*-means) to generate pseudo labels, which are then used to train the loss function and fine-tune the feature extractor. However, those pseudo labels are easily mistaken due to the limitations of the clustering algorithms. Instead, more recent works [4,17,21,24] tend to leverage contrastive learning [6] to train unlabelled instances, with no need for generating their pseudo labels any more. Given any unlabelled image, the contrastive learning strategy treats its augmented views as positive samples, and all the other images as negative ones. Yet, it may mistake potential positive samples from the same classes as negative, as shown in the left of Fig. 1. In addition, the number of positive samples in vanilla contrastive learning is very insufficient, which hinders the model from exploring semantic relations between labelled and unlabelled classes, and weakens the knowledge transfer for NCD and GCD tasks.

To this end, we propose positive-augmented contrastive learning (PACL), which aims to alleviate inaccurate and insufficient positive samples caused by the vanilla counterpart. As shown in the right of Fig. 1, PACL constructs some pseudo-positive relations between instances in addition to positive and negative ones. Specifically, we calculate the similarity scores between images to achieve pseudo-positive samples based on a trained model, like RankStat [7], UNO [5] and GCD [21]. Then, we mine and augment positive samples from labelled and unlabelled images based on the statistical information of the similarity scores.

By the use of the positive and pseudo-positive samples, we formulate a positive-augmented contrastive learning loss, and provide additional theoretical analysis to prove its effectiveness. In a word, PACL helps to identify the visual cues shared between labelled and unlabelled classes, thereby easing the knowledge transfer for novel class discovery. For instance, when "brown bear" acts as a pseudo-positive sample for "black bear", it makes the model learn the same shape cues between the two classes. Likewise, when we construct a pseudo-positive relation between "polar bear" and "dog", the model learns to identify they have the same color. Consequently, we propose a simple yet general feature rectification approach based on PACL. Our rectification approach can be seamlessly built on top of existing NCD and GCD models, so as to rectify their representation learning in an on-the-fly fashion.

Our contributions can be summarized as follows:

1) we propose positive-augmented contrastive learning (PACL), which can mine richer positive/pseudo-positive samples and augment their importance in the training loss. We provide a theoretical analysis of how to define PACL.
2) We develop a feature rectification approach based on PACL, which is the first to investigate how to rectify the feature representation already learned from existing NCD and GCD models.
3) Extensive experiments on generic and fine-grained datasets demonstrate that, our approach benefits improving the performance of several state-of-the-art NCD and GCD methods, without loss of generality.

2   Related Work

Novel and Generalized Class Discovery. Based on the assumption that labelled and unlabelled images are from disjoint classes, current methods on NCD normally use a shared backbone and individual classifiers for recognizing labelled and unlabelled classes [5,7,8,14,26]. In this way, NCD acts as a special case under semi-supervised learning setting [15,20]. In order to strengthen the discrimination of feature representation, RankStat [7] proposes to pre-train the model via self-supervised learning with labelled and unlabelled data. Likewise, UNO [5] unifies the objective of labelled and unlabelled data by generating one-hot pseudo labels for unlabelled images. Considering the fact that the unlabelled images may come from either labelled or unlabelled classes, GCD [21] eliminates the above-mentioned disjoint assumption and builds a simple baseline for generalized class discovery. Another assumption is semantic similarity between labelled and unlabelled data, which supports the success of NCD and GCD [14], but only a few methods take full advantage of this assumption. Motivated by contrastive learning, some methods treat two data augmentation views of one image sample as a positive pair [4,10,21,24], and further enforce their predictions to be consistent [7,10,25–27]. However, these approaches exploit few positive samples, and neglect most semantic information between labelled and unlabelled samples. Instead, we propose to augment positive samples via statistical information, while defining and exploring pseudo-positive samples based on their similarity relationships. The experimental results verify the advantage of our positive-augmented method.

Contrastive Learning. It has been widely used for self-supervised and semi-supervised learning scenarios [6,11,16]. Typically, contrastive learning encourages pulling features of a positive pair together, and at the same time pushing features of a negative pair apart. The positive or negative pairs are normally defined as a binary contrastive label. For example in GCD [21], the positive pair of unlabelled data are two augmentation views of each image, while labelled samples which belong to the same class are positive pairs. In [26], the positive pairs are from the five nearest samples. In addition, the potential positive pairs are mostly ignored, which leads to the imbalance problem between positive and negative pairs in contrastive representation learning. Moreover, binary contrastive relationships are unable to reflect semantic similarity between samples, which is important for representation learning in NCD and GCD. Recently, relaxed contrastive loss (RCL) [11] has been proposed to exploit pairwise similarities between samples for better knowledge transfer. Motivated by [11] but differently, our work is the first to broaden the proportion of potential positive samples for NCD/GCD, while augmenting their importance in contrastive learning.

3   Proposed Approach

This section elaborates positive-augmented contrastive learning (PACL), followed by theoretical analysis. Then, we introduce how to exploit PACL to rectify feature representation already learned by existing NCD and GCD models.

3.1   Problem Definition

Given an image set D from known classes C^k and unknown classes C^u, where C^k, C^u are known a priori, we divide D into a labelled split $D^{lab} = \{(x_i, y_i)\}_{i=1}^{N^{lab}}$ and an unlabelled split $D^{unlab} = \{(x_i, y_i)\}_{i=1}^{N^{unlab}}$. The class labels in D^{lab} and D^{unlab} are denoted with $\{y_1, ..., y_{N^{lab}}\} \in C^{lab}$ and $\{y_1, ..., y_{N^{unlab}}\} \in C^{unlab}$, respectively. The standard NCD task assumes unlabelled images only come from unknown classes, i.e., $C^{unlab} = C^u$, while GCD considers unlabelled data from both known and unknown classes, resulting in $C^{unlab} = C^u \cup C^k$. For both NCD and GCD, they make $C^{lab} = C^k$. The aim is to generate class assignments $\hat{y}$ for unlabelled images from C^{unlab} classes.

3.2   Positive-Augmented Contrastive Learning

Contrastive learning has become an essential and effective technique to make the model learn from unlabelled data. However, its performance on feature representation is significantly limited by inaccurate and insufficient positive samples. As a result, the model may be biased on negative samples, while overlooking semantic information from potential positive samples. To overcome this problem, we propose a novel positive-augmented contrastive learning (PACL) which can enhance the representation learning of positive pairs and explore pseudo-positive relations between labelled and unlabelled data based on the similarity score in contrastive learning. With this aim, we define a positive set X_i^+ for anchor image x_i. X_i^+ is expected to contain more potential positive samples, no matter they are labelled or unlabelled. We define a threshold β for selecting potential positive samples in X_i^+ by

$$X_i^+ = \{x_j | \forall j \in \{1, ..., N\}, \ w_{ij}^s < \beta\}, \tag{1}$$

where $w_{ij}^s \in (0, 1]$ indicates the similarity score calculated with the features derived from a source feature extractor f^s. Since using class labels to define the threshold may overlook the unlabelled positive samples, we instead define the threshold β with the average feature distance of the nearest $\frac{N}{C-1}$ samples, with no need of using any label information. Therefore, we define β with

$$\beta = \frac{1}{k} \sum_{x_j \in top_k w_{ij}^s} w_{ij}^s, \qquad k = \frac{N}{C-1}, \tag{2}$$

where N is the number of input images and C is the class number in the dataset. Through X_i^+, we augment the loss proportion of positive pairs during contrastive learning. Different from the conventional contrastive loss, our PACL loss becomes

$$\mathcal{L}(X) = (C-1) \cdot \frac{1}{N} \sum_{i=1}^N \sum_{j \in X_i^+} l(x_i, x_j) + \frac{1}{N} \sum_{i=1}^N \sum_{j \notin X_i^+} l(x_i, x_j). \tag{3}$$

It can be noted that, we further weigh the loss term of X_i^+ by multiplying a coefficient $C - 1$, so as to enhance the representation learning of positive pairs. Additional theoretical analysis on the coefficient $C - 1$ will be given in Sect. 3.3. Here, we compute $l(x_i, x_j)$ with relaxed contrastive loss [11], which implicitly defines pseudo-positive samples by replacing binary contrastive relationships with real-valued similarity scores. The loss cost of $l(x_i, x_j)$ we use becomes

$$l(x_i, x_j) = w_{ij}^s {d_{ij}^t}^2 + (1 - w_{ij}^s)[\delta - d_{ij}^t]_+^2, \tag{4}$$

where $[\cdot]_+$ is a hinge function that is determined by a distance margin δ.

3.3 Theoretical Analysis for PACL

To reveal the unbalance issue between positive and negative pairs in contrastive representation learning, it is necessary to define and compare the gradients with respect to the positive and negative pairs

Definition 1. δ_i^+, δ_i^-. *Giving a batch of image samples from C classes, that is, $\mathcal{B} = \{(x_i, y_i)\}_{i=1}^N \in \mathbb{R}^C$, we define the gradient δ_i for any x_i with*

$$\delta_i = \sum_{x_j \in \mathcal{B}} \frac{\partial \mathcal{L}(X)}{\partial d_{ij}^t} = N \frac{\sum_{x_j \in \mathcal{B}} \frac{\partial \mathcal{L}(X)}{\partial d_{ij}^t}}{N} = N \left(\frac{\partial \mathcal{L}(X)}{\partial d_{ij}^t} \right)_{mean}.$$

Then we define $\mathcal{B}^+ = \{(x_j, y_j)|x_j \in \mathcal{B}, y_j = y_i\}, \mathcal{B}^- = \{(x_j, y_j)|x_j \in \mathcal{B}, y_j \neq y_i\}$, which represents samples from the same class and other classes. Hence, we have

$$\delta_i^{+/-} = N^{+/-} \Big(\frac{\partial \mathcal{L}(X)}{\partial d_{ij}^t}\Big)^{+/-}_{mean},$$

where δ_i^+ is the gradient of samples from the same class as x_i, and δ_i^- is the opposite. The numbers N^+ and N^- count the amount of samples for two cases.

Theorem 1. $\mathbb{E}[\delta_i^+ : \delta_i^-]$ in contrastive loss is negatively correlated with the total class number C.

Proof. We decompose the gradient of positive and negative pairs:

$$\mathbb{E}[\delta_i^+ : \delta_i^-] = \mathbb{E}\{[N^+ : N^-][\Big(\frac{\partial \mathcal{L}(X)}{\partial d_{ij}^t}\Big)^+_{mean} : \Big(\frac{\partial \mathcal{L}(X)}{\partial d_{ij}^t}\Big)^-_{mean}]\}$$

$$= \frac{1}{C-1} \cdot \mathbb{E}[\Big(\frac{\partial \mathcal{L}(X)}{\partial d_{ij}^t}\Big)^+_{mean} : \Big(\frac{\partial \mathcal{L}(X)}{\partial d_{ij}^t}\Big)^-_{mean}].$$

For one certain x_i, its positive samples are from class y_i, while negative samples are from the remaining $C-1$ classes. Under random sampling, the expected ratio of positive and negative samples is their proportional ratio in the whole dataset, i.e., $\mathbb{E}\{[N^+ : N^-] = \frac{1}{C} : \frac{C-1}{C} = \frac{1}{C-1}$. Meanwhile, the value of the remaining expectation is conditioned on the feature extractor and learning strategy used in NCD or GCD methods. As a result, negative samples can dominate the gradient of the contrastive loss, especially when the number of classes C is large. Moreover, in an open-world scenario, the gradient of positive samples will be gradually diluted as C becomes increasing over time. Based on these observations, we define a coefficient $C-1$ in Eq. (3) to weigh the loss term, which helps to alleviate the imbalance issue between positive and negative samples in contrastive learning.

3.4   PACL for NCD and GCD

PACL can act as a simple and general method to rectify the feature representation learned by any NCD and GCD models. In other words, PACL is an on-the-fly feature rectification that is performed on top of a trained feature extractor. After fine-tuning it with our PACL loss, the rectified feature representation allows to correct some mistaken pseudo labels and transfer knowledge between positive and pseudo-positive samples with respect to unlabelled images. Figure 2 depicts the overall pipeline for our approach. Specifically, let f^s be a source feature extractor that has been trained by any NCD or GCD methods, while f^t be a target feature extractor whose representations are rectified by our PACL. Then, we use f^s to generate a feature distance matrix D^s, a similarity matrix W and a threshold β, and use f^t to generate a feature distance matrix D^t. Afterward, we rectify feature representation with the proposed positive-augmented contrastive loss. Hence, the similarity matrix W with f^s is formulated as

$$W = \Big\{w_{ij}^s \Big| i, j \in [1, .., N], w_{ij}^s = e^{-d_{ij}^s{}^2} \in (0, 1]\Big\}. \tag{5}$$

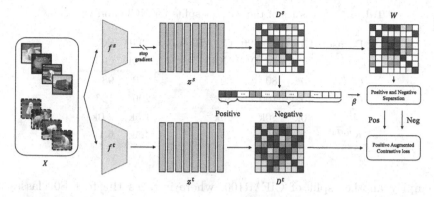

Fig. 2. Overall pipeline of our PACL. We use feature z^s and z^t to generate Euclidean distance matrix D^s and D^t respectively. Then, we use D^s to generate a similarity matrix W. A threshold β is calculated based on D^s, which controls the division of positive and negative pairs. The model is optimized by the PACL loss we develop.

As f^s is the model trained with existing methods, the feature distances in source embedding space can measure the similarity relation between samples. In order to increase the discrimination of feature z^t, we employ batch-level relative distance to calculate the distance matrix D^t, which is given by:

$$D^t = \left\{ \hat{d_{ij}^t} \middle| i, j \in [1, .., N], \hat{d_{ij}^t} = \frac{d_{ij}^t}{\frac{1}{n}\frac{1}{n}\sum_{i=1}^{n}\sum_{j=1}^{n} d_{ij}^t} \right\}. \tag{6}$$

It can be noted that, we use the average feature distance to normalize d_{ij}^t, and remove l_2 normalization, which releases the learned representation to the whole embedding space, rather than in the unit hypersphere. Moreover, considering maintaining the representation learning within labelled and unlabelled data respectively, we jointly fine-tune the model with the proposed loss in labelled data, unlabelled data, and their union. Finally, the overall loss for our feature rectification approach becomes

$$\mathcal{L} = \alpha \mathcal{L}^{label} + \beta \mathcal{L}^{unlabel} + \gamma \mathcal{L}^{all}, \tag{7}$$

where α, β, γ are weight parameters. Here, we refer to the $\mathcal{L}^{label}$, $\mathcal{L}^{unlabel}$ and $\mathcal{L}^{all}$ as the proposed PACL loss with respect to labelled data, unlabelled data, and their union, respectively.

4 Experiments

4.1 Experimental Setup

Datasets. We conduct experiments on a generic image classification dataset CIFAR100 [13], and two fine-grained classification datasets including CUB [22] and Stanford Cars [12]. Following [5,7,21], we take the first 50% classes in each dataset as known, and the remaining 50% classes as unknown. Apart from that,

Table 1. Statistics of the dataset splits for NCD and GCD.

Dataset	CIFAR100-20		CIFAR100-50	CUB	SCars
	NCD	GCD	NCD	GCD	GCD
C^{lab}	80	80	50	100	98
C^{unlab}	20	100	50	200	196
N^{lab}	40k	20k	25k	1.5k	2.0k
N^{unlab}	10k	30k	25k	4.5k	6.1k

we employ another split of CIFAR100, where it takes the first 80 classes as known classes [5] and the other 20 classes being unknown. For clear clarification, we denote them as CIFAR100-50 and CIFAR100-20, which indicates 50 and 20 unknown classes respectively. Regarding GCD, unlabelled training data consists of 50% of the images from C^k and all of the images from C^u.

Evaluation Metrics. Following NCD methods [7], we evaluate their performance with a classification accuracy for labelled images and a clustering accuracy (ACC) for unlabelled images. As for GCD methods [21], it is common to evaluate the performance with the ACC metric for samples from unlabelled training set D^{unlab}. The ACC metric is formulated as $ACC = \frac{1}{N^u} \sum_{i=1}^{N^u} \mathbb{1}(y_i = perm^*(\hat{y}_i))$, where N^u is number of samples from unknown classes, and $perm^*$ is the optimal permutation between predicted cluster assignment $\hat{y}_i$ and ground truth y_i. We report the accuracy results with 'All', 'Old' and 'New' class subsets, which represent all samples, samples from known classes, and samples from unknown classes respectively. Following [5], we evaluate NCD approaches with *task-aware* and *task-agnostic* protocols, which means that instances from known and unknown classes are evaluated separately or jointly, respectively.

Implementation Details. Following the methods in the literature, we use ResNet-18 [9] and ViT-B-16 [3] backbones. For all the models, our method only fine-tunes their final block in the backbone with the proposed PACL loss. We use SGD with momentum [18] as the optimizer. For NCD and GCD, the batch size is set to {1024,256}, and the initial learning rate is set to {0.001,0.00001} which is decayed with a cosine annealed schedule, respectively. The weight coefficients in Eq. (7) are set to $\{\alpha, \beta, \gamma\} = min\{C^{lab}, C^{unlab}\}/\{C^{lab}, C^{unlab}, C^{lab} \cup C^{unlab}\}$, and the distance margin δ in Eq. (4) is set with 1 for NCD and 0.9 for GCD. Like standard practice in contrastive learning, we generate two randomly augmented views for all samples. We fine-tune the model with 10/40 epochs on CIFAR100-20/CIFAR100-50 for NCD setting, and with 20/100/5 epochs on CIFAR100-20/CUB/SCars in the case of GCD setting.

4.2 PACL for Novel Category Discovery

For novel category discovery (NCD), we implement our PACL on top of two representative and state-of-the-art baselines: RankStats [7] and UNO [5]. Table 2 reports the accuracy results on CIFAR100 dataset in terms of three class sets

Table 2. Compared results on CIFAR100 For NCD task. RankStats+ represents the one trained with incremental classifier, and UNO-v2 is the improved version beyond UNO. The test set is evaluated with both *task-agnostic* and *task-aware* protocols, while the train set follows a *task-aware* protocol only.

Method	CIFAR100-20					CIFAR100-50				
	task-agnostic			*task-aware*		*task-agnostic*			*task-aware*	
	Test			Test	Train	Test			Test	Train
	All	Old	New	New	New	All	Old	New	New	New
RankStats+	68.3	71.2	56.8	75.8	75.2	55.3	69.7	40.9	49.1	44.1
RankStats+ w/ RCL	70.0	71.5	64.1	75.1	77.3	57.3	68.7	**46.5**	48.4	49.8
RankStats+ w/ PACL	**70.8**	**72.1**	**65.7**	**76.2**	**77.4**	**59.0**	**73.0**	46.0	**49.7**	**50.4**
UNO-v2	73.6	73.9	72.6	83.5	**85.1**	66.4	75.6	**57.4**	60.3	61.5
UNO-v2 w/ RCL	71.0	70.9	71.5	81.0	82.2	63.0	68.2	57.9	59.7	62.0
UNO-v2 w/ PACL	**74.3**	**74.5**	**73.6**	**84.2**	**85.1**	**67.7**	**78.6**	57.0	**61.7**	**63.1**

Table 3. Compared results on three datasets for GCD task.

Method	CIFAR100-20			CUB			SCars		
	All	Old	New	All	Old	New	All	Old	New
GCD	68.2	75.4	53.9	52.9	51.7	53.5	35.6	**46.5**	30.4
GCD w/ RCL	68.7	77.1	51.8	54.6	55.4	**54.2**	37.8	44.2	34.7
GCD w/ PACL	**70.8**	**78.0**	**56.5**	**55.2**	**57.6**	54.0	**38.3**	45.6	**34.8**
XCon	73.2	79.1	61.5	54.4	55.2	53.9	37.7	43.2	35.1
XCon w/ RCL	73.5	79.1	**62.4**	54.5	59.2	52.1	40.0	44.4	37.9
XCon w/ PACL	**74.7**	**81.2**	61.7	**56.3**	**59.8**	**54.6**	**40.8**	**45.1**	**38.7**

("All", "Old" and "New"). In addition, the experiments are conducted under both task-agnostic and task-aware settings. Overall, we can see that PACL improves the baseline methods by a promising margin over almost all of the metrics, except the "New" accuracy on the test set of CIFAR100-50. It implies the effectiveness of our feature rectification method. To further verify the advantage of our PACL, we also implement relaxed contrastive loss (RCL) to rectify the baselines. It can be noted that, RCL improves RankStats+ on all subsets of CIFAR100-20, while declining the performance of UNO-v2. Instead, our PACL achieves consistent improvements on the two baselines, which further proves the necessity and superiority of exploring more positive samples.

4.3 PACL on Generalized Category Discovery

For generalized category discovery (GCD), we choose two solid baselines: GCD [21] and XCon [4]. We report the results in Table 3, which shows the accuracy performance on "All", "Old" and "New" classes of unlabelled train set [4,21,24]. In particular for the "All" accuracy, the baseline with PACL significantly out-

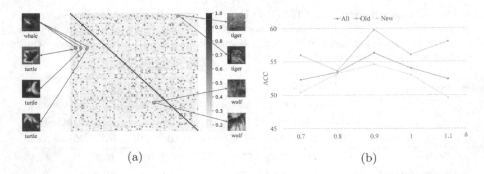

(a)   (b)

Fig. 3. (a) Visualization of the similarity matrix W and several positive-augmented samples in X^+. (b) Hyper-parameter tuning of δ for XCon w/ PACL on CIFAR100-20.

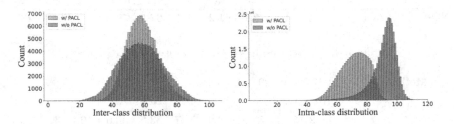

Fig. 4. Distribution of inter-class and intra-class feature distances according to XCon w/o and w/ PACL on CUB dataset.

performs the original baseline and its counterpart with RCL. Although several results achieved by RCL are higher than those by PACL, it should not negate the importance of positive-augmented contrastive learning. Last but not least, PACL witnesses more significant accuracy gains on fine-grained datasets (*i.e.*, CUB and SCars), where positive samples are easier to be incorrectly identified.

4.4 Additional Study

In addition to the compared results above, we conduct the following experiments to provide more insights on PACL:

1) Recall that the set of positive-augmented samples X^+ is constructed based on the similarity matrix W defined in Eq. (5). In Fig. 3(a), we illustrate the W matrix when performing XCon with PACL on CIFAR100-20. It can be seen that richer positive samples allow being used in our contrastive learning. In addition, we tune the hyper-parameter δ in Eq. (4), and the results in Fig. 3(b) show it should be set to 0.9 in GCD.
2) GCD on fine-grained datasets is challenged by small inter-class discrimination and large intra-class variation. We aim to compare the distributions of inter-class and intra-class feature distance between XCon without and with PACL on CUB dataset, as shown in Fig. 4. We can see that, using PACL in the model

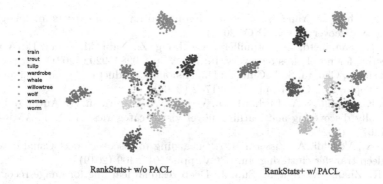

Fig. 5. Illustration of t-SNE feature embeddings according to RankStats+ w/o and w/ PACL on CIFAR100-20 setting for NCD task.

avoids the inter-class distribution being smooth, while benefits the intra-class distribution becomes more smooth.

3) Qualitatively, we visualize the t-SNE embeddings based on our rectified features. As shown in Fig. 5, the feature embeddings with PACL become more discriminative in the space. In particular, the "wolf" and "willow tree" classes are separated from "trout", "whale" and "worm".

5   Conclusion

In this paper, we have proposed positive-augmented contrastive learning that rectifies the feature representation already learned by NCD and GCD models. We mine positive and pseudo-positive samples with statistical information based on similarity relations between samples. Then, we theoretically formulate a PACL loss, which can augment the importance of positive samples in contrastive learning. Experimental results on generic and fine-grained classification datasets demonstrate the consistent improvements achieved by PACL. In future work, we are interested in extending PACL with noise-robust strategies to maintain more reliable positive samples, benefiting knowledge transfer in NCD and GCD.

Acknowledgements. This work was supported in part by the NSF of China under Grant Nos. 62102061 and 62272083, and in part by the Liaoning Provincial NSF under Grant 2022-MS-137 and 2022-MS-128, and in part by the Fundamental Research Funds for the Central Universities under Grant DUT21RC(3)024.

References

1. Cheng, B., et al.: Panoptic-deeplab: a simple, strong, and fast baseline for bottom-up panoptic segmentation. In: CVPR, pp. 12475–12485 (2020)
2. Deng, J., Dong, W., Socher, R., Li, L.J., Li, K., Fei-Fei, L.: Imagenet: a large-scale hierarchical image database. In: CVPR, pp. 248–255 (2009)
3. Dosovitskiy, A., et al.: An image is worth 16x16 words: transformers for image recognition at scale. In: ICLR (2021)

4. Fei, Y., Zhao, Z., Yang, S., Zhao, B.: Xcon: learning with experts for fine-grained category discovery. In: BMVC (2022)
5. Fini, E., Sangineto, E., Lathuilière, S., Zhong, Z., Nabi, M., Ricci, E.: A unified objective for novel class discovery. In: ICCV, pp. 9284–9292 (2021)
6. Hadsell, R., Chopra, S., LeCun, Y.: Dimensionality reduction by learning an invariant mapping. In: CVPR, pp. 1735–1742 (2006)
7. Han, K., Rebuffi, S.A., Ehrhardt, S., Vedaldi, A., Zisserman, A.: Autonovel: Automatically discovering and learning novel visual categories. IEEE TPAMI 44(10), 6767–6781 (2021)
8. Han, K., Vedaldi, A., Zisserman, A.: Learning to discover novel visual categories via deep transfer clustering. In: ICCV, pp. 8401–8409 (2019)
9. He, K., Zhang, X., Ren, S., Sun, J.: Deep residual learning for image recognition. In: CVPR, pp. 770–778 (2016)
10. Jia, X., Han, K., Zhu, Y., Green, B.: Joint representation learning and novel category discovery on single-and multi-modal data. In: ICCV, pp. 610–619 (2021)
11. Kim, S., Kim, D., Cho, M., Kwak, S.: Embedding transfer with label relaxation for improved metric learning. In: CVPR, pp. 3967–3976 (2021)
12. Krause, J., Stark, M., Deng, J., Fei-Fei, L.: 3D object representations for fine-grained categorization. In: ICCV Workshop, pp. 554–561 (2013)
13. Krizhevsky, A., Hinton, G., et al.: Learning multiple layers of features from tiny images (2009)
14. LI, Z., Otholt, J., Dai, B., Hu, D., Meinel, C., Yang, H.: A closer look at novel class discovery from the labeled set. In: NeurIPS (2022)
15. Oliver, A., Odena, A., Raffel, C.A., Cubuk, E.D., Goodfellow, I.: Realistic evaluation of deep semi-supervised learning algorithms. In: NeurIPS (2018)
16. Oord, A.v.d., Li, Y., Vinyals, O.: Representation learning with contrastive predictive coding. arXiv:1807.03748 (2018)
17. Pu, N., Zhong, Z., Sebe, N.: Dynamic conceptional contrastive learning for generalized category discovery. In: CVPR (2023)
18. Sutskever, I., Martens, J., Dahl, G., Hinton, G.: On the importance of initialization and momentum in deep learning. In: ICML, pp. 1139–1147 (2013)
19. Tan, M., Pang, R., Le, Q.V.: Efficientdet: scalable and efficient object detection. In: CVPR, pp. 10781–10790 (2020)
20. Tarvainen, A., Valpola, H.: Mean teachers are better role models: weight-averaged consistency targets improve semi-supervised deep learning results. In: NeurIPS (2017)
21. Vaze, S., Han, K., Vedaldi, A., Zisserman, A.: Generalized category discovery. In: CVPR, pp. 7492–7501 (2022)
22. Wah, C., Branson, S., Welinder, P., Perona, P., Belongie, S.: The caltech-ucsd birds-200-2011 dataset (2011)
23. Wang, H., Zhu, Y., Adam, H., Yuille, A., Chen, L.C.: Max-deeplab: end-to-end panoptic segmentation with mask transformers. In: CVPR, pp. 5463–5474 (2021)
24. Wen, X., Zhao, B., Qi, X.: Parametric classification for generalized category discovery: a baseline study. arXiv:2211.11727 (2022)
25. Yang, M., Zhu, Y., Yu, J., Wu, A., Deng, C.: Divide and conquer: compositional experts for generalized novel class discovery. In: CVPR, pp. 14268–14277 (2022)
26. Zhong, Z., Fini, E., Roy, S., Luo, Z., Ricci, E., Sebe, N.: Neighborhood contrastive learning for novel class discovery. In: CVPR, pp. 10867–10875 (2021)
27. Zhong, Z., Zhu, L., Luo, Z., Li, S., Yang, Y., Sebe, N.: Openmix: reviving known knowledge for discovering novel visual categories in an open world. In: CVPR, pp. 9462–9470 (2021)

CE²: A Copula Entropic Mutual Information Estimator for Enhancing Adversarial Robustness

Lin Liu, Cong Hu[✉], and Xiao-Jun Wu

School of Artificial Intelligence and Computer Science, Jiangnan University, Wuxi, China
conghu@jiangnan.edu.cn

Abstract. Deep neural networks are vulnerable to adversarial examples, which exploit imperceptible perturbations to mislead classifiers. To improve adversarial robustness, recent methods have focused on estimating mutual information (MI). However, existing MI estimators struggle to provide stable and reliable estimates in high-dimensional data. To this end, we propose a *Copula Entropic MI Estimator* (CE²) to address these limitations. CE² leverages copula entropy to estimating MI in high dimensions, allowing target models to harness information from both clean and adversarial examples to withstand attacks. Our empirical experiments demonstrate that CE² achieves a trade-off between variance and bias in MI estimates, resulting in stable and reliable estimates. Furthermore, the defense algorithm based on CE² significantly enhances adversarial robustness against multiple attacks. The experimental results underscore the effectiveness of CE² and its potential for improving adversarial robustness.

Keywords: Adversarial robustness · Copula entropy · Mutual information estimation

1 Introduction

Deep neural networks (DNNs) have propelled advancements in computer vision tasks [5,11,12,14,32], demonstrating remarkable performance in areas such as image classification [10,38], object detection [33,34,39] and face recognition [7, 8]. Albeit triumphing on predictive performance, recent literature has shown that DNNs are vulnerable to adversarial attacks [2,6,24,29]. In an adversarial attack, undetectable but targeted perturbations, which can drastically degrade the performance of a model, are added to input samples. Such attacks have imposed serious threats to the safety and robustness of technologies enabled by

This work was supported in part by the National Natural Science Foundation of China (Grant No. 62006097, U1836218), in part by the Natural Science Foundation of Jiangsu Province (Grant No. BK20200593) and in part by the China Postdoctoral Science Foundation (Grant No. 2021M701456).

DNNs. In order to improve the robustness of DNNs against adversarial attacks, researchers have focused on understanding the dependence between the target model's output and input adversarial examples. However, this particular aspect remains largely unexplored in the literature.

Recently, mutual information (MI) has gained renewed interest in the field of computer vision, inspired by the *Infomax principle* [15]. MI is an entropy-based measure that quantifies the dependence between two variables. In the context of deep learning, MI is used to assess the relationship between features and labels, with higher MI values indicating greater information about the labels with the features. While estimating MI in high-dimensional spaces is a notoriously difficult task, in practice researchers often rely on maximizing tractable lower bounds on MI [1,20,22,25]. Nonetheless, these distribution-free high-confidence lower bounds require exponentially increasing sample sizes [18,21,23], making them inadequate for providing low-variance, low-bias estimates [25,31]. These challenges become fatal in adversarial training, as uncertain and fluctuating MI estimates can hinder the target model's ability to capture meaningful and relevant information about the input-output relationship. Biased estimates can further lead to sub-optimal training and provide misleading guidance. In addition, directly leveraging MI between the input adversarial example and its corresponding output (called *standard MI*) has limitations in improving classification accuracy and adversarial robustness. Adversarial examples contain both natural and adversarial patterns, causing the standard MI to reflect a mixed dependence of the output on these patterns. Utilizing the standard MI of adversarial examples to guide the target model may inadvertently reinforce the dependence between the output and the adversarial pattern, resulting in misclassification from the correct label to an incorrect one [9].

To this end, we propose a *Copula Entropic mutual information Estimator* (CE^2) on the basis of the adversarial training manner. CE^2 is designed specifically to estimate high-dimensional MI and aims to enhance adversarial robustness. By leveraging copula entropy, it offers a method for explicitly estimating the MI between the input and corresponding output in two distinct patterns. This approach provides several advantages, including more stable and reliable estimates of the input-output relationship and an effective trade-off between variance and bias, which is crucial in the face of adversarial attacks. Empirical experiments show that CE^2 outperforms existing estimators in synthetic experiments and produce realistic estimates of MI in real-world datasets. The main contributions in this paper are as follows:

- We investigate copula entropy to fit the high-dimensional MI more precisely. Instead of directly or implicitly modeling the marginals of $P_{x,y}$, CE^2 decouples the computation of MI from marginal distributions, effectively reducing both variance and bias of estimates.
- To improve adversarial robustness, a defense algorithm based on CE^2 is proposed. It guides the target model to prioritize the natural pattern while reducing attention to the adversarial pattern, thereby improving the model's resilience against adversarial attacks.

- Empirical experiments, which encompass various adversarial attacks, demonstrate the effectiveness of the proposed defense algorithm in improving adversarial robustness.

2 Background

2.1 Copula Entropy and MI

Copula is the theory on representation of dependence relationships [26]. According to Sklar theorem [30], any probabilistic distribution can be represented as a copula function with marginal functions as its inputs. Thus, we can construct bivariate distributions with given uni-variate marginal F_1 and F_2 by using copula C_F [26], such that

$$F(x, y) = C_F (F_1(x), F_2(y)) \tag{1}$$

Based on this representation, Ma and Sun [16] proposed a mathematical concept for statistical independence measurement, named copula entropy, and then proved the equivalence between MI and copula entropy.

Definition 1. Let X be a two dimensional random variable with copula density $c(u, v)$. Copula entropy of X is defined as

$$H_c(X) = - \iint_{u,v} c(u, v) \ln c(u, v) du dv \tag{2}$$

Kullback Leibler information:

$$D_{KL}(F, F_1 F_2) = - \int_0^1 \int_0^1 c(u, v) \ln c(u, v) du dv \tag{3}$$

In this case, Kullback Leibler information is called mutual information.

Theorem 1. *Mutual information of the random variable is equal to the negative entropy of their corresponding copula function.*

$$D_{KL}(F, F_1 F_2) = -H_c(X) \tag{4}$$

Copula entropy enjoys several properties which an ideal statistical independence measure should have, such as multivariate, symmetric, non-negative (0 if independence), invariant to monotonic transformation, and equivalent to correlation coefficient in Gaussian cases.

2.2 Adversarial Attacks and Defenses

Adversarial noise can be crafted by optimization-based attacks, such as PGD [17], AA [3], CW [2] and DDN [27]. Besides, some attacks, such as FWA [36], focus on mimicking non-suspicious vandalism by exploiting the geometry and spatial information. These attacks constrain the perturbation boundary by a small norm-ball $\|\cdot\|_p \leq \epsilon$, so that their adversarial instances can be perceptually similar to natural instances.

The issue of adversarial attacks has spurred the development of adversarial defenses, with a major focus on enhancing adversarial robustness through adversarial training [4,17,37]. These methods augment training data with adversarial examples and employ a min-max formulation to train the target model. However, they do not explicitly measure the dependence between adversarial examples and corresponding outputs. Furthermore, certain data pre-processing based methods aim to remove adversarial noise by learning denoising functions or feature-squeezing functions [13,19]. Nonetheless, these methods may encounter challenges such as human-observable loss and residual adversarial noise, which can impact the final prediction.

To avoid the problems above, our proposed method utilizes natural MI and adversarial MI to train an adversarially robust classification model, providing an effective and comprehensive defense against adversarial attacks.

3   Methodology

3.1   Motivation

Estimating MI is core to many problems in machine learning, but at the same time, accurately bounding MI in high dimensions remains a challenging problem. Several recent methods, such as NWJ [20], DV [1], f-divergence [28] and MIAT [40], have attempted to compute gradients of lower bounds on MI with respect to the parameters θ of stochastic encoders $p_\theta(y \mid x)$. These methods aim to avoid direct estimation of MI by introducing flexible parametric distributions or critics parameterized by neural networks to approximate the unknown densities $(p(y), p(y \mid x))$ or density ratios involved $(p(x \mid y)/p(x) = p(y \mid x)/p(y))$.

However, these estimators have shown limitations in providing low-variance, low-bias estimates when dealing with high-dimensional data. Theoretical findings of McAllester & Stratos [18] and empirical experiments conducted by Ben Poole [25] have shed light on these limitations. Specifically, it has been demonstrated that any distribution-free high-confidence lower bound on MI estimated from N samples cannot exceed $O(\ln N)$. In the context of adversarial training, high variance leads to uncertainty and fluctuations in MI estimates, while high bias results in estimated MI deviating from the true MI. Consequently, unstable and unreliable estimates can lead to sub-optimal training and provide misleading information about the input-output relationship.

To address the trade-off between variance and bias in MI estimation, we propose a novel approach based on copula entropy, named CE^2 (Copula Entropic MI Estimator), which allows for more precise estimation of high-dimensional MI. Unlike existing methods that rely on modeling the marginals of the joint distribution, CE^2 decouples the computation of MI from marginal distributions. This decoupling is motivated by the observation that the properties of the marginals are irrelevant to accurately estimate MI. By separating the relevant entropy from the irrelevant entropy, CE^2 effectively reduces the number of implicit quantities that contribute to the final MI estimates. This reduction mitigates the sensitivity of the estimator to choices, such as encoder parameters or architectural configurations. As a result, CE^2 provides more reliable and stable estimates of MI, which in turn enhances the adversarial robustness of target models.

3.2   Copula Entropic MI Estimator

The copula entropy can be expressed as an expectation over the copula density c

$$h(c) = -\mathbb{E}_c[\log_2 c(U)] \tag{5}$$

where $U = (U_1, ..., U_d)$ denotes a random vector from the copula space.

This expectation can then be approximated by the empirical average over a large number of d-dimensional samples $u_j = ((u_j)_1, ..., (u_j)_d)$ from the random vector U:

$$-\mathbb{E}_c[\log_2 c(U)] \approx \widehat{h_k} := -\frac{1}{k} \sum_{j=1}^{k} \log_2(c(u_j)) \tag{6}$$

By the strong law of large numbers, $\widehat{h_k}$ converges almost surely to $h(c)$.

Hence, we can describe the process of our copula entropic MI estimator (CE^2) as follows.

- Sort the N samples of two random variables X and Y, respectively, as $x_{1:N}$, $y_{1:N}$;
- Approximate the empirical distribution function of the original distribution by probability integration, i.e. $\hat{F}(x) = 1/N \sum_{i=1}^{N} I_{[x_i \leq x]}$;
- Estimate the copula function between X and Y: $C(u, v \mid \theta)$;
- Calculate the mutual information:

$$I(X, Y) = \int_0^1 \int_0^1 \log C(u, v)/(u \cdot v) dcdf(X) dcdf(Y)$$

- Use parameter estimation method: maximum likelihood estimation or canonical estimation.

Theoretically, CE^2 introduced above offers the distinct advantage of being independent from the marginal distributions, which in turn enhances its robustness to potential irregularities that may exist within those marginals. This key feature sets it apart from density-dependent methods, like MINE [1] and NWJ [20], which struggle with marginal irregularities.

Note that adversarial examples contain natural and adversarial patterns [40]. The MI between the output and the natural pattern is called natural MI, and the MI between the output and the adversarial pattern is called adversarial MI. Let E_{ϕ_n} denote the CE^2 network for the natural MI and E_{ϕ_a} denote the CE^2 network for the adversarial MI. During training, The optimization mechanism for them is, minimizing the natural MI of adversarial examples and the adversarial MI of natural samples. In addition, to estimate MI more accurately, we select samples that correctly predicted by the target model and the corresponding

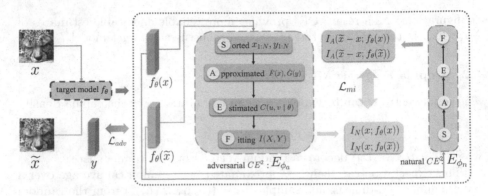

Fig. 1. The overview of the defense algorithm based on CE^2.

adversarial samples are wrongly predicted, to train the estimation networks. The optimization goals are as follows:

$$\widehat{\phi}_n = \underset{\phi_n \in \Phi_N}{\arg\max}[E_{\phi_n}(f_{\theta_0}(X')) - E_{\phi_n}(f_{\theta_0}(\widetilde{X}'))],$$

$$\widehat{\phi}_a = \underset{\phi_a \in \Phi_A}{\arg\max}[E_{\phi_a}(f_{\theta_0}(X')) - E_{\phi_a}(f_{\theta_0}(\widetilde{X}'))] \tag{7}$$

where Φ_N and Φ_A denote the sets of model parameters, f_{θ_0} denotes the pre-trained target model. X' is the selected data:

$$X' = \arg_X[\delta(f_{\theta_0}(X)) = Y, \delta(f_{\theta_0}(\widetilde{X})) \neq Y] \tag{8}$$

where δ is the operation that transforms the logit output into the prediction label.

Based on the CE^2 networks above, we develop an adversarial defense algorithm to enhance adversarial robustness of the target model. We aim to guide the target model towards emphasizing the natural pattern in the input while reducing its focus on adversarial pattern. We provide the overview of the adversarial defense method in Fig. 1.

During the training process, the defense algorithm based on CE^2 adopts an optimization strategy that simultaneously maximizes the natural MI of input adversarial examples and minimizes their adversarial MI. The optimization goal for the target model is as follows:

$$\hat{\theta} = \underset{\arg\max}{\theta \in \Theta}[E_{\hat{\phi}_n}(f_\theta(\widetilde{X})) - E_{\hat{\phi}_a}(f_\theta(\widetilde{X}))] \tag{9}$$

Thus the loss function is formulated as:

$$\begin{aligned}
\mathcal{L}_{mi}(\theta) = \frac{1}{m}\sum_{i=1}^{m}\{&\mathcal{L}_{cos}(E_{\hat{\phi}_n}(f_\theta(\tilde{x}'_i)), E_{\hat{\phi}_n}(f_\theta(x'_i))) \\
&+ \mathcal{L}_{cos}(E_{\hat{\phi}_a}(f_\theta(\tilde{x}'_i)), E_{\hat{\phi}_a}(f_\theta(x'_i))) \\
&+ [E_{\hat{\phi}_a}(f_\theta(\tilde{x}'_i)) - E_{\hat{\phi}_n}(f_\theta(\tilde{x}'_i))]\}
\end{aligned} \tag{10}$$

Algorithm 1. CE^2 defense algorithm

Input: Target model $f_\theta(\cdot)$ parameterized by θ, natural CE^2 $E_{\hat{\phi}_n}$, adversarial CE^2 $E_{\hat{\phi}_a}$, batch size n, and the perturbation budget ϵ.

Output: optimized θ

1: **repeat**
2:     Read mini-batch $\mathcal{B} = \{x_i\}_{i=1}^n$ from training set;
3:     **for** $i = 1$ to n **do**
4:         Use PGD-10 to generate adversarial examples $\tilde{x}_i$ at the given perturbation budget ϵ for x_i;
5:         Forward-pass x_i and $\tilde{x}_i$ through $f_\theta(\cdot)$ and obtain $f_\theta(x_i)$ and $f_\theta(\tilde{x}_i)$;
6:         Select adversarial examples according to Equation (8);
7:     **end for**
8:     Calculate $\mathcal{L}_{all}$ using Equation (11) and optimize θ;
9: **until** training converged

---

where m is the number of the selected data X' and $\mathcal{L}_{cos}$ is the cosine similarity-based loss function, i.e., $\mathcal{L}_{cos} = \|1 - sim(a,b)\|_1$, $sim(\cdot, \cdot)$ denotes the cosine similarity measure.

The optimization strategy above can be exploited together with the adversarial training manner $\mathcal{L}_{adv}(\theta)$, which we use the cross-entropy loss between the adversarial outputs and the ground-truth labels: $\mathcal{L}_{adv}(\theta) = -\frac{1}{n}\sum_{i=1}^n [y_i \cdot \log_2(\sigma(h_\theta(\tilde{x}_i)))]$. n is the number of training samples and σ denotes the *softmax* function. The overall loss function for training the target model is as follows (α is a trade-off hyperparameter):

$$\mathcal{L}_{all}(\theta) = \mathcal{L}_{adv}(\theta) + \alpha \cdot \mathcal{L}_{mi}(\theta) \tag{11}$$

We perform adversarial training through two main procedures: generating adversarial examples and optimizing the target model parameters. The details of the overall procedure are presented in Algorithm 1.

4   Experiments

4.1   Experiment Setups

Datasets. The effectiveness of out defense algorithm is verified on two popular benchmark datasets, i.e., *CIFAR-10* and *CIFAR-100*. *CIFAR-10* has 10 classes of images including 50,000 training images and 10,000 test images. *CIFAR-100* has 100 classes of images including 50,000 training images and 10,000 test images.

Model Architectures. We use a ResNet-18 as the target model for all the datasets. For the MI estimation network, we utilize the same neural network as MIAT [40]. The estimation networks for the natural MI and the adversarial MI have same model architectures.

Attack Settings. Adversarial data for evaluating defense models is crafted by applying state-of-the-art attacks. These attacks are divided into two categories:

L_∞-norm attacks and L_2-norm attacks. The L_∞-norm attacks include PGD [17], AA [3] and FWA [36]. The L_2-norm attacks include PGD, CW [2] and DDN [27]. Among them, the AA attack algorithm integrates three non-target attacks and a target attack. Other attack algorithms are utilized as non-target attacks. The iteration number of PGD and FWA is set to 40 with step size $\epsilon/4$. The iteration number of CW and DDN is set to 20 respectively with step size 0.01. For *CIFAR-10* and *CIFAR-100*, the perturbation budgets for L_∞-norm attacks and L_2-norm attacks are $8/255$ and $\epsilon = 0.5$ respectively.

Defense Settings. The adversarial training data for L_∞-norm and L_∞-norm is generated by using L_∞-norm PGD-10 and L_2-norm PGD-10 respectively. The step size is $\epsilon/4$ and the perturbation budget is $8/255$ and 0.5 respectively. The epoch number is set to 100. For fair comparisons, all the methods are trained using SGD with momentum 0.9, weight decay 2×10^{-4}, batch-size 128 and an initial learning rate of 0.1, which is divided by 10 at the 75-th and 90-th epoch. In addition, we set $\alpha = 5$ for our default algorithm.

4.2 Effectiveness of MI Estimator

To gain a better understanding of CE^2, we compare it with existing estimators in a standard synthetic setting based on correlated Gaussians. Following Poole *et al.* [25], $X, Y \in \mathbb{R}^d$ are random variables, where (X_i, Y_i) are standard normal with correlation $\rho \in [-1, 1]$. It can be checked that $I(X, Y) = -(d/2) \ln 1 - \rho^2$. The following lower-bound estimators are compared: MINE [1], NWJ [20], and CPC [22]. We refer the reader to Poole *et al.* for a detailed exposition of these estimators. The results are presented in Fig. 2.

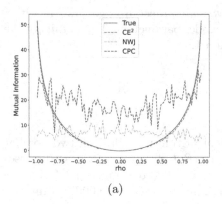
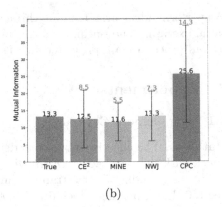

(a)   (b)

Fig. 2. The performances of estimating MI in different setting, while (a) indicates how well CE^2 fits the true MI when d is 128, ρ is from -1.0 to 1.0, (b) compares the estimating deviations of four estimators from the true MI, when ρ is 0.8, d is from 2.0 to 50.0.

Table 1. MI values estimated by four estimators, when the true MI is known, the dimension is 4, 12, 23, 34 and 45.

dim	True	CE^2	MINE	NWJ	CPC
4	2.04	3.02	4.17	**2.81**	5.77
12	6.13	**6.95**	4.47	5.09	9.51
23	11.75	10.94	10.11	**11.77**	18.52
34	17.37	**17.49**	18.75	22.38	38.33
45	22.99	**24.26**	16.02	19.46	31.25

Based on the experimental results in Table 1, CE^2 can estimate and fit the true MI in high dimensions precisely. The performances of four estimators can be contrasted by calculating the standard deviations as a measure of variance and the means as a measure of bias. A smaller standard deviation indicates greater precision and stability in the estimation method, while a larger standard deviation suggests higher estimation errors and variability. Therefore, based on the standard deviation, CE^2 performs better with smaller estimation error and variance. A smaller mean indicates lower bias, meaning the estimation is closer to the true value on average. From the perspective of the mean, CE^2 exhibits the lowest bias and is closest to the true value. Considering precision, variance and bias, CE^2 performs well with smaller variance and bias compared to the other methods.

4.3   Robustness Evaluation and Analysis

To demonstrate the effectiveness of our adversarial defense algorithm, we evaluate the adversarial accuracy using white-box and black-box adversarial attacks, respectively, and compare the results with three representative adversarial training methods, i.e., Standard AT [17], MART [35] and MIAT [40].

White-Box Attacks. In the white-box settings, all attacks can access the architectures and parameters of target models. We evaluate the robustness by exploiting five types of adversarial attacks for *CIFAR-10* and *CIFAR-100*: L_∞-norm PGD, AA, FWA attacks and L_2-norm PGD, CW, DDN attacks. The average natural accuracy and the average adversarial accuracy of defenses are shown in Table 2, and the most successful defense is shown with **bold.**

The results show that our method (CE^2) can achieve better robustness compared with MIAT. The performance of CE^2 on the natural accuracy is competitive (83.86% vs. 83.41% in *CIFAR-10*, 62.35% vs. 61.43% in *CIFAR-100*), and it provides more gains on adversarial accuracy (e.g. 6.60% against CW in *CIFAR-10*, 5.14% against DDN in *CIFAR-100*). Compared with MART and standard AT, the results show that our proposed MI estimator can estimating and maximizing mutual information more precisely and help improve adversarial robustness.

Table 2. Adversarial accuracy (%) of defense methods against white-box attacks.

dataset	Method	L_∞-norm Attack				L_2-norm Attack			
		None	PGD-40	AA	FWA-40	None	PGD-40	CW	DDN
CIFAR-10	Standard AT	83.39	42.38	39.01	15.44	83.97	61.69	30.96	29.34
	MART	78.21	**50.23**	43.96	**25.56**	81.29	58.36	28.41	27.13
	MIAT	83.41	44.79	39.26	15.67	84.35	62.38	34.48	32.41
	CE^2	**83.86**	49.99	**46.68**	15.83	**88.04**	**67.35**	**41.08**	**36.45**
CIFAR-100	Standard AT	**58.03**	22.20	20.62	10.20	62.29	36.23	18.52	16.68
	MART	53.40	**31.34**	22.03	12.27	**62.78**	**43.34**	15.02	**22.16**
	MIAT	56.01	26.24	21.71	11.70	61.43	38.29	14.29	12.57
	CE^2	56.82	31.20	**22.09**	**12.67**	62.35	36.23	**20.24**	17.71

Black-Box Attacks. Block-box adversarial examples are crafted by attacking a surrogate model. We use a VGG-19 as the surrogate model. The surrogate models and defense models are trained separately. We use Standard AT method to train the surrogate model and use several attacks to generate adversarial test data in *CIFAR-100*. The performance of our defense method is reported in Table 3. Among the defense methods, CE^2 demonstrates competitive performance against Standard AT, MART, and MIAT. Specifically, CE^2 achieves 2.53% adversarial accuracy higher than MART against FWA attack and a 1.34% adversarial accuracy higher than MIAT against PGD-L_2.

Table 3. Adversarial accuracy (%) of defense methods against black-box attacks on *CIFAR-100*.

Method	L_∞-norm Attack				L_2-norm Attack			
	None	PGD-40	AA	FWA-40	None	PGD-40	CW	DDN
Standard AT	**58.03**	**43.56**	50.11	32.01	62.29	53.04	56.64	55.23
MART	53.40	40.10	46.67	30.12	**62.78**	53.47	56.14	**57.08**
MIAT	56.01	41.98	50.18	31.53	61.43	52.21	55.18	55.08
CE^2	56.82	42.16	**50.28**	**32.65**	62.75	**53.55**	**56.78**	55.91

5 Conclusion

In this paper, we propose CE^2, a copula entropic MI estimator that effectively balances variance and bias of estimates, even in high-dimensional data. By capturing stable and reliable MI from both clean and adversarial examples, our defense method based on CE^2 enhances the target model's robustness against adversarial attacks. The empirical results demonstrate the effectiveness of our approach in enhancing adversarial robustness against multiple attacks. In future, we will focus on developing more efficient methods for estimating MI and optimize CE^2 further to enhancing its performance against stronger attacks.

References

1. Belghazi, M.I., et al.: Mutual information neural estimation. In: ICML. Proceedings of Machine Learning Research, vol. 80, pp. 530–539. PMLR (2018)
2. Carlini, N., Wagner, D.A.: Towards evaluating the robustness of neural networks. In: IEEE Symposium on Security and Privacy, pp. 39–57. IEEE Computer Society (2017)
3. Croce, F., Hein, M.: Reliable evaluation of adversarial robustness with an ensemble of diverse parameter-free attacks. In: ICML. Proceedings of Machine Learning Research, vol. 119, pp. 2206–2216. PMLR (2020)
4. Ding, G.W., Lui, K.Y.C., Jin, X., Wang, L., Huang, R.: On the sensitivity of adversarial robustness to input data distributions. In: ICLR (Poster). OpenReview.net (2019)
5. Duan, Y., Lu, J., Zheng, W., Zhou, J.: Deep adversarial metric learning. IEEE Trans. Image Process. **29**, 2037–2051 (2020)
6. Goodfellow, I.J., Shlens, J., Szegedy, C.: Explaining and harnessing adversarial examples. In: ICLR (Poster) (2015)
7. Hu, C., Li, Y., Feng, Z., Wu, X.: Attention-guided evolutionary attack with elastic-net regularization on face recognition. Pattern Recogn. 109760 (2023)
8. Hu, C., Xu, H.Q., Wu, X.J.: Substitute meta-learning for black-box adversarial attack. IEEE Sig. Process. Lett. **29**, 2472–2476 (2022). https://doi.org/10.1109/LSP.2022.3226118
9. Ilyas, A., Santurkar, S., Tsipras, D., Engstrom, L., Tran, B., Madry, A.: Adversarial examples are not bugs, they are features. In: NeurIPS, pp. 125–136 (2019)
10. Krizhevsky, A., Sutskever, I., Hinton, G.E.: Imagenet classification with deep convolutional neural networks. In: NIPS, pp. 1106–1114 (2012)
11. Li, H., Wu, X., Kittler, J.: MDLatLRR: a novel decomposition method for infrared and visible image fusion. IEEE Trans. Image Process. **29**, 4733–4746 (2020)
12. Li, X., Wang, W., Hu, X., Yang, J.: Selective kernel networks. In: CVPR, pp. 510–519. Computer Vision Foundation/IEEE (2019)
13. Liao, F., Liang, M., Dong, Y., Pang, T., Hu, X., Zhu, J.: Defense against adversarial attacks using high-level representation guided denoiser. In: CVPR, pp. 1778–1787. Computer Vision Foundation/IEEE Computer Society (2018)
14. Lin, S., et al.: Towards optimal structured CNN pruning via generative adversarial learning. In: CVPR, pp. 2790–2799. Computer Vision Foundation/IEEE (2019)
15. Linsker, R.: Self-organization in a perceptual network. Computer **21**(3), 105–117 (1988)
16. Ma, J., Sun, Z.: Mutual information is copula entropy. CoRR abs/0808.0845 (2008)
17. Madry, A., Makelov, A., Schmidt, L., Tsipras, D., Vladu, A.: Towards deep learning models resistant to adversarial attacks. In: ICLR (Poster). OpenReview.net (2018)
18. McAllester, D., Stratos, K.: Formal limitations on the measurement of mutual information. In: AISTATS. Proceedings of Machine Learning Research, vol. 108, pp. 875–884. PMLR (2020)
19. Naseer, M., Khan, S.H., Hayat, M., Khan, F.S., Porikli, F.: A self-supervised approach for adversarial robustness. In: CVPR, pp. 259–268. Computer Vision Foundation/IEEE (2020)
20. Nguyen, X., Wainwright, M.J., Jordan, M.I.: Estimating divergence functionals and the likelihood ratio by convex risk minimization. IEEE Trans. Inf. Theory **56**(11), 5847–5861 (2010)

21. Noshad, M., Zeng, Y., III, A.O.H.: Scalable mutual information estimation using dependence graphs. In: ICASSP, pp. 2962–2966. IEEE (2019)
22. van den Oord, A., Li, Y., Vinyals, O.: Representation learning with contrastive predictive coding. CoRR abs/1807.03748 (2018)
23. Paninski, L.: Estimation of entropy and mutual information. Neural Comput. **15**(6), 1191–1253 (2003)
24. Papernot, N., McDaniel, P.D., Jha, S., Fredrikson, M., Celik, Z.B., Swami, A.: The limitations of deep learning in adversarial settings. In: EuroS&P, pp. 372–387. IEEE (2016)
25. Poole, B., Ozair, S., van den Oord, A., Alemi, A.A., Tucker, G.: On variational bounds of mutual information. In: ICML. Proceedings of Machine Learning Research, vol. 97, pp. 5171–5180. PMLR (2019)
26. Ravens, B.: An introduction to copulas. Technometrics **42**(3), 317 (2000)
27. Rony, J., Hafemann, L.G., Oliveira, L.S., Ayed, I.B., Sabourin, R., Granger, E.: Decoupling direction and norm for efficient gradient-based L2 adversarial attacks and defenses. In: CVPR, pp. 4322–4330. Computer Vision Foundation/IEEE (2019)
28. Rubenstein, P.K., Bousquet, O., Djolonga, J., Riquelme, C., Tolstikhin, I.O.: Practical and consistent estimation of f-divergences. In: NeurIPS, pp. 4072–4082 (2019)
29. Shi, Y., Liao, B., Chen, G., Liu, Y., Cheng, M., Feng, J.: Understanding adversarial behavior of DNNs by disentangling non-robust and robust components in performance metric. CoRR abs/1906.02494 (2019)
30. Sklar, M.J.: Fonctions de repartition a n dimensions et Leurs Marges (1959)
31. Song, J., Ermon, S.: Understanding the limitations of variational mutual information estimators. In: ICLR. OpenReview.net (2020)
32. Tian, C., Xu, Y., Li, Z., Zuo, W., Fei, L., Liu, H.: Attention-guided CNN for image denoising. Neural Netw. **124**, 117–129 (2020)
33. Tong, J., Chen, T., Wang, Q., Yao, Y.: Few-shot object detection via understanding convolution and attention. In: Yu, S., et al. (eds.) PRCV 2022. LNCS, vol. 13534, pp. 674–687. Springer, Cham (2022). https://doi.org/10.1007/978-3-031-18907-4_52
34. Wang, M., Deng, W.: Deep face recognition: a survey. Neurocomputing **429**, 215–244 (2021)
35. Wang, Y., Zou, D., Yi, J., Bailey, J., Ma, X., Gu, Q.: Improving adversarial robustness requires revisiting misclassified examples. In: ICLR. OpenReview.net (2020)
36. Wu, K., Wang, A.H., Yu, Y.: Stronger and faster Wasserstein adversarial attacks. In: ICML. Proceedings of Machine Learning Research, vol. 119, pp. 10377–10387. PMLR (2020)
37. Zhang, H., Yu, Y., Jiao, J., Xing, E.P., Ghaoui, L.E., Jordan, M.I.: Theoretically principled trade-off between robustness and accuracy. In: ICML. Proceedings of Machine Learning Research, vol. 97, pp. 7472–7482. PMLR (2019)
38. Zhang, W., Gou, Y., Jiang, Y., Zhang, Y.: Adversarial VAE with normalizing flows for multi-dimensional classification. In: Yu, S., et al. (eds.) PRCV 2022. LNCS, vol. 13534, pp. 205–219. Springer, Cham (2022). https://doi.org/10.1007/978-3-031-18907-4_16
39. Zhao, Z., Zheng, P., Xu, S., Wu, X.: Object detection with deep learning: a review. IEEE Trans. Neural Netw. Learn. Syst. **30**(11), 3212–3232 (2019)
40. Zhou, D., et al.: Improving adversarial robustness via mutual information estimation. In: ICML. Proceedings of Machine Learning Research, vol. 162, pp. 27338–27352. PMLR (2022)

Two-Step Projection of Sparse Discrimination Between Classes for Unsupervised Domain Adaptation

Jianhong Xie and Lu Liang(✉)

Guangdong University of Technology, Guangzhou, China
`2112105034@mail2.gdut.edu.cn`

Abstract. In the past few years, researchers have developed domain adaptive (DA) techniques which aim to address the domain shift between training and testing sets. However, most existing unsupervised domain adaptation techniques only use a projection matrix to train the classifier, which do not impose any restrictions on the same class samples. In this paper, we introduce a novel approach to unsupervised domain adaptation, called two-step projection of sparse discrimination between classes for unsupervised domain adaptation (TSPSDC). Unlike existing methods that use a single matrix for classifier learning, TSPSDC leverages two-step projection learning and integrates inter-class sparsity constraints to extract domain-invariant features from the class level. Specifically, 1) aiming to mitigate any adverse consequences arising from domain shift, distribution alignments are implemented to decrease the distribution disparity between the source and target domains in the shared subspace. 2) Our approach involves integrating two types of regularization: manifold regularization and inter-class sparsity regularization. The feature representation exhibits the same row sparsity structure within each class. Simultaneously, the distance between similar classes is minimized as a result. The resulting feature representation is thus advantageous for achieving highly discriminative representations. Extensive experimental results on three different sets of data demonstrate that the proposed approach outperforms many contemporary techniques for domain adaptation.

Keywords: Domain adaptation · Domain shift · Distribution alignment

1 Introduction

Traditional statistical machine learning algorithms are based on the assumption that sample data are independently and identically distributed [1]. However, there are not many scenarios that satisfy the independent identical distribution in nature, but more scenarios with different distributions. Traditional machine learning methods cannot be applied well in scenarios with different distributions. In the field of computer vision, this kind of problem is called domain shift. Figure 1 shows some images with the same semantic information but different distributions. It can be seen from Fig. 1 that even though the images belong to the same class, they have different distributions. In this

© The Author(s), under exclusive license to Springer Nature Singapore Pte Ltd. 2024
Q. Liu et al. (Eds.): PRCV 2023, LNCS 14428, pp. 175–186, 2024.
https://doi.org/10.1007/978-981-99-8462-6_15

case, if these images are used to train a classifier, it is conceivable that the effectiveness of the trained classifier demonstrated on other images will be greatly reduced. To solve such problems, an effective way is usually to extract enough label information from the specified domain and then use this label information to train the classifier [2–4]. However, it is known that obtaining labeled samples is very time-consuming and labor-intensive. An effective model can be trained alternatively from a source domain with the same task as the target domain. There should be enough labeling information to guide the generation of classifiers in the target domain. The challenge of this approach is to overcome the domain bias problem to achieve cross-domain knowledge transfer.

Fig. 1. Several images with the same semantic information have different domains.

In order to solve the above problem, numerous unsupervised domain adaptation (UDA) algorithms have been proposed in the past decades. Generally speaking, UDA algorithms can be categorized into two types, instance-based adaptation and feature-based adaptation [5, 6]. In order to minimize the distribution dissimilarity between the source and target domains, instance-based adaptation algorithms typically adjust the sample weights in the source domain. Feature-based adaptation algorithms are more flexible. They strive to find a general representation that renders the two domains more similar. This facilitates classification via conventional machine learning algorithms.

Despite the proven effectiveness of current UDA algorithms, there remain three issues with the existing methods. (1) Most existing methods assume that there exists a common subspace in which the source and target domains can have the same data distribution. To learn the classifier, they often use only one projection matrix to transform the data. In fact, it is not reasonable to use only one projection matrix. When data is acquired through a single projection matrix transformation, the resulting data appears quite coarse. A simple one-step projection will bring a lot of information compression, which cannot fully utilize the labeled data in the source domain. This greatly reduces the performance of the classifier. (2) Many existing approaches focus on domain alignment solely from either the statistical or geometrical aspect. They fail to incorporate both aspects together to obtain a better alignment of the domain distribution. Statistical attributes and geometric arrangements complement each other in reality. Utilizing them simultaneously produces superior results compared to using them independently. (3)

Prior methods had primarily concentrated on acquiring local discriminative information to learn discriminative features. They disregarded global discriminative information at the class-level. The occurrence known as class confusion serves as a reminder to take into account global discriminative information.

To overcome these issues, in this paper, we propose a novel method, called Two-step projection of sparse discrimination between classes (TSPSDC) for unsupervised domain adaptation, which can effectively solve the above issues, as illustrated in Fig. 2. Extensive experiments were conducted on cross-domain object to demonstrate the effectiveness of the proposed TSPSDC.

The main contributions of this paper can be summarized as follows:

(1) To address the problem of limited performance of classifiers trained by one projection matrix, we propose a two-step projection method to make the acquisition of a common subspace more reasonable and robust.
(2) By considering both statistical and geometric properties in the domain distribution, we align for the conditional and marginal distributions and use manifold regularization, which allows for a more efficient use of the geometric structure of the domain distribution.
(3) To attain the discriminative information of the class-level, a regularization technique referred to as inter-class sparsity is incorporated in the training process. This regularization enforces each class to have an identical row sparsity structure in the learned feature representation.

The rest of the paper is organized as follows. Section 2 provides an overview of relevant research. Our framework is outlined in Sect. 3. The comparison results obtained from actual data sets are presented in Sect. 4. Ultimately, pertinent conclusions are drawn in Sect. 5.

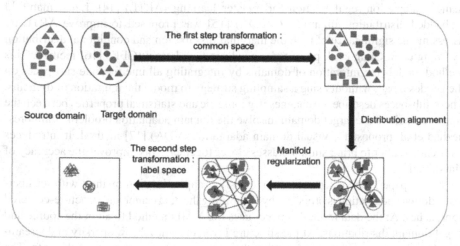

Fig. 2. Flow chart of our method. First, a two-step projection transformation matrix is used and after the learned transformation matrix, distribution alignment and manifold regularization are performed on the common subspace. Finally, row sparsity is achieved on the label space.

2 Related Work

In this section, we discuss prior research in domain adaptation and relevant methods for this paper.

2.1 Distribution Matching

MMD (Maximum Mean Discrepancy) is commonly employed to quantify distribution dissimilarity. Long et al. introduced JDA (Joint Distribution Adaptation) [13], a method that addresses both conditional and marginal distribution. JDA focuses on discovering a shared feature space while neglecting diverse conditional relationships. Additionally, Long et al. introduced a TJM [8] approach to tackle distribution mismatch through sample selection. JGSA (Joint Geometrical and Statistical Alignment) [7] is proposed to enhance within-class compactness and between-class separability. Wang et al. extend JDA with the introduction of BDA (Balanced Distribution Adaptation), which bridges the gap between conditional and marginal distribution. The DICD (Domain Invariant and Class Discriminative) [9] method extracts discriminative information at the class level from both the source and target domains. Although DICD achieves satisfactory performance, it lacks utilization of structural information. Zhang and Wu introduced JPDA [10], a variant method that aims to minimize joint probability differences.

2.2 Structure Maintenance

The authors of JDA also introduced manifold regularization. The purpose is to preserve the local geometric structure of the data to eliminate the differences between domain distributions. This, together with joint distribution adaptation, resulted in a new method named Adaptation regularization for transfer learning (ARTL) [14]. Later, manifold embedded distribution alignment (MEDA) [15] was proposed to improve ARTL by assessing the significance of both the marginal distribution and conditional distribution by Wang et al. Gong et al. proposed a geodesic flow kernel (GFK) [16] method. This method models the migration of domains by integrating all intermediate subspaces on the geodesic path without using a sampling strategy to model the migration of domains. These subspaces describe the changes in geometric and statistical properties between the source domain to the target domain to solve the domain adaptation problem. Tahmoresnezhad et al. proposed a Visual domain adaptation (VDA) [17] method. It introduces intra-class scatter to bring similar classes closer together. This improves the accuracy of classification.

In this paper, a two-step projection classifier approximation method with unsupervised domain adaptation is proposed by combining the data center and structure-centered approaches. At the data level, this paper uses the MMD method to align the source and target domain distributions. At the structure level, class-level row sparsity and streamwise learning are combined to maintain the global and local structure of the data. The common subspace can be learned by these means. In addition, to improve the discriminative power, a classifier containing two projection matrices is learned to fit the source domain data samples to their labels.

3 Two-Step Projection of Sparse Discrimination Between Classes (TSPSDC) Method

3.1 Notation

In this paper, we focus on unsupervised domain adaptation. The datum of the source domain and target domain are respectively denoted as (X_s, Y_s) and (X_t, Y_t), where $X_s \in R^{m \times n_s}$ and $X_t \in R^{m \times n_t}$ are the samples in the source domain and the target domain, respectively. $Y_s \in R^{C \times n_s}$ and $Y_t \in R^{C \times n_t}$ are the binary label matrix of the source domain and the target domain, respectively. Since we work on unsupervised domain adaptation, $Y_t \in R^{C \times n_s}$ is unavailable in this work. n_s and n_t represent the number of the source domain and target domain, respectively. The dimensionality of the original samples is denoted as m and the number of classes is denoted as C. The projection matrix from the original feature space to the common subspace is denoted as $A \in R^{m \times d}$, where d is the dimensionality of the common subspace and $X = [X_s, X_t]$ represents the combination of source domain and target domain samples. The transformation matrix is denoted as $W \in R^{C \times d}$. $\|A\|_p$ represents l_p-norm of matrix.

3.2 Model Formulation

The goal of domain adaptation is to learn a domain-invariant classifier that can be used to predict Y_t for the target domain using samples from both domains. To achieve this goal, we train a standard classifier in the source domain, which provides the basis for subsequent transitions to the target domain. In addition, we also improve the migration ability of the classifier from three aspects: 1) distribution alignment, 2) local discriminative information, 3) structural consistency of class level.

Structural Risk Minimization Principle. TSPSDC's initial goal is to acquire an adaptable classifier that is capable of accurately categorizing source samples. Initially, we establish a standard classifier using labeled source samples. By following the structural risk minimization principle, our objective is to lessen the source empirical error in the process. The formula is as follows:

$$\min_{A,W} \|WA^T X_s - Y_s\|_F^2 + \alpha \|A\|_F^2 \tag{1}$$

It is worth noting that we use a two-step projection method to narrow the source domain data to the label. By doing so, we can prevent data information loss and retain source domain information better. It is better than the structural risk minimization method that only uses one projection.

Distribution Alignment. Maximum Mean Difference (MMD) is a statistical test based on kernel function to determine whether two given distributions are the same [11]. The basic idea is: for two samples with different distributions, find a continuous function from the sample space to the regenerated core Hilbert space (RKHS), then calculate the mean of each sample on the function, and compare the distance between them. If the distributions are the same, then their mean over the function should be close, and vice versa. Thus, MMD can be used as a measure of the distance between two distributions.

For marginal distribution alignment, we project all data into Hilbert space and then align them closer.

$$\min_A \left\| \frac{1}{n_s} \sum_{i=1}^{n_s} A^T x_i - \frac{1}{n_t} \sum_{j=1}^{n_t} A^T x_j \right\|_F^2 = \min_A tr(A^T X M_0 X^T A) \qquad (2)$$

where

$$M_0 = \begin{bmatrix} \frac{1}{n_s^2} 1_{n_s \times n_s} & -\frac{1}{n_s n_t} 1_{n_s \times n_t} \\ -\frac{1}{n_s n_t} 1_{n_t \times n_s} & \frac{1}{n_t^2} 1_{n_t \times n_t} \end{bmatrix} \qquad (3)$$

For conditional distribution alignment, we project data by class into Hilbert space and then align them closer.

$$\min_A \sum_{k=1}^C \left\| \frac{1}{n_s} \sum_{i=1}^{n_s} A^T x_i - \frac{1}{n_t} \sum_{j=1}^{n_t} A^T x_j \right\|_F^2 = \min_A \sum_{k=1}^C tr(A^T X M_k X^T A) \quad (4)$$

where

$$(M_k)_{ij} = \begin{cases} \frac{1}{\left(n_s^{(k)}\right)^2}, & if\ x_i, x_j \in D_s^{(k)}; \\ -\frac{1}{n_s^{(k)} n_t^{(k)}}, & if\ x_i \in D_s^{(k)}, x_j \in D_t^{(k)}; \\ -\frac{1}{n_s^{(k)} n_t^{(k)}}, & if\ x_i \in D_t^{(k)}, x_j \in D_s^{(k)}; \\ \frac{1}{\left(n_t^{(k)}\right)^2}, & if\ x_i, x_j \in D_t^{(k)}; \\ 0, & otherwise. \end{cases} \qquad (5)$$

We can combine (2) and (4) as follow:

$$\min_A tr(A^T X M_0 X^T A) + \min_A \sum_{k=1}^C tr(A^T X M_k X^T A) = \min_A \sum_{k=0}^C tr(A^T X M_k X^T A) \qquad (6)$$

Local Discriminative Information. Discriminability is a key criterion that characterizes the superiority of feature representations to enable domain adaptations. Manifold regularization is a widely used method to extract local discriminative features. According to the manifold assumption, if two points are close in the intrinsic geometry, then the corresponding labels are similar [19]. Following this assumption, the manifold regularization is computed as

$$\sum_{i,j=1}^{n_s+n_t} W_{ij} \|A^T x_i - A^T x_j\|_2^2 = \min_A tr\left(A^T X L X^T A\right) \qquad (7)$$

where W_{ij} is the graph affinity matrix between x_i and x_j. $L = I - G^{-\frac{1}{2}} W G^{-\frac{1}{2}}$ is the graph Laplacian matrix, G is diagonal matrix with $G_{ii} = \sum_{j=1}^{n_s+n_t} W_{ij}$, W is defined as

$$W_{ij} = \begin{cases} \cos(x_i, x_j), & if\ x_i \in kNN(x_j)\ or\ x_j \in kNN(x_i); \\ 0, & otherwise. \end{cases} \qquad (8)$$

where $kNN(x_i)$ is the set of k-nearest neighbors of x_i.

Structural Consistency of Class Level. Studies suggest that ensuring consistency of structure can align the data from the source and target domains in the feature space or decision space, which can reduce the domain gap and enhance the classifier's performance in the target domain [20]. The expression is as follows:

$$\sum_{c=1}^{C} \| WA^T X_s^c \|_{2,1} \tag{9}$$

Introducing the above formula is able to make the transformed feature $WA^T X_s$ have a common sparsity structure in each class. Since the label Y_s possesses remarkable class sparsity traits that provide inherent distinctiveness among samples, we employ them for classification purposes.

3.3 Overall Objective and Optimization Algorithm

By combing (1), (6), (7) and (9), the final objective function of the proposed method is obtained as follows:

$$\min_{A,W} \| WA^T X_s - Y_s \|_F^2 + \alpha \|A\|_F^2 + \sum_{k=0}^{C} \beta tr(A^T XM_k X^T A) + \gamma tr\left(A^T XLX^T A\right)$$
$$+ \sum_{c=1}^{C} \| WA^T X_s^c \|_{2,1}$$
$$\text{s.t. } W^T W = I \tag{10}$$

The introduction of the constraint function in our model serves the purpose of preventing trivial solutions.

1) Update A:

To make the optimization problem separable, an extra variable Q is introduced:

$$\min_{A} \alpha \|A\|_F^2 + \beta tr\left(A^T XMX^T A\right) + \gamma tr\left(A^T XLX^T A\right)$$
$$\text{s.t. } Q = WA^T \tag{11}$$

For solving A, other variables are fixed and the following convex model is obtained:

$$\min_{A} \alpha \|A\|_F^2 + \beta tr\left(A^T XMX^T A\right) + \gamma tr\left(A^T XLX^T A\right) + \delta \|Q - WA^T\|_F^2 \tag{12}$$

Taking the derivation of (12) with respect to A and setting it to zero, we have

$$\left((\alpha + \delta)I + X(\beta M + \gamma L)X^T\right)A = \delta Q^T W \tag{13}$$

$$\Downarrow$$

$$A = \left((\alpha + \delta)I + X(\beta M + \gamma L)X^T\right)^{-1}\left(\delta Q^T W\right) \tag{14}$$

2) Update W:

Fixing variables Q and A, variable W can be obtained by minimizing the following formula:

$$\min_W \|Q - WA^T\|_F^2$$

$$\text{s.t. } W^TW = I \tag{15}$$

By setting the derivative of (15) to zero, the following Equation can be obtained:

$$\min_W \text{tr}\left(QAW^T\right) \tag{16}$$

We denote $B = QA$. Under the orthogonal constraint of W, we make a singular value decomposition of B. Then we get the solution for W:

$$W = UV^T \tag{17}$$

where U and V are the singular value for B.

3) Update Q:

By fixing variables A and W, Q can be obtained by minimizing the following objective:

$$\min_Q \|QX_s - Y_s\|_F^2 + \sum_{c=1}^{C} \|QX_s^c\|_{2,1}$$

$$\text{s.t. } Q = WA^T \tag{18}$$

Let $D^c = QX_s^c$, the problem is equivalent to

$$\min_Q \|QX_s - Y_s\|_F^2 + \sum_{c=1}^{C} \|D^c\|_{2,1}$$

$$\text{s.t. } Q = WA^T, D^c = QX_s^c \tag{19}$$

Then, the above equation becomes solving C subproblems on D^c:

$$\min_{D^c} \mu \|D^c\|_F^2 + \lambda \|D^c - QX_s^c\|_F^2 \tag{20}$$

For D^c, referring to the theorems and optimization in the literature [21–23], we can obtain the following equation:

$$[D^c]_{:,i} = \begin{cases} [QX_s^c]_{:,i} \frac{\|[QX_s^c]_{:,i}\|_2 - \frac{\mu}{\lambda}}{\|[QX_s^c]_{:,i}\|_2}, & \text{if } \|[QX_s^c]_{:,i}\|_2 > \frac{\mu}{\lambda} \\ 0, & \text{otherwise.} \end{cases} \tag{21}$$

For Q, we have:

$$\min_Q \|QX_s - Y_s\|_F^2 + \sum_{c=1}^{C} \|D^c - QX_s^c\|_F^2 + \|Q - WA^T\|_F^2 \tag{22}$$

At last, we get the formula of Q:

$$Q = \left(Y_s X_s^T + \sum_{c=1}^{C} D^c X_s^{cT} + WA^T \right) \left(X_s X_s^T + \sum_{c=1}^{C} X_s^c X_s^{cT} + I \right)^{-1} \quad (23)$$

The proposed algorithm is summarized in Algorithm 1.

Algorithm 1 Two-step projection of sparse discrimination between classes (TSPSDC)

Input: Data $X = [X_s, X_t]$, domain label Y_s, manifold subspace dimension $d = 20$, maximum iterations $T = 20$, regularization parameters α, β, γ, δ and neighbors k

Output: Q, W, A.

   While not converged **do**

      1: Update A by using (14);

      2: Update W by using (17);

      3: Update Q by using (23);

   End while

---

4 Experiments

4.1 The Cross-Domain Experiment

PIE Dataset, COIL Dataset and Office + Caltech10 (Surf). In the experiment of PIE dataset, the parameters α, β, γ and δ are set as 0.005, 0.05, 0.001, 0.01 respectively. The proposed method demonstrates greater accuracy than the comparison methods, as evidenced by the data. In the experiment of COIL dataset, the parameters α, β, γ and δ are set as 0.01, 0.01, 0.001 and 0.05, respectively. The comparison of methods shows that TSPSDC obtains the highest average accuracy, reaching 99.83%. In the experiment of Office + Caltech10 (surf), the parameters α, β, γ and δ are set as 0.01, 0.01, 0.001, 0.05 respectively. Figure 3 displays the outcomes of PIE, COIL and Office + Caltech10 (surf) dataset experiment. As attested by the data, the proposed approach showcases superior precision in comparison to the methods being compared.

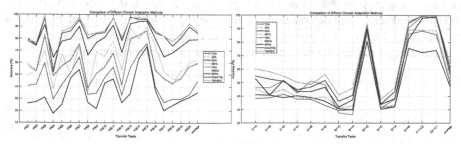

Fig. 3. The results of PIE dataset, COIL dataset and Office + Caltech10 (surf) dataset.

From the above experiments, we can draw the following conclusions. 1) Compared with TCA [12], JDA [13] and GFK [16], TSPSDC can make full use of geometric structure information using manifold regularization. 2) Compared with ARTL [14], VDA [17], MEDA [15] and SPDA [18], TSPSDC can effectively generate discriminative feature using the inter-class sparsity constrain. 3) Compared with ICS_RTSL [24], TSPSDC adopts two-step projection method which prevents the loss of data information, thus better retains the information contained in the data sample of the source domain.

4.2 Analysis of the Convergence

To demonstrate the convergence of the suggested approach, several cross-domain datasets were subjected to three experiments. The outcomes of these experiments are depicted in Fig. 4, from which we can arrive at the following conclusions: firstly, during the iterative process, all the curves exhibit an upward trend and eventually reach stability, providing evidence for the convergence of the proposed method. Secondly, after around ten iterations, nearly all the curves tend to stabilize. This denotes that the proposed method possesses a high convergence rate.

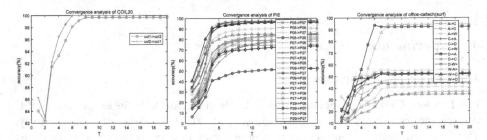

Fig. 4. The results of convergence experiments.

4.3 The Parameter Analysis

The proposed method consists of four parameters, namely α, β, γ and δ. To conduct parameter sensitivity experiments, this paper selected three cross-domain datasets, with the range of α, β, γ and δ set as {1e−3, 5e−3, 1e−2, 5e−2, 0.1, 0.5, 1, 5, 10, 100, 1000}. The experimental outcomes are depicted in Fig. 5 and Fig. 6. Figure 6 indicates that our method is robust to parameters γ and δ. However, for parameters α and β, the optimal values differ among different image databases as can be seen on Fig. 5.

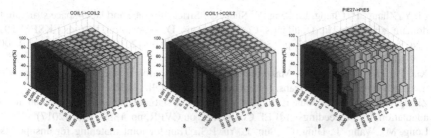

Fig. 5. The results of parameter sensitivity experiments.

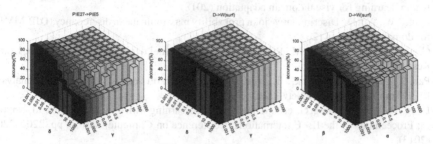

Fig. 6. The results of parameter sensitivity experiments.

5 Conclusion

In this paper, we have proposed a novel transfer learning framework, namely two-step projection of sparse discrimination between classes for unsupervised domain adaptation (TSPSDC). TSPSDC is trained to develop a domain-invariant classifier within a discriminative feature space. Specifically, TSPSDC utilizes MMD to reduce both marginal and conditional distributions, while simultaneously acquiring local and global discriminative characteristics through manifold regularization and inter-class sparsity constraints, respectively. In addition, TSPSDC uses a two-step projection technique to ensure that sample information is not lost during transformation. This technique helps to better explore the correct common subspace, thereby maximizing the prediction structure of the classifier. We conducted experiments comparing image data sets and found that our algorithm outperformed the current leading algorithm.

References

1. Lu, Y., Luo, X., Wen, J., et al.: Cross-domain structure learning for visual data recognition. Pattern Recogn.Recogn. **134**, 109127 (2023)
2. Huang, L.Q., Liu, Z.G., Dezert, J.: Cross-domain pattern classification with distribution adaptation based on evidence theory. IEEE Trans. Cybern. (2021)
3. Deng, W., Zheng, L., Sun, Y., et al.: Rethinking triplet loss for domain adaptation. IEEE Trans. Circuits Syst. Video Technol. **31**(1), 29–37 (2020)
4. Lee, J.H., Lee, G.: Unsupervised domain adaptation based on the predictive uncertainty of models. Neurocomputing **520**, 183–193 (2023)

5. Li, Y., Zhang, P., Cheng, L., Peng, Y., Shen, C.: Strict subspace and label-space structure for domain adaptation. In: Douligeris, C., Karagiannis, D., Apostolou, D. (eds.) KSEM 2019. LNCS (LNAI), vol. 11775, pp. 301–313. Springer, Cham (2019). https://doi.org/10.1007/978-3-030-29551-6_26
6. Xia, H., Jing, T., Ding, Z.: Maximum structural generation discrepancy for unsupervised domain adaptation. IEEE Trans. Pattern Anal. Mach. Intell. (2022)
7. Zhang, J., Li, W., Ogunbona, P.: Joint geometrical and statistical alignment for visual domain adaptation. In: Proceedings of IEEE Conference on CVPR, pp. 1859–1867 (2017)
8. Long, M., Wang, J., Ding, G., Sun, J., Yu, P.S.: Transfer joint matching for unsupervised domain adaptation. In: Proceedings of IEEE Conference on CVPR, pp. 1410–1417 (2014)
9. Li, S., Song, S., Huang, G., Ding, Z., Wu, C.: Domain invariant and class discriminative feature learning for visual domain adaptation (2018)
10. Zhang, W., Wu, D.: Discriminative joint probability maximum mean discrepancy (DJP-MMD) for domain adaptation (2020)
11. Gretton, A., Borgwardt, K.M., Rasch, M.J., et al.: A kernel two-sample test. J. Mach. Learn. Res. 13(1), 723–773 (2012)
12. Pan, S.J., Tsang, I.W., Kwok, J.T., et al.: Domain adaptation via transfer component analysis. IEEE Trans. Neural Netw.Netw. 22(2), 199–210 (2010)
13. Long, M., Wang, J., Ding, G., et al.: Transfer feature learning with joint distribution adaptation. In: Proceedings of the IEEE International Conference on Computer Vision, pp. 2200–2207 (2013)
14. Long, M., Wang, J., Ding, G., et al.: Adaptation regularization: a general framework for transfer learning. IEEE Trans. Knowl. Data Eng.Knowl. Data Eng. 26(5), 1076–1089 (2013)
15. Wang, J., Feng, W., Chen, Y., et al.: Visual domain adaptation with manifold embedded distribution alignment. In: Proceedings of the 26th ACM International Conference on Multimedia, pp. 402–410 (2018)
16. Gong, B., Shi, Y., Sha, F., et al.: Geodesic flow kernel for unsupervised domain adaptation. In: 2012 IEEE Conference on Computer Vision and Pattern Recognition, pp. 2066–2073. IEEE (2012)
17. Tahmoresnezhad, J., Hashemi, S.: Visual domain adaptation via transfer feature learning. Knowl. Inf. Syst.. Inf. Syst. 50, 585–605 (2017)
18. Xiao, T., Liu, P., Zhao, W., et al.: Structure preservation and distribution alignment in discriminative transfer subspace learning. Neurocomputing 337, 218–234 (2019)
19. Belkin, M., Niyogi, P., Sindhwani, V.: Manifold regularization: a geometric framework for learning from labeled and unlabeled examples. J. Mach. Learn. Res. 7(11) (2006)
20. Wen, J., Xu, Y., Li, Z., et al.: Inter-class sparsity based discriminative least square regression. Neural Netw.Netw. 102, 36–47 (2018)
21. Yang, J., Yin, W., Zhang, Y., et al.: A fast algorithm for edge-preserving variational multichannel image restoration. SIAM J. Imag. Sci.Imag. Sci. 2(2), 569–592 (2009)
22. Liu, G., Lin, Z., Yan, S., et al.: Robust recovery of subspace structures by low-rank representation. IEEE Trans. Pattern Anal. Mach. Intell.Intell. 35(1), 171–184 (2012)
23. Yang, L., Zhong, P.: Robust adaptation regularization based on within-class scatter for domain adaptation. Neural Netw.Netw. 124, 60–74 (2020)
24. Yang, L., Lu, B., Zhou, Q., et al.: Unsupervised domain adaptation via re-weighted transfer subspace learning with inter-class sparsity. Knowl.-Based Syst. 110277 (2023)

Enhancing Adversarial Robustness via Stochastic Robust Framework

Zhenjiang Sun, Yuanbo Li, and Cong Hu[✉]

School of Artificial Intelligence and Computer Science, Jiangnan University,
Wuxi, China
conghu@jiangnan.edu.cn

Abstract. Despite deep neural networks (DNNs) have attained remarkable success in image classification, the vulnerability of DNNs to adversarial attacks poses significant security risks to their reliability. The design of robust modules in adversarial defense often focuses excessively on individual layers of the model architecture, overlooking the important inter-module facilitation. To this issue, this paper proposes a novel stochastic robust framework that employs the Random Local winner take all module and the random Normalization Aggregation module (RLNA). RLNA designs a random competitive selection mechanism to filter out outputs with high confidence in the classification. This filtering process improves the model's robustness against adversarial attacks. Moreover, we employ a novel balance strategy in adversarial training (AT) to optimize the trade-off between robust accuracy and natural accuracy. Empirical evidence demonstrates that RLNA achieves state-of-the-art robustness accuracy against powerful adversarial attacks on two benchmarking datasets, CIFAR-10 and CIFAR-100. Compared to the method that focuses on individual network layers, RLNA achieves a remarkable 24.78% improvement in robust accuracy on CIFAR-10.

Keywords: Adversarial robustness · Adversarial training · Local winner take all

1 Introduction

Despite DNNs have achieved remarkable success in the field of computer vision [18,20,26,29], it still has serious security concerns due to the vulnerability of DNNs to adversarial examples [2,15,16,27]. Adversarial examples are misclassified by adding an elaborate, imperceptible adversarial perturbation to an natural image [33].

Much endeavor has been devoted to enhancing the adversarial robustness of DNNs [11]. Among the various methods explored, AT [12,22] stands out as one of

This work was supported in part by the National Natural Science Foundation of China (Grant No. 62006097, U1836218), in part by the Natural Science Foundation of Jiangsu Province (Grant No. BK20200593) and in part by the China Postdoctoral Science Foundation (Grant No. 2021M701456).

Q. Liu et al. (Eds.): PRCV 2023, LNCS 14428, pp. 187–198, 2024.
https://doi.org/10.1007/978-981-99-8462-6_16

the most effective approaches, which incorporates powerful adversarial examples into the training process to encourage correct classification by the target model. AT significantly improves the robustness of the target model, but this enhancement in robustness is often accompanied by a trade-off in terms of reduced natural classification accuracy and increased training time cost. Additionally, recent research has highlighted the crucial role of architecture and module design in enhancing robustness [13]. Random Normalization Aggregation (RNA) [7] introduces randomness by modifying the normalization layer to enhance robustness. Nevertheless, excessive randomness in the model prolongs training time and induces training instability. Moreover, RNA's excessive emphasis on the normalization layer tends to neglect the importance of the activation layer. Consequently, RNA fails to strike an optimal balance between natural classification accuracy and robust accuracy. Thus, it remains a pressing challenge to achieve an effective trade-off between these two factors [19, 32].

In this paper, we are committed to studying the improvement of robustness by introducing different robust components to the target model. To enhancing adversarial robustness, we propose a novel stochastic robust framework named RLNA to defense against adversarial attacks. The randomness is introduced into the model structure by introducing RNA module and local winner take all (LWTA) [25] module. This dual stochasticity enables the target model to possess a more diverse structure, so it is more difficult for an attacker to obtain valid information about the model. RLNA designs a random competitive selection mechanism to filter out outputs with high confidence in the classification. To balance the natural classification accuracy and robust accuracy, the loss function is divided into natural loss and adversarial loss during the AT. Therefore, RLNA significantly improves the robustness of the target model while achieving improved trade-offs between natural and robust accuracy. We demonstrate that RLNA outperforms state-of-the-arts remarkably on several benchmark datasets. Our contributions are as follows:

- We propose a stochastic robust framework named RLNA to demonstrate the effects of different basic components on robustness.
- RLNA divides the loss function into natural loss and robust loss during AT.
- Extensive experiments have been conducted to demonstrate the superiority of RLNA on different benchmark datasets and networks, and obtain state-of-the-arts robust results.

2   Related Work

2.1   Adversarial Attack

Although DNNs have achieved great success in the field of deep learning, they are extremely vulnerable to adversarial attacks. Consequently, researchers have proposed a series of attack methods [28]. Fast Gradient Sign Method (FGSM) [12] generates adversarial examples base on the gradient of the model along the direction of gradient ascent. Madry et al. propose a multi-step version of FGSM called project gradient descent (PGD) [22] in order to solve the nonlinear internal maximization problem. Moosavi-Dezfooli et al. propose a simple and effective hyperplane classification-based attack method Deepfool [23]. In addition, Carlini-Wagner et al. propose an optimization-based powerful attack method C&W [3] to evaluate the robustness of deep neural networks. Dong et al. propose a class of momentum-based iterative algorithms to enhance adversarial attacks which called MIFGSM [9]. Croce et al. combine two new versions of PGD attack (APGD-CE, APGD-DLR) with FAB [4] and Square [1] attack to propose the powerful AutoAttack [5].

2.2   Adversarial Defense

In the past few years, researchers have devoted significant efforts to enhance the robustness of target models [6,7].

Adversarial Training. AT adds powerful adversarial examples to the process of model training and encourages the model to classify them correctly [22]. Zhang et al. propose to divide the loss function into natural loss and robust loss in the AT process and use KL regularization for the latter thus achieving a good trade-off between natural and robust accuracy [32]. Li et al. find a new example called collaborative example and add it to the AT called squeeze training [19].

Stochastic Network. Recently, increased attention has been given to investigating the impact of the basic components of DNNs on adversarial robustness, including activation functions [30], operations [8], and neural structures [21]. Dong et al. investigate the impact of different normalization methods on robustness and propose to aggregate different normalization methods into RNA [7] module. Panousis et al. introduce randomness into LWTA module to improve adversarial robustness [24].

3   Methodology

Inspired by the LWTA mechanism, we propose a novel stochastic robust framework. Section 3.1 provide a comprehensive description of our model. In Sect. 3.2,we elaborate on our proposed method, highlighting its unique characteristics and advantages. Additionally, Sect. 3.2 introduce a novel balance loss function.

3.1   Model Definition

Let us assume an input image x_i and the label y_i comes from $S = \{(x_i, y_i)\}_{i=1}^n$, where $x_i \in \mathcal{X}$ and $y_i \in \mathcal{Y} = \{0, \ldots, C-1\}$. The target model is a convolutional network designed for image classification. It is denoted as $\hat{y} = f_\mathbf{w}(x_i)$, where $\hat{y}$ is the predicted label of the target model, $\mathbf{w}$ represents the model parameters. For simplicity, we refer to $f_\mathbf{w}(x)$ as $f(\cdot)$ in the following. The function $f(\cdot)$ consists of $\mathbf{K}$ hidden units and weight matrix $\boldsymbol{W} \in \mathbb{R}^{I \times K}$. Each hidden unit K within a layer first performs an inner product operation:

$$h_k = \boldsymbol{w}_k^T \boldsymbol{x} = \sum_{i=1}^{I} w_{ik} \cdot x_i \tag{1}$$

where $h_k \in \mathcal{H}$, $\mathcal{H}$ is the distribution of the inner product of each neuron. The output of the current layer is formed by the LWTA module $\sigma(\cdot)$, Subsequently, a normalization method is randomly selected from the RNA module $g \in \mathcal{G}$ to process the data. $\mathcal{G}$ is the aggregation of different normalization methods. As a result, the output vector of the final fully connected layer is obtained by concatenating the processed outputs of each hidden unit, such that $y = [y_1, ..., y_K]$, where $y_k = g(\sigma(h_k))$. The entire classification task can be described as the objective network $f_\mathbf{w}(x) \in \mathcal{F}$ complex optimizes the loss function $\mathcal{L}$ for the input $\mathcal{X}$ and the label $\mathcal{Y}$. The optimization objective can be expressed as:

$$f(\cdot) = \underset{f \in \mathcal{F}}{\mathrm{argmin}} \mathbb{E}_{x,y \sim \mathcal{D}}[\mathcal{L}(f(x), y)]. \tag{2}$$

where $\mathcal{D}$ denotes the data distribution, $\mathcal{L}$ represents the Cross-entropy loss. The formula can be described as:

$$CrossEntropy = \mathbb{E}_{x \sim \mathcal{X}}[-\log Q(x)] \tag{3}$$

3.2   Method

Our method advocates for the utilization of LWTA layer instead of the conventional nonlinear activation function layer. RLNA designs a random competitive selection mechanism to filter out outputs with high confidence in the classification, resulting in a sparse output. RLNA aims to enhance the model's performance by striking a better balance between natural and robust accuracy. Each LWTA layer contains of U linear competing units and each linear competing unit contains two neural units. The output of each linear competing unit is processed using one-hot coding after passing through the Softmax layer. Specifically, the output of each winner is 1 and is pass to the next layer, the rest of the output is 0. Obviously, this leads to a sparse output, since each block has only one winner.

The possibility of two attacks visiting the same path is negligible under the perspective of baseline if the random space is huge. Therefore, leveraging the base component to generate a larger random space can be an effective defense against adversarial attacks. Inspired by baseline, we introduce randomness to the input of the linear competitive unit, sampling a fraction of linear units each time the input passes through the activation layer. As the network deepens, this sampling creates a large random space, which exponentially increases in size. This random space makes the structure of the target model more diverse and makes it challenging for existing attack methods to extract meaningful information from the target model. We randomly sample a path p_a, perform a white-box attack to obtain the adversarial sample X_a. During the AT process, X_a is fed to

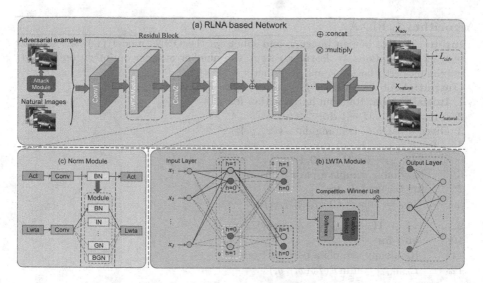

Fig. 1. The illustration of RLNA. The random framework consists of the random local winner take all module and the random normalization aggregation module (RLNA). The LWTA module uses One-Hot coding to process logits in order to obtain a unique winning unit, and the RNA module randomly selects one of the multiple normalization methods from a huge random space to process the data.

another random path p_b to form a black-box adversarial training. In addition, the sparse output formed by the LWTA layer results in a significant reduction in the amount of data the network needs to process, which significantly improves the training speed.

Novel Formulation of Adversarial Training. Given a target network h and inputs x, y, the adversarial example is defined as $\tilde{x} = x + \delta$. Maximize the classification loss in order to misclassify the network as:

$$\tilde{x} = \underset{\|\tilde{x}-x\|_p \leqslant \epsilon}{\operatorname{argmax}} \mathcal{L}(f(\tilde{x}), y) \tag{4}$$

where δ is the carefully craft adversarial perturbation, constrained by its l_p-norm. ϵ is an upper bound on the perturbation size. AT process can be described as a min-max optimization problem in [22], using the labels of the adversarial and natural examples to calculate the loss. The objective function is defined as follow:

$$\min_{\mathbf{w}} \mathbb{E}_{(x,y)\sim\mathcal{D}}[\max_{\delta\in\epsilon} \mathcal{L}(f_{\mathbf{w}}(x+\delta), y)] \tag{5}$$

where ϵ denotes the perturbation distribution, x denotes the natural image, and δ denotes the adversarial perturbation, $\tilde{x}$ denotes the adversarial example. $f_{\mathbf{w}}(.)$ denotes the target network and $L(f_{\mathbf{w}}(x), y)$ denotes the loss function of the target network. The difference is that we introduce the information of natural

---

Algorithm 1. RLNA

Input: A set of benign image x and labels y, the DNNs model $f(\cdot)$, the number of training iterations M, learning rate η, the number of inner optimization steps T, maximum perturbation ϵ, and step factor α, Path set $\mathcal{P}$

Output: A robust classifier $f(\cdot)$ parameterized by $\mathbf{w}$

1: **Initialization:** Random initialization for $f(\cdot)$
2: **for** $m = 0, ..., M-1$ **do**:
3: Given a batch of training data $\{(x_i, y_i)\}_{i=1}^{k}$
4:    **for** $k = 1, ..., K-1$ **do**:
5:      $\tilde{x}_i \leftarrow x_i + 0.001 \cdot \mathcal{N}(0, \mathbf{I})$
6:      Randomly sample a path p_a from $\mathcal{P}$
7:      **for** $t = 0, ..., T$ **do**:
8:         $g_{in} = \max\limits_{\tilde{x} \in \mathbb{B}(x, \epsilon)} \text{CE}(f_{\mathbf{w}}(\tilde{x}_i), y_i)$
9:         $\tilde{x}_i \leftarrow \mathbf{\Pi}_{\tilde{x} \in \mathbb{B}(x, \epsilon)}(\tilde{x}_i - \alpha \cdot \text{sign}(\nabla_{\tilde{x}_i} g_{in}))$
10:     **end for**
11:      Randomly sample a path p_b from $\mathcal{P}$
12:      $\mathcal{L}_{natural} = \min\limits_{x \in \mathcal{X}} \mathcal{L}(f_{\mathbf{w}}(x_i), y_i)$
13:      $\mathcal{L}_{adv} = \max\limits_{\tilde{x} \in \mathbb{B}(x, \epsilon)} \mathcal{L}(f_{\mathbf{w}}(\tilde{x}_i), y_i)$
14:      $\mathcal{L} = \min\limits_{\mathbf{w}} \mathbf{E}_{(x,y) \sim \mathcal{D}} \{\beta \mathcal{L}_{natural} + \mathcal{L}_{adv}\}$
15:    **end for**
16: **end for**

---

examples in the training process. Our novel formulation for AT can be defined as:

$$\min_{\mathbf{w}} \mathbb{E}_{(x,y) \sim \mathcal{D}} \left\{ \beta \cdot \mathcal{L}(f(x), y) + \max_{\tilde{x} \in \mathbb{B}(x, \epsilon)} \mathcal{L}(f_{\mathbf{w}}(\tilde{x}), y) \right\} \tag{6}$$

where $\mathbb{B}(x, \epsilon)$ is the range of data distribution for generating the adversarial examples and β is an important hyperparameter. Compared with Vanilla AT, the most distinct difference is that the loss term in AT is divided into natural loss and robust loss, and use the hyperparameter β to adjust the weight of natural examples in AT, so that the focus of the model is put more on the natural examples. As shown in Fig. 2, we achieve higher robustness with fewer training rounds compared to the baseline approach, which demonstrates the efficiency of our training method. Empirical evidence shows that reinforcing the importance of natural examples during the AT process has a great effect on improving the performance of the model. Enhancing the weights of the natural examples will correct some of the misconceptions of the model about natural examples during AT. Experimental results show that this achieves higher natural accuracy and performs better in the trade-off between natural and robust accuracy.

4 Experiments

To evaluate our method, we conducte experiments on two benchmarking datasets and two models, *i.e.*, CIFAR-10 [17], CIFAR-100 [17], and ResNet18 [14] and WideResNet34-1 [31].

4.1 Adversarial Defenses Under White Box Attacks

Competitive Models. To verify the impact of our method on adversarial robustness, we compare it with several state-of-the-art AT methods and empirically demonstrate the effectiveness of our method. RLNA is compared with the following baselines: (1) PGD-AT [22], (2) TRADES [32], (3) RNA [7], and (4) ST [19]. In the AT phase, the same attack strategy with the same number of iterations is used.

Evaluation. We choose some different attack methods to evaluate our model, including FGSM [12], PGD [22], C&W [3], MIFGSM [10], and AutoAttack (AA) [5]. Among the AA include APGD-CE [5], APGD-DLR [5], FAB [4] and Square [1]. The maximum perturbation ϵ strength is set to $8/255$ and the number of attack iteration steps iters to 10, and the rest are consistent with the default settings of AT. Among the main evaluation criteria are natural accuracy and robust accuracy.

Implementation Details. We first train the models WideResNet34-1 [31] and ResNet18 [14] on two datasets, CIFAR-10 [17], CIFAR-100 [17]. These two datasets contain 50K training images and 10K test images from 10/100 categories, each 32×32 in size. For the target model, SGD optimizer with a momentum set to 0.9. And the initial learning rate is set to 0.1 with a piecewise decay learning rate scheduler. Weight decay is set to 5×10^{-4}. In the AT setting, the attack strategy is PGD-10, where ϵ is set to $8/255$ and step size is set to $2/255$. The batchsize of all comparison methods is set to 128. The experiments are performed on one RTX3090 GPU.

Our results are shown in Tables 1 and 2. We implemente methods in [PGD-AT,TRADES,RNA,ST] under the same experimental setup as they are also based on AT to improve model robustness. Our method achieve state-of-the-art model results on both datasets. For example, using PGD-20 attack on ResNet18 model with AT, our method surpass 24.78% on the CIFAR-10 dataset relative to baseline. On the CIFAR-100 dataset we outperform the baseline by 24.17%.

CIFAR10 Setup. Output numclass of the fully connected layer is set to 10, maximum perturbation ϵ strength to $8/255$ and the number of attack iteration steps iters to 10, step size α is set to $2/255$. Randomtype is set to Batch Normalization, and gntype is set to Group Normalization for the normalization layer. To evaluate the robust accuracy, PGD (white-box) attack with 10 iterations is applied and the step size is 0.003.

Table 1. Test robustness(%) on CIFAR-10 dataset using ResNet18 and WideResNet34. Number in bold indicates the best.

Model	Method	Natural	FGSM	PGD-20	C&W	MIFGSM	AutoAttack
ResNet18	PGD-AT	85.39	54.89	45.50	49.72	51.31	42.45
	TRADES	82.49	57.96	56.10	78.84	55.68	53.40
	RNA	84.58	63.70	59.27	83.93	60.88	64.54
	ST	83.10	59.51	54.62	51.43	57.67	50.50
	RLNA(ours)	**85.32**	**84.41**	**84.05**	**84.31**	**84.09**	**84.48**
WideResNet34	PGD-AT	84.04	50.51	41.55	41.37	47.23	53.97
	TRADES	84.92	60.86	55.33	81.82	58.85	52.50
	RNA	85.36	66.02	61.87	85.80	62.62	67.88
	ST	85.08	62.30	56.94	55.01	60.12	53.71
	RLNA(ours)	**86.68**	**85.47**	**85.83**	**86.03**	**85.87**	**86.19**

Table 2. Test robustness(%) on CIFAR-100 dataset using ResNet18 and WideResNet34. Number in bold indicates the best.

Model	Method	Natural	FGSM	PGD-20	C&W	MIFGSM	AutoAttack
ResNet18	PGD-AT	60.90	26.85	21.41	54.08	24.78	19.76
	TRADES	53.82	30.18	27.38	48.62	28.88	23.48
	RNA	56.79	36.76	35.55	56.86	34.00	42.12
	ST	58.44	33.35	30.53	26.70	31.80	25.61
	RLNA(ours)	**60.96**	**58.94**	**59.72**	**60.40**	**59.52**	**59.13**
WideResNet34	PGD-AT	56.85	24.84	19.64	50.91	22.87	17.00
	TRADES	57.95	33.33	30.68	52.93	32.02	27.25
	RNA	60.57	37.87	36.04	60.21	35.58	42.43
	ST	57.71	31.88	27.80	27.86	32.77	26.95
	RLNA(ours)	**63.35**	**61.70**	**62.41**	**63.35**	**62.30**	**62.62**

4.2   Hyperparameter Selection

The parameter β is an important hyperparameter in our method. We demonstrate the sensitivity of our robust classifier performance to the hyperparameter through extensive numerical experiments on two datasets, CIFAR-10 and CIFAR-100. For both datasets, the cross-entropy loss is minimized in Eq. 6.

As shown in Table 3, We observe that as the hyperparameter β increases, the natural accuracy improves significantly, followed by the robust accuracy, which does not drop significantly. It is important to note that the difference between natural and robust accuracy is smaller on the CIFAR10 dataset than on the CIFAR100 dataset, probably because the classification task is simpler in CIFAR10. Empirically, when β is selected between [0, 4], our method is able to train models with high accuracy and high robustness. In the following experiments, the hyperparameter β is set to 4.

4.3   Ablation Study

In this section we will illustrate the effectiveness of our method from two parts. One is the novel training loss function, the other is the novel network structure.

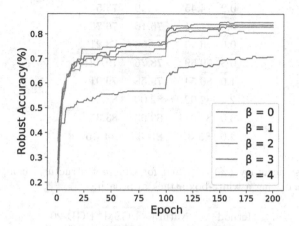

Fig. 2. Extensive experiments is conducted on the value of the hyperparameter β based on Eq. 6.

Novel Adversarial Training Loss. In order to keep the model's attention is not overly focus on the adversarial examples during AT, we try to divide the loss function term into natural accuracy and robust accuracy during AT. As shown in Fig. 2, the natural accuracy is significantly improved and the robust accuracy is not significantly decreased after adding our method. Not adding the natural loss term in the AT phase and using only the adversarial loss leads to a serious reduction in the natural accuracy, which fundamentally limits the upper limit of robustness. This experiment demonstrates that adding information from natural examples during AT helps the model learn more useful information. The natural accuracy is also improved to some extent by controlling the weights of the natural loss terms.

Novel Structure. We advocate introducing the competition mechanism of LWTA into baseline, expecting the model to select more expressive neurons through competition.

As shown in Fig. 2, introducing LWTA on top of baseline in the AT will make the model focus more on the adversarial examples, which will cause the model to incorrectly train the adversarial examples as natural examples excessively. This will cause the label confusion problem, making the model think that the label of the adversarial example is the correct label, resulting in a lower natural

Table 3. Sensitivity of hyperparameter β based on ResNet18 on the CIFAR-10 dataset. Number in bold indicates the best.

β	Natural	FGSM	PGD-20
0.1	73.81	72.48	72.6
0.2	74.45	73.59	73.5
0.4	77.04	76.16	76.75
0.6	79.15	77.56	78.21
0.8	79.59	78.76	79.52
1.0	80.54	79.58	79.91
2.0	82.92	82.00	82.4
3.0	84.01	82.93	83.35
4.0	**85.32**	**83.5**	**84.05**

Table 4. Robust accuracy (%) obtained for different structural components on the CIFAR-10 dataset using a white-box attack approach.

Method	Natural	FGSM	PGD-20
RNA	84.58	63.7	59.27
Ours(Bibn)	**86.02**	71.85	73.58
Ours	85.32	**84.41**	**84.05**

accuracy of the final trained model. To solve this problem, the information from the natural examples is added and their weights is increased.

As shown in Table 4, this indicates that weight update of the normalization layer is a key factor in AT to improve the robustness of the model. It is empirically demonstrated that using a single normalization layer to generate adversarial examples and training to update the parameters will produce good robustness in AT. Conversely, if an additional auxiliary normalization layer is introduced into the adversarial training to generate adversarial examples and used only during the attack (Bibn). This operation reduces the accuracy of robustness to some extent, but improves natural accuracy. This means that we can use the operation of introducing an additional normalization module to improve the upper limit of the robust model's image recognition performance, and the experiments show that this approach is widely applicable to existing models.

5   Conclusion

In this paper, we investigate the issue of defending against imperceptibly small adversarial perturbations in the context of computer vision. The interplay among various robust components within the model is primarily focused on by our research. Specifically, the combination of competing activation and random normalization layers produces a large number of different paths that essentially prevent attackers from successfully attacking the model to improve model robustness. Balance loss is minimized to establish an exact balance between robustness and accuracy. Extensive experiments on two benchmark datasets demonstrates that RLNA achieves higher robust accuracy than the state-of-the-art adversarial defense method. In the future, we will continue to explore the complex relationships between the robust components of the model and hope that this work can provide a new research direction for adversarial defense.

References

1. Andriushchenko, M., Croce, F., Flammarion, N., Hein, M.: Square attack: a query-efficient black-box adversarial attack via random search. In: Vedaldi, A., Bischof, H., Brox, T., Frahm, J.-M. (eds.) ECCV 2020. LNCS, vol. 12368, pp. 484–501. Springer, Cham (2020). https://doi.org/10.1007/978-3-030-58592-1_29
2. Cai, Z., Song, C., Krishnamurthy, S., Roy-Chowdhury, A., Asif, S.: Blackbox attacks via surrogate ensemble search. In: NeurIPS (2022)
3. Carlini, N., Wagner, D.A.: Towards evaluating the robustness of neural networks. In: IEEE Symposium on Security and Privacy, pp. 39–57. IEEE Computer Society (2017)
4. Croce, F., Hein, M.: Minimally distorted adversarial examples with a fast adaptive boundary attack. In: ICML. vol. 119. Proceedings of Machine Learning Research, pp. 2196–2205. PMLR (2020)
5. Croce, F., Hein, M.: Reliable evaluation of adversarial robustness with an ensemble of diverse parameter-free attacks. In: ICML, vol. 119. Proceedings of Machine Learning Research, pp. 2206–2216. PMLR (2020)
6. Dhillon, G.S.: Stochastic activation pruning for robust adversarial defense. In: ICLR (Poster). OpenReview.net (2018)
7. Dong, M., Chen, X., Wang, Y., Xu, C.: Random normalization aggregation for adversarial defense. In: Oh, A.H., Agarwal, A., Belgrave, D., Cho, K., (eds.) Advances in Neural Information Processing Systems (2022)
8. Dong, M., Wang, Y., Chen, X., Xu, C.: Towards stable and robust addernets. In: NeurIPS, pp. 13255–13265 (2021)
9. Dong, Y., et al.: Boosting adversarial attacks with momentum. In: Proceedings of the IEEE/CVF Conference on Computer Vision and Pattern Recognition (CVPR), pp. 9185–9193 (2018)
10. Dong, Y.: Boosting adversarial attacks with momentum. In: CVPR, pp. 9185–9193. Computer Vision Foundation. IEEE Computer Society (2018)
11. Duan, Y., Jiwen, L., Zheng, W., Zhou, J.: Deep adversarial metric learning. IEEE Trans. Image Process. **29**, 2037–2051 (2020)
12. Goodfellow, I., Shlens, J., Szegedy, C.: Explaining and harnessing adversarial examples. In: International Conference on Learning Representations (ICLR) (2015)
13. Guo, M., Yang, Y., Xu, R., Liu, Z., Lin, D.: When nas meets robustness: In search of robust architectures against adversarial attacks. In: Proceedings of the IEEE/CVF Conference on Computer Vision and Pattern Recognition (CVPR) (June 2020)
14. He, K., Zhang, X., Ren, S., Sun, J.: Deep residual learning for image recognition. In: CVPR, pp. 770–778. IEEE Computer Society (2016)
15. Cong, H., Xiao-Jun, W., Li, Z.-Y.: Generating adversarial examples with elastic-net regularized boundary equilibrium generative adversarial network. Pattern Recognit. Lett. **140**, 281–287 (2020)

16. Cong, H., Hao-Qi, X., Xiao-Jun, W.: Substitute meta-learning for black-box adversarial attack. IEEE Signal Process. Lett. **29**, 2472–2476 (2022)
17. Krizhevsky, A.: Learning multiple layers of features from tiny images (2009)
18. Li, H., Xiao-Jun, W., Kittler, J.: Mdlatlrr: a novel decomposition method for infrared and visible image fusion. IEEE Trans. Image Process. **29**, 4733–4746 (2020)
19. Li, Q., Guo, Y., Zuo, W., Chen, H.: Squeeze training for adversarial robustness (2023)
20. Li, X., Wang, W., Hu, X., Yang, J.: Selective kernel networks. In CVPR, pp. 510–519. Computer Vision Foundation/IEEE (2019)
21. Li, Y., Yang, Z., Wang, Y., Xu, C.: Neural architecture dilation for adversarial robustness. In: NeurIPS, pp. 29578–29589 (2021)
22. Madry, A., Makelov, A., Schmidt, L., Tsipras, D., Vladu, A.: Towards deep learning models resistant to adversarial attacks. In: ICLR (Poster). OpenReview.net (2018)
23. Moosavi-Dezfooli, S.-M., Fawzi, A., Frossard, P.: Deepfool: a simple and accurate method to fool deep neural networks. In: CVPR, pp. 2574–2582. IEEE Computer Society (2016)
24. Panousis, K.P., Chatzis, S., Theodoridis, S.: Stochastic local winner-takes-all networks enable profound adversarial robustness. CoRR, abs/ arXiv: 2112.02671 (2021)
25. Srivastava, R.-K., Masci, J., Gomez, F.J., Schmidhuber, J.: Understanding locally competitive networks. In: ICLR (Poster) (2015)
26. Tong, J., Chen, T., Wang, Q., Yao, Y.: Few-Shot Object Detection via Understanding Convolution and Attention. In: Yu, S., et al. (eds.) Pattern Recognition and Computer Vision. PRCV 2022. LNCS, vol 13534. Springer, Cham (2022). https://doi.org/10.1007/978-3-031-18907-4_52
27. Vakhshiteh, F., Nickabadi, A., Ramachandra, R.: Adversarial attacks against face recognition: a comprehensive study. IEEE Access **9**, 92735–92756 (2021)
28. Wang, G., Yan, H., Wei, X.: Enhancing Transferability of Adversarial Examples with Spatial Momentum. In: Yu, S., et al. (eds.) Pattern Recognition and Computer Vision. PRCV 2022. LNCS, vol. 13534. Springer, Cham (2022). https://doi.org/10.1007/978-3-031-18907-4_46
29. Wang, M., Deng, W.: Deep face recognition: a survey. Neurocomputing **429**, 215–244 (2021)
30. Xie, C., Tan, M., Gong, B., Yuille, A.L., Le, Q.V.: Smooth adversarial training. CoRR, abs/ arXiv: 2006.14536 (2020)
31. Zagoruyko, S., Komodakis, N.: Wide residual networks. In: BMVC. BMVA Press (2016)
32. Zhang, H., Yu, Y., Jiao, J., Xing, E.P., Ghaoui, L.E., Jordan, M.I.: Theoretically principled trade-off between robustness and accuracy. In: ICML, vol. 97. Proceedings of Machine Learning Research, pp. 7472–7482. PMLR (2019)
33. Zhang, W., Gou, Y., Jiang, Y., Zhang, Y.: Adversarial VAE with Normalizing Flows for Multi-Dimensional Classification. In: Yu, S., et al. (eds.) Pattern Recognition and Computer Vision. PRCV 2022. LNCS, vol 13534. Springer, Cham (2022). https://doi.org/10.1007/978-3-031-18907-4_16

Pseudo Labels Refinement with Stable Cluster Reconstruction for Unsupervised Re-identification

Zhenyu Liu[1,2], Jiawei Lian[1,2], Jiahua Wu[1,2](✉), Da-Han Wang[1,2], Yun Wu[1,2], Shunzhi Zhu[1,2], and Dewu Ge[3]

[1] School of Computer and Information Engineering, Xiamen University of Technology, Xiamen 361024, China
[2] Fujian Key Laboratory of Pattern Recognition and Image Understanding, Xiamen 361024, China
{liuzy,lianjw}@s.xmut.edu.cn, salmon2wu@gmail.com,
{wangdh,ywu,szzhu}@s.xmut.edu.cn
[3] Xiamen KEYTOP Communication Technology Co., Xiamen 361024, China

Abstract. Most existing unsupervised re-identification uses a clustering-based approach to generate pseudo-labels as supervised signals, allowing deep neural networks to learn discriminative representations without annotations. However, drawbacks in clustering algorithms and the absence of discriminatory ability early in training limit better performance seriously. A severe problem arises from path dependency, wherein noisy samples rarely have a chance to escape from their assigned clusters during iterative training. To tackle this challenge, we propose a novel label refinement strategy based on the stable cluster reconstruction. Our approach contains two modules, the stable cluster reconstruction (SCR) module and the similarity recalculate (SR) module. It reconstructs more stable clusters and re-evaluates the relationship between samples and clearer cluster representatives, providing complementary information for pseudo labels at the instance level. Our proposed approach effectively improves unsupervised reID performance, achieving state-of-the-art performance on four benchmark datasets. Specifically, our method achieves 46.0% and 39.1% mAP on the challenging dataset VeRi776 and MSMT17.

Keywords: Re-identification · Unsupervised learning · Label refinement

1 Introduction

Re-identification aims to train a neural network to select similar samples from many candidate samples with different camera angles and backgrounds. In recent years, supervised methods [1–3] have shown remarkable accuracy and efficiency as deep learning algorithms have improved. However, expensive annotation hinders the practical application of deep learning techniques. Large amounts of quality labeled data often must be made available for real-world applications.

© The Author(s), under exclusive license to Springer Nature Singapore Pte Ltd. 2024
Q. Liu et al. (Eds.): PRCV 2023, LNCS 14428, pp. 199–211, 2024.
https://doi.org/10.1007/978-981-99-8462-6_17

Therefore, more deep-learning researchers have become interested in better unsupervised methods.

Traditional unsupervised re-identification methods [4–6] use the clustering algorithm (e.g., DBSCAN [7]) to generate labels. These pseudo-labels completely replace expensive manual labeling work but leave some non-negligible drawbacks. Samples of the same class may be more focused on inherent noise labels, resulting in stagnation of network performance improvement. For example, as shown in Fig. 1(a), samples from the same cluster are drawn to the exact center regardless of their original class. Even after the next generations of clustering, these noisy samples will be close to the cluster center as the correctly clustered samples. The deep neural network pulls the samples according to the labels, but making it harder to distinguish between different samples.

In recent research, a contrast-learning-based approach [8] further integrates the pseudo-label assignment phase with the neural network training phase, achieving satisfactory results on several datasets. It further optimizes the feature space of the intra-class by linking the label assignment and training phases by using contrast loss, such as infoNCE loss [9]. However, this approach does not tackle the noise problem. The incorrect labels continue to influence network training from start to finish. Refining pseudo-labels is the key to optimizing the unsupervised re-identification algorithm.

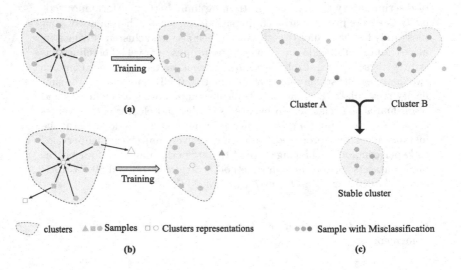

Fig. 1. (a) Noise samples in traditional unsupervised training. (b) Contrast learning with refinement label in representation level. (c) stable cluster reconstruction from two information sources

To address this problem, we propose a label refinement strategy called pseudo-label refinement with stable cluster reconstruction(LRCR), which includes a label refinement module and a stable cluster reconstruction module. Our approach is employed to reduce the harmful effects of noisy labels and to progressively guide the training process to separate noisy labels from correct labels. In the first step, we improve the accuracy of pseudo-labels from the perspective of cluster representation by the stable cluster reconstruction (SCR) module. Specifically, our approach makes the cluster representation more reliable by considering the consistency of the inference results and the pre-assigned pseudo-labels. We use an Intersection over Union (IoU) criterion to measure the consistency of sample labels generated at different times, where a larger IoU score implies a higher degree of similarity between the two categories. After that, the unstable samples were eliminated to create stable clusters like Fig. 1(c), reducing the impact of noisy labels on training. In the second stage, the refinement labels is reassigned using a proposed similarity recalculation (SR) module that is based on the similarity between the stable cluster representation and each sample. This refinement label is utilized to construct a contrast loss for separating noisy samples. Both labels are used in this process to direct the network.

Our contributions are summarized as followed:

- We propose a novel two-step pseudo-labels refinement strategy for unsupervised Re-ID. It complements the information at the instance level, allowing for the progressive separation of noise labels.
- We propose a module that generates refinement labels from a cluster representation perspective, which adds information to the original cluster-based approach and improves the network's performance.
- We propose a reconstruction module that filters out more stable and trustworthy samples and reconstructs the cluster. It explores the relationships of label information from different times and further enhances label refinement work through reconstructive clusters.
- We experimented extensively on several re-identification datasets and achieved performance improvements of 3.0%, 1.1%, 7.7%, and 8.7% on the market1501, Duke, MSMT17, and VeRi detests, respectively, compared to the baseline model.

2   Related Work

Unsupervised re-identification intends to train a deep neural network without labels to search most similar images in a gallery. There are two types of methods depending on the proportion of unlabeled data. The first category is unsupervised domain adaptation (UDA) [5,6,8,10,11], which does not use labels in the target domain. UDA attempts to transfer the knowledge learned in the source domain to the target domain. The other category we focus on is purely unsupervised learning (USL) for re-ID [4,11–16], where label information is completely excluded in training. In USL, clustering methods are commonly used to generate pseudo-labels, but these labels tend to be noisy. These inaccurate pseudo-labels are the primary challenge in unsupervised learning re-identification.

Contrastive learning in USL reID is inspired by self-supervised learning [17,18], which optimizes the network with category prototypes in the memory bank. This approach has achieved good performance and has become an important area of research for purely unsupervised re-identification. With the help of contrast learning techniques, SPCL [8] creatively integrates unsupervised re-ID work and creates a stronger connection between the pseudo-labels and the cluster. Take it a step further, Cluster-Contrast [19] further improves the memory bank and momentum update methods, increasing unsupervised re-identification performance. ICE [14] generates more robust soft labels with instance-level contrast loss.

Label refinement aims to correct pseudo-labels by extracting potential knowledge or differences in the image. To improve domain adaptation performance, MMT [11] builds two collaborative networks and compares two asynchronous networks to produce more reliable soft labels. RLCC [20] learns possible associations between different classes from temporal ensembling. It refines pseudo-labels with temporally propagated and ensembled pseudo-labels. PPLR [15] complements the original pseudo-labels with potential information from fine-grained local features. It exploits the reliable complementary relationship between global and local features to improve network performance. In contrast to previous work, our work searches for stable samples from a temporal integration perspective, providing refinement information on the level of category representation. This novel approach has yet to be attempted in previous studies.

3 Proposed Approach

Define as an unlabelled re-ID dataset $X = x_1, x_2, ...,x_N$, where x_i represents the i-th image in the dataset. The USL re-ID task aims to train a deep neural network $f_\theta(x)$ to extract an image feature embedding $f = f_\theta(x)$. We expect features of the same ID to be naturally close in feature space, with features of different IDs being further apart.

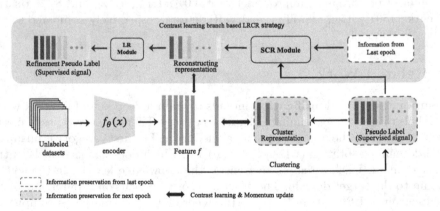

Fig. 2. The proposed framework of label refinement with stable cluster construction (LRCR) Strategy. The upper part of the dark color is based on our LRCR strategy, and the lower part is the baseline based on contrast learning.

Our proposed label refinement strategy consists of two main steps. First, we reconstruct clusters based on stable samples, and then we generate refined labels for the categories by reconstructing representative features of the clusters. As shown in the Fig. 2, after the pseudo-labels have been assigned, we use stable cluster reconstruction (SCR) module to reconstruct another cluster representation. Then, similarity recalculation (SR) module generates refinement labels based on these clustering representatives as supplementary supervisory signals.

3.1 Revisit USL Re-ID Model Based on Contrast Learning

As shown in the lower part of Fig. 2, contrast-based approach recycles the features f and calculating representative information of these features. The query feature is computed with all the representative features during the training process to obtain a similarity score. Contrast-based approach maximizes the similarity to enable the network to learn the most appropriate feature representation. Compared to the traditional method, the contrast learning approach makes the training process closer to the inquiry process.

There are undoubtedly potential risks associated with cluster-based methods. Noisy labels in the pseudo-labels affect the feature-extracting ability. Besides that, representative features in the memory bank become blurred and lose their uniqueness. All of this leads the clustering algorithm to incorrectly perpetuate the noise in subsequent rounds. We argue that providing purer clusters to the network and refining pseudo-labels based on cluster representatives can effectively mitigate these problems.

3.2 Stable Cluster Reconstruction Module

We first propose a clustering reconstruction module based on stable samples. Due to the poor feature extraction capability in the early stages, the clustering algorithm introduces noise, and these noisy samples contaminate the cluster representation. To address this issue, we re-evaluate all the samples before each training step. During the reconstruction of the cluster centers, unstable samples are discarded, allowing the cluster centers to be refined and better represent the underlying clusters.

Formally, after each label assignment process, we define the pseudo-label of each image in t epoch as $Y_p^t = y_1^t, y_2^t, ...,y_N^t$. Additionally, we use the network $f_\theta(x)$ to extract the features of the sample f before the start of the current training round. These features are multiplied with the clustering representatives from the previous round to obtain the sample query results:

$$\hat{y}_i^{(t-1)} = max(M^{(t-1)} \cdot f_i^t) \tag{1}$$

where $M^{(t-1)}$is the set of clustering representative features in epoch $t-1$. The i-th sample can be considered part of the cluster with the maximum similarity. All query results constitute a query set $\hat{Y}_q^{(t-1)} = \hat{y}_1^{(t-1)}, \hat{y}_2^{(t-1)}, ...,\hat{y}_N^{(t-1)}$. We then determine which clusters are worth refining by calculating the cross-agreement score between the pseudo-label and the query results:

$$C^t(i,j) = \frac{|D(\hat{Y}_q^{(t-1)},i) \cap D(Y_p^t,j)|}{|D(\hat{Y}_q^{(t-1)},i) \cup D(Y_p^t,j)|} \tag{2}$$

where the $D(\hat{Y}_q^{(t-1)}, i)$ is expressed as a slice of the sample of the i-th category in set $\hat{Y}_q^{(t-1)}$, the other symbol $| \cdot |$ calculates the number of instances in the set. The two categories with higher IoU scores are expressed as having a higher correlation, and their intersection remains stable at different times. However, it is not fair to directly assume that all samples in the two categories are related. Since the categories in the previous round may overlap with more than one category in the current round, we only reconstructed the categories with the highest scores:

$$Y_{Rec\_j}^t = D(\hat{Y}_q^{(t-1)}, i) \cap D(Y_p^t, j) > m \tag{3}$$

where $Y_{Rec\_j}^t$ is the largest intersection set for j-th cluster in Y_q^t, and m is the minimum reconstruction threshold. We obtained the reconstructed stable clusters and deposited them in a new independent memory bank:

$$M_{stable\_j}^t = \frac{1}{|D(Y_{Rec\_j}^t, k)|} \sum D(Y_{Rec\_j}^t, k) \tag{4}$$

One step further, we can use the reconstructed clusters from the previous round instead of the query results to obtain cleaner and stable clusters. For example, replacing $\hat{Y}_q^{(t-1)}$ in the Eqs. (2,3) with $M_{stable}^{(t-1)}$.

Although the reconstructed clusters are more stable and representative, this may also lead to excluding too many samples. Therefore, instead of directly replacing the original memory bank with the new one, we exclusively use the reconstructed stable clusters for relevant calculations of refinement labels.

3.3   Similarity Recalculation Module

The clustering algorithm groups the image features according to a predefined pattern. The potential risk of this approach is that individual samples are only grouped at the cluster level. This challenge has prompted us to reconsider the relationship between individual samples and groups from a different perspective. Here, we recalculate the similarity between individual samples and all representatives of the group:

$$y_i^t = max(M_{stable}^t \cdot f_i^t) \tag{5}$$

Following Eq. (1). We complemented the association information between individual samples with the stable cluster representative, providing the network with more sophisticated supervised information. Once the refined labels are produced, the new labels and stable clusters are joined together with the original for contrast learning.

3.4   Contrast Learning with Refinement of Labels

Prior research [8] typically implemented cluster-level contrast learning using a tailored InfoNce Loss. Samples are pushed either in the direction of or away from representatives, which can be described by:

$$L_q = -log \frac{exp(f \cdot M_+/\tau)}{\sum_{i=0}^N exp(f \cdot M_i/\tau)} \tag{6}$$

f is the extraction feature, M_+ is the positive cluster representative feature stored in the memory bank, and τ is the temperature hyper-parameter. The loss function 6 provides cluster-level optimization for the network. However, simply providing cluster-level information to the samples cannot separate the noisy samples from the cluster. Therefore, we propose a cluster representation level loss based on the reconstruction cluster:

$$L_{refine} = -log \frac{exp(f \cdot M_{stable+}/\tau)}{\sum_{i=0}^{N} exp(f \cdot M_{stable\_i}/\tau)} \tag{7}$$

The refinement label and stable cluster prefer the association information between the instance-cluster representatives. Ideally, it complements the supplementary information at the instance level. Refinement labels and stable clustering representatives will provide the network with differentiated learning information. The stable cluster's representatives may still differ when the labels and refinement labels are consistent. This phenomenon may lead the network to produce differentiated training results for similar samples. In the opposite condition, this difference will be more pronounced. The new labels and the reconstructed clustering representations provide the network with opposite training signals, hindering the potential noisy labels, thus providing better representation learning. Our method can deliver signals in the opposite direction, so it is necessary to have a hyper-parameter α to balance the losses:

$$L_{total} = (1 - \alpha)L_q + \alpha L_{refine} \tag{8}$$

4 Experiments

4.1 Datasets and Evaluation Metrics

Table 1. Statistics of the datasets used in the experimental section.

Dataset	total images	total IDs	train images	gallery images	query IDs	query images
Market-1501	32,668	1,501	12,936	19732	750	3368
DukeMTMC	36,441	1,110	16,522	17,661	702	2228
MSMT 17	126,441	4,101	32,621	11,659	3060	82,161
VeRi-776	51,003	776	12936	11579	201	1678

Datasets: We evaluated our approach on four widely used re-identification datasets, including three different pedestrian datasets and one vehicle dataset. Market-1501 [21], DukeMTMC-reID [22], and MSMT17 [23] are widely used pedestrian re-identification datasets, and VeRi-776 [24] is a vehicle re-ID dataset. Among them, MSMT17 is difficulty achieving high performance compared to other pedestrian datasets due to its diversity of scenes, lighting, and clothing. In addition, the vehicle dataset is equally challenging in unsupervised re-ID due to minor inter-class differences and more enormous viewpoint differences. Table 1 presents the statistical differences between the datasets.

Evaluation Metrics: The re-identification method uses mean average precision (mAP) and Top-1 scores for comparison. For a fair comparison, we do not use post-processing techniques such as re-ranking [25] in the query process.

4.2   Implementation Details

We used ResNet-50, pre-trained on ImageNet, as the backbone network. The model settings are based on Cluster-Contrast [19]. We resize the image sizes to 256×128 for all datasets. At the label assignment stage, we use DBSCAN to generate pseudo-labels. The maximum distances were set to 0.35, 0.6, 0.67, and 0.6 for Market-1501, DukeMTMC-reID, MSMT17, and VeRi datasets, respectively. At each round of label assignment, the memory bank is initialized with the average cluster center, and it will be updated with a momentum of 0.1 after each iteration. The temperature factors τ are 0.05 for all datasets. We used the Adam optimizer with a weight decay of 5×10^{-4}. The initial learning rate was 3.5×10^{-4} and the basis degree λ_0 was 1.0. The learning rate decreases by 0.1 after every 20 epochs, and the training ends at the 50th epoch.

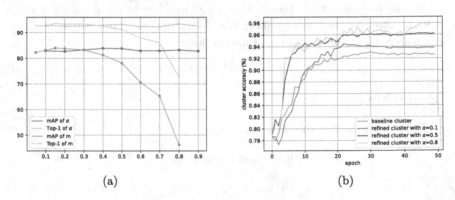

(a)   (b)

Fig. 3. (a) The impact of hyper parameter α and m on Market1501. (b) Clustering accuracy of the pseudo-labels obtained by the clustering algorithm on Market1501.

4.3   Comparison with State-of-the-Arts

Table 2 presents a comparison between the proposed method and other state-of-the-art USL re-ID methods. Our method achieves 84.0% mAP on Market1501 and 73.4% mAP on DukeMTMC-reID. In particular, the ISE method [26] generate support samples from actual samples and their neighbouring clusters in the embedding space. The results on the MSMT dataset indicate that our method has better generalization performance. On the two larger datasets, MSMT17 and VeRi, our method obtains better results, achieving 46.0% and 39.1%, respectively. The experimental results on four datasets showed a steady improvement

Table 2. Comparison of the proposed method to other state-of-the-art models on Market-1501, DukeMTMC, MSMT17, and VeRi776 dataset. The bold font represents the best value in each metric.

Model		Market		DukeMTMC		MSMT		VeRi	
		mAP	Top-1	mAP	Top-1	mAP	Top-1	mAP	Top-1
BUC [13]	AAAI-19	38.3	66.2	27.5	47.4	–	–	–	–
MMT(UDA) [11]	ICLR-20	71.2	87.7	63.1	76.8	–	–	–	–
SPCL [8]	NeurIPS-20	73.1	88.1	–	–	19.1	42.3	36.9	79.9
RLCC [20]	CVPR-21	77.7	90.8	69.2	83.2	27.9	56.5	39.6	83.4
ICE [14]	ICCV-21	79.5	92.0	67.2	81.3	29.8	59.0	–	–
Cluster-Contrast [19]	ACCV-22	82.1	92.3	72.6	84.9	27.6	56.0	40.3	84.6
PPLR [15]	CVPR-22	81.5	92.8	–	–	31.4	61.1	41.6	**85.6**
SECRET(UDA) [10]	AAAI-22	83.0	92.9	69.2	82.0	33.0	62.0	–	–
SECRET(USL)	AAAI-22	81,0	92.6	63.9	77.9	31.3	60.4	–	–
ISE [26]	CVPR-22	**84.7**	**94.0**	–	–	35.0	64.7	–	–
LRCR	this paper	83.8	93.2	**73.4**	**85.5**	**39.1**	**65.2**	**46.0**	82.9

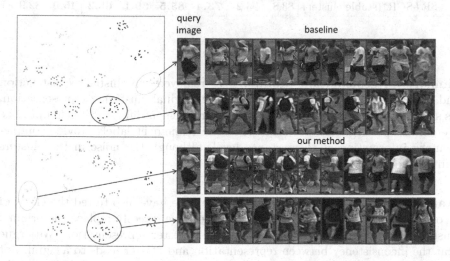

Fig. 4. T-SNE visualization of 10 random samples on Market-1501 and query results.

of our method compared to the baseline approach. In Fig. 4, our method obtains more correct matches in the query results for the ten random samples. Identically, the t-SNE result also reveals that our method has a more compact clustering result than the baseline.

4.4   Ablation Studies

We investigated the effects of our proposed design on the Market-1501 dataset. The experiments include the effects of hyper-parameters and modules. In each experiment, all settings were kept constant except the one mentioned.

Table 3. Performance (%) on ablation experiment over different modules. * denoted as using different information as the source of reconstruction

Method	Market		Duke		MSMT		VeRi	
	mAP	Top1	mAP	Top1	mAP	Top1	mAP	Top1
baseline	79.6	91.0	65.7	79.6	27.5	56.1	36.3	78.4
+SR	81.7	92.5	71.8	84.2	28.8	56.2	38.0	79.7
+SCR	55.0	78.3	57.0	74.9	17.1	40.1	28.4	67.2
+SR+SCR*(label)	82.1	92.8	72.6	84.6	37.2	63.3	45.3	**83.4**
+SR+SCR*(query)	**84.0**	**93.6**	73.1	85.4	37.7	64.8	45.7	83.3
+SR+SCR*(stable cluster)	83.8	93.2	**73.4**	**85.5**	**39.1**	**65.2**	**46.0**	82.9

Similarity Recalculation Module. We removed the cluster reconstruction and modified the refinement labels with the original clustering representation. As shown in row 2 of Table 3, the generated refinement labels still help optimize the details' performance. We argue that the refinement labels provide complementary information at the instance level, although the noise in the clusters blurred the cluster representation.

Stable Cluster Reconstruction Module. We have also tested the effect of reconstruction in isolation. In row 3 of Table 3, we directly replace the original cluster representation with the reconstructed cluster representation. We argue that the inconsistency between representation and cluster leads to a significant degradation in performance. Except for the current epoch of pseudo-labels, additional information is required to reconstruct the cluster. In the rest of the Table 3, We tried three different kinds of information used to reconstruct the cluster. These include the pseudo-labels from the previous epoch, the multi-class query results after the previous epoch training process, and the reconstructed clusters from the previous epoch. All the different information sources contributed to the results. We obtained the best results using multi-class query results on Market1501 and previously reconstructed clusters on the other datasets.

The Proportion of Refinement Strategy. Our method helps to improve the accuracy of pseudo labels. Figure 3(b) show the cluster accuracy of $\alpha = 0.1$, $\alpha = 0.5$ and $\alpha = 0.8$. When $\alpha = 0.8$, the pseudo-label is close to the ground truth but reduces the number of available samples by half, resulting in severe performance degradation. If only refinement labels are trained (set α over 0.9), the clustering algorithm cannot even generate enough categories, causing training to crash. The refinement label provides different information from the cluster, and it is this information that helps to separate the noise labels.

5 Conclusion

This paper proposes a label refinement strategy based on stable reconstruction clusters. The strategy comprises stable cluster reconstruction and representative-based similarity recalculation modules. Initially, stable clusters are reconstructed across generations, making the cluster representation explicit and generating refined labels. Our label refinement strategy provides the deep neural network with instance-level complementary information and progressively separates the noise in the pseudo labels during iterative training. Experimental results demonstrate that our method outperforms the baseline approach and achieves state-of-the-art performance on four datasets.

Acknowledgement. This work is supported by Industry-University Cooperation Project of Fujian Science and Technology Department (No. 2021H6035), the Science and Technology Planning Project of Fujian Province (No. 2021J011191), and Fujian Key Technological Innovation and Industrialization Projects (No. 2023XQ023), and Fu-Xia-Quan National Independent Innovation Demonstration Project (No. 2022FX4).

References

1. Ming, Z., et al.: Deep learning-based person re-identification methods: a survey and outlook of recent works. Image Vis. Comput. **119**, 104394 (2022)
2. He, S., Luo, H., Wang, P., Wang, F., Li, H., Jiang, W.: TransReID: transformer-based object re-identification. In: Proceedings of the IEEE/CVF International Conference on Computer Vision, pp. 15013–15022 (2021)
3. Chen, Y., Wang, H., Sun, X., Fan, B., Tang, C., Zeng, H.: Deep attention aware feature learning for person re-identification. Pattern Recogn. **126**, 108567 (2022)
4. Wang, D., Zhang, S.: Unsupervised person re-identification via multi-label classification. In: Proceedings of the IEEE/CVF Conference on Computer Vision and Pattern Recognition, pp. 10981–10990 (2020)
5. Fu, Y., Wei, Y., Wang, G., Zhou, Y., Shi, H., Huang, T.S.: Self-similarity grouping: a simple unsupervised cross domain adaptation approach for person re-identification. In: proceedings of the IEEE/CVF International Conference on Computer Vision, pp. 6112–6121 (2019)
6. Zhai, Y., et al.: AD-Cluster: augmented discriminative clustering for domain adaptive person re-identification. In: Proceedings of the IEEE/CVF Conference on Computer Vision and Pattern Recognition, pp. 9021–9030 (2020)
7. Ester, M., Kriegel, H.P., Sander, J., Xu, X., et al.: A density-based algorithm for discovering clusters in large spatial databases with noise. In: KDD, vol. 96, pp. 226–231 (1996)

8. Ge, Y., Zhu, F., Chen, D., Zhao, R., et al.: Self-paced contrastive learning with hybrid memory for domain adaptive object re-id. Adv. Neural. Inf. Process. Syst. **33**, 11309–11321 (2020)

9. Oord, A.v.d., Li, Y., Vinyals, O.: Representation learning with contrastive predictive coding. arXiv preprint arXiv:1807.03748 (2018)

10. He, T., Shen, L., Guo, Y., Ding, G., Guo, Z.: Secret: Self-consistent pseudo label refinement for unsupervised domain adaptive person re-identification. In: Proceedings of the AAAI Conference on Artificial Intelligence, vol. 36, pp. 879–887 (2022)

11. Ge, Y., Chen, D., Li, H.: Mutual mean-teaching: Pseudo label refinery for unsupervised domain adaptation on person re-identification. arXiv preprint arXiv:2001.01526 (2020)

12. Lin, Y., Xie, L., Wu, Y., Yan, C., Tian, Q.: Unsupervised person re-identification via softened similarity learning. In: Proceedings of the IEEE/CVF Conference on Computer Vision and Pattern Recognition, pp. 3390–3399 (2020)

13. Lin, Y., Dong, X., Zheng, L., Yan, Y., Yang, Y.: A bottom-up clustering approach to unsupervised person re-identification. In: Proceedings of the AAAI Conference on Artificial Intelligence, vol. 33, pp. 8738–8745 (2019)

14. Chen, H., Lagadec, B., Bremond, F.: Ice: inter-instance contrastive encoding for unsupervised person re-identification. In: Proceedings of the IEEE/CVF International Conference on Computer Vision, pp. 14960–14969 (2021)

15. Cho, Y., Kim, W.J., Hong, S., Yoon, S.E.: Part-based pseudo label refinement for unsupervised person re-identification. In: Proceedings of the IEEE/CVF Conference on Computer Vision and Pattern Recognition, pp. 7308–7318 (2022)

16. Lian, J., Wang, D.H., Du, X., Wu, Y., Zhu, S.: Exploiting robust memory features for unsupervised reidentification. In: Yu, S., et al. Pattern Recognition and Computer Vision. PRCV 2022. Lecture Notes in Computer Science, vol. 13535, pp. 655–667. Springer, Cham (2022). https://doi.org/10.1007/978-3-031-18910-4_52

17. He, K., Fan, H., Wu, Y., Xie, S., Girshick, R.: Momentum contrast for unsupervised visual representation learning. In: Proceedings of the IEEE/CVF Conference on Computer Vision and Pattern Recognition, pp. 9729–9738 (2020)

18. Caron, M., Misra, I., Mairal, J., Goyal, P., Bojanowski, P., Joulin, A.: Unsupervised learning of visual features by contrasting cluster assignments. Adv. Neural. Inf. Process. Syst. **33**, 9912–9924 (2020)

19. Dai, Z., Wang, G., Yuan, W., Zhu, S., Tan, P.: Cluster contrast for unsupervised person re-identification. In: Proceedings of the Asian Conference on Computer Vision, pp. 1142–1160 (2022)

20. Zhang, X., Ge, Y., Qiao, Y., Li, H.: Refining pseudo labels with clustering consensus over generations for unsupervised object re-identification. In: Proceedings of the IEEE/CVF Conference on Computer Vision and Pattern Recognition, pp. 3436–3445 (2021)

21. Zheng, L., Shen, L., Tian, L., Wang, S., Wang, J., Tian, Q.: Scalable person re-identification: a benchmark. In: Proceedings of the IEEE International Conference on Computer Vision, pp. 1116–1124 (2015)

22. Ristani, E., Solera, F., Zou, R., Cucchiara, R., Tomasi, C.: Performance measures and a data set for multi-target, multi-camera tracking. In: Hua, G., Jégou, H. (eds.) ECCV 2016. LNCS, vol. 9914, pp. 17–35. Springer, Cham (2016). https://doi.org/10.1007/978-3-319-48881-3_2

23. Wei, L., Zhang, S., Gao, W., Tian, Q.: Person transfer GAN to bridge domain gap for person re-identification. In: Proceedings of the IEEE Conference on Computer Vision and Pattern Recognition, pp. 79–88 (2018)

24. Liu, X., Liu, W., Mei, T., Ma, H.: A deep learning-based approach to progressive vehicle re-identification for urban surveillance. In: Leibe, B., Matas, J., Sebe, N., Welling, M. (eds.) ECCV 2016. LNCS, vol. 9906, pp. 869–884. Springer, Cham (2016). https://doi.org/10.1007/978-3-319-46475-6_53
25. Zhong, Z., Zheng, L., Kang, G., Li, S., Yang, Y.: Random erasing data augmentation. In: Proceedings of the AAAI Conference on Artificial Intelligence, vol. 34, pp. 13001–13008 (2020)
26. Zhang, X., et al.: Implicit sample extension for unsupervised person re-identification. In: Proceedings of the IEEE/CVF Conference on Computer Vision and Pattern Recognition, pp. 7369–7378 (2022)

Ranking Variance Reduced Ensemble Attack with Dual Optimization Surrogate Search

Zhichao He and Cong Hu$^{(\boxtimes)}$

School of Artificial Intelligence and Computer Science, Jiangnan University, Wuxi, China
conghu@jiangnan.edu.cn

Abstract. Deep neural networks have achieved remarkable success, but they are vulnerable to adversarial attacks. Previous studies have shown that combining transfer-based and query-based attacks in ensemble attacks can generate highly transferable adversarial examples. However, simply aggregating the output results of various models without considering the gradient variance among different models will lead to a suboptimal result. Moreover, fixed weights or inefficient methods for model weight updates can result in excessive query iterations. This paper proposes a novel method called Ranking Variance Reduced Ensemble attack with Dual Optimization surrogate search (RVREDO) as an enhanced ensemble attack method for improving adversarial transferability. Specifically, RVREDO mitigate the influence of gradient variance among models to improve the attack success rate of generated adversarial examples during the attack process. Simultaneously, a dual optimization weight updating strategy is employed to dynamically adjust the surrogate weight set, enhancing the efficiency of perturbation optimization. Experimental results on the ImageNet dataset demonstrate that our method outperforms previous state-of-the-art methods in terms of both attack success rate and average number of queries.

Keywords: Adversarial attack · Adversarial transferability · Ensemble attack

1 Introduction

Deep neural networks (DNNs) have achieved remarkable performance in various computer tasks [6,16,22,24,29]. However, their susceptibility to adversarial examples, which involve imperceptible perturbations added to original images [8–10], poses a significant security threat to DNNs and their real-world applications. Adversarial examples can deceive the target model and cause misclassification of

---

This work was supported in part by the National Natural Science Foundation of China (Grant No. 62006097, U1836218), in part by the Natural Science Foundation of Jiangsu Province (Grant No. BK20200593) and in part by the China Postdoctoral Science Foundation (Grant No. 2021M701456).

the original images. To address this vulnerability, research on adversarial attacks in DNNs focuses on enhancing the security, robustness, and transferability of models, leading to advancements in adversarial training and defense strategies [25].

Adversarial attacks can be divided into two categories: white-box attacks and black-box attacks. In black-box attacks, the adversary does not have access to the information of the target model, including its training process and parameter details. Black-box attacks can be further classified into transfer-based attacks and query-based attacks. The former requires a large number of queries, while the latter suffers from low attack success rates. In order to achieve high attack success rates with fewer queries, recent works have combined transfer-based and query-based attack methods, such as BASES [1]. Compared to previous state-of-the-art methods [4, 11, 18, 21], BASES achieves higher attack success rates with very few query iterations. Although BASES is simple and efficient, its weight update strategy does not take into account the impact of different models on perturbation generation, resulting in excessive query consumption. Moreover, traditional ensemble attack can result in a decrease in attack success rates due to the diverse optimization paths among different models.

To address the limitations mentioned above, we propose a novel approach called Ranking Variance Reduced Ensemble Attack with Dual Optimization Surrogate Search (RVREDO). This approach aims to enhance the adversarial transferability of ensemble attacks. More specifically, in our attack framework, a dual optimization strategy is proposed for weight updates. This approach dynamically adjusts the weights of the models by leveraging the output of the ensemble model. In targeted attacks, when the adversarial example has a low probability of being classified as the target class in the white-box model, it suggests that certain robust features need to be disrupted further. By increasing the weight of that model, the generation of perturbations targeting those features is enhanced. Conversely, for models that have already been successfully attacked, their weights are decreased to maintain a balanced weight distribution. Furthermore, inspired by the work of Xiong et al. [26], the method of ranking variance reduction in ensemble attacks is introduced. In contrast to traditional ensemble attacks, this approach incorporates M internal iterations while preserving the external ensemble gradients. During each internal iteration, the "unbiased gradient" is calculated and accumulated by selecting a white-box model in a ranked order. Adversarial examples updated using these carefully crafted gradients exhibit a higher likelihood of successful transfer to the target black-box model. The contributions of this paper are as follows:

- We propose an effective black-box attack method called RVREDO. This method achieves high success rates in targeted attacks by reducing the gradient variance among models.
- RVREDO utilizes a simple yet efficient weight update strategy to improve query efficiency. This strategy can be combined with ensemble attack methods, further reducing the number of queries required.
- Extensive experiments are conducted on ImageNet dataset, demonstrating the superiority of RVREDO in terms of attack success rate and average number of queries.

2   Related Work

Query-Based Attacks. This attack strategy involves querying the target model and utilizing its feedback to construct adversarial examples. In order to minimize the complexity of queries, certain methods aim to reduce the dimensionality of the search space or leverage transferable priors and surrogate models for query generation [7,11,15,20,21]. For instance, TREMBA [11] trains a generator to initially generate adversarial examples that deceive the source white-box model and then conducts further exploration in low-dimensional spaces. ODS [21] focuses on maximizing diversity in the outputs of the target model to generate more varied perturbations. GFCS [18] performs the search along the direction of the substitute gradient and switches to ODS if the search is unsuccessful. Simulator Attack [19] reduces query complexity by transferring a majority of queries to the simulator.

Transfer-Based Attacks. The fundamental concept behind transfer-based attacks is to utilize local surrogate models to create perturbations and exploit the transferability of these perturbations to deceive the target black-box model. Li *et al.* [14] propose a method to enhance the transferability of gradient-based targeting attacks by identifying and constraining two characteristics of white-box targeting attacks. AA [13] aims to improve the transferability of generated perturbations across different models by perturbing the feature space of the model. FDA [12] focuses on modeling layer-wise and class-wise feature distributions of a white-box model and utilizes this information to modify the label of an adversarial image. Additionally, [21,28,30] train surrogate models in a data-free manner for transferable attacks.

Ensemble Attacks. Liu *et al.* [17] discovered that simultaneously attacking multiple models can enhance the transferability of the attack. By leveraging the differences and diversities among various models, ensemble attacks can overcome the limitations of individual models and offer more effective attack methods [2,26]. MGAA [27] randomly selects a subset of surrogate models from an ensemble and performs meta-training and meta-testing steps to reduce the disparity between white-box and black-box gradient directions. BASES [1] introduces a two-layer optimization framework that iteratively updates the adversarial perturbation based on the linear combination of fixed surrogate models and adjusts the weights of each white-box model according to the query feedback from the target model.

3   Method

3.1   Preliminaries

Adversarial attacks involve iteratively adding imperceptible noise to the original image, resulting in an adversarial example denoted as $x^* = x + \delta$ [23]. The objective is to generate an adversarial perturbation δ for a benign example x to deceive the model $f_\theta(\cdot)$, where θ represents the unknown model parameter. This

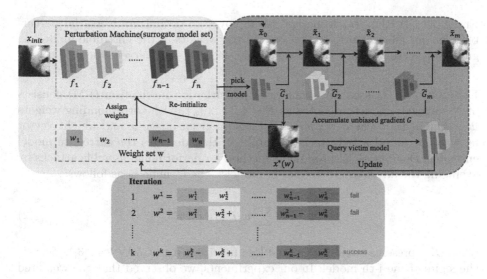

Fig. 1. The structure of RVREDO. (**Top**)We first define a perturbation machine consisting of n surrogate models and its weight set w. The initial example x_{init} are inputted into the perturbation machine, and the Ranking Variance Reduced Ensemble attack is employed to optimize the adversarial examples $x^*(w)$. (**Bottom**) Dual Optimization weight update strategy is utilized to effectively optimize the perturbations by searching for the optimal weight set. This process continues until the attack succeeds or the maximum number of queries is reached.

is known as a black-box adversarial attack. To maintain human imperceptibility, a threshold is set on the ℓ_p norm of the perturbation, such that $||\delta||_p \leq \epsilon$. Untargeted attacks aim to generate a adversarial example with a predicted result different from the true class, while targeted attacks seek to match a pre-defined class. The generation of adversarial examples can be formulated as the following optimization problem:

$$\begin{cases} f(x^*) \neq y \text{ for untargeted attack }, \\ f(x^*) == y^* \text{ for targeted attack }, \end{cases} \text{s.t.} ||\delta||_p \leq \varepsilon \qquad (1)$$

Here, $||\delta||_p$ represents the ℓ_p norm of the perturbation, $f(x^*)$ denotes the output of the target model for the adversarial example x^*, y represents the class of the original image and y^* is the desired target class. The objective is to find the minimal perturbation δ that satisfies the attack objective while keeping the perturbation within the chosen norm constraint.

3.2   Perturbation Machine with RVRE

The perturbation machine is defined as an ensemble of surrogate models, with its weight set represented by w. Using this ensemble, queries are generated to evaluate the target model, as illustrated in Fig. 1. The Ranking Variance Reduced

Ensemble (RVRE) attack method is employed to optimize adversarial perturbations. For efficient attacks, we update the weights of the ensemble models to guide the perturbation towards effectively deceiving the target model.

Ensemble on Losses. Assuming our perturbation mechanism (PM) consists of N surrogate models, denoted as $F = \{f_1, ..., f_n\}$, with corresponding weights $w = [w_1, ..., w_n]$, where $\sum_1^n w_i = 1$. For any given image x and weight vector w, we aim to find a adversarial example x^* that deceives the surrogate model ensemble. Dong et al. [5] proposed a method for performing ensemble attacks by averaging the adversarial losses of N models. The formula is as follows:

$$J(\boldsymbol{x}, y) = \sum_{i=1}^{N} w_i \mathcal{L}\left(f_i(x), y^*\right) \tag{2}$$

where $\mathcal{L}$ represents some adversarial loss function (e.g., C&W loss [3]). $f_i(x)$ is the logits of the i-th model. In our experiments, we observed that the weighted loss formulation outperformed the weighted probabilities and logits formulations, leading us to specifically empoly it.

Loss Function. We primarily use margin loss, which has been proven to be effective in C&W [3] attacks. The problem is formulated as follows:

$$\mathcal{L}\left(f(x), y^*\right) = \max\left(\max_{j \neq y^*} f(x)_j - f(x)_{y^*}, -\kappa\right) \tag{3}$$

where the parameter κ is a margin or threshold value. If the difference between the highest non-target logit and the target logit is larger than κ, the loss becomes zero. Otherwise, the loss is negative, penalizing the model for not sufficiently separating the target class from other classes.

Ranking Variance Reduced Ensemble Attack. The concept of variance reduction in ensemble attacks was introduced to address the issue of different optimization paths resulting from variations in the structures of models involved in traditional ensemble attacks. To mitigate the influence of gradient variance among different models, we incorporated an internal loop with M iterations. As shown in Algorithm 1, we compute the gradients g^{ens} of multiple models by backpropagating the weighted loss. This value remains unchanged during the M iterations in the internal loop. Subsequently, based on the adversarial losses of the external examples on all surrogate models, ranked in descending order, we sequentially select a surrogate model and compute the unbiased gradient $\tilde{g}$. We use the accumulated unbiased gradient to update the internal adversarial examples, while the gradient accumulated in the last internal loop is utilized to optimize the external adversarial examples. By reducing the gradient variance among different models, the unbiased gradient enhances the transferability of the generated perturbations.

Algorithm 1. Perturbation Machine with RVRE

Input: A initial example x_{init}(original or perturbed image) and it's target class $y^*(y^* \neq y$ if untargeted attack), N surrogate models $\mathcal{F} = \{f_1, ..., f_n\}$, and it's corresponding weights set $w = [w_1, ..., w_n]$, external step size α, internal step size β, number of outer update steps T, number of inner update steps M, the perturbation norm ℓ_∞ and the bound ϵ.
The loss founction L as equation 3.
Output: Adversarial example $x^*(w)$
1: Initialize $G_0 = 0$, $x_0^* = x_{init}$
2: **for** $t = 0$ to $T - 1$ **do:**
3:    Get the ensemble loss $J_{ens} = \sum_{i=1}^{N} w_i \cdot L_i(f(x_t^*), y^*)$
4:    Calculate the gradient of the ensemble model $g_t^{ens} = \nabla_x J_{ens}$
5:    Initialize $\tilde{G}_0 = 0$, $\tilde{x}_0 = x_t^*$
6:    Get the model index set: $list = argsort\{L_1, ..., L_n\}$
7:    **for** $m = 0$ to $M - 1$ **do:**
8:       Pick the model index $k = list[m\%N]$, and it's corresponding loss L_k
9:       $\tilde{g}_m = w_k \cdot (\nabla_x L_k (f(\tilde{x}_m), y^*) - \nabla_x L_k (f(x_t^*), y^*)) + g_t^{ens}$
10:      $\tilde{G}_{m+1} = \tilde{G}_m + \tilde{g}_m$
11:      Update $\tilde{x}_{m+1} = Clip_x^\epsilon \left\{ \tilde{x}_m - \beta \cdot \text{sign}\left(\tilde{G}_{m+1}\right) \right\}$
12:   **end for**
13:   $G_{t+1} = G_t + \tilde{G}_M$
14:   Update $x_{t+1}^* = Clip_x^\epsilon \{x_t^* - \alpha \cdot \text{sign}(G_{t+1})\}$
15: **end for**
16: return $x^*(w)$

3.3 Dual Optimization Weight Update Strategy

Many ensemble attack methods simply average the weights of surrogate models or generate queries using a random weight allocation. These methods overlook the impact of different surrogate models on perturbation generation, leading to low attack success rates and excessive query iterations. To overcome this issue, we propose a Dual Optimization (DO) weight update strategy that optimizes the surrogate weight set based on the output results of different models.

The basic idea is as follows: if the adversarial loss of an adversarial example is high in a particular surrogate model, it suggests that certain features of the current adversarial example are robust to that model. In order to deceive the target model, we increase the weight of that model to guide the generation of perturbations and disrupt those robust features. Conversely, for models with lower adversarial loss, we decrease their weights while maintaining the sum of all model weights equal to 1. The optimization objective can be formulated as follows:

$$\mathbf{w} = \arg\min_{\mathbf{w}} \mathcal{L}_\mathbf{v} (f_\mathbf{v} (x^*(w)), y^*) \tag{4}$$

where $f_\mathbf{v}$ represents the target black-box model and $\mathcal{L}_\mathbf{v}$ represents the adversarial loss of the target model. The objective is to find the optimal weights w that minimize the adversarial loss between the predictions of the target black-box model $f_{\mathbf{v}}$ on the adversarial examples $x^*(w)$ and the desired target labels y^*.

$$w_{n_1} = argmax\,[L_1, ..., L_N]$$
$$w_{n_2} = argmin\,[L_1, ..., L_N] \tag{5}$$

Initially, we set the weights of the ensemble models to equal values of $1/n$ and generate an initial adversarial example $x^*(w)$. Based on the feedback obtained by querying the target model, we determine whether to update the model weights. If the attack fails, we select the model indices w_{n_1} and w_{n_2} that correspond to the maximum and minimum adversarial loss values of the adversarial example among the ensemble models, as shown in Eq. 5. We then update w_{n_1} as $w_{n_1} + \eta$ and w_{n_2} as $w_{n_2} - \eta$, where η represents the step size. Next, we input the adversarial example into the PM with the updated weights for further perturbation. This process continues until the attack succeeds or reaches the maximum allowed number of queries.

4   Experiments

4.1   Experiment Setup

Surrogate and Victim Models. In our method, both the surrogate models and the target black-box models are pretrained image classification models from the PyTorch torchvision library. In the experiments, we selected three different model architectures as the target black-box models: VGG19, DenseNet121, and ResNeXt50. As for the surrogate models, we chose a set of 20 models, including VGG-16-BN, ResNet-18, SqueezeNet-1.1, GoogleNet, MNASNet-1.0, DenseNet-161, EfficientNet-B0, RegNet-y-400, ResNeXt-101, Convnext-Small, ResNet-50, VGG-13, DenseNet-201, Inception-v3, ShuffleNet-1.0, MobileNet-v3-Small, Wide-ResNet-50, EfficientNet-B4, RegNet-x-400, and VIT-B-16. These models have diverse architectures, which is helpful for constructing an efficient ensemble attack.

Dataset. We primarily used a dataset of 1000 ImageNet images from the NeurIPS-17 Adversarial Attack and Defense Competition. This dataset consists of carefully selected target classes and is highly suitable for testing our attack method.

Baseline. As BASES [1] has shown excellent performance, our experiments mainly compare with it. In addition, two other powerful methods, TREMBA [11] and GFCS [18], are also included in our comparison.

Query Budget. In the real world scenarios, most commercial models are charged based on the number of queries made. Therefore, we set a very limited maximum number of queries of 50 for both our method and BASES. To provide a fair comparison with GFCS and TREMBA, we set their maximum number of queries to 500.

Table 1. Average number of queries and attack success rate of different attack methods.

method	VGG19		DenseNet121		ResNext50	
	Targeted	Untargeted	Targeted	Untargeted	Targeted	Untargeted
TREMBA	60.45	2.03	45.98	4.9	71.52	4.19
	88.1%	99.7%	89.9%	99.5%	85.4%	98.9%
GFCS	124.9	13.88	87.03	16.07	107.28	16.4
	85.5%	99.8%	93.2%	99.3%	89.1%	99.7%
BASES	3.05	1.2	1.68	1.13	1.79	1.18
	96.2%	99.9%	99.6%	99.9%	**99.8%**	**100%**
RVREDO	**2.58**	**1.1**	**1.56**	**1.1**	**1.62**	**1.1**
	97.2%	**100%**	**99.8%**	**100%**	99.7%	**100%**

Hyper-parameters Setup. We evaluated our method under ℓ_∞, with commonly used perturbation budgets of $\ell_\infty \leq 16$. In all experiments, we set the number of external iterations T to 10 and the number of internal iterations M to 8. The step sizes for both the internal update α and the external update β are set to $3*1.6$. The weight update step size, η, is set to 0.005.

4.2 Attack Effect Analysis

We conducted a comprehensive comparison of our attack method with BASES [1], TREMBA [11], and GFCS [18]. Since our attack method builds upon BASES, the attack settings are generally consistent. However, TREMBA requires training a generator for each target label, which can be time-consuming, particularly when testing against 1,000 target classes from ImageNet. To ensure a fair comparison with TREMBA, we utilized the pre-trained generators from TREMBA for six target labels (0, 20, 40, 60, 80, 100). These generators were employed to attack the same target model, and we averaged the attack results and number of queries. Additionally, GFCS is also an ensemble attack method, so we integrated 20 identical surrogate models and fine-tuned the hyperparameters to achieve optimal performance. As illustrated in Table 1, our method yielded nearly the best results for both targeted and untargeted attacks.

Figure 2 compares the black-box targeted attack performance of the four methods on three different target models. It can be observed that our method achieves higher attack success rates with fewer number of queries compared to BASES. When the number of queries approaches 50, both methods achieve nearly perfect attack performance, but our method still outperforms BASES. TREMBA and GFCS show a significant decrease in attack performance when the number of queries is limited. Even when the number of queries is set to four times that of our method, RVREDO achieves high success rates exceeding of 34.2% and 15.5% compared to TREMBA and GFCS, respectively, in experiments conducted on ResNeXt-50. In the real world, our method is more efficient than these two methods.

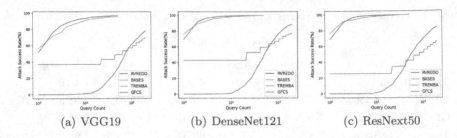

(a) VGG19   (b) DenseNet121   (c) ResNext50

Fig. 2. Four attack methods were compared in targeted attacks against three victim models, with a perturbation budget of $\ell_\infty \leq 16$. We set the maximum number of queries to 50 for both our method and BASES, while GFCS and TREMBA are set to 200.

Table 2. Ablation Study by cutting of different components.

	VGG19		DenseNet121		ResNext50	
	Targeted	Untargeted	Targeted	Untargeted	Targeted	Untargeted
RVREDO	**2.579**	**1.1**	**1.56**	**1.1**	**1.62**	**1.1**
	97.2%	**100%**	**99.8%**	**100%**	99.7%	**100%**
w/o DO	2.88	1.12	1.74	1.16	1.77	**1.1**
	96.3%	**100%**	99.7%	**100%**	99.7%	**100%**
w/o RVRE	2.583	1.17	1.63	1.14	1.67	1.13
	96.6%	99.9%	99.5%	99.9%	**99.9%**	100%
BASES	3.05	1.2	1.68	1.13	1.79	1.18
	96.2%	99.9%	99.6%	99.9%	99.8%	**100%**

4.3   Ablation Study

In this section, we will investigate the impact of the RVRE and DO on the experimental results. "w/o RVRE" indicates the traditional ensemble attack method optimization for perturbation. "w/o DO" denotes the utilization of the surrogate weight searching method from BASES, where the model weights are updated sequentially every two queries. To ensure a fair experimental comparison, we employed the same attack settings on three black-box target models. The experimental results are presented in Table 2.

The experimental results indicate that both RVRE and DO can enhance the transferability of adversarial examples. Specifically, in the targeted attack experiments on the VGG19, RVRE achieves 4 more successful attacks compared to the baseline, while reducing the average number of queries by approximately 0.5. DO achieves one additional successful attack and reduces the average number of queries by 0.2. The combination of RVRE and DO leads to a 1% improvement in attack success rate, with the average number of queries reduced to 2.579.

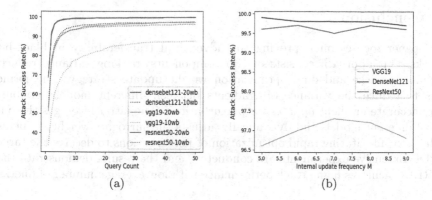

(a)   (b)

Fig. 3. left: Comparison of targeted attack success rate with different number of ensemble models N. right: the targeted attack success rate on different internal update frequency M.

4.4   Comparison on Hyper-parameters

In this analysis, we investigate the influence of two critical hyperparameters, namely the ensemble model size (N) and the internal update frequency (M), on the performance of our method.

We consider the number of ensemble surrogate models as a form of training data, analogous to the training process of models. Ensemble of multiple models helps improve the transferability of adversarial examples. Here, we choose N as 10 and 20 to evaluate the performance of our method on three black-box models. As shown in Fig. 3(a), when N is set to 20, we achieve nearly perfect attack success rates on all target models within the 50-query limit. The attack performance is compromised when N is reduced to 10. This is particularly evident in the case of VGG19, where the attack success rate drops from 97.2% to 87.5%. In theory, a larger number of ensemble models leads to better transferability of generated perturbations. However, excessive ensemble models result in increased computational complexity. Therefore, in our experiments, we primarily set the ensemble model count to 20.

In Fig. 3(b), we investigate the influence of the internal update frequency on the transferability of adversarial examples, specifically focusing on experiments conducted on VGG19. Given that DenseNet121 and ResNext50 have already achieved near-optimal performance, we analyze the results obtained with VGG19. The experiments reveal that the best attack performance is attained when M is set to 7. Increasing M further helps mitigate the impact of inter-model variance on the perturbations, thereby improving their transferability. However, continued increases in M result in a decline in performance, likely due to overfitting of the perturbations to the ensemble models during multiple iterations, which ultimately diminishes their attack effectiveness.

5   Conclusion

This paper focuses on improving the adversarial transferability in black-box attacks. Our approach consists of two components: ranking variance reduced ensemble attacke and dual optimization weight update strategy. The former aims to reduce the variance of gradients between different models, enabling more accurate gradient updates for perturbations. The latter leverages the output of different models and dynamically adjusts the surrogate weight set before each query, facilitating rapid optimization of perturbations to deceive the target model. Extensive experiments are conducted, and the results demonstrate that RVREDO achieves high attack performance with low average number of queries.

References

1. Cai, Z., Song, C., Krishnamurthy, S., Roy-Chowdhury, A., Asif, S.: Blackbox attacks via surrogate ensemble search. In: NeurIPS (2022)
2. Chen, H., Zhang, Y., Dong, Y., Zhu, J.: Rethinking model ensemble in transfer-based adversarial attacks. CoRR abs/2303.09105 (2023)
3. Chen, J., Wu, X., Guo, Y., Liang, Y., Jha, S.: Towards evaluating the robustness of neural networks learned by transduction. In: ICLR. OpenReview.net (2022)
4. Cheng, S., Dong, Y., Pang, T., Su, H., Zhu, J.: Improving black-box adversarial attacks with a transfer-based prior. In: NeurIPS, pp. 10932–10942 (2019)
5. Dong, Y., et al.: Boosting adversarial attacks with momentum. In: Proceedings of the IEEE Conference on Computer Vision and Pattern Recognition (CVPR) (2018)
6. Duan, Y., Lu, J., Zheng, W., Zhou, J.: Deep adversarial metric learning. IEEE Trans. Image Process. **29**, 2037–2051 (2020)
7. Guo, C., Gardner, J.R., You, Y., Wilson, A.G., Weinberger, K.Q.: Simple black-box adversarial attacks. In: ICML, Proceedings of Machine Learning Research, vol. 97, pp. 2484–2493. PMLR (2019)
8. Hu, C., Li, Y., Feng, Z., Wu, X.: Attention-guided evolutionary attack with elastic-net regularization on face recognition. Pattern Recogn. **143**, 109760 (2023)
9. Hu, C., Wu, X., Li, Z.: Generating adversarial examples with elastic-net regularized boundary equilibrium generative adversarial network. Pattern Recognit. Lett. **140**, 281–287 (2020)
10. Hu, C., Xu, H.Q., Wu, X.J.: Substitute meta-learning for black-box adversarial attack. IEEE Signal Process. Lett. **29**, 2472–2476 (2022). https://doi.org/10.1109/LSP.2022.3226118
11. Huang, Z., Zhang, T.: Black-box adversarial attack with transferable model-based embedding. In: ICLR. OpenReview.net (2020)
12. Inkawhich, N., Liang, K.J., Carin, L., Chen, Y.: Transferable perturbations of deep feature distributions. In: ICLR. OpenReview.net (2020)
13. Inkawhich, N., Wen, W., Li, H.H., Chen, Y.: Feature space perturbations yield more transferable adversarial examples. In: 2019 IEEE/CVF Conference on Computer Vision and Pattern Recognition (CVPR), pp. 7059–7067 (2019). https://doi.org/10.1109/CVPR.2019.00723
14. Li, M., Deng, C., Li, T., Yan, J., Gao, X., Huang, H.: Towards transferable targeted attack. In: Proceedings of the IEEE/CVF Conference on Computer Vision and Pattern Recognition, pp. 641–649 (2020)

15. Li, S., et al.: Adversarial attacks on black box video classifiers: leveraging the power of geometric transformations. In: NeurIPS, pp. 2085–2096 (2021)
16. Lin, S., et al.: Towards optimal structured CNN pruning via generative adversarial learning. In: CVPR, pp. 2790–2799. Computer Vision Foundation/IEEE (2019)
17. Liu, Y., Chen, X., Liu, C., Song, D.: Delving into transferable adversarial examples and black-box attacks. In: ICLR (Poster). OpenReview.net (2017)
18. Lord, N.A., Müller, R., Bertinetto, L.: Attacking deep networks with surrogate-based adversarial black-box methods is easy. In: ICLR. OpenReview.net (2022)
19. Ma, C., Chen, L., Yong, J.: Simulating unknown target models for query-efficient black-box attacks. In: CVPR, pp. 11835–11844. Computer Vision Foundation/IEEE (2021)
20. Suya, F., Chi, J., Evans, D., Tian, Y.: Hybrid batch attacks: finding black-box adversarial examples with limited queries. In: USENIX Security Symposium, pp. 1327–1344. USENIX Association (2020)
21. Tashiro, Y., Song, Y., Ermon, S.: Diversity can be transferred: output diversification for white- and black-box attacks. In: NeurIPS (2020)
22. Tian, C., Xu, Y., Li, Z., Zuo, W., Fei, L., Liu, H.: Attention-guided CNN for image denoising. Neural Netw. **124**, 117–129 (2020)
23. Wang, G., Yan, H., Wei, X.: Enhancing transferability of adversarial examples with spatial momentum. In: Yu, S., et al. Pattern Recognition and Computer Vision. PRCV 2022. LNCS, vol. 13534. Springer, Cham (2022). https://doi.org/10.1007/978-3-031-18907-4_46
24. Wang, M., Deng, W.: Deep face recognition: a survey. Neurocomputing **429**, 215–244 (2021)
25. Xie, C., Wu, Y., van der Maaten, L., Yuille, A.L., He, K.: Feature denoising for improving adversarial robustness. In: CVPR, pp. 501–509. Computer Vision Foundation/IEEE (2019)
26. Xiong, Y., Lin, J., Zhang, M., Hopcroft, J.E., He, K.: Stochastic variance reduced ensemble adversarial attack for boosting the adversarial transferability. In: CVPR, pp. 14963–14972. IEEE (2022)
27. Yuan, Z., Zhang, J., Jia, Y., Tan, C., Xue, T., Shan, S.: Meta gradient adversarial attack. In: ICCV, pp. 7728–7737. IEEE (2021)
28. Zhang, J., Li, B., Xu, J., Wu, S., Ding, S., Zhang, L., Wu, C.: Towards efficient data free blackbox adversarial attack. In: 2022 IEEE/CVF Conference on Computer Vision and Pattern Recognition (CVPR), pp. 15094–15104 (2022). https://doi.org/10.1109/CVPR52688.2022.01469
29. Zhang, W., Gou, Y., Jiang, Y., Zhang, Y.: Adversarial VAE with normalizing flows for multi-dimensional classification. In: Yu, S., et al. (eds.) Pattern Recognition and Computer Vision. PRCV 2022. LNCS, vol. 13534. Springer, Cham (2022). https://doi.org/10.1007/978-3-031-18907-4_16
30. Zhou, M., Wu, J., Liu, Y., Liu, S., Zhu, C.: DaST: data-free substitute training for adversarial attacks. In: CVPR, pp. 231–240. Computer Vision Foundation/IEEE (2020)

Performance Evaluation
and Benchmarks

PCR: A Large-Scale Benchmark for Pig Counting in Real World

Jieru Jia[1,2,3]($\boxtimes$), Shuorui Zhang[1,2,3], and Qiuqi Ruan[4]

[1] Institute of Big Data Science and Industry, Shanxi University, Taiyuan, China
jierujia@sxu.edu.cn
[2] Engineering Research Center for Machine Vision and Data Mining of Shanxi
Province, Taiyuan 030006, China
[3] School of Information and Technology, Shanxi University, Taiyuan 030006, China
[4] Institute of Information Science, Beijing Jiaotong University, Beijing 100044, China

Abstract. Automatic pig counting with pattern recognition and computer vision techniques, despite its significance in intelligent agriculture, remains to be a relatively unexplored area and calls for further study. In this paper, we propose a large-scale image-based Pig Counting in Real world (PCR) dataset, covering a variety of real-world scenarios and environmental factors. The dataset consists of two subsets, i.e., PartA captured on real-world pig pens and PartB collected from the Internet, with center point annotations of pig torsos in 4844 images. Moreover, we develop an automatic pig counting algorithm based on weakly-supervised instance segmentation, which can output a single segmentation blob per instance via the proposed Segmentation-Split-Regression (SSR) loss, utilizing point-level annotations only. Experiments show that the proposed algorithm achieves state-of-the-art counting accuracy and exhibits superior robustness against challenging environmental factors. The dataset and source codes are available at https://github.com/jierujia0506/PCR.

Keywords: pig counting · computer vision · instance segmentation · deep neural network

1 Introduction

Counting the number of pigs in a barn is a critical task in modern large-scale farming management. Accurate pig counting can greatly assist in pig management and early detection of abnormal events such as missing pigs. However, up to now, this task still relies heavily on manual labor, which is prone to errors and inefficient. Moreover, frequent trading and the birth of new pigs result in significant fluctuations in the number of pigs, making the task even more challenging for humans. Therefore, there is a urgent need to automate pig counting using pattern recognition and computer vision techniques.

Supplementary Information The online version contains supplementary material available at https://doi.org/10.1007/978-981-99-8462-6_19.

It is fascinating to see artificial intelligence meets agriculture. For example, computer vision algorithms have been exploited to count oil palm trees [5], cranberries [1], banana bunches [23], estimate animal poses [14], to name a few. A few attempts have also been made in pig counting [4,20]. However, the task remains quite challenging due to factors such as heavy pig overlapping, occlusions, changes in illumination and variations in viewpoint, as shown in Fig. 1. Generally speaking, the problem of pig counting is still far from being solved with limited works in the literature and the inferior accuracy. Furthermore, there lacks publicly available large-scale benchmarks and datasets for pig counting. Although there are some existing animal counting datasets such as Animal-Drone [29], CIW [2] and Aerial Elephant Dataset [13] et al., they primarily focus on wildlife species and are captured from drone views, which may be difficult to access for farmers. Currently, there is no publicly available flat-view dataset specifically designed for pig counting.

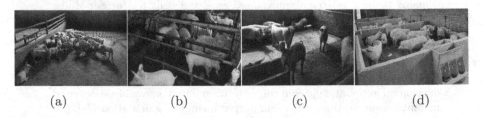

(a)   (b)   (c)   (d)

Fig. 1. Illustrations of pig counting challenges, including (a) overlapping of pigs, (b) occlusions, (c) illumination changes, (d) variations in viewpoint.

To address the lack of publicly available datasets for pig counting, we collect a large-scale dataset named PCR (**P**ig **C**ounting in **R**eal World), which consists of two subsets, i.e., PCR-PartA and PCR-PartB. Specifically, images in PCR-PartA are shot on real-world pig pens, while PCR-PartB is collected from Internet. PartA contains 1860 images with 13,172 center point annotated pigs and PartB consists of 2984 images with 27,453 annotated pigs. To sum up, the PCR dataset contains 4844 images with 40,625 labeled pigs in diverse scenes. The images are taken under different viewpoints, lighting conditions and occlusions, which closely mimic the challenges that may be encountered in real-world scenarios. In summary, the images in PCR-PartA reflects the actual conditions in farms and adds to the authenticity of the dataset, while the addition of internet-collected images in PCR-PartB further broadens the dataset, capturing a wider range of scenarios and increasing the diversity of the dataset.

In addition, we put forward a novel solution to address the pig counting task based on weakly-supervised instance segmentation, which only requires point-level annotations of the center of each pig. Concretely, we design a novel Segmentation-Split-Regression (SSR) loss function, which combines multiple loss terms to guide the model in learning accurate instance segmentation and counting. The Segmentation loss encourages the model to predict the semantic label of each pixel while the Split loss separates overlapping instances to output a single segmentation blob per instance. The Regression loss directly predicts the correct number of pigs in the image, which is an essential component for accurate pig counting. The whole network is trained in an end-to-end manner with the proposed SSR loss. Extensive experiments are carried out on the PCR dataset and the proposed algorithm achieves the state-of-the-art counting accuracy and exhibits superior robustness against the challenging environmental factors. To summarize, the main contributions of this paper are:

- We collect and annotate a large-scale image-based pig counting dataset, i.e., Pig Counting in Real world (PCR), which can greatly benefit the research community by providing a valuable resource for training and evaluating pig counting algorithms.
- We propose a practical and cost-effective pig counting approach based on weakly-supervised instance segmentation via the proposed Segmentation-Split-Regress (SSR) loss, which can output a single segmentation blob per instance given point-level annotations, thus sparing the requirement of exhaustive bounding-box or pixel-level annotation.
- Extensive experiments on PCR dataset are conducted, demonstrating that the proposed method can achieve state-of-the-art performance with evidently small Mean Absolute Error and Mean Square Error.

2 Related Work

2.1 Existing Animal Counting Datasets

There exist some animal based counting datasets, for example, a thermal infrared video dataset is proposed in [25] to count bats within the given bounding boxes. Arteta et al. [2] propose a large-scale Counting in the Wild (CIW) dataset of penguins in Antarctica, which contains 500 thousand images in over 40 different sites. The Aerial Elephant Dataset [13] includes 2,101 images with 15,511 African bush elephants for the aerial object detection task. Rey et al. [17] collects 654 images of animals detected in African Savanna with UAVs and annotate the animals with convex hull polygons. Shao et al. [18] construct two subsets of pasture aerial images for the cattle detection and counting task. The AnimalDrone [29] dataset contains 53,644 frames with over 4 million object annotations.

The majority of the aforementioned datasets have focused on wildlife species such as elephants, yaks, and others, but none of them are specifically dedicated to pig counting. Furthermore, most of these datasets are collected using drones or unmanned aerial vehicles (UAVs), where only the backs of pigs can be seen. Moreover, these collection devices are not easily accessible for farmers in real-life scenarios. As a result, existing datasets are not suitable for the pig counting task in agricultural and farming automation. To the best of our knowledge, the PCR dataset is the first large-scale specifically designed pig counting dataset, filling a significant gap in the research literature.

2.2 Object Counting Methods

In recent years, numerous methods have been introduced for counting various types of objects, with a particular focus on crowd counting. These methods can be broadly categorized into four main branches: regression-based, density map estimation-based, detection-based, and segmentation-based approaches.

Regression Based Methods: Early approaches in object counting utilize regression-based algorithms, which involve mapping extracted image features to corresponding output numbers through machine learning. For example, Glance [3] learns to count using only image-level labels and is accurate when the object count is small. However, regression-based methods have gradually fallen out of favor with the rise of deep learning, as they require manual feature design, which can be time-consuming and challenging.

Density Based Methods: Density map estimation-based methods require annotated object positions in the image, and the ground-truth density map is obtained through Gaussian kernel convolution. Machine learning methods, such as deep learning, are then utilized to estimate density maps for training samples. For example, MCNN [27] uses filters of different sizes to generate density maps with improved robustness to scale changes. CSRNet [9] employs dilated convolutions to yield larger receptive fields. Wang et al. [22] introduce a Spatial Fully Convolutional Network with a spatial encoder that uses a sequence of convolutions in four directions. Zhou et al. [28] design a double recursive sparse self-attention module to capture long-distance dependencies. Zand et al. [26] propose a multiscale and multitask approach based on point supervision for crowd counting and person localization. OT-M [11] presents a parameter-free crowd localization method which alternates between transport plan estimation and point map updating.

Despite their effectiveness and the exciting progress in recent years, this line of methods are not suitable for counting pigs, as they assume objects have fixed sizes and shapes, which clearly does not hold in the case of pig counting. Pigs can vary greatly in sizes and shapes, making it challenging to accurately estimate density maps using Gaussian kernels with fixed sizes.

Detection Based Methods: Counting by detection is a straightforward approach that involves first detecting the objects of interest and then outputting their location and class to determine the count. There are numerous well-developed object detection algorithms, such as YOLO [15], RetinaNet [10], and Faster R-CNN [16]. However, these methods may not be suitable for counting pigs in dense scenes where objects are heavily occluded, as shown in Fig. 1(a). Moreover, these object detection methods typically rely on bounding-box labels, which can be more expensive to obtain compared to our point-level annotations.

Segmentation Based Methods: These methods share a similar principle as detection-based methods but utilize segmentation blobs to represent the instances instead of the complete bounding boxes. The intuition is that predicting the exact size and shape of objects may not be necessary for counting tasks, and a rough segmentation blob per instance would suffice. LCFCN [8] is a representative method that has shown promising performance in counting various types of objects, including people, cars, and animals.

However, LCFCN has limitations in handling highly overlapped and mutually occluded objects, especially when objects are only a few pixels apart, as shown in Fig. 1(a). To address this limitation, the loss function in LCFCN is improved in this paper by directly imposing regression loss on the error between the predicted count and the ground-truth counts. This can mitigate the issue of over-prediction of counts due to too many tiny segmentation regions, which can occur due to over-segmentation. The effectiveness of the proposed improvement is verified through extensive experiments.

2.3 Pig Counting Methods

As for the pig counting task, Tian et al. [20] present a modified Counting network based on ResNeXt to output the density maps. The approach belongs to the density map estimation-based methods, therefore the performance are highly sensitive to the quality of ground-truth density maps and the choice of kernel parameters, which can be challenging to determine accurately. Chen et al. [4] put forward a network to detect five keypoints of pig body parts and associate the keypoints to identify individual pigs. However, the method fails to perform well when only single-point annotation is available. A semantic segmentation and counting network based on DeepLab V3 is proposed in [12], where pixel-level annotation is required.

Different from previous methods, we present a novel and efficient solution for pig counting based on weakly-supervised instance segmentation, where only center point annotation of each instance is needed. Unlike Chen et al. [4], which requires keypoints detection and association, the proposed method can work well for a single image with limited field of view, which is common in practical scenarios. In addition, the proposed method doesn't require the tedious manual tuning of the kernel parameters as in [20] or the full per-pixel annotation in [12].

3 PCR Dataset

3.1 Data Collection

The PCR dataset consists of two subsets, i.e., PCR-PartA and PCR-PartB. The images of PCR-PartA are captured on site in Beijing, Hebei and Shanxi Provinces of China. These images are taken by pig owners or employees of a farming insurance company with various types of cameras, including a Canon EOS 90D SLR camera, surveillance cameras, and mobile phone cameras. As a result, the images in PCR-PartA have significantly different resolutions and shooting perspectives, reflecting the diversity of real-world scenarios. The collection environment includes both indoor and outdoor settings, with varying weather conditions. Meanwhile, the pigs in the images range from newborn piglets to those weighing over 200 kg, representing a wide range of pig ages and sizes commonly encountered in pig farming. In a nutshell, the collected images contain different image resolutions, shooting perspectives, environmental factors, and pig characteristics that can affect pig appearances, which makes it a challenging and realistic benchmark for evaluating pig counting approaches. Importantly, proper permissions have been obtained for data usage, and privacy concerns have been addressed.

The images in PCR-PartB are collected from the Baidu website using specific keywords such as "pig herds," "pig pens," and "hogs." After a careful selection process, 2984 valid piggery images are chosen. These images are selected based on their similarity in shooting scenes and perspectives with PCR-PartA, ensuring consistency and comparability between the two subsets. The inclusion of PCR-PartB adds further diversity to the dataset.

3.2   Data Pruning and Pre-processing

Since the raw collected images vary dramatically in sizes, resolutions and visual quality et al., several strategies have been employed to clean and preprocess the collected images: (1) We remove duplicated and redundant images (Some images are submitted to us multiple times); (2) We remove extremely blurred images that are even difficult to recognize the pigs. Finally, 1860 qualified images are selected into the PCR-PartA dataset. Similar strategies are applied on the PartB images to ensure that only high-quality and relevant images are included.

3.3   Data Annotation

In order to label the data, we have compared several kinds of annotations, including image-level labels, points, bounding-boxes and full pixel-level annotations. Among these methods, center point annotation was chosen as a suitable mediation between counting accuracy and annotation cost.

Specifically, a Matlab-based GUI is developed to obtain the coordinates of the center point of each pig body through a single click. Pigs that appeared in the image as less than 10% of the whole body area are considered invalid and not labeled. Each image is annotated by three groups of annotators separately, and the results are cross-validated between the groups. If the difference between the three annotation results is within the given threshold, they are included in the dataset; if the difference exceed the acceptable threshold, they are re-annotated until the difference is less than the given threshold. In total, 4844 valid images are obtained with a total of 40,625 pigs labeled, of which each image contains at least 1 pig and at most 52 pigs, with an average of 8.39 pigs per image.

4   Proposed Approach

4.1   Network Architecture

The proposed network architecture is an encoder-decoder based convolutional neural network, as shown in Fig. 2. The input image, denoted as $X \in \mathbb{R}^{H \times W \times 3}$, is passed through an encoder to obtain feature embeddings $E \in \mathbb{R}^{H' \times W' \times C}$, where $H' \times W'$ denotes the spatial dimensions and C is the number of channels. The encoder leverages pre-trained models such as ResNet [6], HRNet [19], or other similar architectures to extract meaningful features. The feature embeddings are then passed through the decoder to progressively recover the spatial information. To preserve fine-grained spatial details, skip connections are utilized, where the decoder features are concatenated or element-wise summed with the corresponding encoder features. Finally, a classification layer with softmax activation is added after the decoder to obtain the pixel-wise class probabilities, which form the segmentation map $S \in \mathbb{R}^{H \times W \times c}$, where c represents the number of classes. This segmentation map is then compared with the ground-truth annotations to calculate the SSR loss.

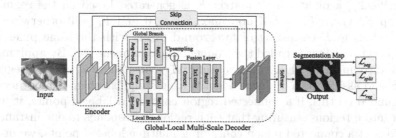

Fig. 2. The proposed network for pig counting. The input image is fed into the network to get the segmentation map, which is then exploited to compute the loss functions.

4.2   The SSR Loss Function

With the proposed network, we can obtain the segmentation map $S \in R^{H \times W \times c}$ where each element S_{ic} denotes the probability that pixel $i \in H \times W$ belongs to category c (for the pig counting task, there are only two categories, pig and background, hence $c = \{0, 1\}$). Y represents the ground-truth point annotation which has label c at the center location of each pig instance and zero elsewhere.

The proposed SSR (Semantic Segmentation Regression) loss function is defined by combining three different components: the segmentation loss, split loss, and regression loss. These components work together to enable the model to accurately predict pig counts in images. Despite the fact that the SSR loss cannot accurately predict the size and shape of the objects, it can flexibly output segmentation blobs for objects of different sizes and get the accurate location and count of the instances. The overall SSR loss function is defined by

$$\mathcal{L}_{SSR}(S, Y) = \mathcal{L}_{seg}(S, Y) + \mathcal{L}_{split}(S, Y) + \mathcal{L}_{reg}(S, Y) \tag{1}$$

Segmentation Loss. Guided by positive point annotations, $\mathcal{L}_{seg}$ aims to correctly predict the semantic label of each pixel in the image. Let $y_{c=1}$ and $y_{c=0}$

denote the positive and negative ground truth points derived from Y respectively, which indicates the presence or absence of pig instances, the segmentation loss can be defined as

$$\mathcal{L}_{seg}(S,Y) = -\sum_{y_{c=1}} \log(S) - \sum_{y_{c=0}} (1 - \log(S)) \qquad (2)$$

Split Loss. Utilizing segmentation loss alone would not suffice for the counting task since the segmented regions usually contain multiple instances. Therefore, it is necessary to split the segmentation regions containing more than one point-annotations to ensure that each blob corresponds to only one pig instance. The split loss penalizes blobs that contain multiple instances, forcing the model to separate them into distinct blobs during training.

Specifically, a binary mask matrix F is generated based on the segmentation map S, where $F_i = 1$ if arg $\max_k S_{ik} > 0$ and $F_i = 0$ otherwise. The image regions in F consisting of foreground pixels with the same pixel value and adjacent positions is called the Connected Component [24]. By applying the connected components algorithm, we can find connected areas in the foreground mask F and assign them a unique identity to distinguish them from others.

Grounded on this, if a connected region contains n labeled points, it has to be split into n regions, ensuring that each region corresponds to one instance. In order to split a connected region containing multiple labeled points, we need to first find the boundary between the instances and encourage the model to predict the boundaries as background. Specifically, we set the ground-truth annotations within each connected region as seeds and apply the watershed segmentation algorithm [21] to obtain n segments and the set of pixels belonging to the boundaries U. Therefore, the loss derived from watershed segmentation can be computed as

$$\mathcal{L}_{split}(S,Y) = -\sum_{i \in U} \alpha_i \log(S_{i0}) \qquad (3)$$

where U denotes the set of pixels representing the boundaries, S_{i0} stands for the probability that pixel i belongs to the background class and α_i refers to the number of point-annotations in the region that pixel i resides.

Regression Loss. For the sake of accurate counting, the regression loss based on Huber loss (i.e., smoothed L1 loss) is introduced to guide the model to directly output the correct number of instances. The count prediction, denoted as $\tilde{g}$, is obtained from the number of connected components derived from F, and the ground-truth number g is obtained from Y. Hence, the regression loss is defined as the Huber loss between the count prediction $\tilde{g}$ and the ground-truth count g:

$$\mathcal{L}_{reg} = \begin{cases} 0.5(\tilde{g} - g)^2 & if \ |\tilde{g} - g| < 1 \\ |\tilde{g} - g| - 0.5 & otherwise \end{cases} \qquad (4)$$

The loss term is deduced by setting the parameter δ in the original Huber loss to 1, i.e., when the error between the true and predicted count is greater than 1, it is minimized by L1 loss, which is less sensitive to outliers and can help avoid missing minimum values. When the error is less than 1, it is minimized by L2 loss, which provides a more stable gradient.

Albeit simple, the incorporation of the regression loss helps to make the SSR loss function more robust to outliers and reduces the issue of over-prediction due to over-segmentation, where the segmented regions are too small and numerous. Hence, the model is encouraged to accurately predict the count of pig instances in the image, making the model more suitable for accurate counting tasks.

5 Experiments

5.1 Implementation Details and Evaluation Setup

We evaluate the proposed method on the PCR dataset. To be specific, the overall 4844 images in PCR-PartA and PCR-PartB are mixed together as the training inputs. Among the 4844 images, 3874 images are randomly selected as the training set and 485 as validation set. The remaining 485 images are used for testing, following the ratio of 0.8:0.1:0.1. As for image resolutions, we first resize all the images to 576×768 and then randomly crop a patch of 384×512 for training.

All the experiments are conducted on a workstation with 2.10 GHz Intel(R) Xeon(R) Silver 4216 CPU, and one NVIDIA GeForce RTX 2080 GPU card. The network is trained with Adam optimizer [7] for 100 epochs ($\beta_1 = 0.99$, $\beta_2 = 0.999$, weight decay $5e - 4$) with the initial learning rate set as $1e - 5$. The batch size is set to 1 since the watershed segmentation algorithm does not support parallel processing of multiple images. Random horizontal flipping (with a probability of 0.5), color jitter and random erasing are applied to augment the training images. In addition, normalization with mean (0.485, 0.456, 0.406) and standard deviation (0.229, 0.224, 0.225) is applied on both the training and testing images for processing. HRNet [19] is chosen as the default encoder owing to its ability to maintain high resolution spatial details.

For evaluation metrics, we adopt the commonly used counting metrics Mean Absolute Error (MAE) and Root Mean Square Error (RMSE) to evaluate the performance of various methods. MAE indicates the accuracy of the counting algorithm while RMSE measures the robustness of methods and represents the degree of dispersion in the differences. They are formally defined as:

$$MAE = \frac{1}{N} \sum_{i=1}^{N} |g_i - \tilde{g}_i| \tag{5}$$

$$RMSE = \sqrt{\frac{1}{N} \sum_{i=1}^{N} |g_i - \tilde{g}_i|^2} \tag{6}$$

where N is the number of test images, and g and $\tilde{g}$ are the ground-truth and predicted counts respectively.

5.2   Results of the Proposed Approach

Figure 3 presents the results of the proposed approach on the PCR dataset, where the predicted segmentation blobs in the third row are assigned with random colors for better visualization. The heatmaps in the fourth row are obtained by representing the magnitude of blobs in the form of red colors. From Fig. 3, it can be observed that superior counting accuracy and robustness can be achieved in various situations such as overlapping (first column), occlusions (second column), illumination changes (third column) and viewpoint influence (last column).

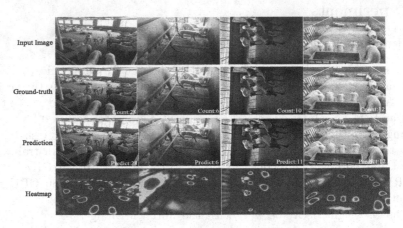

Fig. 3. Qualitative results of the proposed method on PCR dataset. From top to down are testing images, ground-truth, prediction masks and the heatmap. In the heatmap, the red region represents the larger probability that the pixel belongs to pigs, and for the region with lower probability, the color is blue. (Color figure online)

5.3   Comparisons Against State-of-the-Art

To verify the effectiveness of the proposed method, we conduct a comparison with other approaches on PCR dataset in Table 1. To the best of our knowledge, PigCount [20] is the only work that performs pig counting with single point annotation so far. The method in [4] needs to label five keypoints and [12] needs full pixel-level annotation. Therefore, comparisons with them wouldn't be totally fair. In addition, we make a comparison with other object counting methods, including density map estimation-based MCNN [27] et al. and prevalent object detection approaches YOLO [15], RetinaNet [10], Faster R-CNN [16]. Finally, a comparison with the representative segmentation-based method LCFCN [8] is conducted, which is also the baseline of the proposed method.

From Table 1, the following observations can be made: (1) The proposed method outperforms [20] by a considerable margin, with MAE lower than [20] by 1.38 and RMSE lower by 1.82, displaying much better accuracy and robustness.

This can be attributed to the fact that [20] heavily depends on the quality of the ground-truth density map and requires manual tuning of Gaussian kernel parameters, which may not be optimal when the target scales vary dramatically in the case of pig counting. (2) Similarly, the density map estimation-based crowd counting methods MCNN [27] et al. also yield inferior performance in the pig counting task where the size of the targets change significantly, due to the intrinsic defects of density maps. (3) Object detection-based methods, such as YOLO [15] et al. also perform poorly in highly congested scenes in pig counting, as these methods rely on bounding-box representations that are difficult to obtain when objects are heavily overlapped. Moreover, the commonly used Non-Maximum Suppression (NMS) strategy in object detection assigns a single label to overlapping objects, resulting in a large number of false negatives and a much lower predicted count. (4) The proposed method significantly and consistently improves the performance of LCFCN [8], with lower MAE and MSE by 0.63 and 0.71, respectively. This indicates that the improved SSR loss can bring substantial performance improvements. It is worth noting that the encoder of LCFCN is also replaced with HRNet to ensure a fair comparison. The SSR loss helps in circumventing the problem of overly high predicted counts caused by over-segmentation, leading to notable performance boosts.

Table 1. Comparison with state-of-the-art counting methods on PCR dataset.

Category	Method	MAE	RMSE
Density-map	PigCount [20]	1.89	2.83
Density-map	MCNN [27]	2.68	3.32
	CSRNet [9]	2.42	2.76
	SFCN [22]	2.12	2.50
	MSCCL [26]	1.86	2.04
Object detection	YOLO [15]	1.64	2.37
	RetinaNet [10]	1.53	2.21
	Faster R-CNN [16]	2.23	4.27
Segmentation	LCFCN [8]	1.14	1.72
	Ours	0.51	1.01

5.4   Ablation Analysis

To evaluate the contribution of each component in the proposed SSR loss, we separately add each part to the loss function and compare the performance in Table 2. We can see that training the model only with the segmentation loss $\mathcal{L}_{seg}$ results in a deteriorated counting performance with a large MAE and RMSE. This stems from the fact the segmentation loss can only induce a single region containing multiple instances, which is inadequate for the counting task. When the instance split loss $\mathcal{L}_{split}$ is introduced, the model is encouraged to predict blobs that contain no more than one point-annotation, which contributes notably to the increase of counting accuracy. However, there still exist some false positive predictions mainly caused by over-segmentation and false negative results due to the extremely highly overlapping issue, rendering a predicted count deviated from the ground-truth by a considerable margin. Therefore, the introduced regression loss $\mathcal{L}_{reg}$ leads to a further performance gain by penalizing the model from outputting large errors between the predicted and true counts. Overall, every component in the proposed SSR loss is essential, and they complement each other well. It is the joint effort of these components that makes the proposed loss function effective.

Table 2. Ablation study on the impact of different components in SSR loss.

$\mathcal{L}_{seg}$	$\mathcal{L}_{split}$	$\mathcal{L}_{reg}$	MAE	RMSE
✓			3.28	4.23
✓	✓		0.97	1.63
✓	✓	✓	0.51	1.01

6   Conclusion

In this paper, we have introduced a comprehensive framework for pig counting in real-world scenarios. The largest image-based pig counting dataset (PCR) is collected and annotated, which provides a valuable benchmark for pig counting research. The dataset contains 4844 images with 40,625 labeled pigs. Furthermore, we have proposed a strong pig counting baseline based on weakly-supervised instance segmentation using point supervision. The novel Segmentation-Split-Regression (SSR) loss function is developed to minimize the discrepancy between the predicted and ground-truth pig counts, while also ensuring accurate pixel-wise segmentation and blob splitting. Extensive evaluations have been conducted to validate the effectiveness of the proposed approach. We believe the benchmark dataset and proposed approach can contribute to the advancement of pig counting research. In the future, we will try to improve our network with better loss functions and data augmentation methods. Adapting this approach to count other animal crowds for agriculture and wildlife protection is another future avenue to invest.

Acknowledgment. This work was supported by the National Natural Science Foundation of China (62106133).

References

1. Akiva, P., Dana, K.J., Oudemans, P.V., Mars, M.: Finding berries: segmentation and counting of cranberries using point supervision and shape priors. In: CVPR Workshops, pp. 219–228 (2020)
2. Arteta, C., Lempitsky, V., Zisserman, A.: Counting in the wild. In: Leibe, B., Matas, J., Sebe, N., Welling, M. (eds.) ECCV 2016. LNCS, vol. 9911, pp. 483–498. Springer, Cham (2016). https://doi.org/10.1007/978-3-319-46478-7_30
3. Chattopadhyay, P., Vedantam, R., Selvaraju, R.R., Batra, D., Parikh, D.: Counting everyday objects in everyday scenes. In: CVPR, pp. 4428–4437 (2017)
4. Chen, G., Shen, S., Wen, L., Luo, S., Bo, L.: Efficient pig counting in crowds with keypoints tracking and spatial-aware temporal response filtering. In: ICRA, pp. 10052–10058 (2020)
5. Chowdhury, P.N., Shivakumara, P., Nandanwar, L., Samiron, F., Pal, U., Lu, T.: Oil palm tree counting in drone images. Pattern Recognit. Lett. **153**, 1–9 (2022)
6. He, K., Zhang, X., Ren, S., Sun, J.: Deep residual learning for image recognition. In: CVPR, pp. 770–778 (2016)
7. Kingma, D.P., Ba, J.: Adam: A method for stochastic optimization. In: ICLR (2015)
8. Laradji, I.H., Rostamzadeh, N., Pinheiro, P.O., Vazquez, D., Schmidt, M.: Where are the blobs: counting by localization with point supervision. In: Ferrari, V., Hebert, M., Sminchisescu, C., Weiss, Y. (eds.) ECCV 2018. LNCS, vol. 11206, pp. 560–576. Springer, Cham (2018). https://doi.org/10.1007/978-3-030-01216-8_34
9. Li, Y., Zhang, X., Chen, D.: CSRNet: dilated convolutional neural networks for understanding the highly congested scenes. In: CVPR, pp. 1091–1100 (2018)
10. Lin, T.Y., Goyal, P., Girshick, R.B., He, K., Dollár, P.: Focal loss for dense object detection. IEEE Trans. Pattern Ana. Mach. Intell **42**, 318–327 (2020)
11. Lin, W., Chan, A.B.: Optimal transport minimization: crowd localization on density maps for semi-supervised counting. In: CVPR, pp. 21663–21673 (2023)
12. Liu, C., Su, J., Wang, L., Lu, S., Li, L.: LA-DeepLab V3+: a novel counting network for pigs. Agriculture **12**(2), 284 (2022)
13. Naudé, J.J., Joubert, D.: The aerial elephant dataset: a new public benchmark for aerial object detection. In: CVPR Workshops (2019)
14. Rao, J., Xu, T., Song, X., Feng, Z., Wu, X.: KITPose: keypoint-interactive transformer for animal pose estimation. In: Yu, S., et al. (eds.) Pattern Recognition and Computer Vision. PRCV 2022. LNCS, vol. 13534. Springer, Cham. https://doi.org/10.1007/978-3-031-18907-4_51
15. Redmon, J., Divvala, S.K., Girshick, R.B., Farhadi, A.: You only look once: unified, real-time object detection. In: CVPR, pp. 779–788 (2016)
16. Ren, S., He, K., Girshick, R.B., Sun, J.: Faster R-CNN: towards real-time object detection with region proposal networks. IEEE Trans. Pattern Ana. Mach. Intell **39**, 1137–1149 (2015)
17. Rey, N., Volpi, M., Joost, S., Tuia, D.: Detecting animals in African savanna with UAVs and the crowds. Remote Sens. Environ. **200**, 341–351 (2017)
18. Shao, W., Kawakami, R., Yoshihashi, R., You, S., Kawase, H., Naemura, T.: Cattle detection and counting in UAV images based on convolutional neural networks. Int. J. Remote Sens. **41**, 31–52 (2020)
19. Sun, K., Xiao, B., Liu, D., Wang, J.: Deep high-resolution representation learning for human pose estimation. In: CVPR, pp. 5686–5696 (2019)

20. Tian, M., Guo, H., Chen, H., Wang, Q., Long, C., Ma, Y.: Automated pig counting using deep learning. Comput. Electron. Agric. **163**, 104840 (2019)
21. Vincent, L.M., Soille, P.: Watersheds in digital spaces: an efficient algorithm based on immersion simulations. IEEE Trans. Pattern Ana. Mach. Intell **13**, 583–598 (1991)
22. Wang, Q., Gao, J., Lin, W., Yuan, Y.: Learning from synthetic data for crowd counting in the wild. In: CVPR, pp. 8190–8199 (2019)
23. Wu, F., et al.: Detection and counting of banana bunches by integrating deep learning and classic image-processing algorithms. Comput. Electron. Agric. **209**, 107827 (2023)
24. Wu, K., Otoo, E.J., Shoshani, A.: Optimizing connected component labeling algorithms. In: SPIE Medical Imaging (2005)
25. Wu, Z., Fuller, N.W., Theriault, D.H., Betke, M.: A thermal infrared video benchmark for visual analysis. In: CVPR Workshops, pp. 201–208 (2014)
26. Zand, M., Damirchi, H., Farley, A., Molahasani, M., Greenspan, M., Etemad, A.: Multiscale crowd counting and localization by multitask point supervision. In: ICASSP (2022)
27. Zhang, Y., Zhou, D., Chen, S., Gao, S., Ma, Y.: Single-image crowd counting via multi-column convolutional neural network. In: CVPR, pp. 589–597 (2016)
28. Zhou, B., Wang, S., Xiao, S.: Double recursive sparse self-attention based crowd counting in the cluttered background. In: PRCV (2022)
29. Zhu, P., Peng, T., Du, D., Yu, H., Zhang, L., Hu, Q.: Graph regularized flow attention network for video animal counting from drones. IEEE Trans. Image Process. **30**, 5339–5351 (2021)

A Hierarchical Theme Recognition Model for Sandplay Therapy

Xiaokun Feng[1,2], Shiyu Hu[1,2], Xiaotang Chen[1,2,3], and Kaiqi Huang[1,2,3(✉)]

[1] School of Artificial Intelligence, University of Chinese Academy of Sciences, Beijing, China
{fengxiaokun2022,hushiyu2019}@ia.ac.cn
[2] Institute of Automation, Chinese Academy of Sciences, Beijing, China
[3] Center for Excellence in Brain Science and Intelligence Technology, Chinese Academy of Sciences, Beijing, China
{xtchen,kaiqi.huang}@nlpr.ia.ac.cn

Abstract. Sandplay therapy functions as a pivotal tool for psychological projection, where testers construct a scene to mirror their inner world while psychoanalysts scrutinize the testers' psychological state. In this process, recognizing the theme (*i.e.*, identifying the content and emotional tone) of a sandplay image is a vital step in facilitating higher-level analysis. Unlike traditional visual recognition that focuses solely on the basic information (*e.g.*, category, location, shape, *etc.*), sandplay theme recognition needs to consider the overall content of the image, then relies on a hierarchical knowledge structure to complete the reasoning process. Nevertheless, the research of sandplay theme recognition is hindered by following challenges: (1) Gathering high-quality and enough sandplay images paired with expert analyses to form a scientific dataset is challenging, due to this task relies on a specialized sandplay environment. (2) Theme is a comprehensive and high-level information, making it difficult to adopt existing works directly in this task. In summary, we have tackled the above challenges from the following aspects: (1) Based on carefully analysis of the challenges (*e.g.*, small-scale dataset and complex information), we present the **HIST** (**HI**erarchical **S**andplay **T**heme recognition) model that incorporates external knowledge to emulate the psychoanalysts' reasoning process. (2) Taking the split theme (a representative and evenly distributed theme) as an example, we proposed a high-quality dataset called **SP²** (**S**and**P**lay **SP**lit) to evaluate our proposed method. Experimental results demonstrate the superior performance of our algorithm compared to other baselines, and ablation experiments confirm the importance of incorporating external knowledge. We anticipate this work will contribute to the research in sandplay theme recognition. The relevant datasets and codes will be released continuously.

Keywords: Sandplay therapy · Sandplay theme recognition · Visual recognition

---

Supplementary Information The online version contains supplementary material available at https://doi.org/10.1007/978-981-99-8462-6_20.

Q. Liu et al. (Eds.): PRCV 2023, LNCS 14428, pp. 241–252, 2024.
https://doi.org/10.1007/978-981-99-8462-6_20

1   Introduction

Sandplay therapy functions as a pivotal tool for psychological projection, where individuals construct a scene to mirror their inner world while psychoanalysts scrutinize the individual's psychological state [1,2]. In this process, recognizing the *theme* (*i.e.*, identifying the content and emotional tone) of a sandplay image is an important step in facilitating higher-level analysis. As shown in Fig. 1, generating a sandplay image and recognizing its theme always include several steps – a client should first construct a sandplay work based on inner thoughts, and the psychoanalyst then analyzes the theme by synthesizing basic semantic information and high-level semantic information (*e.g.*, psychological knowledge).

Although identifying the theme of a sandplay image is an important and valuable issue, the current visual recognition algorithms are unable to adapt well to this task.

Early visual recognition research studies mainly focus on identifying basic semantic information in images, such as category [3], location [4], shape [5], *etc.* Recently, several researchers have shifted their focus to recognizing high-level visual information like emotions [6,7]. The emotion recognition task enables computer systems to process and comprehend emotional information conveyed by humans, and facilitates natural human-computer interactions [8]. However, unlike existing visual recognition tasks that focuses solely on basic semantic information or emotion, the sandplay theme recognition task is proposed for a more challenging and specialized application scenario. It aims to identify the content and emotional tone expressed in a sandplay image, which can support higher-level tasks such as understanding the psychological state of the creator like psychologists. Therefore, a well-designed sandplay theme recognition method should execute the above analyzing process like experts. It should first consider the overall content of the image, then rely on a hierarchical structure to utilize various knowledge (*e.g.*, basic semantic information, emotion information, and even external knowledge from psychologists), and finally complete the reasoning process.

Nevertheless, two challenges hinder the research of designing a suitable model for the sandplay theme recognition task:

(1) **Obtaining enough high-quality data samples is difficult. ❶** Unlike general scene visual recognition tasks that can easily collect numerous data samples from the Internet [6,9], the sandplay theme recognition task relies on a specialized sandplay environment. ❷ Besides, existing research commonly annotates data samples through crowd-sourcing platforms like Amazon Mechanical Turk, while the annotation of sandplay image should be generated by psychology experts. Thus, gathering high-quality and enough sandplay images paired with expert analyses to form a scientific dataset is challenging.

(2) **A sandplay image's theme information is complex, intensifying the difficulty of recognition.** Existing works have focused on recognizing basic semantics or emotions, but they lack a hierarchical framework for comprehensive understanding, making them difficult to adopt directly in the sandplay theme recognition task.

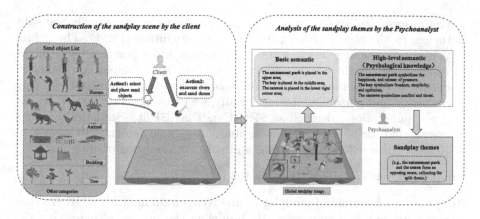

Fig. 1. Framework of sandplay themes recognition process on 3D electronic sandplay platform. Firstly, the client constructs a sandplay scene based on inner thoughts. Then, the psychoanalyst analyzes the sandplay themes by synthesizing basic semantic information and high-level semantic information (*i.e.*, psychological knowledge).

In this paper, we address the difficulties mentioned above and carry out our work, aiming to accomplish the challenging sandplay theme recognition task:

(1) **Design a hierarchical theme recognition model.** Based on carefully analyzing the challenges of the sandplay theme recognition task (*e.g.*, small-scale datasets and complex information), we present a recognition model named **HIST** (**HI**erarchical **S**andplay **T**heme recognition) that incorporates external knowledge. In light of the analysis process of psychologists, our proposed model comprises two fundamental steps. Firstly, we focus on perceiving the basic semantics of the image, which specifically refers to extracting the categorical information of the objects present in the image. Secondly, based on the perceived categorical information, we incorporate the external knowledge by indexing the corresponding high-level attribute information. Then, our model recognizes the sandplay theme by leveraging the above information. Specifically, to evaluate the effects of small-scale datasets caused by the characteristics of sandplay scenario, we also employ several training strategies to enhance the learning capacity of the model.

(2) **Propose a high-quality dataset to train and evaluate the proposed HIST model.** ❶ We take psychological sandplay as our experimental environment, and collect data samples from a 3D electronic platform (Fig. 1). Specifically, we invite a substantial number of testers to participate in sandplay test and collect data samples following the sandplay analysis process, which ensure the professionalism and scientificity of the data samples. According to statistics, each sandplay sample contains an average of 15 sand objects (selected from 494 sand categories), which reflects the diversity of sandplay samples. After screening and sorting, we collect 5,000 samples. ❷ We engage professional psychoanalysts to annotate the theme of the sandplay sample. Without affecting the integrity of the sandplay anal-

ysis process, we select one of the representative themes, namely the split theme, as the object of our psychological recognition (Fig. 1). The definition of split theme refers to a state of isolation and separation between the various parts of the whole sandplay scene (the split samples are shown in Fig. 3). It reflects the inner integration of the tester and is related to many emotional and personality issues. Finally, we construct a dataset with split theme annotations, denoted as SP^2 (SandPlay SPlit). Statistical analysis reveals that the acquisition of each sandplay sample demands an average of 10 min, which reflects the expense and time consumption of obtaining it.

In general, our contributions are as follows:

- We propose a hierarchical sandplay theme recognition model (*i.e.*, HIST) that incorporates external knowledge. By modeling the process of perceiving the basic semantics of images and incorporating corresponding high-level attribute knowledge, our model emulates the analytical ability of psychologists (Sect. 3).
- Based on the 3D electronic sandplay platform, we construct a dataset named SP^2. All the sandplay samples are carefully collected and annotated, aiming to propose a high-quality dataset for recognizing theme in the sandplay environment (Sect. 4). Besides, experimental results indicate the excellent performance of our model, and the ablation experiment demonstrates the importance of introducing external knowledge (Sect. 5).

We anticipate this work will contribute to the research in sandplay theme recognition. The relevant datasets and codes results will be released continuously.

2   Related Work

The recognition of sandplay themes can be regarded as an image recognition task, wherein the sandplay image serves as the input data, and the sandplay theme represents the output. In this section, we firstly introduce existing image recognition tasks to highlight the characteristics of sandplay theme recognition task (Sect. 2.1). Then we introduce the relevant background knowledge in the field of psychological sandplay (Sect. 2.2).

2.1   Image Recognition Tasks

Recognizing semantic information from image data is a fundamental research problem in the field of computer vision. Various research tasks aim to model different aspects of human abilities by focusing on different semantic information. For instance, basic image classification [3] tasks aim to emulate human visual perception abilities, while image emotion recognition tasks [9,10] aim to capture human emotional cognitive abilities. Although these tasks center around different forms of semantic information, their research [9–11] typically utilizes data collected from the Internet (*e.g.*, TV shows, social networks, *etc.*) and relies on crowd-sourcing platforms (like Amazon Mechanical Turk) for semantic label annotations. Consequently, the modeling object of these tasks is primarily centered around ordinary people.

In the context of sandplay theme recognition, the data samples are sampled from the specialized sandplay environment, and the annotation of sandplay themes requires the expertise of psychologists who employ hierarchical cognitive analysis. Through the sandplay theme recognition task, we can explore modeling the hierarchical analytical capabilities of psychoanalysts.

2.2   Projection Test and Sandplay Therapy

Projective test [12] is a well-known and widely used psychoanalysis test in which client offer responses to ambiguous stimuli (*e.g.*, words, images, *etc.*), so as to reveal the hidden conflicts or emotions that client project onto the stimuli. The deeply held feelings and motivations are often not verbalized or even be aware by client, while these subjective psychological semantics can be detected and analyzed through projecting onto the stimuli.

The visual stimulus is the main carrier of projection test, such as Rorschach inkblot image [13], house-tree-person painting and sandplay image [1]. The image carrier of inkblot test is only 10 inkblot pictures, and the painting image in house-tree-person test can only contain 3 elements (*i.e.*, house, tree, person). On the other hand, sandplay therapy usually contains hundreds of miniature figures, and they can be combined and placed in any way. The high degree of interactivity ensures a diverse range of sandplay samples, thus making it an ideal research scene for our work.

However, despite the potential of sandplay to generate numerous data samples, a large-scale sandplay dataset is currently unavailable. For the purpose of using existing data-driven deep neural network models, we invite a substantial number of testers to collect samples. As a result, we build a sandplay dataset consisting of 5,000 diverse samples.

Furthermore, according to the sandplay theme theory developed by Mitchell and Friedman [2], around 20 themes have been identified to encode psychological states (see A.1). This work takes the split theme as the research label, mainly considering the following two reasons: (1)the positive and negative sample size of the split is relatively balanced (see A.3), making it advantageous for model training; (2) split theme exhibits connections with various psychological symptoms, including some common emotional issues (*e.g.*, depression, negative study experiences) and personality problems (*e.g.*, compulsion, paranoia, marginalization, aggression), which reflects the practical application value of the sandplay. For annotation and research on other themes, we leave them for future work.

3   Method

Based on the preceding analysis, sandplay theme recognition task encounters two distinct challenges: small-scale datasets and the intricate processing of complex information (*i.e.*, sandplay themes). In this case, we propose the HIST model, and the framework is shown in Fig. 2. Corresponding to the hierarchical analysis process of psychologists, we model three hierarchical processes: the perception of

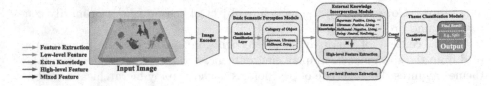

Fig. 2. Overview of the HIST model. Correspondingly to the hierarchical analysis process of psychoanalysts, this model consists of basic semantic perception module, external knowledge incorporation module and theme classification module.

basic semantics (Sect. 3.1), the incorporation of external knowledge (Sect. 3.2), and the final theme classification (Sect. 3.3).

3.1   Basic Semantic Perception Module

Perceiving basic semantic information (such as sand object's category, bounding box, *etc.*) from images is a crucial step for recognizing the theme information. In this process, our model focuses on the category information of the sand objects within the image, and we employ existing multi-label classification [14] techniques to handle this task.

Firstly, we adopt the common ResNet [15] backbone as the image encoder. Given the image $I \in \mathbf{R}^{3 \times 960 \times 540}$, we feed it into the image encoder and obtain the corresponding feature vector $F_i \in \mathbf{R}^{1 \times L}$, where the length of the feature vector $L = 1,000$. To alleviate the limitations imposed by the small-scale sandplay dataset, we utilize image augmentation techniques to expand the dataset's size. Additionally, we leverage the pre-trained model to extract the initial image features (see Sect. 5.1).

Then, F_i is fed into the fully connected layers (accompanied by the *Relu* activation function) to obtain the multi-label classification feature vector $F_{cl} \in \mathbf{R}^{1 \times N_{cl}}$. Here, N_{cl} represents the summary of sand objects' categories, and $N_{cl} = 494$ in our sandplay environment.

Finally, we perform multi-label classification based on F_{cl}. We apply the *Sigmoid* activation function on F_{cl} to obtain F_{clp}. Each element f_{clp}^i in F_{clp} represents the probability of the model classifying the i_{th} sand object. We construct the classification error L_{cl} based on the true labels $Y_c = \{y^1, y^2, ..., y^{N_{cl}}\}$.

$$L_{cl} = -\frac{1}{N_{cl}} \sum_{i=1}^{N_{cl}} [p^i y^i log f_{clp}^i + (1 - y^i) log(1 - f_{clp}^i)], p^i = \frac{neg_i}{pos_i} \qquad (1)$$

where p^i means the weight of the i_{th} object. In order to reduce the long tail distribution problem [16] between different objects, we define p^i as the ratio between i_{th} object's negative sample number neg_i and positive sample number pos_i.

3.2   External Knowledge Incorporation Module

For the sandplay theme recognition task, we consider the psychological attributes of each sand object as external knowledge. According to sandplay therapy, the number of these psychological attributes (denoted as L_h) is fixed, which means that the psychological attributes of each object can be represented by a one-dimensional vector $k_h \in \mathbf{R}^{1 \times L_h}$. In scenarios with N_{cl} objects, we can utilize a feature vector $K_h \in \mathbf{R}^{N_{cl} \times L_h}$ to encode the external knowledge. We set $L_h = 7$ (see Sect. 4.2) in subsequent experiments.

We employ object category indexing to incorporate the external knowledge K_h into our model. By leveraging F_{clp}, we utilize the probability of each object's existence f_{clp}^i, to weight the corresponding high-level semantic vector k_h^i. This process results in the construction of the vector $K_w \in \mathbf{R}^{N_{cl} \times L_h}$.

$$k_w^i = f_{clp}^i \times k_h^i \tag{2}$$

Then, in order to comprehensively analyze the weighted high-level semantic information, we employ self-attention operations to achieve high-level feature extraction. Specifically, we use a transformer encoder [17] module to process and obtain $F_{we} \in \mathbf{R}^{1 \times N_{cl}}$, consider it as the integrated external knowledge feature information.

3.3   Theme Classification Module

Building upon F_{we} and F_i, we can integrate the basic semantic information and external knowledge information of objects. Firstly, we employ the fully connected layers (accompanied by the *Relu* activation function) to further process F_i, obtaining F_i'. Next, we concatenate the F_i' with F_{we}, resulting in the final feature vector, denoted as F_u.

$$F_u = concat\{F_{we} + Relu(Fc(F_I))\} \tag{3}$$

Finally, we feed F_u into the fully connected layers (accompanied by the *Relu* activation function) to obtain the theme category vector $F_t \in \mathbf{R}^{1 \times 2}$ (*i.e.*, recognition for a specific sandplay theme can be regarded as a binary classification task). Based on groundtruth labels Y_t, we construct the classification error L_t using the CrossEntropy loss function.

Based on L_{cl} and L_t. We compute the final error L_a by the relative weight coefficient w_{cl}. L_a is used for backpropagation:

$$L_a = L_t + w_{cl} \times L_{cl} \tag{4}$$

4   SP^2 Dataset

Considering the challenge of gathering high-quality sandplay images paired with theme annotation from psychoanalysts, we take the representative split theme as the example, and construct the SP^2 dataset. In this section, we will firstly present the process of dataset construction (Sect. 4.1). Then, we present the psychological attribute information for each sand object (Sect. 4.2), which can be considered as external knowledge employed by psychoanalysts to achieve hierarchical analysis.

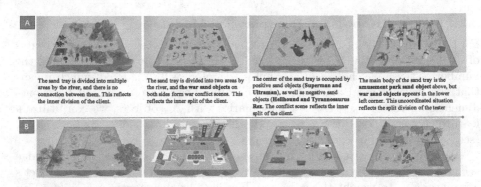

Fig. 3. Examples from the SP^2 dataset. Panel A represents sandplay samples with the split theme, accompanied by explanatory descriptions for the judgment (bold font indicates the concerned sand objects during judgment); Panel B represents sandplay samples without the split theme.

4.1 Dataset Construction

Sampling of Sandplay Samples. We invited a substantial number of testers to participate in the sandplay test and obtained corresponding sandplay samples. In order to ensure the authenticity, we only collected a sandplay sample from each tester. In the end, we sorted out 5,000 sandplay samples.

Labeling of Sandplay Samples. For each sandplay sample, we provide the basic semantic label (*i.e.*, the name and bounding box of each sand object) and the high-level sandplay theme label (*i.e.*, split label). For the former, with the help of the 3D electronic sandplay platform, we can directly obtain it from the terminal; For the latter, we engaged psychoanalysts to discern split theme through binary classification in order to ensure the objectivity of the split label. Initially, we selected 200 samples and each sample was labeled by five psychoanalysts. Through discussions and deliberations, we established consistent criteria for the recognition of split theme. Then, in order to improve annotation efficiency, each sample in the remaining data is labeled by one psychoanalyst.

Format of Sandplay Samples. A sandplay sample consists of a global image $I \in \mathbf{R}^{3 \times 960 \times 540}$, a set of basic semantic labels, and a split label.

In addition, we conducted statistics on the number of positive and negative samples of split label (positive samples mean the sandplay has the split theme). Among SP^2, there are 2,303 positive samples and 2,697 negative samples, which indicates that the number of positive and negative samples is relatively balanced.

4.2 Psychological Attributes of Sand Objects

During the analysis of sandplay images (as depicted in Fig. 1), psychoanalysts integrate the psychological attributes of each sand object to make their final assessments. For instance, the boy represents a positive object, whereas cannon represents a negative object. These psychological attributes associated with the sand objects can be considered as the external knowledge. Drawing from the principles of sandplay analysis, we have organized 7 key psychological attributes for each sand object, including polarity, life attribute, spiritual/material attribute, static/dynamic attribute, *etc*. For more details, please refer to A.2.

5 Experiments

Table 1. Comparison with the state-of-the-art models on SP^2 dataset. The best and second-best results are marked in **bold** and underline.

Model	Acc	F1
AlexNet	0.717	0.676
VGG	0.727	0.707
ResNet	0.742	<u>0.727</u>
ViT	0.731	0.712
MldrNet	0.723	0.710
zhang 2020	0.729	0.707
BiGRU	0.734	0.712
PadNet	<u>0.745</u>	0.722
HIST (Ours)	**0.790**	**0.765**

Table 2. Results of ablation experiments. The best and second-best results are marked in **bold** and underline.

Variants	Acc	F1
ViT-based	0.761	0.745
Without-L_{cl}	<u>0.764</u>	<u>0.750</u>
Without-K_h	0.747	0.723
HIST (Ours)	**0.790**	**0.765**

In this section, we conduct experiments on the SP^2 dataset for evaluation.

5.1 Experimental Setup

Implementation Details. We employ ResNet-50 [15] as the visual encoder, and we initialize it using the model weights pretrained on ImageNet [11]. We firstly reshape the input sandplay image $I \in \mathbf{R}^{3 \times 960 \times 540}$ into $I' \in \mathbf{R}^{3 \times 224 \times 224}$ to match the input of the model. We use basic image rotation and flipping operations for data augmentation. The batch size is set to 64, and the learning rate is set to $1e-3$. We use SGD optimizer, and the relative weight coefficient w_{cl} is set to 2.

Datasets and Metrics. The SP^2 dataset is divided into training, testing and validation sets in the 8:1:1 ratio (see A.3 for detailed information). We evaluate the performance of different models by Accuracy and F1 metric.

5.2   Comparison with the State-of-the-Art

Following the two recent authoritative works [6,18], we compare our model with representative image emotion recognition state-of-the-art methods which are publicly available and adaptable to sandplay images, including BiGRU [19], MldrNet [20], Zhang et al. [18] and PadNet [21]. Additionally, we compare with some classic image classification models, including AlexNet [22], VGG [23], ResNet [15], and ViT [17].

During training our proposed model, two types of supervised information(sand object category and split label) are used. So for the sake of fairness, we provide these two types of supervised information in a multi-task format when training these baseline models. Specifically, a sand object classification head is added in the final output layer of the model, and the model is trained using both object classification loss and split classification loss.

Based on the results presented in Table 1, it is evident that the performance of our proposed model surpasses that of other models. The essential distinction between our model and other models lies in the incorporation of external knowledge, highlighting the necessity of external knowledge for sandplay theme recognition.

5.3   Ablation Study

Effect of Visual Backbone. The visual backbone in our model is used to encode image. In addition to ResNet [15], ViT [17] serves as another popular choice for a backbone network. Hence, we conduct experiments to evaluate the performance of our model using the ViT backbone network (ViT-based).

Experimental results shown in Table 2 indicate that the ResNet based network outperforms ViT. (which is consistent with the experimental result in Table 1). This discrepancy may be attributed to the dataset size. Because prior studies [24] have shown that ViT has the advantageous performance in large scale datasets, and the dataset size of SP^2 is relatively small.

Effect of Semantic Information. Our proposed models aim to emulate the reasoning process of psychoanalysts by considering both basic semantic information and high-level semantic information. To evaluate the effects of different semantic inputs, we conduct experiments under two settings. Firstly, we evaluate the model's performance when the sand object category information is not provided, which means setting l_{cl} to 0 (Without-L_{cl}). Secondly, we evaluate the model's performance when external knowledge is not incorporated, which achieve that by masking K_h to 1 (Without-K_h).

Experimental results shown in Table 2 indicate that the lack of category information or external knowledge can reduce the performance of the model. Moreover, the lack of external knowledge results in more severe performance degradation. This result once again reflects the importance of incorporation external knowledge.

6   Conclusion and Feature Work

Based on the sandplay therapy, sandplay theme recognition task relies on sandplay images and the corresponding theme annotations provided by psychoanalysts. This task offers an opportunity to explore the modeling of psychoanalysts' hierarchical analysis capabilities. Inspired by the analysis process of psychoanalysts, we propose HIST model. To facilitate the sandplay theme recognition task, we construct SP^2 dataset, focusing on split theme. Our proposed model outperforms existing baseline models on the SP^2 dataset, and ablation experiments demonstrates the significance of incorporating external knowledge.

In this work, we leverage a sandplay-based research environment to highlight the significance of external knowledge in the assessment of sandplay theme which represents high-level psychological semantics. Moving forward, we intend to explore incorporating external knowledge into more general scenarios, such as emotion recognition tasks for common images. By incorporating external knowledge into these scenarios, we aim to enhance the machine's recognition capabilities and facilitate more natural human-computer interactions.

References

1. Roesler, C.: Sandplay therapy: an overview of theory, applications and evidence base. Arts Psychother. **64**, 84–94 (2019)
2. Mitchell, R.R., Friedman, H.S.: Sandplay: Past, Present, and Future. Psychology Press (1994)
3. Lu, D., Weng, Q.: A survey of image classification methods and techniques for improving classification performance. Int. J. Remote Sens. **28**(5), 823–870 (2007)
4. Zou, Z., Chen, K., Shi, Z., Guo, Y., Ye, J.: Object detection in 20 years: a survey. In: Proceedings of the IEEE (2023)
5. Guo, Y., Liu, Y., Georgiou, T., Lew, M.S.: A review of semantic segmentation using deep neural networks. Int. J. Multimed. Inf. Retrieval **7**, 87–93 (2018)
6. Zhao, S., et al.: Affective image content analysis: two decades review and new perspectives. IEEE Trans. Pattern Anal. Mach. Intell. **44**(10), 6729–6751 (2021)
7. Tao, J., Tan, T.: Affective computing: a review. In: Tao, J., Tan, T., Picard, R.W. (eds.) ACII 2005. LNCS, vol. 3784, pp. 981–995. Springer, Heidelberg (2005). https://doi.org/10.1007/11573548_125
8. Picard, R.W.: Building HAL: computers that sense, recognize, and respond to human emotion. In: Human Vision and Electronic Imaging VI, vol. 4299, pp. 518–523. SPIE (2001)
9. You, Q., Luo, J., Jin, H., Yang, J.: Robust image sentiment analysis using progressively trained and domain transferred deep networks. In: Proceedings of the AAAI Conference on Artificial Intelligence, vol. 29 (2015)
10. Jou, B., Chen, T., Pappas, N., Redi, M., Topkara, M., Chang, S.-F.: Visual affect around the world: a large-scale multilingual visual sentiment ontology. In: Proceedings of the 23rd ACM International Conference on Multimedia, pp. 159–168 (2015)
11. Deng, J., Dong, W., Socher, R., Li, L.-J., Li, K., Fei-Fei, L.: ImageNet: a large-scale hierarchical image database. In: 2009 IEEE Conference on Computer Vision and Pattern Recognition, pp. 248–255. IEEE (2009)

12. Otsuna, H., Ito, K.: Systematic analysis of the visual projection neurons of drosophila melanogaster. I lobula-specific pathways. J. Comp. Neurol. **497**(6), 928–958 (2006)

13. Gamble, K.R.: The Holtzman inkblot technique. Psychol. Bull. **77**(3), 172 (1972)

14. Tsoumakas, G., Katakis, I.: Multi-label classification: an overview. Int. J. Data Warehousing Min. (IJDWM) **3**(3), 1–13 (2007)

15. He, K., Zhang, X., Ren, S., Sun, J.: Deep residual learning for image recognition. In: Proceedings of the IEEE Conference on Computer Vision and Pattern Recognition, pp. 770–778 (2016)

16. Zhang, Y., Kang, B., Hooi, B., Yan, S., Feng, J.: Deep long-tailed learning: a survey. IEEE Trans. Pattern Anal. Mach. Intell. (2023)

17. Dosovitskiy, A., et al.: An image is worth 16×16 words: transformers for image recognition at scale. arXiv preprint arXiv:2010.11929 (2020)

18. Zhang, W., He, X., Lu, W.: Exploring discriminative representations for image emotion recognition with CNNs. IEEE Trans. Multimed. **22**(2), 515–523 (2019)

19. Zhu, X., et al.: Dependency exploitation: a unified CNN-RNN approach for visual emotion recognition. In: IJCAI, pp. 3595–3601 (2017)

20. Rao, T., Li, X., Min, X.: Learning multi-level deep representations for image emotion classification. Neural Process. Lett. **51**, 2043–2061 (2020)

21. Zhao, S., Jia, Z., Chen, H., Li, L., Ding, G., Keutzer, K.: PDANet: polarity-consistent deep attention network for fine-grained visual emotion regression. In: Proceedings of the 27th ACM International Conference on Multimedia, pp. 192–201 (2019)

22. Krizhevsky, A., Sutskever, I., Hinton, G.E.: ImageNet classification with deep convolutional neural networks. In: Advances in Neural Information Processing Systems, vol. 25 (2012)

23. Simonyan, K., Zisserman, A.: Very deep convolutional networks for large-scale image recognition. arXiv preprint arXiv:1409.1556 (2014)

24. Raghu, M., Unterthiner, T., Kornblith, S., Zhang, C., Dosovitskiy, A.: Do vision transformers see like convolutional neural networks? In: Advances in Neural Information Processing Systems, vol. 34, pp. 12116–12128 (2021)

Remote Sensing Image Interpretation

Change-Aware Network for Damaged Roads Recognition and Assessment Based on Multi-temporal Remote Sensing Imageries

Jiaxin Chen, Ming Wu[✉], Haotian Yan, Binzhu Xie, and Chuang Zhang

School of Artificial Intelligence, Beijing University of Posts and Telecommunications,
Beijing 100876, China
wuming@bupt.edu.cn

Abstract. Road damage assessment holds tremendous potential in evaluating damages and reducing disaster risks to human lives during emergency responses to unforeseen events. The Change Detection (CD) method detects changes in the land surface by comparing bi-temporal remote sensing imageries. Using the CD method for post-disaster assessment, existing research mainly focuses on building, while in terms of road, both the dataset and methodology need to be improved. In response to this, we propose an innovative multi-tasking network that combines Vision Transformer and UNet (BiTransUNet) for identifying road change areas and damage assessments from bi-temporal remote sensing imageries before and after natural disasters, moreover, propose the first road damage assessment model. Notably, our BiTransUNet comprises three efficient modules: Multi-scale Feature Extraction (MFE) module for extracting multi-scale features, Trans and Res Skip Connection (TRSC) module for modeling spatial-temporal global information, and Dense Cased Upsample (DCU) module for change maps reconstruction. In addition, to facilitate our study, we create a new Remote Sensing Road Damage Dataset, RSRDD, thoughtfully designed to contain 1,212 paired imageries before and after disasters, and the corresponding road change masks and road damage levels. Our experimental results on the proposed RSRDD show that our BiTransUNet outperforms current state-of-the-art approaches. BiTransUNet is also applied on the LEVIR-CD building change detection dataset and achieved the best performance, which demonstrates its compatibility in detecting changes of different important ground objects.

Keywords: Remote sensing · Vision transformer · Damaged road dataset · Change detection

1 Introduction

Road damage assessment is the process of evaluating the remote sensing bi-temporal imageries before and after natural disasters to identify the condition

Q. Liu et al. (Eds.): PRCV 2023, LNCS 14428, pp. 255–266, 2024.
https://doi.org/10.1007/978-981-99-8462-6_21

of roads. This process entails road change detection and assessing the extent of damage, playing a pivotal role in road monitoring, emergency rescue, damage assessment, and subsequent road reconstruction.

Change Detection (CD) represents a crucial aspect of road damage assessment and has been widely adopted in remote sensing applications, such as disaster monitoring [28], urban planning [27], and land cover detection [2]. Traditional pixel-based [9] and object-based [23] algorithms are unable to model complex mapping relationships as they primarily focus on extracting simple texture features. Recently, most mainstream CD algorithms, consisting of Autoencoder-based [17], Recurrent Neural Network-based [21], Convolutional Neural Network-based [14], Generative Adversarial Network-based [15], and Transformer-based [4] methods, are end-to-end networks that complete an all-inclusive framework.

Deep Learning has emerged as an essential tool for damage assessment. Nia et al. [19] proposed a high-performance hierarchical model to assess buildings' damage. Hamdi et al. [11] presented a modified UNet [22] architecture for determining the damaged areas in large forest areas.

However, existing methods have limitations when applied to road damage assessment: 1) building change detection methods are unsuitable for road change detection, unable to capture fine-grained, continuous road changes; 2) roads are more resistant to damage and exhibit continuity compared to buildings, so road damage assessment poses greater challenges compared to building change detection and lacks benchmark datasets in remote sensing.

To tackle the aforementioned issues with change detection and damage assessment, we first propose BiTransUNet, a novel multi-tasking network combining Vision Transformer and UNet. To facilitate and validate our approach, we then introduce an entirely new Remote Sensing Road Damage Dataset (RSRDD), comprising 1,212 pre and post-disaster imagery pairs, and the corresponding road change masks and road damage levels. Our BiTransUNet includes three efficient modules: Multi-scale Feature Extraction (MFE) module, Trans and Res Skip Connection (TRSC) module, and Dense Cased Upsample (DCU) module. MFE aims to extract multi-scale feature maps from the CNN backbone that produce coarse-grained and fine-grained features. To model long-short contextual relations, TRSC is designed to work with hierarchical rich features in four different stages. Finally, we propose DCU to enhance the reconstruction process from features to change maps by leveraging densely connected layers from four different stages.

To sum up, the main contributions of this paper are as follows:

- We introduce a new remote sensing road damage dataset RSRDD, which serves as a benchmark for road damage assessment, complementing previous studies that mainly centered on building change detection.
- An end-to-end multi-tasking network architecture BiTransUNet is proposed for road change detection and road damage assessment, which fully exploits multi-scale features to obtain finer details of change maps.
- Extensive experiments have confirmed the versatility of our proposed framework, which can show excellent performance not only in road damage assessment but also in building change detection.

2   Related Works

2.1   Change Detection Methods Based on Deep Learning

CNN-based methods are widely used in CD because of their efficiency in feature extraction, such as FC-Siam-diff [7] and STANet [5]. Daudt et al. [7] presented three fully convolutional networks based on the UNet model, which was modified in a siamese architecture for the first time. Chen et al. [5] proposed a STANet with spatial-temporal attention to obtain better feature representations.

While CNNs have proven to be effective in change detection, the features extracted solely by CNNs lack global contextual information, making them susceptible to pseudo-changes induced by factors such as noise, shadow, and illumination.

Therefore, many improved methods that integrate transformers are proposed to better model long-term spatial-temporal relationships and enhance the ability to discriminate bi-temporal imageries. Chen et al. [4] proposed an efficient transformer-based model (BiT), transforming spatial information into semantic information for change detection in remote sensing imageries. Bandara et al. [1] proposed a transformer-based Siamese network (ChangeFormer), utilizing a hierarchical transformer encoder in the architecture with a simple MLP decoder.

Although previous methods have demonstrated perfect performance in building change detection, they fall short in identifying the continuity of road changes and capturing fine-grained features.

2.2   Transformer and UNet Architecture

Transformer was first introduced in 2017 [24], which was originally used for long-range dependency modeling in the field of Natural Language Processing (NLP). The attention mechanism of the transformer enables it to efficiently model global contextual information. Beginning with the Vision Transformer (ViT) [8], transformers have shown promise in Computer Vision (CV).

UNet, a seminal architecture proposed in 2015 [22], gained popularity for image segmentation tasks. Researchers have applied the UNet architecture to other fields such as image classification [3] and object detection [26].

TransUNet is a combination of Transformer and UNet architecture. Chen et al. [6] first proposed TransUNet for general medical image segmentation, which encodes strong global context by treating image features as sequences while also utilizing low-level CNN features through a u-shaped hybrid architectural design. Since then, many studies [20,25] presented corresponding TransUNet architecture based on the same framework due to its powerful feature extraction and detail recovery capabilities. These methods combine high-level semantic information with low-level texture information through skip connections and model spatial-temporal relationships with a transformer to generate more accurate change maps.

In contrast to previous TransUNet models, which incorporate global context solely in the bottom stage, our BiTransUNet models long-short contextual relations in four different stages.

3   Method

Our overview architecture is depicted in Fig. 1. We incorporate change detection and damage assessment into a new pipeline. Given a couple of imageries, our model starts with Multi-scale Feature Extraction module (MFE) to obtain multi-scale features of bi-temporal imageries. Then, Trans and Res Skip Connection (TRSC) helps to model spatial-temporal global information of features. Besides, Dense Cased Upsample module (DCU) is applied to enhance the reconstruction process from features to change map. Finally, the resulting features are fed to prediction heads and produce the change map and damage level.

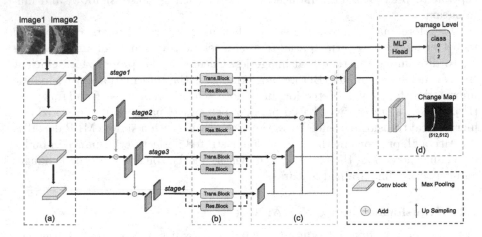

Fig. 1. Architecture of BiTransUNet. (a) Multi-scale Feature Extraction module (MFE); (b) Trans and Res Skip Connection (TRSC); (c) Dense Cased Upsample module (DCU); (d) Prediction Heads for change detection and damage assessment.

3.1   Multi-scale Feature Extraction Module (MFE)

Our intuition is that multi-scale features contain both coarse-grained and fine-grained information, which is better than single-scale features. ResNet [12] is an efficient and portable network that has been widely used by [3–5, 26], so we utilize ResNet-18 as our CNN backbone for feature extraction. We use the first three layers of ResNet-18 and a normal convolutional block to generate feature maps of four different scales. The features extracted at different stages are fed into TRSC to enhance and refine the bi-temporal features.

3.2   Trans and Res Skip Connection (TRSC)

Trans and Res Skip Connection can be divided into two sub-blocks of Trans.Block and Res.Block. **Trans.Block** (Fig. 2 (a)) is initially designed to model spatial-temporal global information of bi-temporal features in different

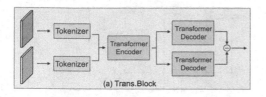 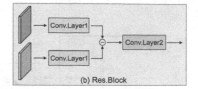

(a) Trans.Block   (b) Res.Block

Fig. 2. Architecture of Trans and Res Skip Connection (TRSC).

stages, which contains high-level semantic information. **Res.Block** (Fig. 2 (b)) consists of two convolutional layers, which aims to obtain the difference map of bi-temporal features and contains low-level geolocation information.

Trans.Block is composed of Tokenizer and Transformer. Inspired by the semantic tokenizer in BiT [4], our tokenizer has the same architecture, which is depicted in Fig. 3 (a). Suppose the height, width, and channel dimension of the feature map are H, W, C. Let $X^1, X^2 \in \mathbb{R}^{HW \times C}$ be the bi-temporal features, and $T^1, T^2 \in \mathbb{R}^{L \times C}$ be the semantic tokens. The semantic tokens $T^i \in \mathbb{R}^{L \times C} (i = 1, 2)$ are computed as:

$$T^i = (A^i)^T X^i = (\sigma(\varphi(X^i)))^T X^i, \tag{1}$$

where $\varphi(\cdot)$ is a pointwise convolution, $\sigma(\cdot)$ denotes the softmax function, A^i means the attention map computed by the pointwise convolution.

Transformer encoder and decoder follow the default architecture of ViT [8] with the pre-norm residual unit in Fig. 3 (b) and (c). At each encoder layer l, the Q(query), K(key), V(value) are maped from the merging of two semantic tokens $T^{(l-1)} \in \mathbb{R}^{2L \times C}$. Then multi-self attention (MSA) will be applied:

$$Attention(Q, K, V) = Softmax(\frac{QK^T}{\sqrt{d}})V, \tag{2}$$

$$MSA(Q, K, V) = Concat(head_1, ..., head_h)W^O, \tag{3}$$
$$where \quad head_j = Attention(QW_j^q, KW_j^k, VW_j^v),$$

where $W_j^q, W_j^k, W_j^v \in \mathbb{R}^{C \times d}$, $W^O \in \mathbb{R}^{hd \times C}$ are parameter matrices of different attention heads. Here h and C represent the number of attention heads and the dimension of embedding tokens and typically we set $d = C/h$.

3.3 Dense Cased Upsample Module (DCU)

For change maps reconstruction, we design a dense cased upsample module (DCU) depicted in Fig. 1 (c). Different from directly upsampling in [6,20,25], dense block [13] has advantages in feature reuse, reducing vanished gradient and efficient memory usage, making it a suitable choice for maintaining the continuity characteristics of roads. In this way, each stage in DCU receives feature maps from all preceding stages, so the coarse-grained features and fine-grained features can be fused for subsequent prediction.

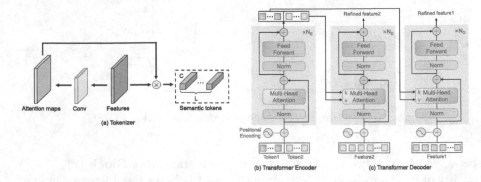

Fig. 3. Detail of Tokenizer, Transformer Encoder and Transformer Decoder.

3.4 Prediction Heads and Loss Function

After the transformer encoder and DCU, two prediction heads are employed for damage assessment and change detection as shown in Fig. 1 (d).

As for damage assessment, class embedding [8] and semantic tokens are concatenated into new tokens in stage 1, and the token length is changed from l to $l + 1$. After the transformer encoder, the class embedding is split and fed into MLP Head to predict the damage level of roads.

As for change detection, the reconstructed features are directly fed into several convolutional layers for upsampling and reducing output channels, then the softmax function is employed to visualize the binary change map of roads.

Considering the positive and negative samples are imbalanced for change detection, the cross-entropy loss function is not suitable for this task, so we utilize focal loss [16] and dice loss [18] for better handling of class imbalance. Besides, we use cross-entropy loss for damage assessment. Formally,

$$L_{focal} = -\sum_{i=1}^{n} y \times \alpha_t (1 - \hat{y})^{\gamma} \log \hat{y}, \tag{4}$$

$$L_{dice} = 1 - \frac{2 \sum \hat{y}y + \epsilon}{\sum \hat{y} + \sum y + \epsilon}, \tag{5}$$

$$L_{ce} = -\frac{1}{n} \sum_{i=1}^{n} y \times \log \hat{y}, \tag{6}$$

where $\hat{y}$ and y are prediction and ground truth, α_t denotes the weighting hyperparameter, γ is a focusing parameter, ϵ is a small constant to prevent division by zero.

The total loss is the sum of all the above three losses, which is formulated as:

$$L_{total} = L_{focal} + L_{dice} + L_{ce}. \tag{7}$$

4 Datasets

4.1 RSRDD

Considering the lack of a benchmark dataset for road damage assessment, we want to fill this gap and provide a new remote sensing road damage dataset (RSRDD). We collected two batches of pre and post-disaster very high-resolution imagery pairs from 6 different regions sitting in several cities in China. Each region has a varied number of image patches and contains a diversity of natural disasters, such as floods, earthquakes, torrential rains, mudslides, and landslides. Besides, our image dataset exhibits temporal variability, spanning a collection timeframe ranging from 2014 to 2022. We located the position of roads with pre-disaster imageries and detected the change areas of roads with post-disaster imageries. Then, we classified road damage into three categories (0/1/2 for no or less/medium/severe damage). Finally, we cut the whole image into small patches with 256×256 pixels, a total of 1212 bi-temporal imagery pairs annotated with change masks and damage levels, and 726/242/244 for training/validation/test respectively. Some examples are shown in Fig. 4 (b).

4.2 LEVIR-CD

LEVIR-CD [5] is a public dataset for building change detection, including 637 pairs of photos from Google Earth at extremely high resolution (1024×1024 pixels). It contains various types of buildings with different shapes and sizes, such as factories, apartments, warehouses, etc. We follow the default dataset split with 256×256 pixels and obtain 7120 training sets, 1024 validation sets, and 2048 test sets, respectively. Figure 4 (a) illustrates some examples of LEVIR-CD.

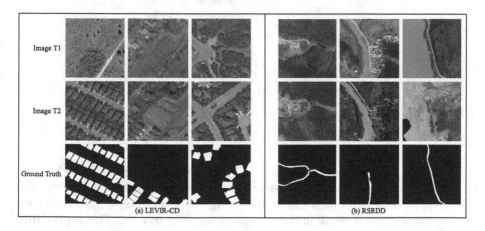

Fig. 4. Bi-temporal imageries and ground truth maps from the LEVIR-CD and the RSRDD datasets.

5   Experiments

5.1   Implemented Details

Our BiTransUNet is implemented using PyTorch and trained on 2 NVIDIA GeForce GTX 1080 Ti GPUs. We apply normal data augmentation to the input imageries, such as flip, rotate, crop, and blur. As for RSRDD, we train BiTransUNet with 8 batch sizes and 400 epochs. To be specific, the initial learning rate is 0.001 and AdamW optimizer is used with $\beta 1 = 0.9$, $\beta 2 = 0.99$, and weight decay is set to 0.0005. For LEVIR-CD, we follow the same setting as [4] to set the initial learning rate and some other parameters of the network.

5.2   Comparison to State-of-the-Art Methods

We conduct experiments on RSRDD and LEVIR-CD to compare the performance of our BiTransUNet with other SOTA methods [1,4,5,10]. Specifically, we demonstrate that BiTransUNet shows better performance in road change detection compared to other methods and can get a good result in road damage assessment on RSRDD. For LEVIR-CD, our BiTransUNet achieves state-of-the-art performance in building change detection tasks.

Table 1. Damage assessment results on RSRDD (0/1/2 for no or less/medium/severe damage). All the scores are described in percentage (%).

Parameter1	Value1	Parameter2	Value2
Accuracy	84.02	-	-
Precision	84.37	Precision_0	86.00
		Precision_1	84.95
		Precision_2	82.18
Recall	82.88	Recall_0	74.14
		Recall_1	82.29
		Recall_2	92.22
F1	**83.38**	F1_0	79.63
		F1_1	83.60
		F1_2	86.91

Table 1 shows the results for road damage assessment as well as refined results for each evaluation metric. The F1-score reaches 83.38, which demonstrates the efficacy of BiTransUNet in precisely discerning levels of damage. Specifically, we can observe a slight increase in the F1-score from F1_0 to F1_2. This is primarily due to the fact that severe damage contains more spatial information compared to less or medium damage, making the network more sensitive to it.

Table 2 shows the comparison results on RSRDD and LEVIR-CD. Obviously, our BiTransUNet is more effective than other methods. Note that BiTransUNet

Table 2. Change detection results on RSRDD and LEVIR-CD. All the scores are described in percentage (%).

Method	RSRDD Pre. / Rec. / F1 / IoU / OA	LEVIR-CD Pre. / Rec. / F1 / IoU / OA
STANet [5]	78.47 / **95.61** / 84.92 / 76.48 / 98.31	94.54 / 83.98 / 88.95 / 80.10 / 98.94
SNUNet [10]	72.13 / 54.43 / 62.04 / 44.97 / 98.54	91.67 / 88.96 / 90.29 / 82.11 / **99.04**
BiT [4]	86.73 / 73.94 / 78.91 / 69.90 / 98.30	89.24 / 89.37 / 89.31 / 80.68 / 98.92
ChangeFormer [1]	86.59 / 81.27 / 83.71 / 75.10 / 98.52	92.05 / 88.80 / 90.40 / 82.48 / **99.04**
BiTransUNet	**90.27** / 88.48 / **89.36** / **82.30** / **99.11**	**95.24** / **93.74** / **94.47** / **89.95** / 98.95

outperforms BiT [4] 10.45 and 12.40 of the F1-score and IoU on RSRDD, which proves that BiTransUNet is more suitable for road change detection. For LEVIR-CD, our method pays more attention to the fine-grained features, which exceeds other methods by a large margin, getting 4.07 and 7.47 improvement in F1-score and IoU compared with ChangeFormer [1]. The results show that BiTransUNet achieves state-of-the-art performance on both two datasets.

5.3 Qualitative Result

Some qualitative examples of BiTransUNet are shown in Fig. 5. The first two columns are bi-temporal imageries and the last column is ground truth, while the remaining columns exhibit the binary change maps that are predicted by several different models. STANet recalls the road change areas as much as possible, but causes a high false positive rate. SNUNet and BiT only detect partial road change regions, resulting in incomplete detection results. ChangeFormer insufficiently captures the continuity of the road. It is readily discernible that BiTransUNet surpasses its counterparts in accurately detecting both the coarse-grained and fine-grained change areas, demonstrating its proficiency and robustness in the face of challenging real-world scenarios.

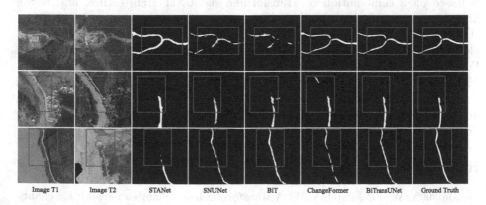

| Image T1 | Image T2 | STANet | SNUNet | BiT | ChangeFormer | BiTransUNet | Ground Truth |

Fig. 5. Qualitative results on RSRDD, and the red boxes are used to highlight the predictions of the different models. (Color figure online)

5.4   Ablation Study

In order to demonstrate the effectiveness of BiTransUNet, we conducted an ablation study on RSRDD.

Table 3. Effect of the loss function, token length, and Res.Block for change detection. All the scores are described in percentage (%).

Parameter	Index	Precision	Recall	F1	IoU
loss function	CE	89.59	85.64	87.43	79.95
	Focal+Dice	**90.27**	**88.48**	**89.36**	**82.30**
token length	4	90.27	**88.48**	89.36	82.30
	8	90.97	87.97	**89.38**	**82.32**
	16	90.38	86.82	88.45	81.18
	32	**91.56**	87.01	89.06	81.93
Res.Block	✓	90.27	**88.48**	**89.36**	**82.30**
	✗	**90.35**	86.45	88.23	80.91

Table 3 exhibits the ablation results for change detection with different parameters. When choosing the cross-entropy loss function, we can observe significant drops in the F1-score and IoU. When employing the Res.Block, both F1-score and IoU get a great enhancement, which indicates that TRSC is critical in our network. Considering the computational complexity of the model, although the model achieves the best performance when setting the token length to 8, we choose 4 because the F1-score and IoU get only 0.02 drops comparatively.

6   Conclusion

In this paper, we introduce a novel multi-tasking network, BiTransUNet, which is based on a combination of Transformer and UNet architecture, and is the first framework for road damage assessment in remote sensing. Our BiTransUNet has the ability to extract multi-scale features and model spatial-temporal global information through three auxiliary modules: MFE, TRSC, and DCU. Furthermore, we present the Remote Sensing Road Damage Dataset (RSRDD), which is the first road damage dataset in remote sensing. Experimental results demonstrate that BiTransUNet achieves state-of-the-art performance in change detection and damage assessment on two related datasets. This paper presents new ideas for future work in this field.

References

1. Bandara, W.G.C., Patel, V.M.: A transformer-based Siamese network for change detection. In: IGARSS 2022–2022 IEEE International Geoscience and Remote Sensing Symposium, pp. 207–210. IEEE (2022)

2. Bolorinos, J., Ajami, N.K., Rajagopal, R.: Consumption change detection for urban planning: monitoring and segmenting water customers during drought. Water Resources Res. **56**(3), e2019WR025812 (2020)
3. Cao, K., Zhang, X.: An improved Res-UNet model for tree species classification using airborne high-resolution images. Remote Sens. **12**(7), 1128 (2020)
4. Chen, H., Qi, Z., Shi, Z.: Remote sensing image change detection with transformers. IEEE Trans. Geosci. Remote Sens. **60**, 1–14 (2021)
5. Chen, H., Shi, Z.: A spatial-temporal attention-based method and a new dataset for remote sensing image change detection. Remote Sens. **12**(10), 1662 (2020)
6. Chen, J., et al.: TransuNet: transformers make strong encoders for medical image segmentation. arXiv preprint arXiv:2102.04306 (2021)
7. Daudt, R.C., Le Saux, B., Boulch, A.: Fully convolutional Siamese networks for change detection. In: 2018 25th IEEE International Conference on Image Processing (ICIP), pp. 4063–4067. IEEE (2018)
8. Dosovitskiy, A., et al.: An image is worth 16x16 words: transformers for image recognition at scale. arXiv preprint arXiv:2010.11929 (2020)
9. El-Hattab, M.M.: Applying post classification change detection technique to monitor an Egyptian coastal zone (Abu Qir Bay). Egypt. J. Remote Sens. Space Sci. **19**(1), 23–36 (2016)
10. Fang, S., Li, K., Shao, J., Li, Z.: SNUNet-CD: a densely connected Siamese network for change detection of VHR images. IEEE Geosci. Remote Sens. Lett. **19**, 1–5 (2021)
11. Hamdi, Z.M., Brandmeier, M., Straub, C.: Forest damage assessment using deep learning on high resolution remote sensing data. Remote Sens. **11**(17), 1976 (2019)
12. He, K., Zhang, X., Ren, S., Sun, J.: Deep residual learning for image recognition. In: Proceedings of the IEEE Conference on Computer Vision and Pattern Recognition, pp. 770–778 (2016)
13. Huang, G., Liu, Z., Van Der Maaten, L., Weinberger, K.Q.: Densely connected convolutional networks. In: Proceedings of the IEEE Conference on Computer Vision and Pattern Recognition, pp. 4700–4708 (2017)
14. Li, H., Wu, K., Xu, Y.: An integrated change detection method based on spectral unmixing and the CNN for hyperspectral imagery. Remote Sens. **14**(11), 2523 (2022)
15. Li, X., Du, Z., Huang, Y., Tan, Z.: A deep translation (GAN) based change detection network for optical and SAR remote sensing images. ISPRS J. Photogramm. Remote. Sens. **179**, 14–34 (2021)
16. Lin, T.Y., Goyal, P., Girshick, R., He, K., Dollár, P.: Focal loss for dense object detection. In: Proceedings of the IEEE International Conference on Computer Vision, pp. 2980–2988 (2017)
17. López-Fandiño, J., Garea, A.S., Heras, D.B., Argüello, F.: Stacked autoencoders for multiclass change detection in hyperspectral images. In: IGARSS 2018–2018 IEEE International Geoscience and Remote Sensing Symposium, pp. 1906–1909. IEEE (2018)
18. Milletari, F., Navab, N., Ahmadi, S.A.: V-Net: fully convolutional neural networks for volumetric medical image segmentation. In: 2016 Fourth International Conference on 3D Vision (3DV), pp. 565–571. IEEE (2016)
19. Nia, K.R., Mori, G.: Building damage assessment using deep learning and ground-level image data. In: 2017 14th Conference on Computer and Robot Vision (CRV), pp. 95–102. IEEE (2017)

20. Pang, L., Sun, J., Chi, Y., Yang, Y., Zhang, F., Zhang, L.: CD-TransUNet: a hybrid transformer network for the change detection of urban buildings using L-band SAR images. Sustainability **14**(16), 9847 (2022)
21. Papadomanolaki, M., Verma, S., Vakalopoulou, M., Gupta, S., Karantzalos, K.: Detecting urban changes with recurrent neural networks from multitemporal sentinel-2 data. In: IGARSS 2019–2019 IEEE International Geoscience and Remote Sensing Symposium, pp. 214–217. IEEE (2019)
22. Ronneberger, O., Fischer, P., Brox, T.: U-Net: convolutional networks for biomedical image segmentation. In: Navab, N., Hornegger, J., Wells, W.M., Frangi, A.F. (eds.) MICCAI 2015. LNCS, vol. 9351, pp. 234–241. Springer, Cham (2015). https://doi.org/10.1007/978-3-319-24574-4_28
23. Tomowski, D., Ehlers, M., Klonus, S.: Colour and texture based change detection for urban disaster analysis. In: 2011 Joint Urban Remote Sensing Event, pp. 329–332. IEEE (2011)
24. Vaswani, A., et al.: Attention is all you need. In: Advances in Neural Information Processing Systems, vol. 30 (2017)
25. Yang, Y., Mehrkanoon, S.: AA-TransUNet: attention augmented TransUNet for nowcasting tasks. In: 2022 International Joint Conference on Neural Networks (IJCNN), pp. 01–08. IEEE (2022)
26. Yuan, L., et al.: Multi-objects change detection based on Res-Unet. In: 2021 IEEE International Geoscience and Remote Sensing Symposium IGARSS, pp. 4364–4367. IEEE (2021)
27. Zhang, Z., Vosselman, G., Gerke, M., Tuia, D., Yang, M.Y.: Change detection between multimodal remote sensing data using Siamese CNN. arXiv preprint arXiv:1807.09562 (2018)
28. Zheng, Z., Zhong, Y., Wang, J., Ma, A., Zhang, L.: Building damage assessment for rapid disaster response with a deep object-based semantic change detection framework: from natural disasters to man-made disasters. Remote Sens. Environ. **265**, 112636 (2021)

UAM-Net: An Attention-Based Multi-level Feature Fusion UNet for Remote Sensing Image Segmentation

Yiwen Cao[1,2], Nanfeng Jiang[1,2], Da-Han Wang[1,2], Yun Wu[1,2(✉)], and Shunzhi Zhu[1,2]

[1] School of Computer and Information Engineering,
Xiamen University of Technology, Xiamen 361024, China
cywen@s.xmut.edu.cn, {wangdh,szzhu}@xmut.edu.cn
[2] Fujian Key Laboratory of Pattern Recognition and Image Understanding,
Xiamen 361024, China
ywu@xmut.edu.cn

Abstract. Semantic segmentation of Remote Sensing Images (RSIs) is an essential application for precision agriculture, environmental protection, and economic assessment. While UNet-based networks have made significant progress, they still face challenges in capturing long-range dependencies and preserving fine-grained details. To address these limitations and improve segmentation accuracy, we propose an effective method, namely UAM-Net (UNet with Attention-based Multi-level feature fusion), to enhance global contextual understanding and maintain fine-grained information. To be specific, UAM-Net incorporates three key modules. Firstly, the Global Context Guidance Module (GCGM) integrates semantic information from the Pyramid Pooling Module (PPM) into each decoder stage. Secondly, the Triple Attention Module (TAM) effectively addresses feature discrepancies between the encoder and decoder. Finally, the computation-effective Linear Attention Module (LAM) seamlessly fuses coarse-level feature maps with multiple decoder stages. With the corporations of these modules, UAM-Net significantly outperforms the most state-of-the-art methods on two popular benchmarks.

Keywords: Semantic segmentation · U-shape architecture · Attention mechanism · Feature fusion · Remote sensing images

1 Introduction

Semantic segmentation of Remote Sensing Images (RSIs) is a crucial task that can assign a semantic label to each pixel in images. It is widely used in a variety of applications, such as precision agriculture [6,15], environmental protection [17,24], and economic assessment [18,25]. Existing RSI semantic segmentation methods mainly rely on convolutional U-shape structure. The typical U-shape

© The Author(s), under exclusive license to Springer Nature Singapore Pte Ltd. 2024
Q. Liu et al. (Eds.): PRCV 2023, LNCS 14428, pp. 267–278, 2024.
https://doi.org/10.1007/978-981-99-8462-6_22

network, namely UNet [16], combines encoder and decoder with skip connections. This structure can fuse different scale features to realize accurate segmentation while preserving more spatial information. Following this route, many UNet-based variants such as ResUnet [26], MAResU-Net [8] and UNetFormer [22] have been developed for this task and obtained great success.

Although UNet-based networks have achieved excellent performance in RSIs semantic segmentation, they still cannot fully meet the strict requirements of remote sensing image applications for segmentation accuracy. Firstly, their capacity to learn long-range dependencies is limited to the localized receptive fields [27]. As a result, such a deficiency in capturing multi-scale information leads to sub-optimal segmentation of objects with different shapes and scales (e.g. cars and buildings). Some studies have tried to address this problem by using atrous convolutions [4,13], feature pyramid structure [2,11,27]. However, these methods still have limitations in modeling long-range dependencies. Secondly, the hierarchical structure of UNet and its upsampling operations will introduce spatial scale differences between low-level and high-level features. Moreover, the direct transmission of low-level features from the encoder to the decoder through skip connections can also result in interference with high-level features, leading to unfavorable segmentation results. Based on the aforementioned analysis, our model design needs to consider the following aspects: 1) How to effectively utilize global long-range information to enhance the discriminative ability of feature extraction. 2) How to make full use of detailed information from low-level features while maintaining semantic accuracy on high-level features.

To address the above issues, we combine attention mechanism with multi-level feature fusion to design a novel UAM-Net for RSI semantic segmentation. Firstly, we introduce a Global Context Guidance Module (GCGM) that incorporates a modified Pyramid Pooling Module (PPM) and Global Context Flows (GCF). This enables the integration of semantic information from the PPM into each stage of the decoder, ensuring sufficient global contextual information. Secondly, we employ a Triple Attention Module (TAM) to narrow feature discrepancies between the encoder and decoder. TAM captures cross-dimensional interactions via a three-branch structure, which can effectively compute attention weights to assign and fuse different level features. Thirdly, we introduce a computation-effective Linear Attention Module (LAM) to seamlessly fuse coarse-level feature maps from GCGM with feature maps in multiple decoder stages. This allows the fusion of global and local information, which effectively improves the ability to capture fine-grained details while maintaining semantic accuracy.

The main contributions can be concluded as follows:

- We introduce a GCGM, which consists of a modified PPM and GCFs, to ensure the incorporation of semantic information at different level decoders. This way can effectively learn global context information for improving segmentation accuracy.
- We employ a TAM to adaptively assign features at different levels, enhancing consistency between the encoder and decoder. In addition, we introduce the LAM to fuse coarse-level feature maps from GCGM to enrich semantic information.

- With the help of these modules, our proposed UAM-Net is of high accuracy and efficiency. Extensive experiments conducted on popular datasets also demonstrate the superiority of our approach over other state-of-the-art methods.

2   Related Work

2.1   Encoder-Decoder Designs

Based on UNet architecture, ResUnet [26] adopted residual units as basic blocks to retain high-level semantic information. Resunet-a [4] incorporated atrous convolution and residual connections. They all introduced short skip connections within one module. This way will have limited ability to capture rich information. In addition, some networks focused on enhancing the learning ability of skip connections. GSN [19] designed an entropy control module to adaptively weighted feature maps to integrate contextual information and details. HCANet [1] replaced the copy and crop operation with designed Compact Atrous Spatial Pyramid Pooling (CASPP) modules in UNet to extract the multi-scale context information. These networks still have details loss on semantic results and ignore the importance of learning long-range information. Unlike them, our proposed network fuses coarse-grained and fine-grained features reasonably while making full use of global contextual information.

2.2   Attention Mechanism

Attention modules have gained widespread adoption in various tasks due to their ability to enhance feature learning. DC-Swin [20] selected the Swin Transformer as the encoder and utilized CNN-based decoders that significantly improved model performance. However, Transformer-based encoders are often of high computational complexity, which limits their suitability for real-time applications. Other popular attention networks [3,5,14,23] have employed spatial attention, channel attention, or a combination of both to capture dependencies among channels and spatial dimensions. While these mechanisms excel at capturing local dependencies, there is a potential drawback of neglecting long-range information, which is crucial for tasks requiring a broader contextual understanding. Differently, our proposed approach introduces the Triple Attention Module (TAM) to capture cross-dimension dependencies. By incorporating TAM, our model can address the limitation of neglecting long-range information by effectively capturing interactions and dependencies across different dimensions.

3   Proposed Method

3.1   Philosophy of Model Design

In the RSIs semantic segmentation tasks, there are several inherent characteristics that need to be considered. Firstly, RSIs often contain objects with varying

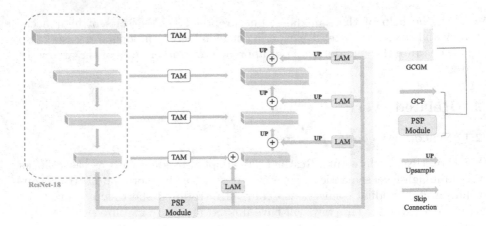

Fig. 1. The overall pipeline of proposed UAM-Net.

scales and complex structures. Secondly, the presence of long-range dependencies is crucial for accurate segmentation, as the context information plays a significant role in understanding the spatial relationships between objects. Lastly, this task requires sufficient semantic information and preservation of fine-grained spatial details to guarantee accurate segmentation.

In this paper, we achieve these goals by introducing the Global Context Guidance Module (GCGM), Triple Attention Module (TAM) [12] and Linear Attention Module (LAM) [9] on a U-shape structure.

3.2   Network Design

As shown in Fig. 1, we adopt ResNet-18 [7] as the backbone. However, different from the original UNet and its variants [8,22,26], we propose a Global Context Guidance Module (GCGM) and put it on the top of the encoder to capture multi-scale context information. To avoid the dilution of high-level semantic information in the later stage of upsampling phase, we aggregate the global prior generated by GCGM with feature maps at each level in the decoder. In the process of feature merging, we further introduce the Linear Attention Module (LAM) to ensure that feature maps at different scales can be seamlessly merged. Furthermore, we add the Triple Attention Module (TAM) to the skip connection for the feature fusion of the encoder and decoder. The descriptions of these modules are as bellow.

3.3   Global Context Guidance Module

We propose a GCGM to address the issues in U-shape architectures. The GCGM consists of two parts: a modified Pyramid Pooling Module (PPM) for capturing global context information and a series of Global Context Flows (GCF) for transferring contextual information to each stage of the decoder. In our approach, the

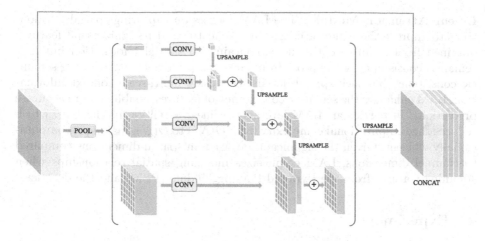

Fig. 2. Structure of modified Pyramid Pooling Module (PPM).

PPM is designed with four levels of bin sizes (1×1, 2×2, 3×3, and 6×6), and the feature maps from these levels are fused as the global prior. Unlike the original PPM, we perform the cascaded upsampling on adjacent feature maps and sum them up to preserve local details. By introducing the GCF, high-level semantic information is effectively delivered to each stage of the decoder, ensuring that global context information is retained during the upsampling phase. Overall, our GCGM modifies the PPM and incorporates GCF to mitigate the disturbance and dilution of high-level features while preserving the global context information during the upsampling process in U-shape architectures.

3.4 Attention Module

Remote sensing images commonly contain objects with large-scale variations and significant intra-class variance, while the inter-class variance tends to be small. It is beneficial to integrate multi-level features to leverage rich information for semantic segmentation. However, challenges arise due to semantic gaps between different features and the potential inclusion of redundant information, which can adversely affect prediction accuracy. Thus, effective feature selection and refinement are of paramount importance in enhancing the segmentation performance. To tackle these issues, this work introduces two attention modules: the Triple Attention Module (TAM) and the Linear Attention Module (LAM).

Triple Attention Module. The TAM addresses the fusion of low-level and high-level features by capturing cross-dimensional relationships. Unlike conventional attention mechanisms that separately model spatial and channel dependencies, TAM considers both spatial and channel relationships simultaneously. It enables a more comprehensive fusion of features captured by the encoder and decoder, enhancing the discriminative ability of the model. Especially, TAM achieves this without introducing significant computational overhead.

Linear Attention Module. The LAM focuses on capturing spatial relationships to improve the semantic consistency of features. It recognizes that features obtained from small receptive fields contain rich spatial information but lack semantic consistency, while those from large receptive fields have strong semantic consistency but lack spatial details. By fusing multi-scale context information, LAM enhances the semantic consistency of features, enabling more accurate predictions. In particular, LAM leverages a linear attention mechanism, which reduces the computational complexity from $O(N^2)$ to $O(N)$, making it computationally efficient. Given the significant variation in spatial dimensions compared to channel dimensions, LAM emphasizes modeling spatial relationships when summing features from the Pyramid Pooling Module (PPM) and the decoder.

4   Experiment

4.1   Datasets

We evaluate the effectiveness of the proposed framework on two public 2D semantic labeling datasets, including Vaihingen and Potsdam, which are provided by the International Society for Photogrammetry and Remote Sensing (ISPRS). The images and corresponding labels in the two popular datasets (Vaihingen and Potsdam) are cropped into 512×512 patches.

4.2   Experimental Setting

To demonstrate the segmentation performance of the proposed UAM-Net, we compare it with seven state-of-the-art methods, including ABCNet (P&RS'21) [10], BANet (Remote Sensing'21) [21], MANet (TGRS'21) [9], DCSwin (GRSL'22) [20], MAResUnet (GRSL'21) [8], and UnetFormer (P&RS'22) [22]. For a fair comparison, these methods are trained, validated, and tested using the source codes provided by the original authors.

All the experiments are implemented with PyTorch on a single TITAN Xp. We adopt ResNet-18 as the backbone and initialize it with the pre-trained model on ImageNet. We train the network using AdamW with batch size 8, learning rate 0.0003, and weight decay 0.0025. The training epoch number is 100 for the Vaihingen dataset and 80 for the Potsdam dataset. For each method, the soft cross-entropy loss function is used as a quantitative evaluation, and the performance on the two datasets is assessed by using the F1 score (F1), Overall Accuracy (OA), and mean Intersection over Union (mIoU).

4.3   Quantitative Results

From Table 1 and Table 2, we can observe that our proposed network outperforms almost all previous state-of-the-art results. Especially, UAM-Net achieves a mIoU of 83.1% on the Vaihingen Dataset, which is the highest among all the listed methods. Additionally, UAM-Net shows competitive results in terms of per-class F1 scores. It achieves the highest F1 scores for three classes: "Low vegetation" (85.0%), "Tree" (90.2%), and "Car" (89.4%). UAM-Net and DCSwin both have an OA of 91.0%, achieving satisfactory overall accuracy among all the methods. Overall, UAM-Net appears to be a promising method for semantic segmentation.

Table 1. Quantitative Results on the Vaihingen Dataset

Method	Per-class F1 Score (%)					Mean F1 (%)	OA (%)	mIoU (%)
	Imp.surf.	Building	Low veg.	Tree	Car			
ABCNet	92.2	94.6	83.7	89.9	87.2	89.5	90.2	81.3
BANet	92.2	94.7	83.5	89.6	87.5	89.5	90.2	81.2
MANet	91.9	95.0	83.4	89.6	87.9	89.6	90.1	81.4
UnetFormer	93.0	95.4	84.4	89.9	88.0	90.1	91.0	82.7
MAResUnet	92.5	95.1	83.3	89.6	87.1	89.5	90.3	81.3
DCSwin	93.3	95.8	84.5	90.1	88.2	90.4	91.1	82.7
LANet	92.4	94.6	84.0	89.7	87.0	89.5	90.3	81.3
UAM-Net	92.9	95.5	85.0	90.2	89.4	90.6	91.0	83.1

Table 2. Quantitative Results on the Potsdam Dataset

Method	Per-class F1 Score (%)					Mean F1 (%)	OA (%)	mIoU (%)
	Imp.surf.	Building	Low veg.	Tree	Car			
ABCNet	92.3	96.5	87.5	89.0	95.6	92.2	90.5	85.7
BANet	92.7	96.0	87.0	88.5	95.7	92.0	90.5	85.4
MANet	93.1	96.2	87.3	88.7	96.1	92.3	90.9	85.9
UnetFormer	93.4	96.9	87.7	88.7	96.5	92.6	91.2	86.5
MAResUnet	92.2	95.9	86.8	87.9	95.4	91.6	90.1	84.8
DCSwin	93.7	97.2	88.5	89.4	96.1	93.0	91.7	87.2
LANet	93.3	96.4	87.4	89.2	96.5	92.6	90.9	86.4
UAM-Net	93.5	96.6	88.0	89.1	96.4	92.7	91.3	86.7

4.4   Qualitative Results

We provide visualizations of the segmentation results for the comparable models in Fig. 3 to demonstrate the effectiveness of our proposed UAM-Net in handling challenging situations. The red boxes highlight regions where UAM-Net outperforms the other methods. In the first row, most networks incorrectly segment some cars as buildings. In contrast, UAM-Net accurately segments these cars as separate objects. This demonstrates the ability of UAM-Net to capture detailed features and make precise distinctions between different object classes. In the second row of Fig. 3, it can be observed that UAM-Net is able to distinguish trees, low vegetation and road with blurred edges in shadows more accurately compared to the other networks. The boundaries between these classes are clearly defined by UAM-Net, indicating its robustness in handling challenging lighting conditions.

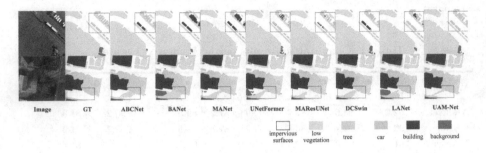

Fig. 3. Qualitative comparisons to previous state-of-the-art methods on the Vaihingen dataset (top) and Potsdam dataset (bottom). Obviously, compared to other methods, our approach is capable of extracting more discriminating features to better segment small objects and confusing objects in shadows.

Table 3. Complexity Comparisons on the Vaihingen Dataset

Method	mIoU (%)	FLOPs (G)	Inference Time (s)
ABCNet	81.3	15.63	0.88
BANet	81.2	13.06	7.00
MANet	81.4	22.20	3.17
UnetFormer	82.7	11.68	0.98
MAResUnet	81.3	25.42	3.30
DCSwin	82.7	70.15	5.84
LANet	81.3	33.24	2.01
UAM-Net	83.4	17.39	0.79

4.5　Efficiency Comparison

For real-time scene tasks, inference speed is a crucial metric to be considered. To measure the complexities of all methods, we further evaluate the average inference time per image and floating point operations per second (FLOPs) on the Vaihingen dataset, and the results are shown in Table 3 and Fig. 4. We can observe that UAM-Net has a FLOPs value of 17.39 billion. It is relatively lower compared to some other methods like DCSwin, which has a higher FLOPs value of 70.15 billion. Furthermore, UAM-Net achieves a low inference time of 0.79 s per image, which is the lowest among all the listed methods. This means that UAM-Net is able to perform real-time tasks efficiently, as it can process images quickly.

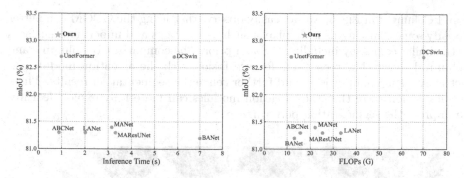

Fig. 4. Comparison with state-of-the-art methods in inference speed and FLOPs

4.6 Ablation Study

In this section, we systematically decompose our approach to examine the impact of each component. We adopt a U-shape architecture with a ResNet-18 backbone as the baseline. All ablation experiments are conducted on the ISPRS Vaihingen and Potsdam dataset, the comparison results are presented in Table 4.

Table 4. Quantitative Results of Different Configurations

Method	Vaihingen			Potsdam		
	Mean F1(%)	OA(%)	mIoU(%)	Mean F1(%)	OA(%)	mIoU(%)
Baseline	88.6	90.2	79.9	92.3	90.6	85.5
Baseline+PPM	89.8	90.5	81.8	92.4	90.8	85.9
Baseline+GCGM (PPM+GCF)	90.0	90.7	82.1	92.4	90.9	86.1
baseline+GCGM+LAM	90.4	90.9	82.7	92.6	91.2	86.5
baseline+GCGM+LAM+TAM	**90.6**	**91.0**	**83.1**	**92.7**	**91.3**	**86.7**

Ablation for Global Context Guidance Module. GCGM enables our network to use global prior knowledge to label pixels. The results of Table 4 show that this configuration can greatly improve the performance of segmentation in terms of all metrics. Besides, the proposed network uses the modified PPM also can bring improvement with 0.9% in mIoU. However, the GCFs only improve the mIoU by 0.3% because the features are fused directly without refinement.

Ablation for Attention Module. As mentioned above, we introduce lightweight attention modules for better feature fusion with fewer model parameters. The former is used for encoder and decoder features, and the latter for features from GCGM and decoder. Table 4 shows that these two modules improve the performance by 0.6% and 0.4% on the Vaihingen dataset, and 0.4% and 0.2% on the Potsdam dataset. By comparing the visualization results in Column c and Column d in Fig. 5, we can easily observe that using the GCGM can better classify some indistinguishable pixels, such as buildings and impervious surfaces with similar colors, by including a global prior. The comparison of column c and column d shows that reasonable feature fusion with the attention module can better segment small objects and further refine the segmentation results. These results demonstrate that the attention modules can more fully exploit features to improve the segmentation performance.

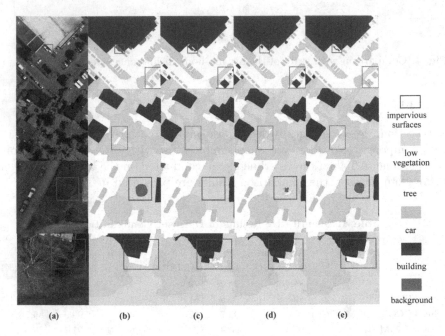

Fig. 5. Visual comparisons for prediction with different configurations of GCGM and attention modules on the Vaihingen dataset and Potsdam dataset. (a) Image; (b) Ground truth; (c) Results of U-shape structure with a ResNet-18 backbone baseline; (d) Results of baseline + GCGM; (e) Results of our proposed UAM-Net.

5   Conclusion

In this paper, we propose an effective RSI semantic segmentation network, namely UAM-Net, which compensates for some shortages of the U-shape network to improve the segmentation performance. We design a GCGM consisting of a modified PPM, which collects global context information, and a series of GCFs to transmit the context features to each decoder stage. Additionally, the TAM and the LAM are adopted to fuse different features reasonably. Both of the attention models are computation-effective. A wide range of experiments shows the effectiveness of UAM-Net, which achieves the best trade-off between complexity and accuracy.

Acknowledgements. This work is supported by Industry-University Cooperation Project of Fujian Science and Technology Department (No. 2021H6035), the Science and Technology Planning Project of Fujian Province (No. 2021J011191), and Fujian Key Technological Innovation and Industrialization Projects (No. 2023XQ023), and Fu-Xia-Quan National Independent Innovation Demonstration Project (No. 2022FX4).

References

1. Bai, H., Cheng, J., Huang, X., Liu, S., Deng, C.: HCANet: a hierarchical context aggregation network for semantic segmentation of high-resolution remote sensing images. IEEE Geosci. Remote Sens. Lett. **19**, 1–5 (2021)
2. Chen, L.C., Papandreou, G., Schroff, F., Adam, H.: Rethinking atrous convolution for semantic image segmentation. arXiv preprint arXiv:1706.05587 (2017)
3. Chen, L., et al.: SCA-CNN: spatial and channel-wise attention in convolutional networks for image captioning. In: Proceedings of the IEEE Conference on Computer Vision and Pattern Recognition, pp. 5659–5667 (2017)
4. Diakogiannis, F.I., Waldner, F., Caccetta, P., Wu, C.: ResUNet-a: a deep learning framework for semantic segmentation of remotely sensed data. ISPRS J. Photogramm. Remote. Sens. **162**, 94–114 (2020)
5. Fu, J., et al.: Dual attention network for scene segmentation. In: Proceedings of the IEEE/CVF Conference on Computer Vision and Pattern Recognition, pp. 3146–3154 (2019)
6. Griffiths, P., Nendel, C., Hostert, P.: Intra-annual reflectance composites from Sentinel-2 and Landsat for national-scale crop and land cover mapping. Remote Sens. Environ. **220**, 135–151 (2019)
7. He, K., Zhang, X., Ren, S., Sun, J.: Deep residual learning for image recognition. In: Proceedings of the IEEE Conference on Computer Vision and Pattern Recognition, pp. 770–778 (2016)
8. Li, R., Zheng, S., Duan, C., Su, J., Zhang, C.: Multistage attention resU-Net for semantic segmentation of fine-resolution remote sensing images. IEEE Geosci. Remote Sens. Lett. **19**, 1–5 (2021)
9. Li, R., et al.: Multi attention network for semantic segmentation of fine-resolution remote sensing images. IEEE Trans. Geosci. Remote Sens. **60**, 1–13 (2021)
10. Li, R., Zheng, S., Zhang, C., Duan, C., Wang, L., Atkinson, P.M.: ABCNet: attentive bilateral contextual network for efficient semantic segmentation of fine-resolution remotely sensed imagery. ISPRS J. Photogramm. Remote. Sens. **181**, 84–98 (2021)
11. Lin, T.Y., Dollár, P., Girshick, R., He, K., Hariharan, B., Belongie, S.: Feature pyramid networks for object detection. In: Proceedings of the IEEE Conference on Computer Vision and Pattern Recognition, pp. 2117–2125 (2017)
12. Misra, D., Nalamada, T., Arasanipalai, A.U., Hou, Q.: Rotate to attend: convolutional triplet attention module. In: Proceedings of the IEEE/CVF Winter Conference on Applications of Computer Vision, pp. 3139–3148 (2021)
13. Oda, H., et al.: BESNet: boundary-enhanced segmentation of cells in histopathological images. In: Frangi, A.F., Schnabel, J.A., Davatzikos, C., Alberola-López, C., Fichtinger, G. (eds.) MICCAI 2018. LNCS, vol. 11071, pp. 228–236. Springer, Cham (2018). https://doi.org/10.1007/978-3-030-00934-2_26
14. Park, J., Woo, S., Lee, J., Kweon, I.S.: BAM: bottleneck attention module. In: British Machine Vision Conference 2018, BMVC 2018, Newcastle, UK, September 3–6, 2018. p. 147. BMVA Press (2018)

15. Picoli, M.C.A., et al.: Big earth observation time series analysis for monitoring Brazilian agriculture. ISPRS J. Photogramm. Remote. Sens. **145**, 328–339 (2018)
16. Ronneberger, Olaf, Fischer, Philipp, Brox, Thomas: U-Net: convolutional networks for biomedical image segmentation. In: Navab, Nassir, Hornegger, Joachim, Wells, William M.., Frangi, Alejandro F.. (eds.) MICCAI 2015. LNCS, vol. 9351, pp. 234–241. Springer, Cham (2015). https://doi.org/10.1007/978-3-319-24574-4_28
17. Samie, A., et al.: Examining the impacts of future land use/land cover changes on climate in Punjab province, Pakistan: implications for environmental sustainability and economic growth. Environ. Sci. Pollut. Res. **27**, 25415–25433 (2020)
18. Tong, X.Y., et al.: Land-cover classification with high-resolution remote sensing images using transferable deep models. Remote Sens. Environ. **237**, 111322 (2020)
19. Wang, H., Wang, Y., Zhang, Q., Xiang, S., Pan, C.: Gated convolutional neural network for semantic segmentation in high-resolution images. Remote Sens. **9**(5), 446 (2017)
20. Wang, L., Li, R., Duan, C., Zhang, C., Meng, X., Fang, S.: A novel transformer based semantic segmentation scheme for fine-resolution remote sensing images. IEEE Geosci. Remote Sens. Lett. **19**, 1–5 (2022)
21. Wang, L., Li, R., Wang, D., Duan, C., Wang, T., Meng, X.: Transformer meets convolution: a bilateral awareness network for semantic segmentation of very fine resolution urban scene images. Remote Sens. **13**(16), 3065 (2021)
22. Wang, L., et al.: UNetFormer: a UNet-like transformer for efficient semantic segmentation of remote sensing urban scene imagery. ISPRS J. Photogramm. Remote. Sens. **190**, 196–214 (2022)
23. Woo, Sanghyun, Park, Jongchan, Lee, Joon-Young., Kweon, In So.: CBAM: convolutional block attention module. In: Ferrari, Vittorio, Hebert, Martial, Sminchisescu, Cristian, Weiss, Yair (eds.) ECCV 2018. LNCS, vol. 11211, pp. 3–19. Springer, Cham (2018). https://doi.org/10.1007/978-3-030-01234-2_1
24. Yin, H., Pflugmacher, D., Li, A., Li, Z., Hostert, P.: Land use and land cover change in inner Mongolia-understanding the effects of China's re-vegetation programs. Remote Sens. Environ. **204**, 918–930 (2018)
25. Zhang, C., et al.: Joint deep learning for land cover and land use classification. Remote Sens. Environ. **221**, 173–187 (2019)
26. Zhang, Z., Liu, Q., Wang, Y.: Road extraction by deep residual U-Net. IEEE Geosci. Remote Sens. Lett. **15**(5), 749–753 (2018)
27. Zhao, H., Shi, J., Qi, X., Wang, X., Jia, J.: Pyramid scene parsing network. In: Proceedings of the IEEE Conference on Computer Vision and Pattern Recognition, pp. 2881–2890 (2017)

Improved Conditional Generative Adversarial Networks for SAR-to-Optical Image Translation

Tao Zhan, Jiarong Bian, Jing Yang, Qianlong Dang, and Erlei Zhang[✉]

College of Information Engineering, Northwest A&F University, Yangling 712100,
Shaanxi, China
{zhantao,jiarong98,yangjinghit,Dangqianlong,
erlei.zhang} @nwafu.edu.cn

Abstract. Synthetic aperture radar (SAR) has the potential to operate effectively in all weather conditions, making it a desirable tool in various fields. However, the inability of untrained individuals to visually identify ground cover in SAR images poses challenges in practical applications such as environmental monitoring, disaster assessment, and land management. To address this issue, generative adversarial networks (GANs) have been used to transform SAR images into optical images. This technique is commonly referred to as SAR to optical image translation. Despite its common use, the traditional methods often generate optical images with color distortion and blurred contours. Therefore, a novel approach utilizing conditional generative adversarial networks (CGANs) is introduced as an enhanced method for SAR-to-optical image translation. A style-based calibration module is incorporated, which learns the style features of the input SAR images and matches them with the style of real optical images to achieve color calibration, thereby minimizing the differences between the generated output and real optical images. Furthermore, a multi-scale strategy is incorporated in the discriminator. Each branch of the multi-scale discriminator is dedicated to capturing texture and edge features at different scales, thereby enhancing the texture and edge information of the image at both local and global levels. Experimental results demonstrate that the proposed approach surpasses existing image translation techniques in terms of both visual effects and evaluation metrics.

Keywords: Optical images · Synthetic Aperture Radar (SAR) ·
SAR-to-optical translation · Conditional Generative Adversarial
Network (CGANs)

1 Introduction

Synthetic aperture radar (SAR) images offer high-resolution radar images with penetrating capabilities, containing valuable information like geomorphological

This work was supported by the National Natural Science Foundation of China (No. 62006188, No. 62103311); Chinese Universities Scientific Fund (No. 2452022341); the QinChuangyuan High-Level Innovation and Entrepreneurship Talent Program of Shaanxi (No. 2021QCYRC4-50).

Q. Liu et al. (Eds.): PRCV 2023, LNCS 14428, pp. 279–291, 2024.
https://doi.org/10.1007/978-981-99-8462-6_23

structures and surface features. Due to these unique characteristics, SAR images have become indispensable in many applications, such as military [22], agriculture [6], environmental monitoring [1], etc. Radar is a valuable remote sensing instrument, whereas its ability to discriminate land cover types and extract meaningful information from SAR images is severely limited by the presence of speckle noise. These obstacles pose significant challenges to the widespread application and promotion of SAR imagery. In contrast, due to the unique characteristics of optical satellites, they have emerged as an indispensable tool for Earth observation. Optical images are generally abundant in color details and information, which makes them well-suited for various applications that require high precision, such as scene classification [25], image denoising [4], and change detection [24]. Consequently, converting SAR images to optical images enhances land type identification by non-experts, and also expands the applicability of SAR images in various domains.

Recently, SAR-to-optical image translation has garnered considerable attention, many methods have been proposed. Conventional techniques generally utilize colorization to translate SAR imagery. Nonetheless, these kinds of methods exclusively allow for the separation of the ground objects while failing to accurately restore the real ground truth. Recently, researchers have proposed many deep learning-based methods for image translation, among which the methods based on adversarial generative networks (GANs) [7] have shown impressive performance. GAN has achieved outstanding performance in image translation tasks, leading to an increasing number of studies utilizing GAN as the fundamental model [3]. Nevertheless, the process of GAN-based image translation generation is typically uncontrollable, therefore, additional constraints must be incorporated into the network to achieve the intended result. Bermudez *et al.* [14] proposed a technique utilizing Pix2Pix [8] networks to translate optical images into SAR images. Similarly, Merkle *et al.* [10] utilized the generated images for image matching and explored the translation of SAR images into optical images. Wang *et al.* [18] converted optical images to SAR images using CycleGAN [27] networks, which can maintain the accuracy and stability of the conversion by performing cyclic consistency training between the two domains. Ma *et al.* [20] proposed a hybrid CGAN [11] coupling global and local features for SAR-to-optical image translation. Nevertheless, this task poses significant challenges as the translation model must effectively extract image information while maintaining precise details amidst strong interference, such as the inevitable speckle noise. Unfortunately, the aforementioned algorithms suffer from contour blurring, texture loss, and color distortion.

To resolve these issues, a novel model for SAR-to-optical image translation is introduced. The main contributions of this work are as follows:

1. A style-based calibration module is introduced to enhance the translation quality and stability by effectively capturing image style features.
2. A multi-scale discriminator is proposed to improve the texture details of generated images by capturing features of different sizes using distinct receptive domains, thereby generating high-quality images with rich details and styles.

3. The impressive results obtained on different datasets have demonstrated that the proposed method is more effective than other existing techniques at preserving image textures and preventing color distortions.

The remaining sections are structured as follows: Sect. 2 presents an overview of some related work. Section 3 provides a detailed description of the specific details of the improved CGAN model. Section 4 introduces the experimental dataset and analyzes the experimental results. Finally, Sect. 5 provides a summary of the paper and presents future work plans.

2   Related Work

2.1   Descriptions of GAN

In 2014, GAN was introduced by Goodfellow *et al.* [7], showcasing its inherent capacity for data generation. Since then, with the continuous progress of deep learning, GAN has become increasingly significant in numerous computer vision applications, including but not limited to image super-resolution [12], style transfer [26], semantic segmentation [23], image compression [21], and object detection [15].

GAN consists of two networks: generator and discriminator. The generator's primary task is to produce synthetic data using random noise as input, whereas the discriminator is responsible for distinguishing between real and fake data. Through adversarial training, the generator learns to produce more realistic data while the discriminator improves its ability to accurately discriminate between real and fake data. Despite being a powerful unsupervised learning method, GAN has limited application due to the lack of direct control over its generated results. To address this issue, CGAN has been proposed. CGAN differs from standard GAN by incorporating additional conditional information into the generator and discriminator, such as labels, text descriptions, or images. This allows CGAN to control the generated content during image generation, resulting in more controllable generated results. Inspired by this, the Pix2Pix [8] image transformation model was developed and validated on multiple datasets to demonstrate its effectiveness. On the framework of Pix2Pix, a modified method dubbed pix2pixHD [19] was proposed in order to better manage high-resolution images and accomplish more precise detail generation. Recently, a novel unsupervised CGAN model called NiceGAN [2] has been developed, which achieves outstanding performance by reusing the image coding process.

2.2   SAR-to-Optical Image Translation

SAR-to-optical image translation is a fascinating and rapidly evolving field within remote sensing research. Over the past years, a significant amount of research has been dedicated to developing effective methods for generating high-quality optical images from SAR images. To cater to the needs of visual recognition, SAR image adjustment techniques are broadly categorized into two types: image improvement algorithms and coloring algorithms.

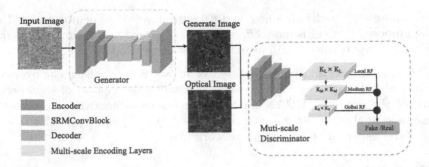

Fig. 1. Framework of the proposed method. RF stands for the receptive field.

However, these traditional techniques have their limitations in achieving superior image quality and interpretation. As a result, researchers have turned their attention to deep learning techniques to overcome these challenges. In particular, the emergence of GAN provides novel directions for researching SAR-to-optical image translation.

Schmitt *et al.* [16] introduced the SEN1-2 dataset, providing a valuable resource for deep learning studies focused on the fusion of SAR and optical images using Pix2Pix for SAR-to-optical image translation. Fu *et al.* [5] introduced a cascaded-residual adversarial network as a novel approach for bidirectional translation between SAR and optical remote sensing images. Wang *et al.* [17] proposed a method for translating SAR images to optical images that makes use of hierarchical latent features. By incorporating a hierarchical feature extractor, this approach enhances the capability to capture information between SAR and optical images. Niu *et al.* [13] provided a solution for cross-modality image translation in remote sensing using the Pix2Pix network.

However, most of these methods still suffer from limitations, including blurred boundaries, loss of spatial information, and color inaccuracies, making it difficult to handle speckle noise and adapt to a diverse range of image styles.

3   Method

This paper introduces an improved CGAN model for SAR-to-optical image translation, as shown in Fig. 1. This section provides a comprehensive explanation of the generator and discriminator architectures, as well as the loss function employed in the model.

3.1   The Generator with Style-Based Recalibration Module

A novel generator framework is introduced, as illustrated in Fig. 2. The input image is encoded using a three-layer convolutional encoder module. Meanwhile, the decoder module, which includes one convolutional layer and two fractionally stridden convolutions, is utilized for image reconstruction. During the translation

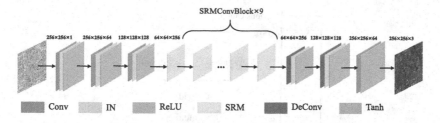

Fig. 2. The architecture of our generator.

process, the transformation module integrates 9 residual blocks to manipulate the feature maps.

The generator framework in this study includes a style-based recalibration module (SRM) [9] within the residual block to enhance the capture of style features in the feature space. The SRM architecture consists of two main components: style pooling and style integration. The style pooling operator aggregates the feature responses across spatial dimensions to extract style features from each channel, enabling a more comprehensive representation of style information from SAR images. The style integration operator employs these style features in channel-level operations to generate style weights that are specific to each sample. These style weights subsequently recalibrate the feature mappings by amplifying or reducing their information, resulting in a unified style representation. It is important to note that the SRMConvBlock, which incorporates SRM, represents the residual block as illustrated in Fig. 3.

The style pooling operation selects the average mean and standard deviation of each feature map as the style feature. Given the input feature map $F \in R^{C \times H \times W}$, C denotes the total number of channels, while H and W represent the height and width respectively. Additionally, f_{ij} denotes the activation value of the c-th channel at position (i, j) in the feature map F. The style feature $T \in R^{C \times 2}$ is calculated as follows:

$$\mu_c = \frac{1}{HW} \sum_{i=1}^{H} \sum_{j=1}^{W} f_{ij} \tag{1}$$

$$\sigma_c = \sqrt{\frac{1}{HW} \sum_{i=1}^{H} \sum_{j=1}^{W} (f_{ij} - \mu_c)^2} \tag{2}$$

$$t_c = [\mu_c, \sigma_c] \tag{3}$$

here, t_c represents a stylized vector that provides a concise representation of the style information for channel c, formed by joining average mean μ_c and standard deviation σ_c.

During style integration, the style features are converted into channel-level style weights, which can adjust the significance of styles related to each channel and facilitate the following feature recalibration. Previous studies have utilized

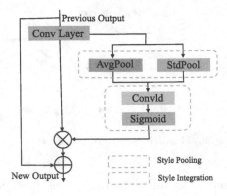

Fig. 3. The architecture of SRMConvBlock.

a fully-connected layer in this module to implement style integration, but such a design leads to a heavy computational burden. To overcome this issue, a method is introduced to perform style integration using a 1-D convolutional layer with fewer parameters, which effectively reduces the computational cost while maintaining the performance of the model. Besides, the style integration module includes a sigmoid activation function. Using the output t_c obtained from style pooling as input, the style integration module computes the style integration weights $S \in R^{C \times 1}$ for each channel as follows:

$$s_c = \sigma(k * t_c) \tag{4}$$

the weight for the cth channel is denoted as s_c, $\sigma(\cdot)$ represents the sigmoid activation function, k donates the convolution kernel, as well as $*$ represents the convolution operation. The input features $F \in R^{C \times H \times W}$ are finally recalibrated by the channel-wise style integration weights $S \in R^{C \times 1}$ to obtain the output, which is represented as:

$$\hat{F} = S \cdot F \tag{5}$$

the output $\hat{F} \in R^{C \times H \times W}$ is obtained by channel-wise multiplication of the style integration weights W and the input features F.

3.2   Multi-scale Discriminator

A multi-scale discriminator method is employed to improve the model's discriminative capacity. The discriminator is composed of an encoder and three compact classifiers that correspond to local, medium, and global RFs at varying scales, respectively. The loss function for the multi-scale discriminator can be formulated as:

$$\mathcal{L}_{D_{(p=Real,Fake)}} = \sum_{q=l,m,g} ((\|K_q - p_q\|_2^2)/3) \tag{6}$$

Table 1. Selected categories of scenes from SEN1-2 dataset.

Scene	Training Set	Testing Set
Vegetation	900	225
Desert	702	176
Residence	840	210
Farmland	1111	278
Mountain	801	180
Others	1536	384
Total	5089	1453

here, the discriminator classifies the results as true or false, and the receptive fields at local, medium, and global scales are denoted by K_l, K_m, and K_g respectively. Their respective sizes are 70×70, 140×140, and 280×280. Furthermore, p_q represents the feature representation of real or generated samples at the corresponding receptive field scale q.

The multi-scale discriminator enhances the texture of generated images by extracting features at multiple scales, thereby improving the accuracy of the network. It also improves the texture of generated images at different scales.

3.3 Loss Functions

The model utilizes two loss functions in the full objective.

Adversarial Loss. The purpose of this loss is to align the distribution of the generated optical images with that of the real optical images, thereby reducing the average pixel difference between them, and it is defined as follows:

$$\min_G \max_D \mathcal{L}_{CGAN}(G, D) = \mathbb{E}[\log(D(y, x))] + \mathbb{E}[\log(1 - D(y, G(x)))] \quad (7)$$

here, G donates the generator, and D is the discriminator. $\mathbb{E}$ donates the calculation of the expected value of a random variable. x donates the SAR image and y donates the optical image. $D(y, G(x))$ is the discriminator's output when it classifies the generated image $G(x)$, which is the result of transforming the SAR image x into the optical domain, along with the optical image y.

L1 Loss. This loss helps the model in capturing the real color distribution and reducing the average pixel difference between the generated and real optical images. It is defined as follows:

$$\mathcal{L}_{L1}(G) = \mathbb{E}[||G(x) - y||_1] \quad (8)$$

Full Objective. In conclusion, the full objective of the method is defined as:

$$\mathcal{L}_{total} = \mathcal{L}_{CGAN}(G, D) + \lambda \cdot \mathcal{L}_{L1}(G) \quad (9)$$

where λ is the hyperparameter that control the importance of $\mathcal{L}_{L1}(G)$.

Table 2. Impact of hyperparameter λ on evaluation metrics for $\mathcal{L}_{L1}$.

λ	PSNR↑	SSIM↑	MSE↓	VIF↑
10	14.9326	0.2893	0.0274	0.3107
20	15.0824	0.3158	0.0265	0.3140
30	15.3679	0.3121	0.0283	0.3122
40	15.6399	0.3369	0.0278	0.3120
50	15.8698	0.3277	0.0289	0.3134
60	15.9764	0.3464	0.0212	0.3155
70	16.0382	0.3353	0.0213	0.3175
80	16.1653	0.3472	0.0228	0.3172
90	16.3824	0.3563	0.0202	0.3186
100	**16.6164**	**0.3754**	**0.0186**	**0.3291**
110	16.6052	0.3649	0.0197	0.3279
120	16.5883	0.3602	0.0189	0.3266

4   Experiments

4.1   Experimental Setup

Dataset. The SEN1-2 dataset comprises SAR images and multispectral optical images acquired globally throughout the year 2017. All images are 256×256 pixels in size with a spatial resolution of 10 meters. The dataset is divided into two parts: Sentinel-1 and Sentinel-2. Sentinel-1 includes fully polarized and monopolized SAR images, while Sentinel-2 consists of multi-spectral and high-resolution optical images.

A total of 5890 SAR-optical image pairs were randomly sampled from the SEN1-2 dataset for training the model. Additionally, 1453 pairs were allocated as test data for evaluation purposes. The selected images represent diverse land types and topography to ensure the generalizability of the proposed method. More detailed information about the dataset can be found in Table 1.

Evaluation Metrics. Four metrics were used to evaluate the proposed model's performance: peak signal-to-noise ratio (PSNR), structural similarity measure (SSIM), mean square error (MSE), and visual information fidelity (VIF). These metrics provide quantitative assessments of the generated image quality from various aspects. Higher PSNR, SSIM, and VIF values, and lower MSE values indicate better image quality.

Implementation Details. The experiments were performed using PyTorch on a computer with an Intel(R) Xeon(R) Platinum 8124M CPU, 64GB of RAM, and an NVIDIA GeForce RTX 3090 graphics card. The network was trained using Adam optimization with a learning rate of 0.0002, β_1 of 0.9, and β_2 of 0.999.

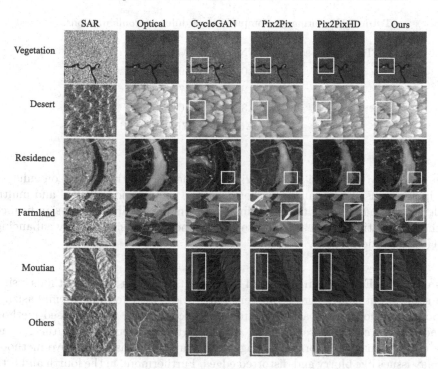

Fig. 4. Visualization results of SAR-to-optical image translation of different models.

The batch size was set to 8, and the model was trained for 200 epochs. The same hardware setup and parameter settings provided by the authors' official code were used for the comparison with other algorithms.

4.2 Experimental Analysis

In this section, the influence of hyperparameters, ablation study, and experimental results will be presented.

Impact of Hyperparameters. Table 2 presents the results of the evaluation metrics for different values of λ. As shown in the table, it can be observed that when λ is set to 100, all evaluation metrics achieve their optimal values. This value achieves a balance between the adversarial loss and the $L1$ loss, enabling the model to generate visually accurate outputs while preserving important input features. Based on these results, the value of λ was set to 100.

Ablation Study. To validate the effectiveness of each component of the proposed method, SRM and multi-scale discriminator, ablation experiments were conducted. All experiments were performed on the same dataset. The results are shown in Table 3. Analyzing the various evaluation metrics in the table reveals

Table 3. Performance comparison of different configurations.

SRM	Multi-scale D	PSNR↑	SSIM↑	MSE↓	VIF↑
×	×	14.0318	0.2696	0.0261	0.2472
✓	×	15.2362	0.3374	0.0201	0.2968
×	✓	15.3279	0.3468	0.0197	0.3042
✓	✓	**16.6164**	**0.3754**	**0.0186**	**0.3291**

that both SRM and multi-scale discriminator have a positive impact on enhancing image translation quality. Furthermore, the combination of SRM and multi-scale discriminator exhibits the optimal performance, affirming the significance of the synergistic effect between these two components in significantly enhancing image translation results.

Qualitative Evaluation. Figure 4 presents the results of all methods using the same input image across different categories, with red boxes emphasizing important regions. Comparing the first and second rows, the proposed method stands out in contour preservation. The desert scene generated in the second row closely mirrors the original image's contours, whereas alternative methods display issues like blurry and distorted edges. Furthermore, in the fourth and fifth rows, the generated images exhibit minimal color disparity from the real images. Notably, the proposed method enhances texture features while maintaining color precision, evident in the third and sixth rows. The riverbank's contours in the third row and the river's color closely resemble the original image. Overall, the proposed approach produces more authentic images with vivid colors, sharper edges, and enhanced textures in contrast to alternative methods.

Quantitative Evaluation. Table 4 presents the evaluation results for different scenarios in the SEN1-2 dataset. The proposed method surpasses other algorithms in quantitative evaluation of the SEN1-2 dataset. It consistently achieves higher PSNR values, up to a maximum improvement of 9.1559, indicating enhanced accuracy in reconstructing pixel intensities. Additionally, it obtains lower MSE values, with a maximum reduction of 0.0823, showcasing better resemblance to the ground truth images. Moreover, the proposed method attains the highest VIF values, with a maximum improvement of 0.4144, emphasizing its superior preservation and reproduction of visible details. Overall, the proposed SAR-to-optical translation model exhibits superior performance across various scenarios, affirming its practical efficacy.

Table 4. Results of evaluation metrics on different scenarios of the SEN1-2 dataset.

Scene	Method	PSNR↑	SSIM↑	MSE↓	VIF↑
Vegetation	CylceGAN	12.8663	**0.1491**	0.0513	0.1997
	Pix2Pix	12.7919	0.1149	0.0469	0.0777
	Pix2PixHD	12.9476	0.1378	0.0419	0.0909
	Ours	**13.9593**	0.1382	**0.0414**	**0.2038**
Desert	CylceGAN	13.4822	0.2681	0.0392	0.4532
	Pix2Pix	11.2846	0.3145	0.0758	0.5584
	Pix2PixHD	16.5809	0.4563	0.0132	0.7392
	Ours	**19.0318**	**0.4670**	**0.0101**	**0.8679**
Residence	CylceGAN	13.0263	0.1260	0.0653	0.0746
	Pix2Pix	12.8637	0.1651	0.0607	0.0498
	Pix2PixHD	13.9332	0.1872	0.0336	0.0632
	Ours	**15.6138**	**0.1884**	**0.0303**	**0.1161**
Farmland	CylceGAN	10.2489	0.1391	0.0990	0.0844
	Pix2Pix	11.0748	0.2148	0.0916	0.0717
	Pix2PixHD	13.9058	**0.3360**	0.0337	0.1169
	Ours	**16.0567**	0.3356	**0.0260**	**0.2481**
Mountain	CylceGAN	9.2646	0.3119	0.0938	0.0319
	Pix2Pix	11.9893	0.2061	0.0516	0.0570
	Pix2PixHD	17.4524	0.3405	0.0118	0.1387
	Ours	**18.4205**	**0.3607**	**0.0115**	**0.2096**

5   Conclusion

This paper introduces a novel improved CGAN model for SAR-to-optical image translation. The proposed model incorporates a style-based calibration module to capture style features and mitigate color distortion in the generated images. Additionally, a multi-scale discriminator is introduced to enhance the sharpness of the image contours. The proposed method is evaluated through comprehensive qualitative and quantitative assessments, demonstrating superior performance compared to existing state-of-the-art methods. Future research will focus on exploring new network architectures to further enhance the quality of image-to-image translation.

References

1. Ardila, J., Laurila, P., Kourkouli, P., Strong, S.: Persistent monitoring and mapping of floods globally based on the iceye sar imaging constellation. In: IGARSS 2022, pp. 6296–6299 (2022)
2. Chen, R., Huang, W., Huang, B., Sun, F., Fang, B.: Reusing discriminators for encoding: towards unsupervised image-to-image translation. In: CVPR 2020, pp. 8165–8174 (2020)
3. Fang, Y., Deng, W., Du, J., Hu, J.: Identity-aware CycleGAN for face photo-sketch synthesis and recognition. PR **102**, 107249 (2020)
4. Feng, L., Wang, J.: Research on image denoising algorithm based on improved wavelet threshold and non-local mean filtering. In: ICSIP 2021, pp. 493–497 (2021)
5. Fu, S., Xu, F., Jin, Y.Q.: Reciprocal translation between SAR and optical remote sensing images with cascaded-residual adversarial networks. Sci. China Inf. Sci. **64**, 122301 (2021)
6. Gonzalez-Audicana, M., Lopez-Saenz, S., Arias, M., Sola, I., Alvarez-Mozos, J.: Sentinel-1 and sentinel-2 based crop classification over agricultural regions of navarre (spain). In: IGARSS 2021, pp. 5977–5980 (2021)
7. Goodfellow, I.J., et al.: Generative adversarial networks (2014). arXiv:1406.2661
8. Isola, P., Zhu, J.Y., Zhou, T., Efros, A.A.: Image-to-image translation with conditional adversarial networks (2018), arXiv:1611.07004
9. Lee, H., Kim, H.E., Nam, H.: Srm: a style-based recalibration module for convolutional neural networks. In: ICCV 2019, pp. 1854–1862 (2019)
10. Merkle, N., Auer, S., Müller, R., Reinartz, P.: Exploring the potential of conditional adversarial networks for optical and SAR image matching. IEEE J. Sel. Top. Appl. Earth Observ. Remote Sens. **11**(6), 1811–1820 (2018)
11. Mirza, M., Osindero, S.: Conditional generative adversarial nets (2014). arXiv:1411.1784
12. Naik, D.A., Sangeetha, V., Sandhya, G.: Generative adversarial networks based method for generating photo-realistic super resolution images. In: ETI 4.0 2021, pp. 1–6 (2021)
13. Niu, X., Yang, D., Yang, K., Pan, H., Dou, Y.: Image translation between high-resolution remote sensing optical and SAR data using conditional GAN. In: PCM 2018, pp. 245–255 (2018)
14. Vishwakarma, D.K.: Comparative analysis of deep convolutional generative adversarial network and conditional generative adversarial network using hand written digits. In: ICICCS 2020, pp. 1072–1075 (2020)
15. Prakash, C.D., Karam, L.J.: It gan do better: GAN-Based detection of objects on images with varying quality. IEEE Trans. Image Process. **30**, 9220–9230 (2021)
16. Schmitt, M., Hughes, L.H., Zhu, X.X.: The SEN1-2 dataset for deep learning in SAR-optical data fusion (2018). arXiv:1807.01569
17. Wang, H., Zhang, Z., Hu, Z., Dong, Q.: SAR-to-optical image translation with hierarchical latent features. IEEE Trans. Geosci. Remote Sens. **60**, 1–12 (2022)
18. Wang, L., et al.: SAR-to-optical image translation using supervised cycle-consistent adversarial networks. IEEE Access **7**, 129136–129149 (2019)
19. Wang, T.C., Liu, M.Y., Zhu, J.Y., Tao, A., Kautz, J., Catanzaro, B.: High-resolution image synthesis and semantic manipulation with conditional GANs. In: CVPR 2018, pp. 8798–8807 (2018)
20. Wang, Z., Ma, Y., Zhang, Y.: Hybrid cgan: coupling global and local features for sar-to-optical image translation. IEEE Trans. Geosci. Remote Sens. **60**, 1–16 (2022)

21. Wu, M., He, Z., Zhao, X., Zhang, S.: Generative adversarial networks-based image compression for consumer photo storage. In: GCCE 2019, pp. 333–334 (2019)
22. Xiao, M., He, Z., Lou, A., Li, X.: Center-to-corner vector guided network for arbitrary-oriented ship detection in synthetic aperture radar images. In: ICGMRS 2022, pp. 293–297 (2022)
23. Yang, H., Zhang, J.: Self-attentive semantic segmentation model based on generative adversarial network. In: AICIT 2022, pp. 1–5 (2022)
24. Zhan, T., Gong, M., Jiang, X., Zhao, W.: Transfer learning-based bilinear convolutional networks for unsupervised change detection. Remote Sens. Lett. **19**, 1–5 (2022)
25. Zhang, E., et al.: Attention-embedded triple-fusion branch CNN for hyperspectral image classification. Remote Sens. **15**(8) (2023)
26. Zhao, L., Jiao, Y., Chen, J., Zhao, R.: Image style transfer based on generative adversarial network. In: ICCNEA 2021, pp. 191–195 (2021)
27. Zhu, J.Y., Park, T., Isola, P., Efros, A.A.: Unpaired image-to-image translation using cycle-consistent adversarial networks (2020). arXiv:1703.10593

A Novel Cross Frequency-Domain Interaction Learning for Aerial Oriented Object Detection

Weijie Weng[1], Weiming Lin[1(✉)], Feng Lin[1], Junchi Ren[2], and Fei Shen[3]

[1] School of Opto-Electronic and Communication Engineering,
Xiamen University of Technology, Xiamen 361024, China
wjweng@s.xmut.edu.cn, {linwm,linfeng}@xmut.edu.cn
[2] China Telecom Co., Ltd., Beijing, China
renjc@chinatelecom.cn
[3] Tencent AI Lab, Shenzhen, China

Abstract. Aerial oriented object detection is a vital task in computer vision, receiving significant attention for its role in remote image understanding. However, most Convolutional Neural Networks (CNNs) methods easily ignore the frequency domain because they only focus on the spatial/channel interaction. To address these limitations, we propose a novel approach called Cross Frequency-domain Interaction Learning (CFIL) for aerial oriented object detection. Our method consists of two modules: the Extraction of Frequency-domain Features (EFF) module and the Interaction of Frequency-domain Features (IFF) module. The EFF module extracts frequency-domain information from the feature maps, enhancing the richness of feature information across different frequencies. The IFF module facilitates efficient interaction and fusion of the frequency-domain feature maps obtained from the EFF module across channels. Finally, these frequency-domain weights are combined with the time-domain feature maps. By enabling full and efficient interaction and fusion of EFF feature weights across channels, the IFF module ensures effective utilization of frequency-domain information. Extensive experiments are conducted on the DOTA V1.0, DOTA V1.5, and HRSC2016 datasets to demonstrate the competitive performance of the proposed CFIL in the aerial oriented object detection. Our code and models will be publicly released.

Keywords: Oriented object detection · Frequency-domain · Interaction

1 Introduction

Aerial oriented object detection [6–8,20] is an increasingly important task in the field of computer vision. It involves identifying and locating targets [19,21,26],

This work was supported by the Natural Science Foundation of Fujian Province of China under Grant 2022J011271, the Foundation of Educational and Scientific Research Projects for Young and Middle-aged Teachers of Fujian Province under Grant JAT200471, as well as the High-level Talent Project of Xiamen University of Technology under Grant YKJ20013R.

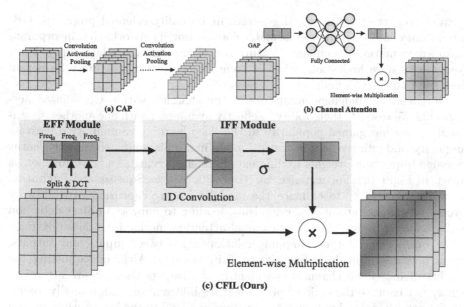

Fig. 1. Comparisons of different approaches for feature extraction from feature maps. We obtain richer frequency domain features by crossing frequency domains and interact fully and efficiently without dimension reduction.

such as vehicles, buildings, or other objects, in aerial images captured from varying heights and angles. The significance of this task is attributed to its broad range of applications, including surveillance, security, and disaster management. However, aerial object detection presents a formidable challenge since it includes objects exhibiting diverse scales and orientations, densely packed distributions amidst complex backgrounds, and variations in target appearance caused by changes in illumination, pose, or occlusion.

Considerable progress is currently being made in detecting rotated objects. Specifically, extensive analysis is being conducted on various representations of rotated objects [6–8,12,20,37] and the development of more appropriate loss functions for these representations [3]. Furthermore, substantial exploration is being undertaken to extract rotated regions of interest [34] and assign labels [16]. Additionally, comprehensive studies are being conducted on the structure of detection networks, including the neck [38] and head [6] of detectors, as well as rotated region proposal networks [34]. However, limited attention is being given to acquiring additional feature information of target objects.

The existing studies on the identification of directional targets in aerial images predominantly employ CNNs for extracting target features [22,32,35,36]. As shown in Fig. 1 (a), CNNs capture powerful image descriptions by utilizing stacked convolutional layers, interspersed with non-linearities and downsampling, to capture hierarchical patterns with global receptive fields [24,27–29]. For example, Oriented R-CNN [34] employs a lightweight fully convolutional network to extract features and generate high-quality oriented proposals. Oriented RepPoints [12] utilizes deformable convolutional networks that incorporate spatial deformations to enhance feature representation and adaptability. However, CNNs have limitations when it comes to extracting feature information from the frequency domain.

Moreover, the independent nature of the channels within CNN-based networks [23, 25] restricts their ability to effectively interact with one another. Channel attention has gained popularity in the deep learning community due to its simplicity and effectiveness in feature modeling. This approach directly learns to assign importance weights to different channels, serving as a powerful tool, as shown in Fig. 1 (b). For instance, SCRDet [39] utilizes Squeeze-and-Excitation Network (SENet) [9] to enhance the model's feature capturing capability by leveraging channel attention mechanisms, leading to improved object detection performance. However, including channel attention mechanisms does not guarantee a direct and optimal mapping relationship between inputs and outputs, which may result in degrading the resulting quality. Although adjusting the weights of individual channels can identify the targets that require attention, it may not capture the entirety of the feature information. Additionally, representing attention weights using scalar values can lead to the loss of valuable and nuanced information.

For that, this paper presents a novel Cross Frequency-domain Interaction Learning (CFIL) for aerial oriented object detection, as shown in Fig. 1 (c). CFIL comprises two key modules: The Extraction of Frequency-domain Features (EFF) module and the Interaction of Frequency-domain Features (IFF) module. EFF extracts frequency-domain information from the input feature maps, enriching the understanding of aerial objects. By incorporating frequency-domain analysis, we capture diverse feature information across different frequencies. IFF facilitates efficient interaction and fusion of frequency-domain feature maps obtained from the EFF module. It enables cross-channel interactions, producing feature weights optimized for the frequency domain. This module ensures effective utilization of frequency-domain information by enabling complete and efficient interaction of EFF feature weights. The frequency-domain weights obtained from the IFF are combined with the time-domain feature maps, resulting in a more robust representation for aerial object detection. This fusion process integrates complementary information from both domains.

This paper makes the following contributions:

- We propose a plug-and-play Cross Frequency-domain Interaction Learning (CFIL) for aerial oriented object detection that addresses the interaction of frequency domain problem.
- We present a new Extraction of Frequency-domain Features (EFF) module to extract frequency-domain information from the feature maps, enhancing the richness of feature information across different frequencies.
- We devise a new Interaction of Frequency-domain Features (IFF) module to facilitate efficient interaction and fusion of the frequency-domain feature maps obtained from the EFF module across channels.

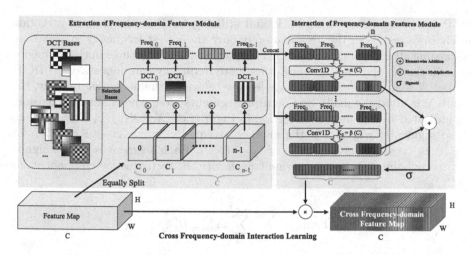

Fig. 2. Pipline of CFIL. The pipline comprises two modules: the EFF module and IFF module, represented by the light blue region and the light purple separately. (Color figure online)

2 Proposed Method

2.1 Extraction of Frequency-Domain Features Module

Let $x^{2D} \in \mathbb{R}^{H \times W}$ be a two-dimensional input, where $x_{i,j}^{2D}$ represents the pixel value at position (i, j) in the input. where i represents the row index and j represents the column index of the input, $i \in \{0, 1, \ldots, H-1\}$, $j \in \{0, 1, \ldots, W-1\}$. The 2D discrete cosine transform (2D DCT) is denoted as Eq. 1, as follows,

$$f_{h,w}^{2d} = \sum_{i=0}^{H-1} \sum_{j=0}^{W-1} x_{i,j}^{2D} \cos\left(\frac{\pi h}{H}\left(i + \frac{1}{2}\right)\right) \cos\left(\frac{\pi w}{W}\left(j + \frac{1}{2}\right)\right), \quad (1)$$

where $f^{2D} \in \mathbb{R}^{H \times W}$ is the 2D DCT frequency spectrum, h and w represent the positions in the frequency spectrum of the 2D DCT, $h \in \{0, 1, \ldots, H-1\}$, $w \in \{0, 1, \ldots, W-1\}$, H is the height of x^{2D}, and W is the width of x^{2D}. $\cos\left(\frac{\pi h}{H}\left(i + \frac{1}{2}\right)\right) \cos\left(\frac{\pi w}{W}\left(j + \frac{1}{2}\right)\right)$ is defined as the basis function for 2D DCT. Let the basis function in Eq. 1 for 2D DCT be denoted as $B_{h,w}^{i,j}$, then the 2D DCT can be written as Eq. 2.

$$f_{h,w}^{2D} = \sum_{i=0}^{H-1} \sum_{j=0}^{W-1} x_{i,j}^{2D} B_{h,w}^{i,j} \quad (2)$$

When both h and w in Eq. 1 are zero, it is equivalent to Global Average Pooling (GAP), defined as $\text{GAP}(x) = \frac{1}{h \cdot w} \sum_{i=0}^{h-1} \sum_{j=0}^{w-1} x_{ij}$. The result of GAP is directly proportional to the lowest frequency component of 2D DCT. This indicates that

GAP is a special case of 2D DCT, and utilizing GAP in the channel attention mechanism results in the preservation of only the lowest frequency information, discarding all components from other frequencies. More features in the frequency domain can be obtained by using other more frequency components of 2D DCT.

Let $X \in \mathbb{R}^{C \times H \times W}$ be the image feature tensor in the networks. Initially, the input X is divided equally into multiple segments along the channel dimension, As indicated by the light blue section in Fig. 2. Let $[X_0, X_1, \ldots, X_{n-1}]$ denote these segments, where n is the number of equally split, $X_i \in \mathbb{R}^{C \times H \times W}$ for $i \in \{0, 1, \ldots, \text{n-1}\}$, and $C_0 = C_1 = \ldots = C_{n-1}$. Each segment is then assigned a corresponding 2D discrete cosine transform (DCT) frequency component, as Eq. 3,

$$\text{Freq}_i = \sum_{h=0}^{H-1} \sum_{w=0}^{W-1} X^i_{:,h,w} B^{u_i, v_i}_{h,w},\tag{3}$$

where u_i and v_i represent the 2D indices of the frequency components corresponding to X_i, while $\text{Freq}_i \in \mathbb{R}^{C_i}$ is the C_i-dimensional vector resulting from aggregation.

As shown in Fig. 2, The Frequency-domain Features Extraction module ultimately produces a complete aggregation vector by concatenating various feature information from different frequency components in the frequency domain, as described in Eq. 4.

$$\text{Freq} = \text{cat}\left([\text{Freq}_0, \text{Freq}_1, \cdots, \text{Freq}_{n-1}]\right).\tag{4}$$

2.2   Interaction of Frequency-Domain Features Module

The EFF Module ultimately yields a vector containing frequency-domain information at different frequencies, introducing additional information into the embedded channels. However, there is a lack of interaction among the different frequency information. Therefore, As indicated by the light purple area in Fig. 2, we present our Interaction of Frequency-domain Features Module which efficient interaction and fusion of the frequency-domain feature maps obtained from the EFF module across channels without channel dimensionality reduction, aiming at guaranteeing both efficiency and effectiveness. We found that two convolutional kernels can achieve the best interactive effect. Calculation based on the number of input channels, we conducted 1D convolution operations on the aggregated vectors obtained from the EFF module using m different kernel sizes, followed by fusion of the m results, as described in Eq. 5,

$$\sum_{l=1}^{m}(\text{Freq} * \text{Conv1D}_{k=2l-1}),\tag{5}$$

where m is the number of 1D convolutions fusions, l represents the index, $l \in \{0, 1, \ldots, m\}$ and k represents the kernal size of the corresponding 1D convolution. We obtained a vector, effective interplay between different frequencies,

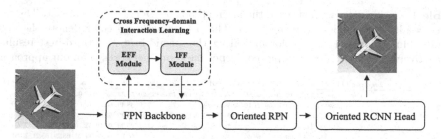

Fig. 3. The overall framework of CFIL. The framework comprises two modules, the EFF module, represented by the light blue region, and the IFF module, represented by the light purple region.

comprising feature weights obtained from the frequency domain. Our IFF module, with convolutional kernels of size 3 and 15, achieves superior results when compared to a Fully Connected (FC) layer, while maintaining significantly lower model complexity. This ensures both efficiency and effectiveness by effectively capturing local cross-channel interaction. The frequency domain feature are ultimately fused with the time-domain feature maps following interaction (Fig. 3).

2.3 Combined with Existing Framework

This paper proposes a novel method called CFIL, incorporated into each Feature Pyramid Network (FPN) upsampling operation. CFIL involves extracting additional oriented object features from the frequency domain and enabling comprehensive interaction. By introducing frequency-domain interaction, we address the limitations of the existing framework and offer a fresh perspective on feature extraction and representation. By integrating these additional oriented object features, CFIL achieves a more comprehensive understanding of objects in the image, significantly improving model accuracy.

3 Experiments and Analysis

3.1 Datasets

DOTA [4,33] is a comprehensive dataset for object detection in aerial images with two released versions: DOTA V1.0 [33] and DOTA V1.5 [4]. DOTA V1.0 comprises 2,806 images and 188,282 instances with annotated oriented bounding boxes and includes 15 object classes. DOTA-v1.5 uses the same images as DOTA V1.0 and also annotates the extremely small instances (below 10 pixels). It has a total of 403,318 instances.

HRSC2016 [14] is a widely-used dataset for the detection of ships with arbitrary orientation, which contains 1,061 images that range from 300 to 1500 pixels and contains 2,976 objects in Google Earth. We employ the mean average precision (mAP) as the evaluation criterion, which is consistent with the PASCAL VOC2007 and VOC2012.

Table 1. Comparison with state-of-the-art methods on DOTA V1.0 dataset. The top three results are highlighted using RGB. The results with red color denote the best results, blue color and green color indicating the second-best and third-best results, respectively. The gray entries employed deeper backbone networks than our approach.

Method	Backbone	PL	BD	BR	GTF	SV	LV	SH	TC	BC	ST	SBF	RA	HA	SP	HC	mAP (%)
ICN [1]	ResNet101	81.36	74.30	47.70	70.32	64.89	67.82	69.98	90.76	79.06	78.20	53.64	62.90	67.02	64.17	50.23	68.16
CAD-Net [40]	ResNet101	87.80	82.40	49.40	73.50	71.10	63.50	76.60	90.90	79.20	73.30	48.40	60.90	62.00	67.00	62.20	69.90
RoI Transformer [4]	ResNet101	88.64	78.52	43.44	75.92	68.81	73.68	83.59	90.74	77.27	81.46	58.39	53.54	62.83	58.93	47.67	69.56
CenterMap-Net [31]	ResNet50	88.88	81.24	53.15	60.65	78.62	66.55	78.10	88.83	77.80	83.61	49.36	66.19	72.10	72.36	58.70	71.74
SCRDet [39]	ResNet101	89.98	80.65	52.09	68.36	68.36	60.32	72.41	90.85	87.94	86.86	65.02	66.68	66.25	68.24	65.21	72.61
FAOD [11]	ResNet101	90.21	79.58	45.49	76.41	73.18	68.27	79.56	90.83	83.40	84.68	53.40	65.42	74.17	69.69	64.86	73.28
FR-Est [5]	ResNet101	89.63	81.17	50.44	70.19	73.52	77.98	86.44	90.82	84.13	83.56	60.64	66.59	70.59	66.72	60.55	74.20
Mask OBB [30]	ResNet50	89.61	85.09	51.85	72.90	75.28	73.23	85.57	90.37	82.08	85.05	55.73	68.39	71.61	69.87	66.33	74.86
Gliding Vertex [37]	ResNet101	89.64	85.00	52.26	77.34	73.01	73.14	86.82	90.74	79.02	86.81	59.55	70.91	72.94	70.86	57.32	75.02
Oriented R-CNN [34]	ResNet101	88.86	83.48	55.27	76.92	74.27	82.10	87.52	90.90	85.56	85.33	65.51	66.82	74.36	70.15	57.28	76.28
Baseline	ResNet50	89.46	82.12	54.78	70.86	78.93	83.00	88.20	90.90	87.50	84.68	63.97	67.69	74.94	68.84	52.28	75.87
Ours	ResNet50	89.78	85.03	62.16	80.85	81.02	84.91	88.63	90.88	87.01	87.75	73.71	70.50	82.67	82.26	71.95	81.32

Table 2. Comparison with state-of-the-art methods on DOTA V1.5 dataset.

Method	Backbone	PL	BD	BR	GTF	SV	LV	SH	TC	BC	ST	SBF	RA	HA	SP	HC	CC	mAP (%)
RetinaNet-OBB [13]	ResNet50	71.43	77.64	42.12	64.65	44.53	56.79	73.31	90.84	76.02	59.96	46.95	69.24	59.65	64.52	48.06	0.83	59.16
FR-OBB [20]	ResNet50	71.89	74.47	44.45	59.87	51.28	69.98	79.37	90.78	77.38	67.50	47.75	69.72	61.22	65.28	60.47	1.54	62.00
MASK RCNN [8]	ResNet50	76.84	73.51	49.90	57.80	51.31	71.34	79.75	90.46	74.21	66.07	46.21	70.61	63.07	64.46	57.81	9.42	62.67
HTC [2]	ResNet50	77.80	73.67	51.40	63.99	51.54	73.31	80.31	90.48	75.12	67.34	48.51	70.63	64.84	64.48	55.87	5.15	63.40
ReDet [7]	ReResNet50	79.20	82.81	51.92	71.41	52.38	75.73	80.92	90.83	75.81	68.64	49.29	72.03	73.36	70.55	63.33	11.53	66.86
Baseline	ResNet50	79.55	81.83	56.56	72.18	52.43	76.95	88.13	90.86	78.77	68.88	62.06	72.93	75.16	66.09	60.90	8.11	68.21
Ours	ResNet50	80.77	81.18	59.49	76.95	67.26	81.88	89.73	90.86	80.81	78.39	70.89	76.29	80.07	76.69	72.53	51.44	75.95

3.2 Implementation Detail

The deep learning model in this study was implemented with consistent details compared to the baseline model. The same hyperparameters and learning rates were used for both models. The training was performed on a single Nvidia RTX 2080Ti with a batch size of 2. The mmdetection platform is used, with ResNet50 as the backbone. Data augmentation techniques such as horizontal and vertical flipping were applied during training. The network was optimized using the SGD algorithm with a momentum of 0.9 and weight decay of 0.0001.

3.3 Comparisons with the State-of-the-Art

As shown in Table 1, CFIL is compared with state-of-the-art (SOTA) methods on the DOTA V1.0 dataset. Firstly, we observe that when using ResNet50 as the backbone network, CFIL achieves the highest mAP performance. In particular, the overall mAP reaches 81.32%. Our ResNet50-based method outperforms the current best ResNet50-based algorithm (baseline), by 2.09% and 7.38% in the SV and BR categories, respectively. This indicates the superiority of the CFIL proposed in our approach. Lastly, we find that when we compare with the use of deeper backbone networks (such as Oriented R-CNN and Gliding Vertex with ResNet101), our proposed module still surpasses it in terms of mAP on ResNet50. This further demonstrates that our algorithm exhibits better generalization capabilities for downstream tasks.

Table 3. Results on HRSC2016 test set. $mAP_{50}(07)$ and $mAP_{50}(12)$ represents the reported results under VOC2007 and VOC2012 mAP metrics.

Method	Backbone	$mAP_{50}(07)$	$mAP_{50}(12)$
R2CNN [10]	ResNet101	73.07	79.73
Rotated RPN [15]	ResNet101	79.08	85.64
RoI Transformer [4]	ResNet101	86.20	–
Gliding Vertex [37]	ResNet101	88.20	–
PIoU [3]	DLA-34	89.20	–
DRN [18]	H-34	–	92.70
CenterMap-Net [31]	ResNet50	–	92.80
DAL [17]	ResNet101	89.77	–
S²ANet [6]	ResNet101	90.17	95.01
R3Det [38]	ResNet101	89.26	96.01
Oriented R-CNN [34]	ResNet101	90.50	96.50
Baseline	ResNet50	90.40	96.50
Ours	ResNet50	90.70	98.47

Table 4. Evaluation of the role of EFF and IFF module on HRSC2016 with ResNet50.

No.	EFF Module	IFF Module	$mAP_{50}(12)$
1	×	×	96.43
2	$n = 1$ (GAP)	×	96.61 (+0.18)
3	✓	×	97.03 (+0.60)
4	$n = 1$ (GAP)	✓	97.22 (+0.79)
5	✓	✓	**98.47** (+2.04)

Comparing CFIL with SOTA approaches on the DOTA V1.5 dataset, we present the results in Table 2. Firstly, we can find that when ResNet50 is used as the backbone, our algorithm gets the highest performance in 15 specific categories and overall mAP, especially the overall mAP reaches 75.95%, and our method is higher than the baseline, for example on SV subclasses by 14.83% on the SV class and 8.83% on the SBF subclass. Particularly, we find that our algorithm has a substantial improvement (39.91%) over the previous algorithm on the container crane (CC) subclass, which demonstrates the superiority of CFIL.

We conducted a comparative analysis of CFIL with state-of-the-art (SOTA) methods using the HRSC2016 dataset. In order to provide a comprehensive evaluation, we present the results using both the VOC2007 and VOC2012 metrics. The experimental results are summarized in Table 3. Firstly, our method based on ResNet50 is higher than the best current algorithm based on ResNet50

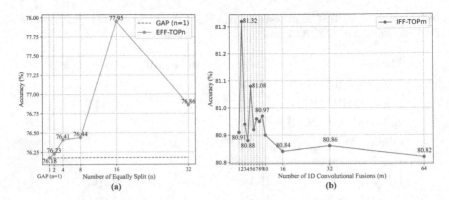

Fig. 4. (a) Accuracy of the different number of equally split, i.e., n in Eq. 4. (b) Accuracy of different numbers of 1D convolutional fusions, i.e., m in Eq. 5.

backbone, Oriented R-CNN. Our method achieves SOTA on VOC2007 and VOC2012 evaluation metrics with 90.07 and 98.47 respectively, which shows the superiority of CFIL, and secondly, we find that when comparing deeper networks, our ResNet50-based method can also outperform them on mAp. This reinforces the fact that our proposed algorithm has better generalization for downstream tasks.

3.4   Ablation Studies

To examine the effectiveness of each component in our proposed method, we perform a series of ablation experiments.

Impact of the Number of Channels. To evaluate the influence of different numbers of channels in EFF module, i.e., n in Eq. 4, we verify the results of using n highest frequency components of the 2D discrete cosine transform (2D DCT) on Dota V1.0 datasets with ResNet50. Due to the total number of feature map channels being 256 in the experiment, n is set to 1, 2, 4, 8, 16, and 32. As discussed in Sect. 3.2, when $n = 1$, it is equivalent to Global Average Pooling (GAP). As shown in Fig. 4 (a), we can observe two phenomena. Firstly, compared to the experiment with $n = 1$ (GAP), experiments that introduce more frequency domain features ($n > 1$) show significant performance gains. This validates our idea of introducing more frequency domain features. Secondly, dividing the feature maps into 16 parts (n=16) and selecting the 16 highest frequency components of the 2D DCT yields optimal performance. By employing this configuration, we utilize the same setting in our approach as well as in all other experiments.

Influence of the Number of 1D Convolutional. As shown in Fig. 4 (b), to assess the impact of varying numbers of 1D convolutional fusions in the IFF module, we examine the outcomes obtained by employing the m highest 1D convolutional fusions (arranged according to different kernel sizes) on the Dota V1.0 datasets with ResNet50. The input incorporates the results of an EFF module with $n = 16$. The best performance is achieved by the two highest 1D convolutional fusions ($m = 2$). In our approach and all other experiments, we utilize this configuration to ensure consistent results. Our algorithm can combine different backbone networks and FPNs in a good and generic way.

Role of Different Settings in EFF and IFF Module. To investigate the effectiveness of the EFF and IFF modules, We compare them respectively against the baseline method. Table 4 shows the experimental results. We set up five groups for the ablative experiments (No. 1–No. 5) on the Dota V1.0 datasets with ResNet50. No. 1 is the baseline. No. 2 serves as the control group for No. 3 and No. 4 experiments. Since the input of the IFF module needs to be a one-dimensional vector, No. 2 output directly utilizes the sigmoid function to learn its weights without any interaction. No. 3 only employs the EFF module with $n = 16$, and its output is directly learned using the sigmoid function for its weights without any interaction. No. 4 utilizes the IFF module with $m = 2$.

Lastly, No. 5 combines the EFF module with $n = 16$ and the IFF module with $m = 2$ simultaneously. Comparison No. 1 and No. 2, the improvement in No. 3 demonstrates the effectiveness of the EFF module, proving the validity of introducing richer features from the frequency domain and integrating temporal and frequency domain features. The improvement in No. 4 demonstrates the effectiveness of the IFF module. No. 5, in comparison to No. 1, shows a significant improvement, providing evidence for the effectiveness of combining the EFF and IFF modules to fully utilize frequency domain features and promote effective interaction.

Universality for Different Backbones. To verify the applicability of our method, a comparison between using CFIL and not using CFIL in different backbone networks, as shown in Fig. 5(a). It can be clearly observed that using CFIL has a great improvement over not using CFIL. Using CFIL, the network with ResNet101 and SENet as the Backbone network increased by 5.39% and 5.38% respectively. This shows that the proposed CFIL can enhance the performance of capturing aerial oriented object features in different backbone networks.

Universality for Different FPNs. We conducted a comparative analysis between employing CFIL and not across various FPNs, as shown in Fig. 5(b). It can be observed that using our method has a great improvement over not using CFIL. Using CFIL, The accuracy of all networks has increased by no less than 5%. This shows that the proposed CFIL can adapt to various FPN variants and enhance the performance of aerial oriented object detection.

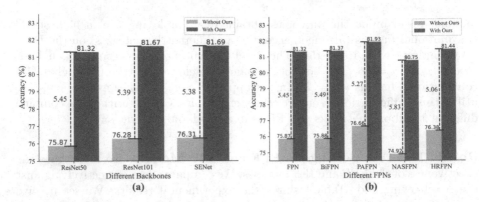

Fig. 5. (a) Accuracy of the Dota V1.0 test set under different backbones. (b) Accuracy of the Dota V1.0 test set under different FPNs. All results are performed with ResNet50-FPN backbone.

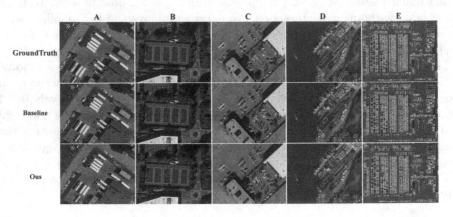

Fig. 6. Visualization results. Compared our method with baseline. The red bounding boxes indicate the presence of error detection and omission detection for ground truth.

3.5   Visualization Results

As shown in Fig. 6, we visualized the results, comparing the results of the baseline and our algorithm with ground truth. In comparison with ground truth all missed detections, wrong detections, etc. are marked with red bounding boxes. It can be observed that our algorithm detects more target objects than baseline, especially in the scenes of light transformation, occlusion, etc. Our algorithm can capture more information about the target objects and detect them correctly.

4   Conclusion

In this paper, we propose CFIL, which includes the EFF module and IFF module, to enhance the feature learning capability of the network through frequency domain feature extraction and comprehensive interaction. Our CFIL is a plug-and-play method that can be inserted into any backbone network with convolutional layers. We conduct extensive experiments on three challenging oriented object detection benchmarks. The experimental results demonstrate that equipped with CFIL, the performance of variously oriented object detectors is significantly improved on commonly used rotation object detection benchmarks. In the future, we will apply our CFIL to more CNN architectures and further explore its potential in the frequency domain.

References

1. Azimi, S.M., Vig, E., Bahmanyar, R., Körner, M., Reinartz, P.: Towards multi-class object detection in unconstrained remote sensing imagery. In: Jawahar, C.V., Li, H., Mori, G., Schindler, K. (eds.) ACCV 2018. LNCS, vol. 11363, pp. 150–165. Springer, Cham (2019). https://doi.org/10.1007/978-3-030-20893-6_10
2. Chen, K., et al.: Hybrid task cascade for instance segmentation. Cornell University (2019)
3. Chen, Z., et al.: PIoU loss: towards accurate oriented object detection in complex environments. In: Vedaldi, A., Bischof, H., Brox, T., Frahm, J.-M. (eds.) ECCV 2020. LNCS, vol. 12350, pp. 195–211. Springer, Cham (2020). https://doi.org/10.1007/978-3-030-58558-7_12
4. Ding, J., Xue, N., Long, Y., Xia, G.S., Lu, Q.: Learning roi transformer for oriented object detection in aerial images. In: 2019 IEEE/CVF Conference on Computer Vision and Pattern Recognition (CVPR) (2020)
5. Fu, K., Chang, Z., Zhang, Y., Sun, X.: Point-based estimator for arbitrary-oriented object detection in aerial images. IEEE Trans. Geosci. Remote Sens. **59**, 4370–4387 (2021)
6. Han, J., Ding, J., Li, J., Xia, G.S.: Align deep features for oriented object detection. IEEE Trans. Geosci. Remote Sens. **60**, 1–11 (2021)
7. Han, J., Ding, J., Xue, N., Xia, G.S.: Redet: a rotation-equivariant detector for aerial object detection. In: Computer Vision and Pattern Recognition (2021)
8. He, K., Gkioxari, G., Dollár, P., Girshick, R.: Mask r-cnn. In: Proceedings of the IEEE International Conference on Computer Vision, pp. 2961–2969 (2017)
9. Hu, J., Shen, L., Albanie, S., Sun, G., Wu, E.: Squeeze-and-excitation networks. IEEE Trans. Pattern Anal. Mach. Intell. (2018)
10. Jiang, Y.Y., et al.: R2cnn: rotational region CNN for orientation robust scene text detection (2017)
11. Li, C., Xu, C., Cui, Z., Wang, D., Zhang, T., Yang, J.: Feature-attentioned object detection in remote sensing imagery. In: International Conference on Image Processing (2019)
12. Li, W., Zhu, J.: Oriented reppoints for aerial object detection. Cornell University (2021)
13. Lin, T.Y., Goyal, P., Girshick, R., He, K., Dollár, P.: Focal loss for dense object detection. Cornell University (2017)
14. Liu, Z., Wang, H., Weng, L., Yang, Y.: Ship rotated bounding box space for ship extraction from high-resolution optical satellite images with complex backgrounds. IEEE Geosci. Remote Sens. Lett. **13**, 1074–1078 (2016)
15. Ma, J., et al.: Arbitrary-oriented scene text detection via rotation proposals. IEEE Trans. Multimedia **20**, 3111–3122 (2018)

16. Ming, Q., Miao, L., Zhou, Z., Song, J., Yang, X.: Sparse label assignment for oriented object detection in aerial images. Remote Sens. **13**(14), 2664 (2021)
17. Ming, Q., Zhou, Z., Miao, L., Zhang, H., Li, L.: Dynamic anchor learning for arbitrary-oriented object detection. In: Proceedings of the AAAI Conference on Artificial Intelligence, pp. 2355–2363 (2022)
18. Pan, X., et al.: Dynamic refinement network for oriented and densely packed object detection. In: Proceedings of the IEEE/CVF Conference on Computer Vision and Pattern Recognition, pp. 11207–11216 (2020)
19. Qiao, C., et al.: A novel multi-frequency coordinated module for sar ship detection. In: 2022 IEEE 34th International Conference on Tools with Artificial Intelligence (ICTAI), pp. 804–811. IEEE (2022)
20. Ren, S., He, K., Girshick, R., Sun, J.: Faster r-cnn: towards real-time object detection with region proposal networks. Cornell University (2015)
21. Shen, F., Du, X., Zhang, L., Tang, J.: Triplet contrastive learning for unsupervised vehicle re-identification. arXiv preprint arXiv:2301.09498 (2023)
22. Shen, F., et al.: A large benchmark for fabric image retrieval. In: 2019 IEEE 4th International Conference on Image, Vision and Computing (ICIVC), pp. 247–251. IEEE (2019)
23. Shen, F., Peng, X., Wang, L., Zhang, X., Shu, M., Wang, Y.: HSGM: a hierarchical similarity graph module for object re-identification. In: 2022 IEEE International Conference on Multimedia and Expo (ICME), pp. 1–6. IEEE (2022)
24. Shen, F., Shu, X., Du, X., Tang, J.: Pedestrian-specific bipartite-aware similarity learning for text-based person retrieval. In: Proceedings of the 31th ACM International Conference on Multimedia (2023)
25. Shen, F., Wei, M., Ren, J.: HSGNET: object re-identification with hierarchical similarity graph network. arXiv preprint arXiv:2211.05486 (2022)
26. Shen, F., Xie, Y., Zhu, J., Zhu, X., Zeng, H.: Git: graph interactive transformer for vehicle re-identification. IEEE Trans. Image Process. **32**, 1039–1051 (2023)
27. Shen, F., et al.: An efficient multiresolution network for vehicle reidentification. IEEE Internet Things J. **9**(11), 9049–9059 (2021)
28. Shen, F., Zhu, J., Zhu, X., Xie, Y., Huang, J.: Exploring spatial significance via hybrid pyramidal graph network for vehicle re-identification. IEEE Trans. Intell. Transp. Syst. **23**(7), 8793–8804 (2021)
29. Tian, Z., Shen, C., Chen, H., He, T.: FCOS: fully convolutional one-stage object detection. In: 2019 IEEE/CVF International Conference on Computer Vision (ICCV) (2020)
30. Wang, J., Ding, J., Guo, H., Cheng, W., Pan, T., Yang, W.: Mask OBB: a semantic attention-based mask oriented bounding box representation for multi-category object detection in aerial images. Remote Sens. **11**, 2930 (2019)
31. Wang, J., Yang, W., Li, H.C., Zhang, H., Xia, G.S.: Learning center probability map for detecting objects in aerial images. IEEE Trans. Geosci. Remote Sens. **59**, 4307–4323 (2021)
32. Wu, H., Shen, F., Zhu, J., Zeng, H., Zhu, X., Lei, Z.: A sample-proxy dual triplet loss function for object re-identification. IET Image Proc. **16**(14), 3781–3789 (2022)
33. Xia, G.S., et al.: DOTA: a large-scale dataset for object detection in aerial images. In: Proceedings of the IEEE Conference on Computer Vision and Pattern Recognition, pp. 3974–3983 (2018)
34. Xie, X., Cheng, G., Wang, J., Yao, X., Han, J.: Oriented R-CNN for object detection. In: Proceedings of the IEEE/CVF International Conference on Computer Vision, pp. 3520–3529 (2021)

35. Xie, Y., Shen, F., Zhu, J., Zeng, H.: Viewpoint robust knowledge distillation for accelerating vehicle re-identification. EURASIP J. Adv. Signal Process. **2021**, 1–13 (2021)
36. Xu, R., Shen, F., Wu, H., Zhu, J., Zeng, H.: Dual modal meta metric learning for attribute-image person re-identification. In: 2021 IEEE International Conference on Networking, Sensing and Control (ICNSC), vol. 1, pp. 1–6. IEEE (2021)
37. Xu, Y., et al.: Gliding vertex on the horizontal bounding box for multi-oriented object detection. IEEE Trans. Pattern Anal. Mach. Intell. **43**, 1452–1459 (2020)
38. Yang, X., Yan, J., Feng, Z., He, T.: R3det: Refined single-stage detector with feature refinement for rotating object. In: Proceedings of the AAAI Conference on Artificial Intelligence, vol. 35, no. 4, pp. 3163–3171 (2022)
39. Yang, X., et al.: SCRDET: towards more robust detection for small, cluttered and rotated objects. In: International Conference on Computer Vision (2019)
40. Zhang, G., Lu, S., Zhang, W.: CAD-Net: a context-aware detection network for objects in remote sensing imagery. IEEE Trans. Geosci. Remote Sens. **57**, 10015–10024 (2019)

DBDAN: Dual-Branch Dynamic Attention Network for Semantic Segmentation of Remote Sensing Images

Rui Che[1], Xiaowen Ma[1], Tingfeng Hong[1], Xinyu Wang[1], Tian Feng[1], and Wei Zhang[1,2](✉)

[1] School of Software Technology, Zhejiang University, Hangzhou 310027, China
cstzhangwei@zju.edu.cn
[2] Innovation Center of Yangtze River Delta, Zhejiang University, Jiaxing 314103, China

Abstract. Attention mechanism is capable to capture long-range dependence. However, its independent calculation of correlations can hardly consider the complex background of remote sensing images, which causes noisy and ambiguous attention weights. To address this issue, we design a correlation attention module (CAM) to enhance appropriate correlations and suppress erroneous ones by seeking consensus among all correlation vectors, which facilitates feature aggregation. Simultaneously, we introduce the CAM into a local dynamic attention (LDA) branch and a global dynamic attention (GDA) branch to obtain the information on local texture details and global context, respectively. In addition, considering the different demands of complex and diverse geographical objects for both local texture details and global context, we devise a dynamic weighting mechanism to adaptively adjust the contributions of both branches, thereby constructing a more discriminative feature representation. Experimental results on three datasets suggest that the proposed dual-branch dynamic attention network (DBDAN), which integrates the CAM and both branches, can considerably improve the performance for semantic segmentation of remote sensing images and outperform representative state-of-the-art methods.

Keywords: Remote Sensing · Semantic Segmentation · Attention Mechanism · Deep Learning

1 Introduction

Recent advances in sensors enable increasingly available high-resolution remote sensing images (RSIs) to be captured worldwide with rich spatial details and semantic contents. In particular, semantic segmentation of RSIs can serve various applications, such as land cover classification [14], road extraction [13], urban planning [28] and so on. The drastic development of deep learning techniques has driven a series of methods based on convolutional neural networks (CNNs) for semantic segmentation given their capability to capture fine-grained local

© The Author(s), under exclusive license to Springer Nature Singapore Pte Ltd. 2024
Q. Liu et al. (Eds.): PRCV 2023, LNCS 14428, pp. 306–317, 2024.
https://doi.org/10.1007/978-981-99-8462-6_25

context information, which benefits feature representation and pattern recognition. The convolution operation with a limited receptive field is designed to extract local patterns, but lacks the power to model global context information or long-range dependence due to its nature.

Introducing the attention mechanism is an effective means to reduce the confusion in predicted classes without losing spatial information. With the global statistics aggregated from the entire image, scene information can be embedded to highlight the features with appropriate correlations or suppress those with erroneous ones. However, the correlation of each query-key pair is calculated independently, ignoring the correlations of other pairs. This may cause erroneous correlations due to the imperfection of feature representation as well as the interference of complex background, and lead to additional background noise and ambiguous attention weights. Therefore, our opinion is towards that the typical attention-based methods [4, 19, 22] can hardly be applied to the semantic segmentation of RSIs directly.

Generally speaking, if a key is highly correlated with a query, its adjacent keys are supposed to have high correlations with the query. Otherwise, correlations might be with noise. Following this observation, we design a correlation attention module (CAM), which extends the general attention mechanism, to refine the correlations via seeking the consensus among all correlation vectors. Meanwhile, achieving the trade-off between the effectiveness of feature embedding and of spatial localization is critical for the semantic segmentation of RSIs. Hence, we embed the CAM into a local dynamic attention (LDA) branch and a global dynamic attention (GDA) branch to extract the information on the local texture details and the global context. In addition, considering the different demands of complex and diverse geographical objects for both local texture details and global context, we devise a dynamic weighting mechanism to fuse the outputs of both branches so as to capture the feature representations on land cover more discriminatively. Experimental results on three datasets suggest the effectiveness of the proposed dual-branch dynamic attention network (DBDAN) for semantic segmentation of RSIs. The contributions of this study can be summarized as follows,

- a correlation attention module that effectively strengthens appropriate relationships while suppressing erroneous ones to promote feature aggregation for visual recognition;
- a dynamic weighting mechanism fusing the information on local texture details and global context to receive more discriminative features in RSIs; and
- a dual-branch dynamic attention network for semantic segmentation of RSIs, whose superiority to other representative methods can be proved by experimental results on three datasets.

2   Related Work

2.1   Semantic Segmentation of RSIs

RSIs are normally captured by sensors from a distance, typically from aircraft or satellites, which often cover diverse and complex ground objects. Semantic segmentation, which assigns a semantic label to each pixel in an image based on the class of the ground object or the region it represents, plays a crucial role in image processing. Deep neural networks have facilitated considerable advancements in semantic segmentation accuracy because of their capacity for automatic extraction of more informative image features and integration of richer context information.

FCN [10] is a pioneering CNN architecture that can effectively perform end-to-end semantic segmentation. After its introduction, many methods have been innovatively developed based on FCN. For example, PSPNet [29] uses a pyramid pooling module to perform multi-scale processing on input images and DeepLab [1] uses dilated convolution to increase the receptive field and retain more detailed information. STLNet [32] utilizes global information in the lower-level feature mapping to effectively extract multi-scale statistical texture features, thereby enhancing texture details.

An adjacent pixel affinity loss [3] is introduced to improve the capability to discriminate semantically heterogeneous regions. HBCNet [24] is porposed to improve the accuracy of semantic segmentation by utilizing boundary information, semantic information classes, and region feature representations. CTMFNet [18], a multi-scale fusion network, integrates CNN and Transformer mechanisms into its backbone encoder-decoder framework. DMF-CLNet [17] extracts intensive multi-scale features based on contrast learning.

2.2   Attention-Based Methods

Attention mechanism refers to the strategy of allocating biased computational resources to highlight the informative parts of processed signals. Non-Local [22] is the first to propose applying self-attention mechanism to the image processing, allowing the individual feature at any position to perceive features at all other positions. Based on this, CBAM [23] and DANet [4] combine spatial attention and channel attention for model features in serial and parallel, respectively. In addition, FLANet [19] and SAPNet [30] reconstruct the self-attention mechanism as a fully attentive approach which can be obtained in a single non-local block.

Semantic segmentation of RSIs benefits a lot from attention-base methods. For example, Li et al. [7] propose the use of deep CAM and shallow spatial attention modules to segment large-scale satellite images. AFNet [25] has designed a multi-path encoder structure, a multi-path attention fusion module, and a refined attention fusion module. The refined attention fusion module is used for semantic segmentation of ultra-high-resolution RSIs. HMANet [15] combines representations from spatial domain, channel domain and class domain to enhance feature

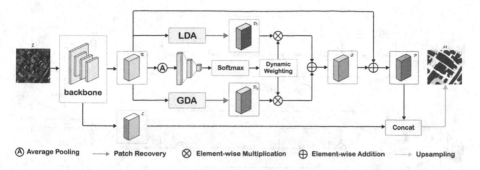

Fig. 1. Architecture of the proposed DBDAN.

extraction. Similarly, Li et al. [8] propose HCANet, which extracts and aggregates background information across levels from pixels, superpixels, and global space, as well as global space. In addition, Liu et al. design an adaptive fusion network [25] to enhance feature pixels with low-level or high-level feature maps using attention modules. LANet [2] introduces a patch-wise attention module to incorporate regional information, and a designed attention embedding module to fuse cross-level attention.

Inspired by the success of attention-based methods above and considering their limitations, we rethink the attention mechanism from a different perspective. There is an inherent limitation of the attention mechanism that the reasoning of spatial correlations is computed independently, which may introduce noises and ambiguity problems due to the complex background of RSIs, resulting in unsatisfactory segmentation performance. Consequently, we propose a dual-branch dynamic attention network for semantic segmentation of RSIs, seeking consensus among all correlation vectors to enhance vision focus and associating local details with global information to obtain enhanced feature representations.

3   Method

3.1   Overview

As shown in Fig. 1, the proposed method takes as input an image $\mathcal{I}$, whose feature representation $\mathcal{R}$ is extracted from the backbone. Two branches, namely, a *local dynamic attention* (LDA) branch and a *global dynamic attention* (GDA) branch, each of which includes a *correlation attention module* (CAM), process $\mathcal{R}$ to reach the local feature representation $\mathcal{D}_l$ and the global feature representation $\mathcal{D}_g$, respectively. The weights of both branches α_l and α_g are generated by a dynamic weighting mechanism from $\mathcal{R}$ to fuse $\mathcal{D}_l$ and $\mathcal{D}_g$ into the refined feature representation $\mathcal{S}$. The element-wise summation of $\mathcal{R}$ and S is then employed to reach the enhanced feature representation $\mathcal{P}$. Finally, the output segmentation map $\mathcal{M}$ is obtained by the concatenation of $\mathcal{P}$ and the low-level feature representation $\mathcal{L}$ from the backbone followed by upsampling.

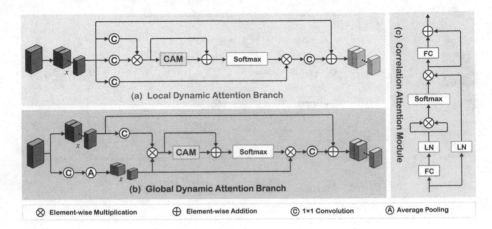

Fig. 2. Structures of key components in our DBDAN: (a) a local dynamic attention (LDA) branch to extract the information on the local texture details, (b) a global dynamic attention (GDA) branch to extract the information on the global context, and (c) a correlation attention module (CAM) embedded in both branches to denoise the feature representations.

3.2   Correlation Attention Module

We briefly review the general attention mechanism in computer vision as follows. An attention function takes as input a query and a set of key-value pairs and outputs a weighted sum of the values. Specifically, the weights of the values are obtained by calling the Softmax function on the scaled dot products of the query and the corresponding keys. Suppose an feature representation $X \in \mathbb{R}^{H \times W \times \hat{C}}$, where H, W and $\hat{C}$ denote X's height, width and number of channels, respectively, 1×1 convolutions W_q, W_k, and $W_v \in \mathbb{R}^{\hat{C} \times C}$ are adopted to transform X into three embeddings $Q \in \mathbb{R}^{H \times W \times C}$, $K \in \mathbb{R}^{H \times W \times C}$, and $V \in \mathbb{R}^{H \times W \times C}$ as,

$$Q = W_q(X), K = W_k(X), V = W_v(X),   \quad (1)$$

where C denotes the number of channels. These embeddings are then flattened to size $N \times C$, where $N = H \times W$ represents the total number of the spatial locations. Afterwards, the attention map $A \in \mathbb{R}^{N \times N}$ involves correlation vectors of all queries, which is calculated via matrix multiplication as,

$$A = Q \times K^T.   \quad (2)$$

For each location in V, the output of the attention function is formulated as,

$$O = (f(A) \times V)W_o + X,   \quad (3)$$

where $f(\cdot)$ represents a function that applies the normalization to A, and $W_o \in \mathbb{R}^{C \times \hat{C}}$ denotes the linear transformation weights to adjust the importance of the attention function and recover the channel number from C to $\hat{C}$.

In the general attention mechanism, however, the correlation of each query-key pair, included by the attention map A, is independently calculated while ignoring its relationship with other query-key pairs. This strategy may introduce erroneous correlations due to the imperfection of feature representation as well as the interference of complex background, and thus cause additional background noise and ambiguous attention weights. These issues negatively impact the feature aggregation in the attention function, which results in a sub-optimal performance for semantic segmentation of RSIs.

To address the above-mentioned issue, we observe that a high query-key correlation should enable the similar relationship between the query and the key's adjacent pixels. Therefore, we design a correlation attention module (CAM) to improve correlation map generation via exploiting the clues in the general attention mechanism. Specifically, our CAM seeks the correlation consistency around each key to enhance the appropriate correlations extracted from the relevant query-key pairs and suppress the erroneous ones from the irrelevant pairs. As a variant of the general attention mechanism, our CAM considers the columns in attention map A as a sequence of correlation vectors and takes them as queries Q', keys K' and values V' to generate a residual attention map. As shown in Fig. 2 (c), our CAM first applies a linear transformation to reduce the dimensions of Q' and K' to $N \times D$ where $D \ll N$ for computational efficiency and then $\overline{Q'}$ and $\overline{K'}$ are generated after normalization. Besides, V' is processed by layer normalization to reach the normalized correlation vector $\overline{V'}$. With $\overline{Q'}$, $\overline{K'}$ and $\overline{V'}$, the residual attention map A' is generated as,

$$A' = (f(\frac{\overline{Q'K'}^T}{\sqrt{C}})\overline{V'})(1 + W'_a), \tag{4}$$

where $W'_a \in \mathbb{R}^{N \times N}$ denotes the linear transformation weights to adjust the aggregated correlations with an identical connection. Consequently, the attention function with our CAM can be formulated as,

$$Attn(Q, K, V) = (f(A + A') \times V)W_o + X, \tag{5}$$

3.3 Local Dynamic Attention Branch

In most cases, a RSI covers a large area and thus includes massive local information (e.g., building rooftops, trees, and utility poles). However, these details are usually represented by a limited number of pixels, and likely to be obscured, blurred, or covered by noise, which causes difficulty in accurate segmentation. To solve this problem, we design the local dynamic attention (LDA) branch to enhance the extraction of local information in a RSI.

As shown in the Fig. 2 (a), the attention function is limited to local patches for feature aggregation since global attention is not suitable for processing high-resolution RSIs. Meanwhile, the CAM is introduced for denoising so that the enhanced feature representation contains accurate and detailed local information. Specifically, the original feature representation $R \in \mathbb{R}^{H \times W \times \hat{C}}$ is first split into patches $X \in \mathbb{R}^{(N_h \times N_w) \times (h \times w) \times \hat{C}}$. Each patch $x_i \in \mathbb{R}^{h \times w \times \hat{C}}$ is processed into $q_i, k_i, v_i \in \mathbb{R}^{h \times w \times C}$ via linear transformation and the enhanced feature $\mathcal{D}_l \in \mathbb{R}^{H \times W \times \hat{C}}$ can be then calculated as follows,

$$D_l = Recover(Attn(q_i, k_i, v_i)), \tag{6}$$

where $Recover(\cdot)$ denotes recovering patches to its original spatial dimension, h and w refer to the height and width of a patch, and $N_h = \frac{H}{h}$ and $N_w = \frac{W}{w}$ represent the number of patches along the dimension of height and of width.

3.4   Global Dynamic Attention Branch

There is a certain spatial dependence between the geographical areas of RSIs. For example, buildings are distributed on both sides of the road and lakes are surrounded by farmland. The LDA branch models the pixel relationships within regions to enhance local features, but ignores the global context. Therefore, we design a global dynamic attention (GDA) branch to model remote dependence between regions, thus enhancing the perception of pixels to the global context.

As shown in Fig. 2 (b), the key of the GDA branch is to apply the CAM in the global dimension to further enrich the feature representation. Specifically, the pooling operation and the linear embedding are applied to the original feature representation $\mathcal{R}$ to generate global descriptors Z. Each global descriptor $z_i \in \mathbb{R}^{N_h \times N_w \times C}$ is processed into $\hat{k}_i, \hat{v}_i$, while $\hat{q}_i \in \mathbb{R}^{h \times w \times C}$ is generated as in the LDA branch. It is noteworthy that $\hat{k}_i, \hat{v}_i$ need extending to the same dimension of $\hat{q}_i$ in advance. The final output $\mathcal{D}_g \in \mathbb{R}^{H \times W \times \hat{C}}$ is calculated as,

$$D_g = Recover(Attn(\hat{q}_i, \hat{k}_i, \hat{v}_i)). \tag{7}$$

3.5   Dual-Branch Feature Fusion

Considering the various demands of complex and diverse geographical objects for both local texture details and global context in RSIs, we propose a dynamic weighting allocation mechanism to adaptively integrate the above-mentioned local and global information and thus effectively capture a more discriminative feature for segmentation. Specifically, the weighted summation is adopted to control the contributions of LDA and GDA branches as,

$$S = f_{1 \times 1}(\alpha_l \times D_l + \alpha_g \times D_g), \tag{8}$$

where $f_{1 \times 1}(\cdot)$ denotes the 1×1 convolution, α_l and α_g respectively refer to the weights of LDA and GDA branches. In particular, both weights are calculated according to the original feature representation R as,

$$\alpha_l, \alpha_g = \sigma(H_d(\delta(H_r(AvgPool(R))))), \tag{9}$$

where σ and δ denote Softmax and ReLU functions, respectively, $H_r \in \mathbb{R}^{\hat{C} \times \frac{\hat{C}}{16}}$ and $H_d \in \mathbb{R}^{\frac{\hat{C}}{16} \times 2}$ represent the 1×1 convolutions for dimension reduction. It is noteworthy that the sum-to-one constraint (i.e., $\alpha_l + \alpha_g = 1$) is employed to compress the kernel space and simplify the learning of both weights.

4   Results

We implement our DBDAN using Python and PyTorch on a workstation with two NVIDIA GTX A6000 graphics cards (96 GB GPU memory in total). For training, the AdamW optimizer with weight decay of 1e-4 is adopted, and the initial learning rate is set to 1e-4 with the poly decay strategy. The training epochs for the Vaihingen, Potsdam, and LoveDA datasets are 250, 100, and 50, respectively, with a batch size of 4 for all. Following previous work [2,6,11,12], we use data augmentation methods such as random scaling (0.5, 0.75, 1.0, 1.25, 1.5), random vertical flipping, random horizontal flipping, and random rotation during the training process. The augmented images are then randomly cropped into patches of 512×512 pixels. During the inference process, data augmentation techniques such as random flipping and multi-scale prediction are used.

4.1   Datasets and Evaluation Metrics

We evaluate the proposed method on the LoveDA dataset, the ISPRS Potsdam dataset, and the ISPRS Vaihingen dataset in three common metrics, namely, average F1 score per class (AF), mean intersection over union (mIoU) and overall accuracy (OA).

The LoveDA dataset [21] contains 5987 fine-resolution optical RSIs (GSD 0.34 m) of size 1024×1024 pixels and includes 7 land-cover classes (i.e., building, road, water, barren, forest, agriculture, and background). We use 2522 images for training, 1669 images for validation, and the remaining 1796 images for testing.

The ISPRS Potsdam dataset [16] contains 38 true orthophoto (TOP) tiles and the corresponding digital surface model (DSMs) collected from a historic city with large building blocks. All images (GSD 0.05 m) are of size 6000×6000 pixels. The ground truth labels comprise six land-cover classes (i.e., impervious surfaces, building, low vegetation, tree, car, and clutter/background). We adopt 24 images for training and the remaining 14 for testing.

The ISPRS Vaihingen dataset [16] contains 33 TOP tiles and the corresponding DSMs collected from a small village. The spatial size varies from 1996×1995 to 3816×2550 pixels, with a GSD of 0.09 m. The ground truth labels comprise the same six ones as the ISPRS Potsdam dataset. We adopt 16 images for training and the remaining 17 for testing.

4.2   Comparison with the State-of-the-Art Methods

We compare our DBDAN with other state-of-the-art methods on three datasets. As shown in Table 1, the proposed method achieves the best performance on all datasets. Specifically, our DBDAN achieves an increase by 1.8% in mIoU on the LoveDA dataset compared to PoolFormer [26], and on the ISPRS Vaihingen dataset compared to ConvNeXt [9]. On the ISPRS Potsdam dataset, the proposed method reaches a less significant improvement of 0.4% in mIoU compared to FLANet [19]. In addition, our DBDAN provides a considerable performance for classes with large variance (e.g., buildings), obtaining an increase by 3% in

Table 1. Comparison between our DBDAN and several state-of-the-art methods on the LoveDA, ISPRS Vaihingen and ISPRS Potsdam datasets. Best results are in bold. Second best results are underscored. All results are in percentage.

Method	LoveDA								ISPRS Vaihingen			ISPRS Potsdam		
	Back	Buil	Road	Water	Barren	Forest	Agri	mIoU	AF	mIoU	OA	AF	mIoU	OA
PSPNet [29]	44.4	52.1	53.5	76.5	9.7	44.1	57.9	48.3	86.47	76.78	89.36	89.98	81.99	90.14
DeepLabv3+ [1]	43.0	50.9	52.0	74.4	10.4	44.2	58.5	47.6	86.77	77.13	89.12	90.86	84.24	89.18
DANet [4]	44.8	55.5	53.0	75.5	17.6	45.1	60.1	50.2	86.88	77.32	89.47	89.60	81.40	89.73
FarSeg [31]	43.1	51.5	53.9	76.6	9.8	43.3	58.9	48.2	87.88	79.14	89.57	91.21	84.36	89.87
OCRNet [27]	44.2	55.1	53.5	74.3	18.5	43.0	60.5	49.9	89.22	81.71	90.47	92.25	86.14	90.03
LANet [2]	40.0	50.6	51.1	78.0	13.0	43.2	56.9	47.6	88.09	79.28	89.83	91.95	85.15	90.84
ISNet [5]	44.4	<u>57.4</u>	<u>58.0</u>	77.5	**21.8**	43.9	60.6	51.9	90.19	82.36	90.52	92.67	86.58	91.27
Segmenter [20]	38.0	50.7	48.7	77.4	13.3	43.5	58.2	47.1	88.23	79.44	89.93	92.27	86.48	91.04
MANet [6]	38.7	51.7	42.6	72.0	15.3	42.1	57.7	45.7	90.41	82.71	90.96	92.90	86.95	91.32
FLANet [19]	44.6	51.8	53.0	74.1	15.8	45.8	57.6	49.0	87.44	78.08	89.60	<u>93.12</u>	<u>87.50</u>	<u>91.87</u>
ConvNeXt [9]	<u>46.0</u>	53.5	56.8	76.1	15.9	**47.5**	61.8	51.2	<u>90.50</u>	<u>82.87</u>	<u>91.36</u>	93.03	87.17	91.66
PoolFormer [26]	45.8	57.1	53.3	**80.7**	<u>19.8</u>	45.6	<u>64.5</u>	<u>52.4</u>	89.59	81.35	90.30	92.62	86.45	91.12
Biformer [33]	43.6	55.3	55.9	79.5	16.9	45.4	61.5	51.2	89.65	81.50	90.63	91.47	84.51	90.17
DBDAN (ours)	**48.4**	**60.4**	**59.7**	<u>80.2</u>	18.6	<u>46.5</u>	**65.3**	**54.2**	**91.57**	**84.64**	**91.74**	**93.28**	**87.86**	**91.94**

Table 2. Ablation study on the feature fusion strategy. Best results are in bold. All results are in percentage.

Strategy	LoveDA	Vaihingen		
	mIoU	AF	mIoU	OA
Addition	53.5	91.19	84.02	91.46
Concatenation	53.3	91.35	84.29	91.55
Dynamic Weighting	**54.2**	**91.57**	**84.64**	**91.74**

mIoU. These findings suggest that the proposed method is superior to the methods for comparison and enables stable performance for semantic segmentation of RSIs.

Figure 3 shows the visualization of several example outputs from our DBDAN, as well as DANet [4], OCRNet [27], FLANet [19], and BiFormer [33]. It is observable that the proposed method not only better preserves the integrity and regularity of ground objects (e.g., vehicles and low vegetation) but also significantly reduces the number of noise pixels. The qualitative and quantitative results above validate the effectiveness and efficiency of our DBDAN in improving the performance for semantic segmentation of RSIs.

4.3   Ablation Study

We conduct ablation experiments regarding the strategy of fusing the features from LDA branch and GDA branch and the patch size in both branches on the LoveDA dataset and the Vaihingen dataset, as both variables significantly influence the effectiveness of the corresponding network.

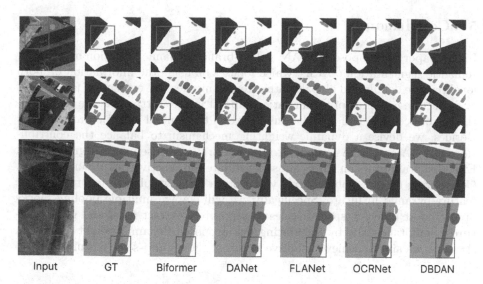

| Input | GT | Biformer | DANet | FLANet | OCRNet | DBDAN |

Fig. 3. Example outputs from our DBDAN and other methods for comparison on the ISPRS Vaihingen dataset (line 1 and 2) and the ISPRS Potsdam dataset (line 3 and 4). Best viewed in color and zoom in.

Table 3. Ablation study on the patch size in both branches. Best results are in bold. All results are in percentage.

Patch Size	LoveDA	ISPRS Vaihingen		
	mIoU	AF	mIoU	OA
8×8	52.9	90.38	82.69	90.75
16×16	53.8	91.37	84.33	91.57
32×32	**54.2**	**91.57**	**84.64**	**91.74**
64×64	53.3	90.82	83.52	91.12

Dual-Branch Feature Fusion Strategy. To demonstrate the effectiveness of our feature fusion strategy, we fix the patch size in LDA and GDA branches to 32×32 and compare the dynamic weighting mechanism with other representative feature fusion strategies, namely, addition and concatenation. As shown in Table 2, our dynamic weighting mechanism achieves a considerable improvement over others.

Patch Size in LDA and GDA Branches. Using the dynamic weighting mechanism, we explore the impact of the patch size in LDA and GDA branches on performance for semantic segmentation, following the previous work [2]. As shown in Table 3, the performance reaches the optimal when the patch size is set to 32×32. Therefore, we adopt the patch size of 32×32 in the final model.

5 Conclusion

In this paper, we propose DBDAN, a dual-branch dynamic attention network for semantic segmentation of RSIs, which focuses on using the information on both local texture details and global context to enhance feature extraction for efficient and effective modeling of pixel relationships. It is noteworthy that we exploratorily introduce pixel-pair correlations based on the general attention mechanism. Besides, we devise a dynamic weighting mechanism to balance the contributions of LDA and GDA branches to the performance of semantic segmentation. Experimental results on three datasets validate the effectiveness of our DBDAN. Considering the large object-scale variation in RSIs, we plan to incorporate the feature pyramid in future. Specifically, we will employ multiple pooling blocks operating at different spatial scales to separately extract local and global features. Simultaneously, by integrating our proposed dynamic weight allocation strategy, we aim to achieve adaptive fusion of these multi-scale features.

References

1. Chen, L.C., Zhu, Y., Papandreou, G., Schroff, F., Adam, H.: Encoder-decoder with atrous separable convolution for semantic image segmentation. In: ECCV, pp. 801–818 (2018)
2. Ding, L., Tang, H., Bruzzone, L.: LANet: local attention embedding to improve the semantic segmentation of remote sensing images. TGARS **59**(1), 426–435 (2020)
3. Feng, Y., et al.: Npaloss: neighboring pixel affinity loss for semantic segmentation in high-resolution aerial imagery. ISPRS Ann. **5**(2), 475–482 (2020)
4. Fu, J., et al.: Dual attention network for scene segmentation. In: CVPR, pp. 3146–3154 (2019)
5. Jin, Z., Liu, B., Chu, Q., Yu, N.: Isnet: integrate image-level and semantic-level context for semantic segmentation. In: ICCV, pp. 7189–7198 (2021)
6. Li, R., et al.: Multiattention network for semantic segmentation of fine-resolution remote sensing images. TGARS **60**, 1–13 (2021)
7. Li, X., et al.: Dual attention deep fusion semantic segmentation networks of large-scale satellite remote-sensing images. IJRS **42**(9), 3583–3610 (2021)
8. Li, X., Xu, F., Xia, R., Lyu, X., Gao, H., Tong, Y.: Hybridizing cross-level contextual and attentive representations for remote sensing imagery semantic segmentation. Remote Sens. **13**(15), 2986 (2021)
9. Liu, Z., Mao, H., Wu, C.Y., Feichtenhofer, C., Darrell, T., Xie, S.: A convnet for the 2020s. In: CVPR, pp. 11976–11986 (2022)
10. Long, J., Shelhamer, E., Darrell, T.: Fully convolutional networks for semantic segmentation. In: CVPR, pp. 3431–3440 (2015)
11. Ma, X., et al.: Sacanet: scene-aware class attention network for semantic segmentation of remote sensing images. arXiv preprint arXiv:2304.11424 (2023)
12. Ma, X., et al.: Log-can: local-global class-aware network for semantic segmentation of remote sensing images. In: ICASSP2023, pp. 1–5. IEEE (2023)
13. Maboudi, M., Amini, J., Malihi, S., Hahn, M.: Integrating fuzzy object based image analysis and ant colony optimization for road extraction from remotely sensed images. ISPRS PRS **138**, 151–163 (2018)

14. Marcos, D., Volpi, M., Kellenberger, B., Tuia, D.: Land cover mapping at very high resolution with rotation equivariant cnns: towards small yet accurate models. ISPRS PRS **145**, 96–107 (2018)
15. Niu, R., Sun, X., Tian, Y., Diao, W., Chen, K., Fu, K.: Hybrid multiple attention network for semantic segmentation in aerial images. TGARS **60**, 1–18 (2021)
16. Rottensteiner, F., et al.: International society for photogrammetry and remote sensing, 2d semantic labeling contest. Accessed 29 Oct 2020. https://www.isprs. org/education/benchmarks
17. Song, M., Li, B., Wei, P., Shao, Z., Wang, J., Huang, J.: DMF-CL: dense multi-scale feature contrastive learning for semantic segmentation of remote-sensing images. In: PRCV 2022, pp. 152–164. Springer, Heidelberg (2022). https://doi.org/10. 1007/978-3-031-18916-6_13
18. Song, P., Li, J., An, Z., Fan, H., Fan, L.: CTMFNet: CNN and transformer multi-scale fusion network of remote sensing urban scene imagery. TGARS **61**, 1–14 (2022)
19. Song, Q., Li, J., Li, C., Guo, H., Huang, R.: Fully attentional network for semantic segmentation. In: AAAI, vol. 36, pp. 2280–2288 (2022)
20. Strudel, R., Garcia, R., Laptev, I., Schmid, C.: Segmenter: transformer for semantic segmentation. In: ICCV, pp. 7262–7272 (2021)
21. Wang, J., Zheng, Z., Ma, A., Lu, X., Zhong, Y.: Loveda: a remote sensing land-cover dataset for domain adaptive semantic segmentation. arXiv preprint arXiv:2110.08733 (2021)
22. Wang, X., Girshick, R., Gupta, A., He, K.: Non-local neural networks. In: CVPR, pp. 7794–7803 (2018)
23. Woo, S., Park, J., Lee, J.Y., Kweon, I.S.: CBAM: convolutional block attention module. In: ECCV, pp. 3–19 (2018)
24. Xu, Y., Jiang, J.: High-resolution boundary-constrained and context-enhanced network for remote sensing image segmentation. Remote Sens. **14**(8), 1859 (2022)
25. Yang, X., et al.: An attention-fused network for semantic segmentation of very-high-resolution remote sensing imagery. ISPRS PRS **177**, 238–262 (2021)
26. Yu, W., et al.: Metaformer is actually what you need for vision. In: CVPR, pp. 10819–10829 (2022)
27. Yuan, Y., Chen, X., Wang, J.: Object-contextual representations for semantic segmentation. In: Vedaldi, A., Bischof, H., Brox, T., Frahm, J.-M. (eds.) ECCV 2020. LNCS, vol. 12351, pp. 173–190. Springer, Cham (2020). https://doi.org/10.1007/ 978-3-030-58539-6_11
28. Zhang, Q., Seto, K.C.: Mapping urbanization dynamics at regional and global scales using multi-temporal DMSP/OLS nighttime light data. Remote Sens. Environ. **115**(9), 2320–2329 (2011)
29. Zhao, H., Shi, J., Qi, X., Wang, X., Jia, J.: Pyramid scene parsing network. In: CVPR, pp. 2881–2890 (2017)
30. Zheng, S., Lu, C., Wu, Y., Gupta, G.: Sapnet: segmentation-aware progressive network for perceptual contrastive deraining. In: WACV, pp. 52–62 (2022)
31. Zheng, Z., Zhong, Y., Wang, J., Ma, A.: Foreground-aware relation network for geospatial object segmentation in high spatial resolution remote sensing imagery. In: CVPR, pp. 4096–4105 (2020)
32. Zhu, L., Ji, D., Zhu, S., Gan, W., Wu, W., Yan, J.: Learning statistical texture for semantic segmentation. In: CVPR, pp. 12537–12546 (2021)
33. Zhu, L., Wang, X., Ke, Z., Zhang, W., Lau, R.W.: Biformer: vision transformer with bi-level routing attention. In: CVPR, pp. 10323–10333 (2023)

Multi-scale Contrastive Learning for Building Change Detection in Remote Sensing Images

Mingliang Xue[1], Xinyuan Huo[1], Yao Lu[2(✉)] (iD), Pengyuan Niu[1], Xuan Liang[1], Hailong Shang[1], and Shucai Jia[1]

[1] School of Computer Science and Engineering,
Dalian Minzu University, Dalian 116650, China
[2] Beijing Institute of Remote Sensing Information, Beijing, China
`yaolu@bjirs.org.cn`

Abstract. Self-supervised contrastive learning (CL) methods can utilize large-scale label-free data to mine discriminative feature representations for vision tasks. However, most existing CL-based approaches focus on image-level tasks, which are insufficient for pixel-level prediction tasks such as change detection (CD). This paper proposes a multi-scale CL pre-training method for CD tasks in remote sensing (RS) images. Firstly, unlike most existing methods that rely on random augmentation to enhance model robustness, we collect a publicly available multi-temporal RS dataset and leverage its temporal variations to enhance the robustness of the CD model. Secondly, an unsupervised RS building extraction method is proposed to separate the representation of buildings from background objects, which aims to balance the samples of building areas and background areas in instance-level CL. In addition, we select an equal number of local regions of the building and background for the pixel-level CL task, which prevents the domination caused by local background class. Thirdly, a position-based matching measurement is proposed to construct local positive sample pairs, which aims to prevent the mismatch issues in RS images due to the object similarity in local areas. Finally, the proposed multi-scale CL method is evaluated on benchmark OSCD and SZTAKI databases, and the results demonstrate the effectiveness of our method.

Keywords: Remote sensing images · Contrastive learning · Change detection

1 Introduction

Remote sensing (RS) image change detection (CD) aims to identify changes in geographical features over time. It has been widely applied in various domains [1]. Recently, a plethora of deep learning-based CD methods in RS have been

---

This work was supported by the Research Foundation of Liaoning Educational Department (Grant No. LJKMZ20220400). Please contact Yao Lu (`yaolu@bjirs.org.cn`) for access to the pre-training dataset.

Q. Liu et al. (Eds.): PRCV 2023, LNCS 14428, pp. 318–329, 2024.
https://doi.org/10.1007/978-981-99-8462-6_26

proposed [2]. These methods, leveraging the potent representation capabilities of convolutional neural networks, have become the predominant strategies [3]. Nevertheless, their reliance on high-quality annotated samples introduces substantial labor and time expenses [4], impeding their practical utility [5].

Recently, self-supervised learning (SSL) has gained attention as an annotation-free approach for extracting discriminative feature representations. Contrastive learning (CL), a potent SSL technique, has rapidly advanced in natural image processing [6,7]. It has also found success in RS image CD tasks [8,9]. However, existing CL approaches primarily concentrate on image-level representation learning, which may not be sufficient for pixel-level classification tasks, such as CD in RS images. To overcome these limitations, [10] proposed a multitask self-supervised approach that learns low-level and high-level features simultaneously, while [11] introduced a global and local matching CL method for effective local-level representation modeling. Furthermore, [4] designs a global-local CL method that demonstrates its effectiveness for fine-grained CD tasks.

Although the above CL methods have achieved remarkable results in RS images, there are still some limitations: Firstly, due to the lack of multi-temporal RS image data [11], most methods rely on random augmentation to simulate real imaging environment changes, aiming to enhance model robustness. However, random augmentation is not able to capture the rich semantic information provided by actual imaging environment changes, limiting the improvement in model robustness. Secondly, RS images are characterized by imbalanced class distributions [12]. However, many CL methods commonly overlook this aspect. These methods obtain the image representations by pooling the entire image feature into a single vector, which may lead to the dominance of background categories in the CL process. Furthermore, in pixel-level CL, there also exists the issue of local background class dominance [11]. Meanwhile, the utilization of distance-based measurement for local instance matching in RS images usually causes mismatches due to the intricate nature of the objects contained. Furthermore, this will result in the location-sensitive characteristics that are especially important for CD tasks [13] not being adequately learned in CL.

In this paper, we propose a multi-scale CL pre-training method for CD tasks in RS images. Firstly, we collect a multi-temporal RS dataset, which can provide temporal variation information for CL pre-training. Secondly, an unsupervised RS building extraction method is designed to separate the buildings from background objects in instance-level CL, which prevents the background class-dominated problem. Furthermore, an equal number of local regions of building and background are selected in pix-level CL, which aims to balance the representation learning of local areas. The local positive sample pairs are then constructed based on position matching, which resolves mismatch issues. The contributions of this paper can be summarized as follows:

1) We construct a multi-temporal urban RS dataset for CL pre-training in CD. The temporal variations information provided by the constructed dataset can be utilized to enhance the robustness of the pre-trained models.
2) A multi-scale CL method is proposed for CD in RS images. The proposed method considers the category distribution imbalance in RS images in both

instance-level and pixel-level CL, which can obtain robust feature representations for CD in RS images.

3) We propose a position-based matching method for the construction of local positive sample pairs in pixel-level CL, which encourages the pre-trained model to learn location-sensitive information for the CD task.

2   Related Works

2.1   Self-supervised Contrastive Learning

Recent SSL progress driven by CL methods [14], as an unsupervised approach [15–17]. SimCLR [6] achieves excellent performance in CL using large batch size. To reduce the need for large batch size, MoCo [18] introduces momentum encoders and negative sample queues to learn effective feature representation. Furthermore, SimSiam [19] acquires representations without relying on negative samples. Recently, VICReg [7] effectively addresses model collapse by using two distinct regularization methods separately. SMoG [20] demonstrates superior results with their group-level CL approach. [21] proposed a dense CL method (DenseCL) that better adapts to dense prediction downstream tasks.

2.2   SSL in Remote Sensing

With SSL advancements, notable research successfully applies them in RS. [8] introduces an SSL signal acquisition method and constructs the publicly available S2MTCP dataset. [9] proposed Seco, a CL method that leverages seasonal variations to effectively utilize unlabeled RS imagery data. Although the mentioned SSL methods show promise, they lack fine-grained information processing for pixel-level classification of different ground objects [4]. Therefore, a series of pretext tasks based on pixel-level CL are proposed [4,10,11], guiding the pre-training model to learn both high-level and low-level image semantic information. However, these methods have some limitations. First, their simulated imaging environment changes fall short of fully capturing the semantic information inherent in complex real-world imaging conditions. Secondly, randomly selecting local regions in pixel-level CL can lead to the dominance of the local background class. Thirdly, matching local instances using distance-based measurements in RS images can result in pairing errors.

3   Methodology

We outline the steps involved in the proposed pre-training method as follows: Firstly, the input two temporal variation images are augmented to generate augmented view pairs. The pairs undergo feature extraction using the Siamese encoder network. The extracted features undergo two processing processes: 1) The mask obtained by the building extraction method is used to pool the extracted features, resulting in the representation vectors with different semantics. These vectors are then fed into the instance-level CL module for representation learning. 2) The features are decoded by the decoder. The mask, obtained by selecting equal amounts of local regions from different classes, is then used to pool the decoded features into representation vectors for different categories. These category-specific vectors are subsequently input into the pixel-level CL module for representation learning. The overall architecture is illustrated in Fig. 1.

3.1 Data Preprocessing and Representation Generation

In data preprocessing, two images $(x^i, i \in \{1, 2\})$ capturing temporal variations are randomly chosen from a set of four RS images taken at three-month intervals at a specific location. Next, the Building Extraction Module extracts the building area M from x^i. Based on the M, we choose an equal number of buildings and background local areas, generating mask K. Meantime, data augmentation is applied to x^1 and x^2, resulting in x^{t1} and x^{t2}. Semantic mask M^{ti} and local region mask K^{ti} are also derived M and K through employing identical geometric augmentation methods as their counterparts in x^{ti}. Next, x^{t1} and x^{t2} are fed into the Siamese encoder to obtain e^{t1} and e^{t2}. In the instance-level CL module, the semantic masks M^{t1} and M^{t2} are respectively sampled to the same size as the corresponding feature e^{ti}. Through the M^{ti} mask, the corresponding feature e^{ti} is decoupled into different semantic representation vectors $e_j^{ti}, j \in \{1, 2\}$, j stands for semantic category(building and background). In the pixel-level CL module, feature e^{ti} is decoded by the decoder to obtain feature d^{ti}, then k^{ti} pools d^{ti} into the representation vector $P_n^{ti}, n \in \{1, 2, ..., 10\}$, where n ranges from 1 to 5 for the buildings area and from 6 to 10 for the background area.

3.2 Instance-Level CL Module

Most existing CL methods obtain image representations by pooling the entire image feature into a single vector. However, in RS images, there exists a characteristic of imbalanced class distributions among categories [12], which may result in the dominance of background classes in the CL process. Therefore, expanding on the building extraction method we have designed, we separate the representation of the building object from the background. This approach effectively addresses the issue of background class dominance in representation learning and simultaneously enhances the learning of semantic relationships between instances in CL. While [22] has alleviated the problem of background class-dominated representation learning to some extent. However, the difference is that the instance-level CL method we design does not rely on any of the labeled data. The instance-level CL architecture is shown on the left side of Fig. 2.

Firstly, we are inspired by SimSiam and adopt a positive example-based CL method. Differently, within the CL process, we decouple the representations of distinct semantic categories. For example, each semantic representation vector e_j^{ti} is projected into the embedding space $z \in R^d$ by the Projector, resulting in vector g_j^{ti}. Next, the vector (e.g. g_j^{t1}.) is mapped to a vector (e.g. h_j^{t1}.) by the predictor, and its cosine-like distance from the corresponding projection vector (e.g. g_j^{t2}.) is calculated. Then, optimize the architecture by minimizing

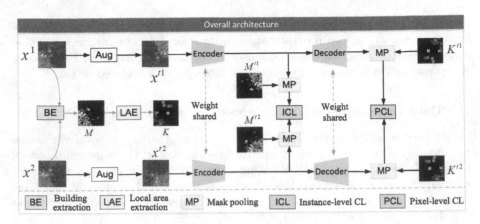

Fig. 1. Overall architecture of the proposed multi-scale contrastive learning method for change detection in remote sensing images.

the negative cosine similarity distance between the projection vector and the corresponding prediction vector. The corresponding loss function is defined as:

$$L_{pu,j}^{ti} = 1 - \frac{1}{2}(C(h_j^{t1}, st(g_j^{t2})) + C(h_j^{t2}, st(g_j^{t1}))) \tag{1}$$

where st represents gradient clipping. SimSiam shows the effectiveness of gradient clipping in preventing model collapse. Finally, the overall similarity loss for the current sample is denoted as $L_{pu} = \frac{1}{2}\sum_{j=1}^2 L_{pu,j}^{ti}$, where j represents different semantic categories. As a result of the decoupling of representations for distinct semantic classes, the CL process inherently avoids being dominated by background classes. Minimizing L_{pu} ensures that the model encodes similar data similarly. By generating similar semantic representations for buildings at the same location but at different times, the model learns robustness to complex imaging environment changes. Additionally, learning similar semantic representations for the background area at the same location and at different times enhances the model's robustness to seasonal variations. Furthermore, to further explore the relationship between different semantic categories, we utilize the cosine similarity loss to separate their representations in a high-dimensional space. According to the two types of semantic representation vectors e_1^{ti} and e_2^{ti} corresponding to the input image x^i, the dissimilarity loss $L_{p,j}^{ti}$ we design is $L_{p,j}^{ti} = 1 + (C(e_1^{ti}, e_2^{ti}))$, where $C(\cdot)$ denotes cosine similarity, and the additional term guarantees a positive loss value. Thus, the overall dissimilarity loss for the current sample is denoted as $L_p = \frac{1}{2}\sum_{i=1}^2 L_{p,j}^{ti}$. Ultimately, the total loss L_g for the current sample in the instance CL model is denoted as $L_g = L_p + \alpha L_{pu}$. The positive coefficient α is used to balance the learning speed of the model between tasks L_p and L_{pu}.

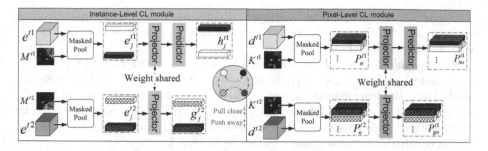

Fig. 2. Instance and pixel-level Contrastive Learning architecture.

3.3   Pixel-Level CL Module

We enhance local-level representations with a pixel-level CL module, advantageous for dense prediction tasks [21]. Additionally, [11] suggests balancing category selection for local regions could lead to further improvements. Therefore, we balance local region selection based on building extraction results. Meantime, due to the intricate spectral responses exhibited by objects in RS images and the image enhancement, there may be a mismatch issue with commonly used distance-based matching methods for local instances. We match local instances at the same location to address this issue. Specifically, we apply the same spatial augmentation from the augmented image pair x^{ti} to the selected local region mask K to obtain K^{ti}. Through K^{ti}, each local instance corresponding to the current augmented image is matched with another augmented image's local instances. Each pair shares the same position in the original image. This matching process forms corresponding dense positive sample pairs (P_n^{t1}, P_n^{t2}). The pixel-level CL architecture is shown on the right side of Fig. 2.

The encoder-decoder network is optimized to ensure dissimilarity between different category local representations and similarity within the same category. Firstly, we introduce the local region positive pair CL method based on matching corresponding positions. For example, each vector representation P_n^{ti} is mapped to the vector Pg_n^{ti} by the Projector. Next, through the Predictor, the projection vector (e.g. Pg_n^{t1}) is mapped to the prediction vector (e.g. Ph_n^{t1}), and the cosine similarity distance between it and the corresponding matching projection vector (e.g. Pg_n^{t2}) is calculated. The corresponding loss function is defined as:

$$L_r = \frac{1}{n} \sum_{n=1}^{10} (1 - \frac{1}{2}(C(Ph_n^{t1}, st(Pg_n^{t2})) + C(Ph_n^{t2}, st(Pg_n^{t1})))) \qquad (2)$$

The CL of local areas from categories building ($n = 1$ to 5) and background ($n = 6$ to 10) is balanced. Next, we introduce the CL for dissimilarity between local representations belonging to different categories. Based on the local representations of different categories, the semantic dissimilarity loss $L_{d,p}^{ti}$ follows.

$$L_{d,p}^{ti} = 1 + \frac{1}{3}(C(P_1^{ti}, P_6^{ti}) + C(P_2^{ti}, P_7^{ti}) + C(P_3^{ti}, P_8^{ti})) \qquad (3)$$

The model learns fine-grained category differences through dissimilarity loss based on semantic representations. The total local dissimilarity loss for the current sample is calculated as $L_d = \frac{1}{2} \sum_{i=1}^{2} L_{d,p}^{ti}$.

Finally, the total loss L_t of the current sample in the pixel-level CL module is calculated as $L_t = L_d + \beta L_r$. By applying a positive coefficient β to weight and balance the learning rate across different tasks. Finally, the overall loss L_a of the current sample is defined as $L_a = (1 - \lambda)L_g + \lambda L_t$. [4] highlight the significance of high-level semantic and low-level detail information for accurate CD. Accordingly, we set a constant value of 0.5 for λ.

4   Experiments

4.1   Datasets

Firstly, we automatically collect a global multi-temporal urban dataset from the Sentinel-2 platform [23]. Sentinel-2 satellite imagery consists of 12 spectral bands with resolutions of 10 m, 20 m, and 60 m, offering a wide coverage and short revisit cycles. More specifically, we use the Google Earth Engine [24] to extract, process, and download a sample of satellite image patches from approximately 24000 cities worldwide. For cities with a population below 0.1 million, we conduct normal distribution sampling within a 1.4 km radius around the city center. Each location was associated with four images, taken three months apart. For cities with a population exceeding 0.1 million, normal distribution sampling is performed with a standard deviation of 2 km around the city center. To mitigate cloud interference, patches with a cloud coverage exceeding 5% are filtered out. The image size was standardized to 256×256 pixels. Finally, approximately 78,000 images are obtained.

Additionally, we evaluate our proposed building extraction module using a test dataset provided in [25], which includes 101 locations across six continents from 2018 to early 2020. The dataset focuses on 60 rapidly developed test sites.

Finally, we evaluate our pre-training method on two CD datasets: 1) OSCD [26], containing 24 sets of Sentinel-2 multispectral images from 2015 and 2018. We adopt Seco's partitioning, using 14 images for training and 10 for validation. 2) SZTAKI [27], an aerial dataset with three subsets: SZADA, TIZADOB, and ARCHIVE. For our evaluation of a small dataset, we focus on the SZADA subset, which consists of 7 pairs of dual-temporal images. The first pair is for testing, cropped into 6 non-overlapping 256 × 256 regions. The remaining 6 pairs form a training set of 168 images with 50% overlap.

4.2   Implementation Details

Building Extraction Method. We accurately extract building areas from images based on factors such as: 1) Buildings display relatively less visual variation compared to the background across different periods, such as seasonal change. 2) Visual disparities between buildings and vegetation regions. 3)

Texture differences between buildings and the background regions. First, after adjusting image brightness, we identify subtle building changes compared to the background across seasons using image differencing and thresholding, approximating building areas as M_S. Next, we adapted the vegetation index [28] to $3 \times g - 2.4 \times r - 1.3 \times b$ for precise vegetation extraction in RS images, obtaining non-vegetation areas as M_T. Furthermore, we employ a novel GLCM-based approach for pixel-level texture segmentation of RS images, refining target area selection based on texture differences among objects. Initially, the grayscale image based on the B channel is extracted from the image. It is then quantized into eight gray levels. For each pixel, GLCMs are computed within its 1-pixel neighborhood for the average values along four directions: $0°$, $45°$, $90°$, and $135°$. Next, contrast texture features are further calculated:

$$Contrast = \sum_i \sum_j p(i,j) \times (i-j)^2 \qquad (4)$$

where i and j represent the rows and columns of the GLCM, $p(i,j)$ represents the current texture map in the corresponding grayscale image. Next, after thresholding, we obtain the final building extraction result M_G, which accurately delineates the building area M in the remote sensing image through a bitwise AND operation with the precomputed masks M_S and M_T.

Training Details. 1) Pre-training details: we employ ResNet-18 [29] as the CL backbone. The decoder consists of bilinear interpolation and convolution modules. The Projector and Predictor structures are 2-layer MLPs with output dimensions of 1024 and hidden dimensions of 1048 and 256. Data augmentation includes color jittering, Gaussian blur, random rotation, and flipping. Stochastic Gradient Descent (SGD) with an initial learning rate of 0.01, weight decay of 0.0005, and momentum of 0.9 served as the optimizer. Pre-training is conducted for 15 epochs with a batch size of 64. 2) Change detection details: we compare our method with random initialization, ImageNet pre-training, Seco [9], SimSiam [19], VICReg [7] and SMoG [20]. We adopt Seco's CD task model architecture. For the OSCD dataset, we use Seco's training parameter settings to ensure fair comparison. For the SZADA dataset, we employ Adam optimizer to train U-net network with frozen backbone weights. The training was done for 50 epochs with an initial learning rate of 0.0001, gradually decreasing by 0.95 per epoch, using a batch size of 32.

4.3   Experiment on Change Detection and Building Extraction

We evaluate pre-training methods on OSCD and SZADA datasets using precision, recall, and F1 score as metrics. Focusing on the F1 score, which combines precision and recall, provides a comprehensive assessment.

In OSCD datasets, based on the results shown in Table 1, our proposed method has achieved superior performance, possibly attributed to the introduction of instance-level CL, which effectively captures the relationships between

Table 1. The fine-tuning results on the OSCD and SZADA datasets.

Pre-training	OSCD			SZADA		
	Precision	Recall	F1	Precision	Recall	F1
Random	**0.705**	0.191	0.294	0.672	0.130	0.218
ImageNet	0.704	0.251	0.362	**0.713**	0.158	0.258
SimSiam [19]	0.657	0.291	0.399	0.539	0.203	0.295
SeCo [9]	0.654	0.380	0.469	0.467	0.216	0.295
VICReg [7]	0.671	0.365	0.459	0.588	0.137	0.222
SMoG [20]	0.674	0.315	0.409	0.356	0.364	0.360
Ours	0.696	**0.395**	**0.488**	0.451	**0.382**	**0.414**

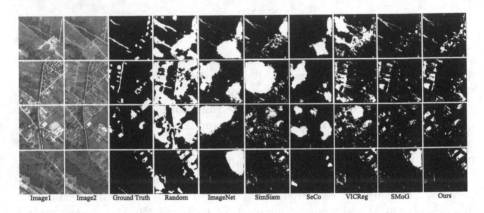

Fig. 3. Visualization results on SZADA dataset.

different land cover classes (e.g., buildings and background). In contrast, most baseline methods overlook the importance of learning such discriminative features. Furthermore, by introducing and improving the pixel-level CL pretext task, our approach enhances the model's ability to analyze and process fine-grained information within images, which aligns perfectly with the requirements of dense CD tasks. In SZADA datasets, based on the results shown in Table 1 and Fig. 3, our method demonstrates superior performance even in the presence of complex environmental factors such as varying lighting and atmospheric conditions, as illustrated in Fig. 3. Furthermore, our model excels at capturing fine-grained information within images. This can be attributed to the introduction and enhancement of pixel-level CL. Then, we evaluate the proposed method, as shown in Table 2. It achieves high accuracy for various ground object types.

4.4 Ablation Studies

We ablatively analyze the key modules of the proposed pretraining approach: utilizing temporal variation information on the pre-training dataset (TV), instance-

Table 2. Accuracy on building extraction datasets

F1	Precision	Recall	CPA(building)	CPA(background)	mPA
0.507	0.431	0.613	0.852	0.613	0.733

Table 3. Ablation experiment on OSCD dataset

Base	TV	IL	PL	CB	F1
✓	×	×	×	×	0.377
✓	✓	×	×	×	0.399
✓	✓	✓	×	×	0.467
✓	✓	✓	✓	×	0.478
✓	✓	✓	✓	✓	0.488

level CL (IL), pixel-level CL (PL), and considering class balance in pixel-level CL (CB). We use SimSiam (Not utilize temporal enhancement in the dataset) as the baseline method. Modules were added incrementally to the baseline for evaluation, and the downstream OSCD dataset results are presented in Table 3. Firstly, integrating the TV module improves the F1 score by 2.2% over random augmentation, as realistic temporal changes introduce imaging condition variations, which will bolster model robustness more effectively. Secondly, implementing the IL module raised the F1 score by 6.8%, because the IL module decouples representations of buildings from backgrounds, overcoming the dominance of background classes inherent in traditional CL methods that condense the entire image into low-dimensional representations. Moreover, the inclusion of the PL module adds a 1.1% F1 score enhancement, driven by the importance of local contextual data in CD [30]. Additionally, adopting the CB module adds a 1% F1 score enhancement, as it balances representation learning across various local areas. In the PL module, our approach introduces a local positive pair representation method based on position matching, achieving a 1% F1 improvement over traditional local feature similarity matching (F1 score: 0.478). This resolves the mismatch in traditional methods due to the unique object complexity of RS images.

5 Conclusion

In this paper, numerous multi-temporal urban change images are sampled for the robustness of the pre-trained model to changes in complex imaging conditions. Furthermore, we decouple the representations of interest and background objects in instance-level CL, which better models the relationship between instances. Simultaneously, considering the characteristics of the downstream dense CD task, we introduce and improve the pixel-level CL module to make the representation learning of local regions more balanced. It overcomes the mismatch problem that may be caused when the traditional local feature similarity matching method is applied to RS images. Finally, extensive experiments demonstrate the effectiveness of our proposed method.

References

1. Zhang, M., Liu, Z., Feng, J., Liu, L., Jiao, L.: Remote sensing image change detection based on deep multi-scale multi-attention siamese transformer network. Remote Sens. **15**(3), 842 (2023)
2. Bai, T., et al.: Deep learning for change detection in remote sensing: a review. Geo-Spatial Inf. Sci. **26**, 262–288 (2022)
3. Cheng, G., Xie, X., Han, J., Guo, L., Xia, G.S.: Remote sensing image scene classification meets deep learning: challenges, methods, benchmarks, and opportunities. IEEE J. Sel. Topics Appl. Earth Obs. Remote Sens. **13**, 3735–3756 (2020)
4. Jiang, F., Gong, M., Zheng, H., Liu, T., Zhang, M., Liu, J.: Self-supervised global-local contrastive learning for fine-grained change detection in VHR images. IEEE Trans. Geosci. Remote Sens. **61**, 1–13 (2023)
5. Zhu, X.X., et al.: Deep learning in remote sensing: a comprehensive review and list of resources. IEEE Geosci. Remote Sens. Maga. **5**(4), 8–36 (2017)
6. Chen, T., Kornblith, S., Norouzi, M., Hinton, G.: A simple framework for contrastive learning of visual representations. In: International Conference on Machine Learning, pp. 1597–1607. PMLR (2020)
7. Bardes, A., Ponce, J., LeCun, Y.: VICReg: variance-invariance-covariance regularization for self-supervised learning. In: International Conference on Learning Representations (2022)
8. Leenstra, M., Marcos, D., Bovolo, F., Tuia, D.: Self-supervised pre-training enhances change detection in sentinel-2 imagery. In: Del Bimbo, A., et al. (eds.) ICPR 2021. LNCS, vol. 12667, pp. 578–590. Springer, Cham (2021). https://doi.org/10.1007/978-3-030-68787-8_42
9. Manas, O., Lacoste, A., Giró-i Nieto, X., Vazquez, D., Rodriguez, P.: Seasonal contrast: unsupervised pre-training from uncurated remote sensing data. In: Proceedings of the IEEE/CVF International Conference on Computer Vision, pp. 9414–9423 (2021)
10. Li, W., Chen, H., Shi, Z.: Semantic segmentation of remote sensing images with self-supervised multitask representation learning. IEEE J. Sel. Topics Appl. Earth Obs. Remote Sens. **14**, 6438–6450 (2021)
11. Li, H., et al.: Global and local contrastive self-supervised learning for semantic segmentation of HR remote sensing images. IEEE Trans. Geosci. Remote Sens. **60**, 1–14 (2022)
12. Gu, X., Li, S., Ren, S., Zheng, H., Fan, C., Xu, H.: Adaptive enhanced swin transformer with u-net for remote sensing image segmentation. Comput. Electr. Eng. **102**, 108223 (2022)
13. Fang, S., Li, K., Shao, J., Li, Z.: SNUNET-CD: a densely connected siamese network for change detection of VHR images. IEEE Geosci. Remote Sens. Lett. **19**, 1–5 (2021)
14. Miyai, A., Yu, Q., Ikami, D., Irie, G., Aizawa, K.: Rethinking rotation in self-supervised contrastive learning: adaptive positive or negative data augmentation. In: Proceedings of the IEEE/CVF Winter Conference on Applications of Computer Vision, pp. 2809–2818 (2023)

15. Wang, H., Yao, M., Jiang, G., Mi, Z., Fu, X.: Graph-collaborated auto-encoder hashing for multiview binary clustering. IEEE Trans. Neural Netw. Learn. Syst. (2023)
16. Wang, H., Peng, J., Fu, X.: Co-regularized multi-view sparse reconstruction embedding for dimension reduction. Neurocomputing **347**, 191–199 (2019)
17. Feng, L., Meng, X., Wang, H.: Multi-view locality low-rank embedding for dimension reduction. Knowl.-Based Syst. **191**, 105172 (2020)
18. He, K., Fan, H., Wu, Y., Xie, S., Girshick, R.: Momentum contrast for unsupervised visual representation learning. In: Proceedings of the IEEE/CVF Conference on Computer Vision and Pattern Recognition, pp. 9729–9738 (2020)
19. Chen, X., He, K.: Exploring simple siamese representation learning. In: Proceedings of the IEEE/CVF Conference on Computer Vision and Pattern Recognition, pp. 15750–15758 (2021)
20. Pang, B., Zhang, Y., Li, Y., Cai, J., Lu, C.: Unsupervised visual representation learning by synchronous momentum grouping. In: Avidan, S., Brostow, G., Cissé, M., Farinella, G.M., Hassner, T. (eds.) ECCV 2022. LNCS, vol. 13690, pp. 265–282. Springer, Cham (2022). https://doi.org/10.1007/978-3-031-20056-4_16
21. Wang, X., Zhang, R., Shen, C., Kong, T., Li, L.: Dense contrastive learning for self-supervised visual pre-training. In: Proceedings of the IEEE/CVF Conference on Computer Vision and Pattern Recognition, pp. 3024–3033 (2021)
22. Chen, H., Zao, Y., Liu, L., Chen, S., Shi, Z.: Semantic decoupled representation learning for remote sensing image change detection. In: IGARSS 2022–2022 IEEE International Geoscience and Remote Sensing Symposium, pp. 1051–1054. IEEE (2022)
23. Drusch, M., et al.: Sentinel-2: esa's optical high-resolution mission for GMES operational services. Remote Sens. Environ. **120**, 25–36 (2012)
24. Gorelick, N., Hancher, M., Dixon, M., Ilyushchenko, S., Thau, D., Moore, R.: Google earth engine: planetary-scale geospatial analysis for everyone. Remote Sens. Environ. **202**, 18–27 (2017)
25. Hafner, S., Ban, Y., Nascetti, A.: Unsupervised domain adaptation for global urban extraction using sentinel-1 sar and sentinel-2 msi data. Remote Sens. Environ. **280**, 113192 (2022)
26. Daudt, R.C., Le Saux, B., Boulch, A., Gousseau, Y.: Urban change detection for multispectral earth observation using convolutional neural networks. In: IGARSS 2018–2018 IEEE International Geoscience and Remote Sensing Symposium, pp. 2115–2118. IEEE (2018)
27. Benedek, C., Szirányi, T.: Change detection in optical aerial images by a multilayer conditional mixed markov model. IEEE Trans. Geosci. Remote Sens. **47**(10), 3416–3430 (2009)
28. Meyer, G.E., Neto, J.C.: Verification of color vegetation indices for automated crop imaging applications. Comput. Electron. Agric. **63**(2), 282–293 (2008)
29. He, K., Zhang, X., Ren, S., Sun, J.: Deep residual learning for image recognition. In: Proceedings of the IEEE Conference on Computer Vision and Pattern Recognition, pp. 770–778 (2016)
30. Ailimujiang, G., Jiaermuhamaiti, Y., Jumahong, H., Wang, H., Zhu, S., Nurmamaiti, P.: A transformer-based network for change detection in remote sensing using multiscale difference-enhancement. Comput. Intell. Neurosci. **2022** (2022)

Shadow Detection of Remote Sensing Image by Fusion of Involution and Shunted Transformer

Yifan Wang[1]([⊠]), Jianlin Wang[1], Xian Huang[1,2], Tong Zhou[1], Wenjun Zhou[1], and Bo Peng[1]

[1] School of Computer Science, Southwest Petroleum University, Chengdu, China
yifan.wang@swpu.edu.cn
[2] Sichuan Xinchuan Aviation Instrument Co., Ltd., Guanghan, China

Abstract. In the field of remote sensing image analysis, this paper addresses the challenges of accurately detecting edge details and small shadow regions through the introduction of STAISD-Net. This end-to-end network synergistically combines the capabilities of CNN and Transformer to enhance shadow detection accuracy. Our network architecture incorporates several innovations. In the encoder, we propose an improved multi-scale asymmetric involution to extract detailed features across multiple scales. Additionally, we improve Shunted Transformer to extract global features, generating four-scale global feature information. In the decoder, we employ bilinear interpolation for upsampling and skip connections for fusing the high-level and low-level image features. Finally, we generate shadow masks by integrating feature maps of different scales. Comprehensive experiments conducted on the Aerial Imagery dataset for Shadow Detection (AISD) validate the effectiveness of STAISD-Net, showing that when compared to several state-of-the-art methods, our method demonstrates superior performance, achieving higher accuracy and the shadow detection results are more consistent with actual shadows.

Keywords: Shadow detection · Remote sensing image · Involution · Shunted Transformer · Multiscale feature

1 Introduction

Shadows in satellite remote sensing images are an inherent phenomenon, cast when direct sunlight encounters obstructions such as vehicles, trees, or architectural structures, resulting in shadows on roads, rooftops, or other surfaces [1]. The presence of shadows introduces substantial complexities in the processing of high-spatial-resolution remote sensing images, thereby necessitating shadow detection as a critical precursor to shadow removal. Numerous research endeavors aim to mitigate the detrimental impacts of shadows in remote sensing images. The ubiquitous presence of shadows in these images leads to information loss and

Q. Liu et al. (Eds.): PRCV 2023, LNCS 14428, pp. 330–342, 2024.
https://doi.org/10.1007/978-981-99-8462-6_27

disrupts grayscale continuity, hindering the interpretation and limiting application potential [2,3]. Tasks of significant importance, such as building identification, super-resolution reconstruction, and object tracking [4,5], require precise shadow detection to minimize the influence of shadows [6,7]. Therefore, the development of an effective shadow detection methodology is of paramount importance in the field of remote sensing image analysis.

Shadow detection in imagery is a specialized task within the broader context of semantic segmentation or pixel classification, essentially constituting a binary classification problem. Unlike natural image datasets, which often contain sparse information, remote sensing images are characterized by their rich information content. However, the inherent complexity of remote sensing images presents unique challenges. These images exhibit significant variations in object size and color texture, even among objects belonging to the same class. This contrasts sharply with natural images, adding an additional layer of complexity to shadow detection. In addition, shadows in remote sensing images are not uniformly distributed but are scattered across the image, displaying a wide range of shapes and sizes. This variability further complicates the task of shadow detection. In the context of remote sensing images, the shadow detection of prominent objects such as vegetation or buildings is typically straightforward due to their large size and distinct features. However, the identification of edge information for smaller objects, e.g. utility poles situated in corners, poses a more significant challenge. This is primarily due to their smaller size and the complex interplay of light and shadow in these regions.

The application of deep learning-based shadow detection methods within the domain of remote sensing images has been constrained due to the scarcity of relevant datasets. It was not until 2020 that the first public dataset in this domain, the Aerial Imagery Dataset for Shadow Detection (AISD), was made available. Most networks for shadow detection are built on the principles of Convolutional Neural Networks (CNNs) and Residual Networks, primarily using an encoder-decoder framework for end-to-end feature extraction and training. Within the encoder, CNNs are leveraged for feature extraction, systematically reducing the resolution of the feature map to enrich the semantic information. Conversely, the decoder employs CNNs to utilize the encoded features as input, decoding them to generate the final segmentation prediction results. However, these methodologies have limitations in identifying prominent shadows and are less effective at segmenting object edges.

In this study, we introduce a pioneering deep learning approach that synergistically integrates CNN and Transformer [8], specifically designed to enhance the efficiency and accuracy of shadow detection in remote sensing images. The primary contributions of this research are three-fold:

1. By harnessing the strengths of both CNN and Transformer, our network retains rich global and local features, thus significantly improving the efficiency and accuracy of shadow detection in remote sensing images.
2. We leverage the Transformer's capability to extract global features from images, overcoming the inherent limitations of CNN's receptive fields that impede the capture of comprehensive global information.

3. We design Multiscale Asymmetric Involution (MAI) modules by substituting conventional convolution with involution to amalgamate multiscale feature information, thereby capturing local image details and forming effective contextual information.

2   Related Work

Currently, there have been various developments in shadow detection methods, which can be mainly divided into three categories: geometry-based methods, threshold-based methods, and machine learning and deep learning-based methods [9].

Geometry-Based Methods: These methods necessitate prior knowledge about the environment and sensors, heavily relying on prior positioning. Additionally, shadow extraction methods based on physical models are computationally complex and have high data requirements, making them challenging to apply in practice [10,11].

Threshold-Based Methods: While simple and efficient, these methods overlook the spatial correlation between pixels, complicating the distinction between dark objects illuminated by sunlight and bright shadow objects [12,13].

Machine Learning-Based Methods: These methods primarily rely on the basic visual features of non-shadow pixels for shadow recognition, but they ignore the spatial correlation between adjacent pixels. Machine learning methods struggle to differentiate between dark objects and bright shadow objects, resulting in poor performance in remote sensing image processing [14].

Deep Learning-Based Methods: In recent years, CNNs [15–17] have been widely used in image processing due to their excellent feature extraction capability and ability to automatically extract image information, overcoming the limitations of traditional methods requiring prior knowledge, cumbersome steps, and poor performance. In the field of semantic segmentation, CNNs, including fully convolutional network (FCN) [18], U-Net [19], and SegNet [20], have achieved commendable results. In recent years, deep learning-based methods have also been introduced to shadow detection in high-resolution images. In 2020, Jin et al. [21] proposed a method combining spatial attention modules and adaptive weighted binary cross-entropy loss function to enhance shadow detection effectiveness. Luo et al. [22] proposed a deep supervised convolutional neural network (DSSDNet) and established a remote sensing shadow detection dataset (AISD) to provide training data for deep learning shadow detection. In 2021, Liu et al. [23] proposed a multiscale space-based shadow detection network with an attention mechanism (MSASDNet) to address the problem of incomplete shadows.

Ufuktepe et al. [24] proposed a shadow mask estimation method that helps capture global contextual information better. However, deep learning-based methods still face challenges in handling shadows of varying scales in remote sensing images, particularly when the shadows are scattered, small, irregular, and randomly distributed.

Although these methods have advanced the state-of-the-art, all of them still inevitably generate wrong edges for small shadow regions because they fail to obtain rich edge details correctly. Besides, they paid no attention to forming effective contextual information, which leads to inaccurate shadow detection as well. Different from these methods, our method demonstrates efficiency and accuracy in this aspect.

3 Methods

3.1 Proposed STAISD-Net

We introduce STAISD-Net, a comprehensive end-to-end framework for shadow detection in remote sensing images, ingeniously integrating CNN [25] and Shunted Transformer. The architecture of the network, illustrated in Fig. 1, is bifurcated into an encoder and a decoder, each encompassing four stages that progressively extract image features from superficial to profound levels.

The encoder of our network adeptly captures high-level semantic features by diminishing the dimensions of the input feature map, thereby facilitating the effective extraction of shadow-specific feature information embedded within the image. Each stage within the encoder amalgamates MAI and Shunted Transformer to concurrently extract local and global shadow image features. Every MAI layer is processed through Batch Normalization (BN) and ReLU activation function, further refining the image by filtering noise and smoothing via a 3×3 filter. The MAI module actualizes multiscale asymmetric involution through a combination of involution, asymmetric convolution blocks, and feature fusion. Notably, involution supplants the traditional pooling layer, thus reducing computational burden while attenuating information loss and preserving the image's

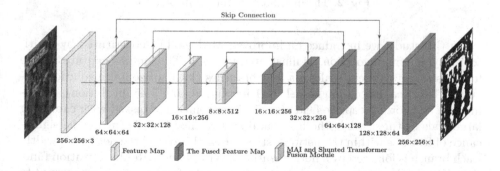

Fig. 1. The architecture of proposed STAISD-Net framework.

low-resolution information. The convolution kernel employed has a size of 3×3, a stride of 2, and a padding layer of 1, thereby effectively extracting intricate local feature details and contextual information from the image.

The decoder of our network combines low-level details and high-level semantic information from different scale feature maps through skip connections. It comprises four stages, each incorporating a fused MAI module. Each fused MAI module is succeeded by a convolution operation with a 3×3 kernel and a stride of 2 to further amplify the feature map. In the terminal module, a 1×1 convolution layer coupled with a sigmoid activation function is employed for thresholding, culminating in the generation of the final binary segmentation image that represents the shadow distribution. The ultimate output of the network is the shadow image.

3.2   Improved Multiscale Asymmetric Involution

The MAI introduces an enhanced asymmetric involution structure. As depicted in Fig. 2, our approach introduces the replacement of conventional convolutions with involution. These involution possess symmetrical and localized properties. The square involution kernel captures features with varying scales. Our framework employs asymmetric involution to approximate the existing square convolution, thus promoting model compression and acceleration. The cross-shaped receptive field of involution reduces the capture of redundant feature information.

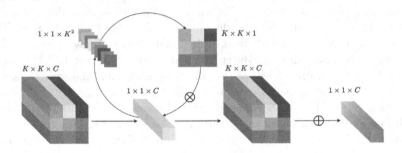

Fig. 2. The internal structure of involution.

In this study, we introduce an improved MAI module constituted by multiscale asymmetric involution, as illustrated in Fig. 3. The module captures image features across different scales via four branches (1×3 horizontal involution kernel, 3×1 vertical involution kernel, 1×1 involution, and 5×5 involution), resulting in a cruciform receptive field. The 5×5 convolution is tasked with capturing large-scale features, while the horizontal and vertical kernels ensure the significance of features within the network structure and broaden the network's width. Each branch is followed by a normalization layer BN and a ReLU activation function. Ultimately, the feature maps generated by the four branches are merged to obtain features encompassing more contextual information.

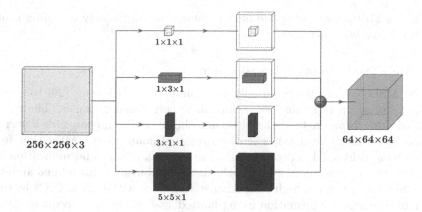

Fig. 3. Improved asymmetric MAI convolution block.

After multiple convolutions and fusions, the cascaded features may contain noise. Consequently, an initial fusion is required, utilizing a 3×3 convolution as a filter to diminish parameter count and prevent overfitting. The formula is expressed as

$$f_{(P,Q)}^m(V) = \text{ReLU}\left(w_{(P,Q)}^m \otimes V + b_{(P,Q)}^m\right), \tag{1}$$

where the input feature is V. $f_{(P,Q)}^m(V)$ represents the feature of $w_{(P,Q)}^m(V)$ following the m-th step convolution kernel of $P \times Q$ dimensions. $b_{(P,Q)}^m(V)$ denotes the bias of the m-th step convolution kernel, and ReLU signifies the rectified linear activation function.

The features after initial fusion are

$$F_{fir}^m(V) = f_{(3,3)}^m\left(C\left(f_{(1,1)}^m(V), f_{(1,3)}^m(V), f_{(3,1)}^m(V), f_{(5,5)}^m(V)\right)\right), \tag{2}$$

where F_{fir}^m denotes the initial feature following the MAI module, and C symbolizes feature fusion.

Subsequently, the initial feature F_{fir}^m applies involution with a kernel of 2 to amplify effective contextual information, thereby obtaining more discriminative features across various scales. Global features from the corresponding layer of the Shunted Transformer are then input and combined with pooling features to acquire more image scale information. After several rounds of convolutions and fusions, cascaded features may exhibit noise. As a result, refinement is performed once more using a 3×3 filter kernel for MAI features, which also reduces feature dimensions and averts overfitting. The improved MAI module formula is expressed as

$$F_{MAI}^m(V) = f_{(3,3)}^m\left(C\left(I\left(F_{fir}^m(V), F_{con}^m\right)\right)\right), \tag{3}$$

where I indicates involution with a stride of 2, padding of 1, and a kernel of 2. F_{con}^m representing the regular features. $F_{MAI}^m(V)$ representing the improved MAI module features. The MAI feature is then input into the following MAI module for further abstraction using the next regular scale, ultimately encoding robust context across various scales.

3.3   Improved Shunted Transformer

Our STAISD-Net integrates an enhanced Shunted Transformer [26] to extract global information from shadow regions in remote sensing images. The refined Shunted Transformer network architecture, shown in Fig. 4, comprises a normalization layer, multi-head self-attention, residual connections, activation functions, and feed-forward network layers. The network employs normalization and regularization techniques to stabilize the data distribution. This refined architecture effectively leverages high-resolution spatial information from CNN features and global semantic information from Shunted Transformers, consequently facilitating the capture of global context information.

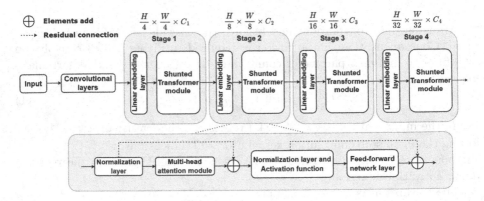

Fig. 4. Improved Shunted Transformer.

The introduction of self-attention modules helps the model better understand the semantic information of different regions in the image. The self-attention structure is hierarchically divided into three branches: query, key, and value.

$$\begin{cases} Q_i = XW_i^Q \\ K_i, V_i = MTA\left(X, r_i\right) W_i^K, MTA\left(X, r_i\right) W_i^V \\ V_i = V_i + LE\left(V_i\right) \end{cases} \tag{4}$$

In the above equation, Q, K, and V represent the query matrix, key matrix, and value matrix, respectively. X signifies the input features. i denotes the number of scales in K and V; MTA (Multiscale Token Aggregation) refers to the feature aggregation module with a downsampling rate of r_i (implemented via convolution with stride). LE represents a depth-wise separable convolution layer.

The updated method for the multi-head self-attention sub-module is as follows:

$$Attention(Q, K, V) = Softmax\left(\frac{QK^T}{\sqrt{d}} + B\right) V. \tag{5}$$

It is utilized to enhance the connections between adjacent pixels in V, thus enabling the model to effectively integrate rich local and global feature information.

3.4   Optimization

In practice, the loss value is calculated by averaging the loss across all samples in the batch. The computed loss value is subsequently utilized to update the model's weights. However, an imbalanced sample distribution, characterized by a large proportion of negative samples, can bias the model toward predicting negative samples, potentially deteriorating the network's detection performance. To address this issue, we employ weighted cross-entropy in our framework to balance the weight difference between positive and negative samples. We precompute the weights from the samples to avoid the computational burden of calculating weights for each batch. The loss function is expressed as follows:

$$Loss = -\left(\frac{1}{N}\sum_{i=1}^{N} y_i \times \log\left(p(y_i) + (1-y_i) \times \log(1-p(y_i))\right)\right), \quad (6)$$

where y_i represents the label value of the i-th pixel in the image, which takes a value of either 0 (non-shadow) or 1 (shadow). $y_i \in \{0,1\}$, indicates the i-th shadow pixel, and $p(y_i)$ is the probability that the i-th pixel is predicted to be a shadow pixel.

4   Experiments

4.1   Dataset and Evaluation Metrics

The network proposed in this paper is evaluated for shadow detection performance on the AISD dataset, which contains 514 remote sensing images with resolutions ranging from 256×256 to 1688×1688 pixels. To enhance the model's generalization ability and enrich the training data, we use data augmentation to process the dataset images (shifting, scaling, rotation, flipping, and sliding window cropping). After this step, we get 12936 images with the same resolution of 256×256, in which 80% are used for training, 10% are used for validation and 10% are used for testing, respectively.

We use *accuracy, precision, recall*, $F1$ score, and BER (Balance Error Rate) as indicators to evaluate the model's performance. The true positive (TP) sample refers to the cumulative count of correctly recognized shadow pixels. The false negative (FN) represents the cumulative count of shadow pixels mistakenly identified as non-shadow pixels. The false positive (FP) indicates the number of non-shadow pixels incorrectly classified as shadow pixels. The true negative (TN) is the accurate identification of non-shadow pixels. BER is a commonly used metric for shadow detection evaluation. The definitions of *accuracy*, *precision*, *recall*, *F*1 score, and *BER* are as follows.

$$Accuracy = (TP + TN)/(TP + TN + FP + FN) \tag{7}$$

$$Precision = TP/(TP + FP) \tag{8}$$

$$Recall = TP/(TP + FN) \tag{9}$$

$$F1 = (2 \times Precision \times Recall) / (Precision + Recall) \tag{10}$$

$$BER = 1 - 0.5 \times (Precision + Recall) \tag{11}$$

Table 1. Quantitative comparison(%) of shadow detection on AISD.

Methods	Accuracy	Precision	Recall	F1	BER
FCN	95.719	95.416	84.487	89.620	10.049
SegNet	94.536	91.774	82.891	87.107	12.668
U-Net	95.058	**95.971**	80.478	87.544	11.776
Swin-T	96.405	90.975	91.903	91.437	8.561
Shunted-T	96.632	93.351	90.832	90.074	7.909
CADNet	96.936	94.037	91.239	92.617	7.362
MSASDNet	96.996	93.455	92.303	92.875	7.121
Ours	**97.112**	92.972	**93.121**	**93.046**	**6.954**

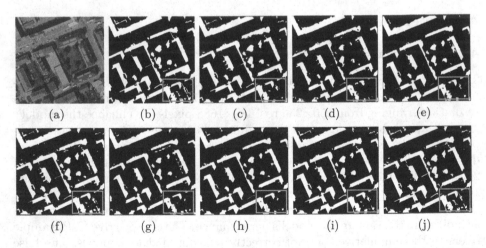

Fig. 5. Comparison of shadow detection results in AISD dataset (a) Input image (b) Ground truth (c) FCN (d) SegNet (e) U-Net (f) Swin Transformer (g) Shunted Transformer (h) CADNet (i) MSASDNet (j) Ours.

4.2 Comparisons

We compare our method with several semantic segmentation networks (See Table 1), including deep learning baselines, such as FCN [18], SegNet [20], and U-Net [19]. Besides, two Transformer networks (Swin Transformer [27] and Shunted Transformer [26]) and two popular shadow detection methods (CADNet [28] and MSASDNet [23]) are also compared.

We use both quantitative and qualitative evaluations to compare our method with other foundational methods. For the quantitative evaluation, given our use of the augmented AISD dataset, we adhere to the standard evaluation method and report the *BER*. We report the values of the *accuracy, precision, recall, F*1 score, and *BER* in Table 1, demonstrating that our method generally outperforms the other seven methods on the AISD dataset. Especially, we have the lowest *BER* (i.e. the best detection result). Our qualitative comparison experiments are depicted in Fig. 5. It is evident that all deep learning models perform well, accurately detecting shadows when the image is simple and the shadow features are distinct. However, when examining the details, our method surpasses others in detecting small shadows and preserving image smoothness. The predicted shadows may appear slightly larger than the actual shadows, leading to a decrease in Precision. Specifically, in Fig. 5(a), for the small rectangular shadow in the lower right, our method's result closely aligns with the ground truth. In total, our method exhibits fewer omissions and achieves more complete shadow detection.

4.3 Ablation Experiments

In order to further evaluate the effectiveness of each module in our STAISD-Net, we conduct ablation experiments. As shown in Table 2, we compare the performance of our network after keeping different modules. The evaluation metrics in Table 2 demonstrate the superior performance of our proposed method with all the modules, while the red boxes in Fig. 6 indicate that the combination of our modules exhibits better detection results in small shadows.

Table 2. Ablation study of our methods (%).

Ablations	Accuracy	Precision	Recall	F1	BER
Shunted Transformer	96.632	**93.350**	90.831	92.073	7.910
Shunted Transformer+Involution	96.902	93.161	91.677	92.413	7.581
Involution+MAI	94.936	91.669	90.223	90.940	9.054
Ours	**97.112**	92.981	**93.121**	**93.051**	**6.949**

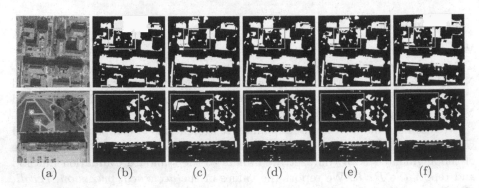

Fig. 6. Ablation experiment results: (a) Input image (b) GroundTruth (c) Shunted Transformer (d) Shunted Transformer+Involution (e) Involution+MAI (f) Ours.

5   Conclusion

We propose an end-to-end shadow detection network for remote sensing images, integrating CNN and Transformer. Our STAISD-Net network successfully mitigates the issue of false positives, achieving significant improvements in both accuracy and efficiency. Through experiments on the AISD dataset, we have validated the effectiveness of our method in detecting shadows of small objects, demonstrating superior detection accuracy. Future work will focus on designing a more lightweight version of the Transformer and CNN fusion network to further enhance both efficiency and accuracy.

Acknowledgment. This work was supported by the Natural Science Foundation of Sichuan, China (No. 2023NSFSC1393, No. 2023NSFSC0504) and the Scientific Research Starting Project of SWPU (No. 2021QHZ001).

References

1. Azevedo, S., Silva, E., Pedrosa, M.: Shadow detection improvement using spectral indices and morphological operators in urban areas in high resolution images. In: International Archives of the Photogrammetry, Remote Sensing and Spatial Information Sciences-ISPRS Archives, pp. 587–592 (2015)
2. Li, H., Zhang, L., Shen, H.: An adaptive nonlocal regularized shadow removal method for aerial remote sensing images. IEEE Trans. Geosci. Remote Sens. **52**(1), 106–120 (2013)
3. Mo, N., Zhu, R., Yan, L., Zhao, Z.: Deshadowing of urban airborne imagery based on object-oriented automatic shadow detection and regional matching compensation. IEEE J. Sel. Topics Appl. Earth Obs. Remote Sens. **11**(2), 585–605 (2018)
4. Luo, S., Shen, H., Li, H., Chen, Y.: Shadow removal based on separated illumination correction for urban aerial remote sensing images. Signal Process. **165**, 197–208 (2019)
5. Ok, A.O.: Automated detection of buildings from single VHR multispectral images using shadow information and graph cuts. ISPRS J. Photogramm. Remote Sens. **86**, 21–40 (2013)

6. Okabe, T., Sato, I., Sato, Y.: Attached shadow coding: estimating surface normals from shadows under unknown reflectance and lighting conditions. In: 2009 IEEE 12th International Conference on Computer Vision, pp. 1693–1700. IEEE (2009)

7. Tolt, G., Shimoni, M., Ahlberg, J.: A shadow detection method for remote sensing images using VHR hyperspectral and lidar data. In: 2011 IEEE International Geoscience and Remote Sensing Symposium, pp. 4423–4426. IEEE (2011)

8. Zheng, S., et al.: Rethinking semantic segmentation from a sequence-to-sequence perspective with transformers. In: Proceedings of the IEEE/CVF Conference on Computer Vision and Pattern Recognition, pp. 6881–6890 (2021)

9. Alvarado-Robles, G., Osornio-Rios, R.A., Solis-Munoz, F.J., Morales-Hernandez, L.A.: An approach for shadow detection in aerial images based on multi-channel statistics. IEEE Access 9, 34240–34250 (2021)

10. Salvador, E., Cavallaro, A., Ebrahimi, T.: Cast shadow segmentation using invariant color features. Comput. Vis. Image Underst. 95(2), 238–259 (2004)

11. Hsieh, J.W., Hu, W.F., Chang, C.J., Chen, Y.S.: Shadow elimination for effective moving object detection by gaussian shadow modeling. Image Vis. Comput. 21(6), 505–516 (2003)

12. Silva, G.F., Carneiro, G.B., Doth, R., Amaral, L.A., de Azevedo, D.F.: Near real-time shadow detection and removal in aerial motion imagery application. ISPRS J. Photogramm. Remote. Sens. 140, 104–121 (2018)

13. Ma, H., Qin, Q., Shen, X.: Shadow segmentation and compensation in high resolution satellite images. In: IGARSS 2008-2008 IEEE International Geoscience and Remote Sensing Symposium, vol. 2, pp. II–1036. IEEE (2008)

14. Zhang, R., Sun, D., Li, S., Yu, Y.: A stepwise cloud shadow detection approach combining geometry determination and SVM classification for MODIS data. Int. J. Remote Sens. 34(1), 211–226 (2013)

15. Zhu, L., et al.: Bidirectional feature pyramid network with recurrent attention residual modules for shadow detection. In: Proceedings of the European Conference on Computer Vision (ECCV), pp. 121–136 (2018)

16. Hu, X., Fu, C.W., Zhu, L., Qin, J., Heng, P.A.: Direction-aware spatial context features for shadow detection and removal. IEEE Trans. Pattern Anal. Mach. Intell. 42(11), 2795–2808 (2019)

17. Le, H., Vicente, T.F.Y., Nguyen, V., Hoai, M., Samaras, D.: A + D net: training a shadow detector with adversarial shadow attenuation. In: Proceedings of the European Conference on Computer Vision (ECCV), pp. 662–678 (2018)

18. Long, J., Shelhamer, E., Darrell, T.: Fully convolutional networks for semantic segmentation. In: Proceedings of the IEEE Conference on Computer Vision and Pattern Recognition, pp. 3431–3440 (2015)

19. Ronneberger, O., Fischer, P., Brox, T.: U-Net: convolutional networks for biomedical image segmentation. In: Navab, N., Hornegger, J., Wells, W.M., Frangi, A.F. (eds.) MICCAI 2015. LNCS, vol. 9351, pp. 234–241. Springer, Cham (2015). https://doi.org/10.1007/978-3-319-24574-4_28

20. Badrinarayanan, V., Kendall, A., Cipolla, R.: SegNet: a deep convolutional encoder-decoder architecture for image segmentation. IEEE Trans. Pattern Anal. Mach. Intell. 39(12), 2481–2495 (2017)

21. Jin, Y., Xu, W., Hu, Z., Jia, H., Luo, X., Shao, D.: GSCA-UNet: towards automatic shadow detection in urban aerial imagery with global-spatial-context attention module. Remote Sens. 12(17), 2864 (2020)

22. Luo, S., Li, H., Shen, H.: Deeply supervised convolutional neural network for shadow detection based on a novel aerial shadow imagery dataset. ISPRS J. Photogramm. Remote. Sens. 167, 443–457 (2020)

23. Liu, D., Zhang, J., Wu, Y., Zhang, Y.: A shadow detection algorithm based on multiscale spatial attention mechanism for aerial remote sensing images. IEEE Geosci. Remote Sens. Lett. **19**, 1–5 (2021)

24. Ufuktepe, D.K., Collins, J., Ufuktepe, E., Fraser, J., Krock, T., Palaniappan, K.: Learning-based shadow detection in aerial imagery using automatic training supervision from 3D point clouds. In: Proceedings of the IEEE/CVF International Conference on Computer Vision, pp. 3926–3935 (2021)

25. Zhang, Y., Zhang, L., Ding, Z., Gao, L., Xiao, M., Liao, W.H.: A multiscale topological design method of geometrically asymmetric porous sandwich structures for minimizing dynamic compliance. Mater. Des. **214**, 110404 (2022)

26. Ren, S., Zhou, D., He, S., Feng, J., Wang, X.: Shunted self-attention via multi-scale token aggregation. In: Proceedings of the IEEE/CVF Conference on Computer Vision and Pattern Recognition, pp. 10853–10862 (2022)

27. Liu, Z., et al.: Swin transformer: hierarchical vision transformer using shifted windows. In: Proceedings of the IEEE/CVF International Conference on Computer Vision, pp. 10012–10022 (2021)

28. Yang, Y., Guo, M., Zhu, Q.: Cadnet: top-down contextual saliency detection network for high spatial resolution remote sensing image shadow detection, pp. 4075–4078 (2021)

Few-Shot Infrared Image Classification with Partial Concept Feature

Jinyu Tan[1] , Ruiheng Zhang[1]([✉]) , Qi Zhang[2], Zhe Cao[1], and Lixin Xu[1]

[1] Beijing Institute of Technology, Beijing 100081, China
{ruiheng.zhang,3120220172,lxxu}@bit.edu.cn
[2] Technical University of Munich, 80333 Munchen, Germany
rachelqi.zhang@tum.de

Abstract. Few infrared image samples will bring a catastrophic blow to the recognition performance of the model. Existing few-shot learning methods most utilize the global features of object to classify infrared image. However, their inability to sufficiently extract the most representative feature for classification results in a degradation of recognition performance. To tackle the aforementioned shortcomings, we propose a few-shot infrared image classification method based on the partial conceptual features of the object. It enables the flexible selection of local features from targets. With the integration of these partial features into the concept feature space, the method utilizes Euclidean distance for similarity measurement to accomplish infrared target classification. The experimental results demonstrate that our proposed method outperforms previous approaches on a new infrared few-shot recognition dataset. It effectively mitigates the adverse effects caused by background blurring in infrared images and significantly improving classification accuracy.

Keywords: Deep learning · Infrared image · Partial feature · Few-shot classification

1 Introduction

In recent years, infrared target recognition has gained increasing attention due to its affordability, high concealment, and ability to penetrate through smoke and other obstacles [1]. With the vigorous development of deep learning techniques, infrared target recognition based on deep learning has gradually become the mainstream in research. These methods employ end-to-end feature learning to analyze deep nonlinear features at multiple levels in the data, demonstrating strong generalization ability and robustness. However, the effectiveness of deep learning largely relies on the availability of massive annotated training data [2]. Acquiring a large-scale infrared image dataset in complex environments incurs high costs, and the size of collected samples is often insufficient to meet the training requirements of the model.

© The Author(s), under exclusive license to Springer Nature Singapore Pte Ltd. 2024
Q. Liu et al. (Eds.): PRCV 2023, LNCS 14428, pp. 343–354, 2024.
https://doi.org/10.1007/978-981-99-8462-6_28

Motivated by the remarkable rapid learning ability observed in humans, the concept of few-shot learning has emerged as a solution to address the afore-mentioned challenges [3–5]. Unlike traditional deep learning methods, few-shot learning enables efficient learning and rapid generalization to new tasks, even with a limited number of samples. Consequently, it has attracted significant research attention in recent years, leading to the proposal of numerous innova-tive solutions. Among these approaches, metric learning stands out as a promi-nent methodology. The fundamental idea behind metric-based few-shot learning methods is to acquire an embedding function that maps the input space to a dis-tinct embedding space, and then leverage the similarity metric for classification within this embedding space. Distinguished methods in this domain encompass prototype networks [6], matching networks [7], and relation networks [8], all of which have demonstrated promising outcomes.

However, these methods primarily rely on global features at the image level for feature embedding, disregarding the crucial utilization of local features for classifying challenging targets. Due to the inherent redundancy in global fea-tures, placing excessive emphasis on them introduces confounding contextual information, ultimately misleading the model and resulting in severe degrada-tion of its recognition performance. In addition to this, the global-based approach results in insignificant label correlations in the case of few samples. In human perception, the parts constitute the whole, where an object consists of multiple parts. Each part is a reusable and learnable concept, and combining these con-cepts determines an object. Given that certain objects are only deterministically distinct from each other in terms of parts with the same semantics, focusing on distinguishing between these concepts contributes to improved recognition performance. It is worth noting that the partial feature attributes possessed by the concepts as well reveal the potential relevance of few label classification.

To this end, stimulated by the experience that human beings perceive objects with multiple concept patterns [9], we propose a Partial Representation-based Concept Network (PRCNet) for few-shot classification in infrared images. Specif-ically, we first extract the concept prototype of the object with the learner, so as to flexibly select the most representative local feature information and eliminate the adverse effects brought by the messy background information of infrared images, then adopt the Euclidean distance as the similarity measure between samples, hence accomplish the classification of the infrared object. The derived approach offers a robust model to recognize distinctive features within a given infrared dataset and increase holistic performance. Upon closer examination, the reasons for this can be attributed to two factors: (a) Partial features are distin-guishable between categories, which leads to more accurate classification ability of the model. (b) Partial features provide condensed and reduced the information compared to global features, making it easier for the model to learn.

In this paper, our main contributions can be summarized as the following three points:

1. We propose a new method for few-shot infrared image classification, which fully leverages the partial feature of the object and resists the disadvantage of messy background.
2. We construct a new infrared image dataset for few-shot learning, which has 17 categories with 50 images each.
3. Sufficient experiments have been conducted on the infrared dataset to demonstrate the performance of the proposed method. Experiments show that it has the advantage in the few-shot infrared image classification.

2   Related Work

2.1   Few-Shot Learning

Few-shot learning has received a great deal of attention in the past few years, with most of the mainstream approaches being based on meta-learning. Meta-learning-based approaches aim to learn a good initialization of the model with a meta-learner using the learn-to-learn paradigm [10,11]. The kernel ideology is that the model acquires meta-knowledge by performing training on a large number of similar tasks, and learns and applies it quickly on the new tasks. This is helpful for the model to adjust parameters and converge quickly under the condition of a few samples. Some representative methods are referred to [12–15]. MAML [12] and its variants [16–18] are designed to acquire favorable model initialization such that the meta-learner allows for rapid adaptation to new tasks. Among the many few-shot learning algorithms based on meta-learning, metric-based methods are widely used. The basic idea of the metric-based few-shot learning method is to learn a feature embedding space, so that the model in this feature embedding space is more effective in measuring the similarity between samples, with higher similarity between samples of the same category and lower similarity between samples of different categories. The common metric-based methods can be divided into three main types: matching network [7] and their variants [19,20], prototype network [6] and their variants [9,21,22], relation network [8] and their variants [23–26].

2.2   Partial-Based Methods

Several few-shot learning methods [9,23,24,27–30] are beginning to use local features for image recognition and classification. DN4 [23] learns image class metrics through the measurement of the cosine similarity between the depth part descriptor of each query instance supporting the class and its neighbors. DeepEMD [27] acquires the global distance by solving the optimization problem of the earth movement distance with many-to-many matching between partial patches. ConstellationNet [30] extends CNNs with a constellation model that clusters and encodes the cell features in a dense partial representation.

3 Problem Definition

Contrasting with traditional image identification, few-shot learning intends to train an effective model to classify categories that have not been seen before in a way that exploits a handful of annotated samples, which implies that the sample label spaces are mutually exclusive for testing and training.

In this paper, a standard few-shot learning setting, the episodic training mechanism [7], is applied to train the model. Supposing the distribution of training and testing tasks is consistent, a series of tasks are randomly selected from the task distribution of the training set, each of which includes a support set and a query set. In this case, the support set separately samples K labelled samples from N various classes, and the query set is composed of q samples per class. Such is the N-way K-shot classification problem in few-shot learning. In the training stage, learning samples of the support set S thereby allows the classification of unknown samples of the query set Q. In a similar way, the task is arbitrarily drawn out of the testing set and subjected to model testing.

4 Method

The model framework proposed in this paper is shown in Fig. 1, which is mainly composed of two modules: feature extraction and similarity measurement. The method enhances the generalization ability of the model via studying along the human interpretable concept dimension. The way to think about it is, assuming that each class of objects is composed of several specific concepts, i.e., the object is regarded as consisting of some partial features, such as the wheels, seat and head of a bicycle.

The concept learner defines concept prototypes along each concept dimension which capture the class-level differences in the metric space of the underlying concepts. The PRCNet leverages a concept embedding function parameterized by a neural network, named concept learner, to learn a unique concept embedding for each concept dimension. Then, the concept prototypes are compared with the concept embeddings to assign concept significance scores along each concept dimension. Finally, the concept learner efficiently integrates information from different concept learners and concept prototypes across concept dimensions to complete the classification task.

4.1 Concept Feature Extraction

Supposing that the samples of each category contain a collection of M concepts as prior knowledge, namely $C = \{\mathbf{c}^j\}_{j=1}^M$, where each concept denoted as $\mathbf{c}^j \in (0,1)^D$ is a vector consisting of 0 and 1, and D stands for the dimensionality of the output. In the individual training task, images are partitioned into the support set $S = \{(\mathbf{x}_i, y_i)\}_{i=1}^{N \times K}$ and the query set $Q = \{(\mathbf{x}_j, y_j)\}_{j=1}^q$, where $\mathbf{x}_i$ and $\mathbf{x}_j$ indicate the samples, and $\mathbf{y}_i$ and $\mathbf{y}_j$ are the corresponding labels. Dividing the sample space into multiple subspaces of predefined concepts and learning an

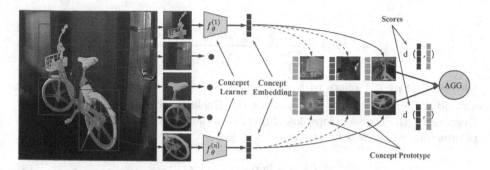

Fig. 1. Architecture of the proposed method. Firstly, the PRCNet leverages the neural network parameterized embedding function to learn the concept embeddings of the samples, then verifies them with the existing concept prototypes to obtain the similarity scores of the concepts in each category, and finally integrates them effectively to achieve the classification.

independent embedding function $f_\theta^{(j)}$ for each concept. For each category n in the support set, each concept learner learns the concept prototype $P_n^{(j)}$ of its corresponding concept, which is calculated as follows:

$$P_n^{(j)} = \frac{1}{|S_n|} \sum_{(\mathbf{x}_i, y_i) \in S_n} f_\theta^{(j)}(\mathbf{x}_i \circ \mathbf{c}^j) \tag{1}$$

where S_n denotes the subset consisting of samples belonging to category n in the support set S, $\circ$ means Hadamard product, and $\mathbf{x}_i$ and $\mathbf{c}^j$ Multiply by bit. Each category n can be represented by the union of M concept prototypes, that is, $\{P_n^{(j)}\}_{j=1}^M$. Given a query sample $\mathbf{x}_q$, corresponding concept embeddings are gained by exploiting the concepts of the query set samples:

$$P_q^{(j)} = f_\theta^{(j)}(\mathbf{x}_q \circ \mathbf{c}^j) \tag{2}$$

4.2   Similarity Metric

Mapping the samples to the concept feature space with embedding function, the concept prototypes of the support set samples and the concept embeddings of the query set samples are acquired. To obtain an efficient similarity measure between samples in the concept feature space, we adopt the Euclidean distance to measure the distance between the concept embedding of the query set samples and the concept prototype of each category and aggregate all the conceptual contributions by summing the distances between the concept embedding and the concept prototype. Finally, the model predicts the probability that the query set sample $\mathbf{x}_q$ belongs to the nth category by converting it to the form of probability through the softmax function.

$$\mathbf{p}_\theta(y = n \mid \mathbf{x}_q) = \frac{exp(-\sum_j d(f_\theta^{(j)}(\mathbf{x}_q \circ \mathbf{c}^j), P_n^{(j)})}{\sum_{n'} exp(-\sum_j d(f_\theta^{(j)}(\mathbf{x}_q \circ \mathbf{c}^j), P_{n'}^{(j)})} \tag{3}$$

where $f_\theta^{(j)}(\mathbf{x}_q \circ \mathbf{c}^j)$ refers to the concept embedding of the query set sample $\mathbf{x}_q$ and $d(f_\theta^{(j)}(\mathbf{x}_q \circ \mathbf{c}^j), P_n^{(j)})$ represents the Euclidean distance metric function. Eventually, the cross-entropy loss function is deployed for updating the model parameters, whose loss function is:

$$L_\theta = -\lg \mathbf{p}_\theta(y = n \mid \mathbf{x}_q) \tag{4}$$

5   Experiment

5.1   Infrared Dataset

The infrared image dataset employed for the experiments in this paper has a total of 17 categories, each containing 50 infrared images. Some of the data are obtained from publicly available thermal image datasets, and others are manually captured (Fig. 2).

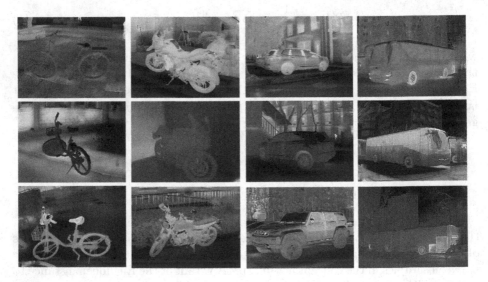

Fig. 2. Display of some images in infrared dataset. Shown here are various infrared images of vehicles taken on the ground.

5.2   Experiment Settings

Randomly selected 10 categories of 17 infrared datasets with a total of 500 infrared images as the training set and the remaining 7 categories including 350 infrared images as the testing set, while all the training tasks are carried out with the 5-way K-shot (where K = 1 or K = 5) strategy. Specifically, after randomly selecting 5 categories from the training dataset, K labelled samples from each category are selected as the support set, 16 of the remaining samples are arbitrarily picked as the query set, and then the category labels of the query set samples are inferred based on the support set. All experiments utilize the Adam optimization algorithm [31] to update the model parameters with an initial learning rate of 0.001. Furthermore, we applied various backbone networks for feature extraction, which are specifically Conv4, Conv6, ResNet10, and ResNet18. During the experiments, the size of all input images is standardized to 84 × 84 pixels, as well as image dithering, regularization and other operations are implemented to achieve sample enhancement.

For the 5-way 1-shot training task, with epoch set to 150 and batch size set to 85, we randomly select 100 meta-tasks from the query set once every 100 training cycles to verify the classification accuracy of the model for sample data. As for the 5-way 5-shot training task, with the epoch of 100 and batch size of 105, we evaluate the correct classification rate of the model for small sample data by selecting 100 meta-tasks randomly from the query set every 100 training cycles. Suppose the sum of testing samples correctly predicted in 100 meta-tasks is N_{right}, then the accuracy rate is calculated as:

$$acc = \frac{N_{right}}{100 \times 80} \tag{5}$$

5.3   Ablation Study

With this IR dataset, we verified the effectiveness of each module on PRCNet. For the purpose of equitable comparison, we consistently implemented the standard ResNet18 as the backbone, employed two few-shot experimental setups of 5-way 1-shot and 5-way 5-shot, and conducted an ablation study for both modules of concept feature extraction and similarity metric. Table 1 shows the results of our ablation study on the infrared dataset. According to the table, for 1-shot, it can be seen that with only the partial concept feature extraction module, the average accuracy of the experiment is 58.42%. If there is only the similarity metric module, the average accuracy of the experiment decreases to 53.74%. The accuracy, when both modules are present, which means that PRCNet is used, however, increases to 75.74%. The trend of accuracy change for the 5-shot experiments is similar to that of the 1-shot. Owing to the fact that the conceptual component of the feature extraction module resists the spurious contextual message in the infrared images, more robust and representative partial characteristics are derived. Alternatively, it contributes to the classification by the similarity measure of Euclidean distance that integrates these conceptual characteristics in the embedding space comparatively.

Other than that, for more visual reflection of the effectiveness of the two modules manifested in the ablation experiment, we visualized the data with the T-SNE, as shown in Fig. 3. There are two few-shot settings, 5-way 1-shot and 5-way 5-shot, displayed in the figure, and from the left to the right are the visualized distributions of the data after the backbone network, partial concept feature extraction, and similarity metric, respectively. It is observed that with the installation of modules, the distribution of similar data is more convergent.

Table 1. Ablation study results on the infrared dataset. The impact of PRC-Net's concept feature extraction module and metric module. ResNet18 was selected as the backbone of the experiment, with a confidence interval of 95%.

Concept Feature	Similarity Metric	Accuracy(%)	
		5-way 1-shot	5-way 5-shot
√		58.42 ± 1.15	68.91 ± 0.96
	√	53.74 ± 1.09	61.46 ± 0.63
√	√	**75.74±1.04**	**78.60±0.38**

5.4   Analysis of Classification Effect

As the proposed PRCNet belongs to the metric-based few-shot learning method, we conduct comparison experiments with ProtoNet [6], MatchingNet [7] and RelationNet [8] on the infrared dataset to validate the effectiveness of this method for target recognition of infrared images. Under the four backbone networks and two few-shot task settings, the experimental results are shown in Table 2.

The PRCNet significantly outperforms the other three typical few-shot learning methods in both the 5-way 5-shot and 5-way 1-shot tasks. The reason is that PRCNet can flexibly introduce the most reliable partial feature attributes into the structured conceptual space of the image. As compared with the other three methodologies for global image mapping, it can remove the adverse influence of cluttered contextual information by ignoring the complex background environment of infrared images.

When it comes to the selection of feature extraction modules, Conv4 performs comparatively well in the 5-way 5-shot task for the prototype network, matching network and relational network, while it shows erratic improvement in the 5-way 1-shot task. For PRCNet, the execution of different backbones is stable in both task settings, with ResNet18 performing optimally, decreasing only 3% in the 5-way 1-shot task compared to the 5-way 5-shot accuracy. Through analysis, we suppose this may be due to the fact that the few-shot learning method for feature mapping against global images can affect the generalization ability of the model if it adopts too deep a structure, while the conceptual learner method based on partial feature representation has considerably mitigation of this impact owing to the higher stability of feature mapping targeting partial features.

Table 2. Comparison to prior work. A quantitative comparativeness of PRCNet with three classical metric-based few-shot learning methods. The average classification accuracy (%) on the testing set with the confidence interval of 95%

Method	Backbone	Accuracy(%)	
		5-way 1-shot	5-way 5-shot
MatchingNet	Conv4	56.55±1.85	74.33±1.23
	Conv6	58.88±1.60	70.98±1.29
	ResNet10	61.25±1.61	70.77±1.27
	ResNet18	58.10±1.65	70.44±1.23
ProtoNet	Conv4	62.32±1.67	74.08±1.09
	Conv6	59.33±1.73	73.45±1.15
	ResNet10	60.62±1.73	75.08±1.09
	ResNet18	57.39±1.65	74.98±1.14
RelationNet	Conv4	53.98±1.66	71.63±1.29
	Conv6	59.58±1.96	70.97±1.22
	ResNet10	56.94±1.73	67.90±1.19
	ResNet18	53.38±1.82	66.02±1.34
Ours	Conv4	**70.04±1.61**	**76.08±0.63**
	Conv6	**70.01±1.24**	**76.86±0.55**
	ResNet10	**71.76±1.23**	**78.26±0.42**
	ResNet18	**75.74±1.04**	**78.60±0.38**

Table 3. Comparison about inference time. Time-consumption tests are performed on the trained model on the testing set with the same requirement and count the computational speed of the model to finish a round of inference.

Method	Backbone	Time(ms)	
		5-way 1-shot	5-way 5-shot
MatchingNet	Conv4	0.33	0.58
	Conv6	0.51	1.81
	ResNet10	0.38	0.67
	ResNet18	0.29	0.69
ProtoNet	Conv4	0.03	0.04
	Conv6	0.05	0.01
	ResNet10	0.04	0.05
	ResNet18	0.04	0.04
RelationNet	Conv4	1.84	2.11
	Conv6	1.84	1.91
	ResNet10	4.00	2.86
	ResNet18	3.10	2.83
Ours	Conv4	10.01	12.91
	Conv6	9.89	15.43
	ResNet10	12.43	14.35
	ResNet18	13.4	18.60

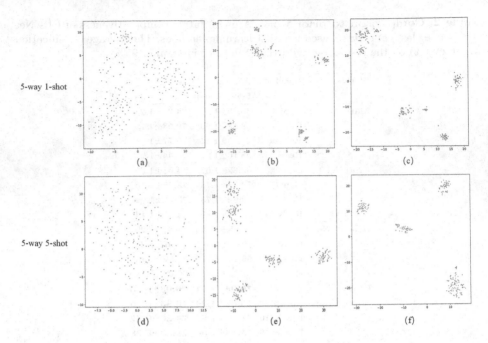

Fig. 3. The T-SNE visualization of ablation experiments. From the query set, 50 samples are selected for each category, and the distribution of the data in the feature embedding is gradually evident by adding modules. It is worth noting that the backbone network serves as ResNet18.

5.5   Real-Time Analysis

In this section, we systematically evaluate the real-time performance of the proposed model on the infrared dataset and compare it with several exemplary few-shot learning classification approaches. To be more specific, with the same hardware conditions, we conduct time-consumption tests of the trained model on the testing set and make statistics on the inference rates of the model in completing a round of tasks, with the results shown in Table 3.

In contrast to the remaining three typical metric-based few-shot learning classification methods, PRCNet consumes the longest time, taking about 15ms to accomplish the 5-way 5-shot task and about 10ms to finish the 5-way 1-shot task. Therefore, the inference process with high accuracy sacrifices the responsiveness time, so it needs to be chosen reasonably according to the actual requirements during the application.

6   Conclusion

In this paper, we propose a partial representation-based concept network approach for infrared image target recognition with few-shot learning. The PRCNet makes use of concept learners to obtain concept prototypes of the support set and concept embeddings of the query set, utilizes Euclidean distance to calculate the similarity of query samples to each category of samples in the support set, and attains the probability that samples belong to a certain category. The experimental results show that the network proposed in this paper performs superiorly, but is time-consuming. For the following study, the model will be modified, such as infrared and visible fusion and data augmentation, in order to improve the inference speed as well as the generalization ability of the model while maintaining or promoting accuracy in the conditions of sample scarcity.

References

1. Jiangrong, X., Fanming, L., Hong, W.: An enhanced single-transmission multi-frame detector method for airborne infrared target detection. Acta Optica Sinica **39**(6), 0615001 (2019)
2. Antonie, M.L., Zaiane, O.R., Coman, A.: Application of data mining techniques for medical image classification. In: Proceedings of the Second International Conference on Multimedia Data Mining, pp. 94–101 (2001)
3. Jankowski, N., Duch, W., Grąbczewski, K.: Meta-Learning in Computational Intelligence, vol. 358. Springer, Heidelberg (2011). https://doi.org/10.1007/978-3-642-20980-2
4. Lake, B.M., Salakhutdinov, R.R., Tenenbaum, J.: One-shot learning by inverting a compositional causal process. Adv. Neural Inf. Process. Syst. **26**, 1–9 (2013)
5. Phaphuangwittayakul, A., Guo, Y., Ying, F.: Fast adaptive meta-learning for few-shot image generation. IEEE Trans. Multimedia **24**, 2205–2217 (2021)
6. Snell, J., Swersky, K., Zemel, R.: Prototypical networks for few-shot learning. Adv. Neural Inf. Process. Syst. **30**, 1–11 (2017)
7. Vinyals, O., Blundell, C., Lillicrap, T., Wierstra, D., et al.: Matching networks for one shot learning. Adv. Neural Inf. Process. Syst. **29** (2016)
8. Sung, F., Yang, Y., Zhang, L., Xiang, T., Torr, P.H., Hospedales, T.M.: Learning to compare: relation network for few-shot learning. In: Proceedings of the IEEE Conference on Computer Vision and Pattern Recognition, pp. 1199–1208 (2018)
9. Cao, K., Brbic, M., Leskovec, J.: Concept learners for few-shot learning. arXiv preprint arXiv:2007.07375 (2020)
10. Thrun, S.: Lifelong learning algorithms. In: Learning to Learn, pp. 181–209. Springer, Heidelberg (1998). https://doi.org/10.1007/978-1-4615-5529-2_8
11. Vilalta, R., Drissi, Y.: A perspective view and survey of meta-learning. Artif. Intell. Rev. **18**, 77–95 (2002)
12. Finn, C., Abbeel, P., Levine, S.: Model-agnostic meta-learning for fast adaptation of deep networks. In: International Conference on Machine Learning, pp. 1126–1135. PMLR (2017)
13. Gidaris, S., Komodakis, N.: Dynamic few-shot visual learning without forgetting. In: Proceedings of the IEEE Conference on Computer Vision and Pattern Recognition, pp. 4367–4375 (2018)
14. Ravi, S., Larochelle, H.: Optimization as a model for few-shot learning. In: International Conference on Learning Representations (2017)
15. Santoro, A., Bartunov, S., Botvinick, M., Wierstra, D., Lillicrap, T.: Meta-learning with memory-augmented neural networks. In: International Conference on Machine Learning, pp. 1842–1850. PMLR (2016)

16. Antoniou, A., Edwards, H., Storkey, A.: How to train your maml. arXiv preprint arXiv:1810.09502 (2018)
17. Jamal, M.A., Qi, G.J.: Task agnostic meta-learning for few-shot learning. In: Proceedings of the IEEE/CVF Conference on Computer Vision and Pattern Recognition, pp. 11719–11727 (2019)
18. Sun, Q., Liu, Y., Chua, T.S., Schiele, B.: Meta-transfer learning for few-shot learning. In: Proceedings of the IEEE/CVF Conference on Computer Vision and Pattern Recognition, pp. 403–412 (2019)
19. Cai, Q., Pan, Y., Yao, T., Yan, C., Mei, T.: Memory matching networks for one-shot image recognition. In: Proceedings of the IEEE Conference on Computer Vision and Pattern Recognition, pp. 4080–4088 (2018)
20. Prol, H., Dumoulin, V., Herranz, L.: Cross-modulation networks for few-shot learning. arXiv preprint arXiv:1812.00273 (2018)
21. Oreshkin, B., Rodríguez López, P., Lacoste, A.: Tadam: task dependent adaptive metric for improved few-shot learning. Adv. Neural Inf. Process. Syst. **31**, 1–11 (2018)
22. Xing, C., Rostamzadeh, N., Oreshkin, B., O Pinheiro, P.O.: Adaptive cross-modal few-shot learning. Adv. Neural Inf. Process. Syst. **32**, 1–11 (2019)
23. Li, W., Wang, L., Xu, J., Huo, J., Gao, Y., Luo, J.: Revisiting local descriptor based image-to-class measure for few-shot learning. In: Proceedings of the IEEE/CVF Conference on Computer Vision and Pattern Recognition, pp. 7260–7268 (2019)
24. Li, W., Xu, J., Huo, J., Wang, L., Gao, Y., Luo, J.: Distribution consistency based covariance metric networks for few-shot learning. In: Proceedings of the AAAI Conference on Artificial Intelligence, vol. 33, pp. 8642–8649 (2019)
25. Hui, B., Zhu, P., Hu, Q., Wang, Q.: Self-attention relation network for few-shot learning. In: 2019 IEEE International Conference on Multimedia & Expo Workshops (ICMEW), pp. 198–203. IEEE (2019)
26. Wu, Z., Li, Y., Guo, L., Jia, K.: Parn: position-aware relation networks for few-shot learning. In: Proceedings of the IEEE/CVF International Conference on Computer Vision, pp. 6659–6667 (2019)
27. Zhang, C., Cai, Y., Lin, G., Shen, C.: Deepemd: few-shot image classification with differentiable earth mover's distance and structured classifiers. In: Proceedings of the IEEE/CVF Conference on Computer Vision and Pattern Recognition, pp. 12203–12213 (2020)
28. Chen, H., Li, H., Li, Y., Chen, C.: Multi-scale adaptive task attention network for few-shot learning. In: 2022 26th International Conference on Pattern Recognition (ICPR), pp. 4765–4771. IEEE (2022)
29. Yuille, J.H.A.K.A.: Compas: representation learning with compositional part sharing for few-shot classification. arXiv preprint arXiv:2101.11878 (2021)
30. Xu, W., Xu, Y., Wang, H., Tu, Z.: Attentional constellation nets for few-shot learning. In: International Conference on Learning Representations (2021)
31. Kingma, D.P., Ba, J.: Adam: a method for stochastic optimization. arXiv preprint arXiv:1412.6980 (2014)

AGST-LSTM: The ConvLSTM Model Combines Attention and Gate Structure for Spatiotemporal Sequence Prediction Learning

Xuechang Wang[1], Hui Lv[1(✉)], and Jiawei Chen[2]

[1] Key Laboratory of Advanced Design and Intelligent Computing, Ministry of Education, School of Software Engineering, Dalian University, Dalian 116622, China
lh8481@tom.com
[2] Department of Computer Science, Old Dominion University, Norfolk, VA 23529, USA

Abstract. Spatiotemporal sequence prediction learning generates one or more frames of images by learning from multiple frames of historical input. Most current spatiotemporal sequence prediction learning methods do not adequately consider the importance of long-term features for spatial reconstruction. Based on the convolutional LSTM (ConvLSTM) network unit, this paper adds a memory storage unit that updates information through the original memory unit in the ConvLSTM unit and uses the same zigzag memory flow as the PredRNN network, which can focus on long-term and short-term spatiotemporal features at the same time. Then, an attention module is proposed to extract important information from the long-term hidden state and aggregate it with the short-term hidden state, expand the temporal feeling field of the hidden state, and propose the attention gate spatiotemporal LSTM (AGST-LSTM) model, which further enhances the model's capacity to catch the spatiotemporal correlation. This paper validates the model through two different prediction tasks. The AGST-LSTM model has competitive performance compared to the comparative model to some degree, as exhibited in experiments.

Keywords: Deep Learning · Spatiotemporal Prediction · ConvLSTM

This work is supported by 111 Project (No. D23006), the National Natural Science Foundation of China (Nos. 62272079, 61972266), the Natural Science Foundation of Liaoning Province (Nos. 2021-MS-344, 2021-KF-11-03, 2022-KF-12-14), the Scientific Research Fund of Liaoning Provincial Education Department (No. LJKZZ20220147), the Scientific Research Platform Project of Dalian University (No. 202301YB02), the State Key Laboratory of Synthetical Automation for Process Industries, the State Key Laboratory of Light Alloy Casting Technology for High-end Equipment (No. LACT-006), the Postgraduate Education Reform Project of Liaoning Province (No. LNYJG2022493), the Dalian Outstanding Young Science and Technology Talent Support Program (No. 2022RJ08).

Q. Liu et al. (Eds.): PRCV 2023, LNCS 14428, pp. 355–367, 2024.
https://doi.org/10.1007/978-981-99-8462-6_29

1   Introduction

For the past few years, deep learning has been extensively applied in a variety of complex real-world scenarios. Deep learning has been implemented for mostly kinds of prediction behaviors, including video prediction [15,16,18,19], traffic prediction [10], power prediction [17], wind speed prediction [4], stock price prediction [6], and more. As a key application direction of predictive learning, the spatiotemporal sequence prediction of multi-frame images after given continuous images has attracted more and more attention in the domain of computer vision. Capturing spatiotemporal correlation is the key to spatiotemporal sequence prediction, and how to better extract temporal and spatial dependence has been the focus of researchers in spatiotemporal sequence prediction in recent years.

Convolutional neural network(CNN) [8] extract spatial features through convolution kernels, while Recurrent Neural Network (RNN) [14,21] capture temporal dependence through their recurrent structures. These two network architectures are naturally suited for spatiotemporal sequence prediction. ConvLSTM [15] is a model that combines convolutional operations with recurrent architectures. ConvLSTM replaces the linear operation in the LSTM [5] by convolutions, so that the network model has the ability to extract spatiotemporal correlations.

The PredRNN [18,19] network adds a spatiotemporal feature memory unit to the ConvLSTM. And the PredRNN paper proposes a spatiotemporal memory flow that enables the spatiotemporal memory state information of the first layer of the current time to be obtained from the top layer of the previous time in a stacked network structure.

In the work of spatiotemporal sequence prediction, short-term characteristics play a major role in the prediction results, but the influence of long-term characteristics on the prediction results cannot be ignored. Long-term features show the main trend of change in the overall sequence change, so that the model can focus on the overall change trend. However, the PredRNN and its mostly variant models fail to grasp the importance of long-term features in image spatial reconstruction. The motivation for our research in this paper is to make the network able to capture long-term features, increase the influence of long-term features on the prediction results, and enhance the ability of the model to capture spatiotemporal correlation. Based on the ConvLSTM model unit, adding a new memory unit with the ability to capture the spatiotemporal characteristics of short-term and long-term memory, and adding the hidden state receptive field of the temporal model are the focus of this paper.

Aiming at the shortcoming of the existing spatiotemporal sequence prediction models that do not pay enough attention to long-term spatiotemporal features, this paper adds a spatiotemporal memory unit that uses the original memory unit to update information and uses a stacked network structure based on the ConvLSTM network unit so that the model has the ability to extract long-term and short-term features to enhance the ability to capture spatiotemporal correlation. In view of the insensitivity of hidden state to long-term features, this paper adds an attention module for the aggregation of long-term hidden state and short-term hidden state, and the temporal receptive field of the hidden state is expanded.

Based on the above, the new spatiotemporal prediction module, GM-LSTM, is proposed in this paper, which adds a new memory unit M that interacts with the original time memory unit C to store spatiotemporal dependence information on the basis of the ConvLSTM network unit. Then, inspired by the MAU [3] network, the attention module is introduced to augment the temporal perception domain of the hidden state and form a final spatiotemporal prediction model, AGST-LSTM. M uses the same spatiotemporal memory flow as the memory unit M in the PredRNN network in the entire AGST-LSTM network model, which enables M to effectively focus on long-term and short-term spatiotemporal features. This paper verifies the availability of the AGST-LSTM network through multiple different datasets, and experiments make it clear that the AGST-LSTM model is more competitive than the comparison model. The central contributions of this article is that:

1. A spatiotemporal memory unit that uses the original memory unit to update information is added to the ConvLSTM model unit, the GM-LSTM module is proposed.
2. An attention module is added to expand the temporal sensing field of hidden states on the basis of the GM-LSTM module, and proposes a new spatiotemporal sequence prediction model, AGST-LSTM.
3. The model proposed in this article exhibits competitive power compared with the comparative model in multiple different datasets.

The organizational structure of the remaining parts of this article is as follows: The related work is introduced in Sect. 2; a detailed introduction to the AGST-LSTM method is provided in Sect. 3; Sect. 4 presents the experimental outcomes of AGST-LSTM compared to the competition models; and Sect. 5 is the conclusion.

2   Related Work

The previous spatiotemporal sequence prediction learning methods are introduced in this section. On account of the exceptional capability of RNN in temporal modeling, RNN and its variant networks were naturally first applied to the field of video prediction. Srivastava et al. [16] use LSTM network to learn video sequences and proposed the FC-LSTM model.

For better extraction of spatiotemporal features, Shi et al. [15] propose the ConvLSTM network based on the LSTM network combined with the convolution operation, which can carry out unified modeling of time and space. Wang et al. [18,19] propose PredRNN network, which proposes a Spatiotemporal LSTM (ST-LSTM) that adds a memory state in the longitudinal data stream compared to the ConvLSTM unit, improving the ability to capture spatiotemporal correlations. Wang et al. [20] used differential works to mitigate the order of non-stationary hidden in spatiotemporal data, and a network named Memory in Memory (MIM) is proposed, which designed a non-stationary module and a stationary module, and cascaded these two modules to take the place of the forgetting gate in the ST-LSTM unit. The RST-LSTM network, is proposed by Luo et al. [12], which solves the puzzle of mismatch and inaccurate intensity prediction caused by convolution through a global reconstruction module and a local reconstruction module.

Chang et al. [3] believe that accurately predicting the motion information between frames is the key to video prediction and propose the Motion-Aware Unit (MAU). Wu et al. [22] propose MotionRNN, and this network adapts to spatiotemporal changes by capturing internal changes in motion.

A novel video prediction model, VPTR, based on Transformer, was proposed by Ye et al. [23,24]. The network proposes an effective local spatiotemporal separation attention mechanism to reduce the complexity of Transformer. The network uses the structure of Transformer for video prediction tasks, but its effect is still lacking compared with the mainstream RNN type prediction models.

The PredRNN series of networks mentioned above do not pay enough attention to long-term spatial features. The method proposed in this article uses data from the original time memory to update information on newly added spatiotemporal memory state, which can obtain long-term spatiotemporal features. Then, inspired by the MAU network, the attention mechanism is used to raise the temporal sensing field of the hidden state and increase the influence of long-term features on the prediction results.

3   Method

This article adds a memory state to the ConvLSTM unit structure to extract long-term spatiotemporal features and construct a Gate Memory LSTM (GM-LSTM) module. Please refer to Sect. 3.1 for details. Inspired by the MAU network model, this paper adds an attention module on the basis of the GM-LSTM module and proposes the AGM-LSTM network module. In the end, the AGST-LSTM network model proposed in this paper consists of stacked AGM-LSTM network units. See Sect. 3.2 for details.

Problem Definition: given a spatiotemporal sequence $X_T = \{X_t \in \mathbb{R}^{C \times H \times W}\}$, where $t = 1, 2, ..., T$, X_t is a frames with C channels, H height and W width, it is possible to predict the sequence $\hat{X}_{T'} = \{\hat{X}_{T+t'} | t' = 1, 2, ..., T'\}$.

3.1   GM-LSTM Module

The GM-LSTM structural unit is designed on the basis of the ConvLSTM network unit, and a new spatiotemporal memory unit M is added compared to the ConvLSTM network. The structure inside the red box in Fig. 1 is the block of GM-LSTM.

To conveniently prove the effectiveness of the GM-LSTM component in spatiotemporal sequence prediction, this section stacks GM-LSTM to form a GST-LSTM network model. Figure 2 shows the specific network structure of AGST-LSTM, the GST-LSTM model is formed by stacking the GM-LSTM structural elements likes AGST-LSTM structure. C_{t-1}^l is the time memory transmitted from the previous time slice t−1 to the current moment in each GM-LSTM layer, and C_t^l is the calculated time memory of the current moment and will be transmitted to the next moment, where l is the network layer. H_{t-1}^l is the hidden state that is transmitted from the previous moment to the current moment. The input of the current moment is X_t if in the first layer, and otherwise it is the hidden state H_t^{l-1} by the GM-LSTM unit in the previous layer in the current moment.

A gate structure for information update is constructed for the newly added spatiotemporal memory state M, which covers an input gate i_t' and an input modulation gate g_t'. After i_t' and g_t' perform Hadamard product on the result, add the Hadamard product result of $1-i_t'$ and the M_t^{l-1} passed from the previous network layer to get the spatiotemporal memory state M_t^l at the current time.

The state transition method of spatiotemporal memory state M_t^l adopts the same zigzag state transfer method as the PredRNN network structure, and the memory state $M_{t-1}^{l=L}$ at the highest level of the prior time slice will be transmitted to the memory state $M_t^{l=1}$ of the first layer at the current time slice, where L is the number of network stacking layers. The gradient disappearance phenomenon is efficiently reduced by the zigzag data transfer method of spatial characteristics between the layers of the network in the stacked network structure, and the network considers the spatial intercorrelation of memory states at different levels and at different times, which helps the model capture the spatiotemporal correlation.

H_t^l is the hidden state at the current time after calculation, and it is obtained by point multiplication with the O_t at the current time after C_t^l and M_t^l are connected in the channel dimension and dimensionally reduced through the nonlinear hyperbolic activation function $tanh$. The specific state transfer method of the GST-LSTM unit is:

$$i_t = \sigma(W_{ix} * X_t + W_{ih} * H_{t-1}^l) \tag{1}$$

$$g_t = \tanh(W_{gx} * X_t + W_{gh} * H_{t-1}^l) \tag{2}$$

$$f_t = \sigma(W_{fx} * X_t + W_{fh} * H_{t-1}^l) \tag{3}$$

$$C_t^l = i_t \odot g_t + f_t \odot C_{t-1}^l \tag{4}$$

$$A_t^l = [C_{t-1}^l, M_t^{l-1}] \tag{5}$$

$$i_t' = \sigma(W_{i'a} * A_t^l) \tag{6}$$

$$g_t' = \tanh(W_{g'a} * A_t^l) \tag{7}$$

$$M_t^l = i_t' \odot g_t' + (1 - i_t') \odot M_t^{l-1} \tag{8}$$

$$O_t = \sigma(W_{ox} * X_t + W_{oh} * H_{t-1}^l + W_{om} * M_t^l + W_{oc} * C_t^l) \tag{9}$$

$$H_t^l = O_t \odot \tanh(W_{1\times1} * [C_t^l, M_t^l]) \tag{10}$$

Here, W is the weight, $\odot$ is the Hadamard product, and $*$ is the convolution operation,σ is the sigmoid function.

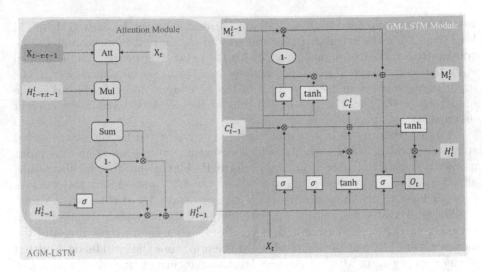

Fig. 1. The specific construction of the AGM-LSTM network unit

3.2   Attention Module

Due to the attention mechanism's ability to pay close attention to key features, this article is inspired by the MAU network structure to design a module that uses attention mechanisms to aggregate hidden state based on importance. The GM-LSTM network unit after adding this module is called the Attention Gate Memory LSTM (AGM-LSTM) network unit, and the stacked multiple AGM-LSTM network units form the AGST-LSTM network model. The attention module structure is inside the green box in the left half of Fig. 1.

Considering that the input at each moment has a primary impact on the hidden state, it is a good choice to use the correlation between inputs to give different weights to the long-term hidden state. Assuming that the number of time slices used is defined as τ, this paper uses the input $X_{t-\tau:t-1}$ of multiple time slices as the long-term input, the input X_t at present as the short-term input, the hidden state $H_{t-\tau:t-1}^l$ of multiple time slices as the extended hidden state, the hidden state H_{t-1}^l transmitted to the current moment at the previous moment as the short-term hidden state, And in the attention mechanism, Q is the long-term input, K is the short-term input, and the long-term hidden state as V. Use the correlation between long-term input and short-term input to extract information from the long-term hidden state to obtain important information in the long-term hidden state. The short-term hidden state is also important for the generation of the prediction image at the current moment, so it is also necessary to aggregate the information extracted from the long-term hidden state and the short-term hidden state H_{t-1}^l through a gate structure to form a new hidden state $H_{t-1}^{l}{}'$ with the same size as the original hidden state H_{t-1}^l. The state switching equation of the attention block is:

$$B_h = \sigma(W_h * H_{t-1}^l) \tag{11}$$

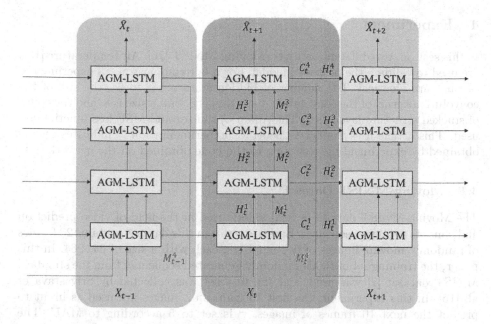

Fig. 2. Main Framework Structure of AGST-LSTM

$$H_{t-1}^{l}{}' = B_h \odot H_{t-1}^{l} + (1 - B_h) \odot sum(H_{t-\tau:t-1}^{l} \odot ((X_{t-\tau:t-1} \odot X_t)/\sqrt{d})) \quad (12)$$

where d is the product of the dimensions of X_t, σ is the sigmoid function, and τ is the length of time. Note that X_t is the input at present if it is in the first layer, otherwise, it is the hidden state H_t^{l-1} passed up by the AGM-LSTM unit at the current moment. The same is true if not in the first layer of the network, $X_{t-\tau:t-1}$ is $H_{t-\tau:t-1}^{l-1}$.

The attention mechanism increases the temporal receptive field of the model's hidden state, increases the persistence of data features, and helps the model extract long-term features.

Afterwards, the new hidden state $H_{t-1}^{l}{}'$ generated by the attention module replaces H_{t-1}^{l} in the GM-LSTM network element for calculation, thus forming a new structural unit, the AGM-LSTM. In Fig. 1, the framework of the AGM-LSTM is shown. Finally, the final AGST-LSTM model is formed by stacking the AGM-LSTM structural elements likes Fig. 2.

AGST-LSTM is composed of stacked AGM-LSTM structural elements. This stacked structure enables the network model to obtain different levels of spatiotemporal correlation, which not only obtains the top-level macro features but also the low-level detailed features for the top-level predicted images. Then the model will vividly depict the future image through the calculation of the detailed and macroscopic features of each layer at the current moment.

4   Experiment

In this section, two different datasets (Moving-MNIST [16], AirTemperature [13]) are used to validate the proposed model. For a fair comparison, all experimental settings are identical. The number of hidden state is set to 64, the size of the convolution kernel of the convolution operation is 5, batch size is 8, and the count of stacked structure layers is 4. The Adam optimizer and MSE loss function are used. This article uses the GTX 2080ti for experimentation. * indicates results obtained by experimenting according to the code obtained on the network.

4.1   Moving-MNIST Dataset

The Moving-MNIST dataset is very widely used in the field of video prediction and consists of image sequences. Each sequence in this dataset contains 20 frames of randomly moving images of two numbers, each with a size of 64×64. In this paper, the training set of 10000 randomly generated sequences from the standard MNIST dataset [7] was used, and the test set was collected by Srivastava et al. [16]. In this experiment, the first 10 frames of images are used as input to predict the next 10 frames of images. τ is set to 5 according to MAU. The comparison results are shown in Table 1 between AGST-LSTM and competition models. Figure 3 is an example of a visualization of the results compared with some comparison models. In this experiment, MSE(Mean Square Error) and SSIM(Structural Similarity Index) were used as evaluation indicators.

Table 1. Assessment results of AGST-LSTM and comparison models on the Moving-MNIST dataset (10 frames by 10 frames).

Method	SSIM	MSE
ConvLSTM [15]	0.707	103.3
PredRNN [18,19]	0.869	56.8
PredRNN-V2 [19]	0.891	48.4
RST-LSTM [12]*	0.907	39.0
LMC-Memory [9]	0.924	41.5
MSPN [11]	0.926	39.5
VPTR [23,24]	0.882	63.6
GST-LSTM (ours)	0.927	31.7
AGST-LSTM (ours)	**0.929**	**31.0**

By studying the results in Table 1, AGST-LSTM model performs better than comparative models. From the consequences in Table 1, The MSE value of AGST-LSTM is 45.4% lower than that of PredRNN, and the SSIM value is increased by 7.0%. And, as can be seen from Fig. 3, the model proposed in this

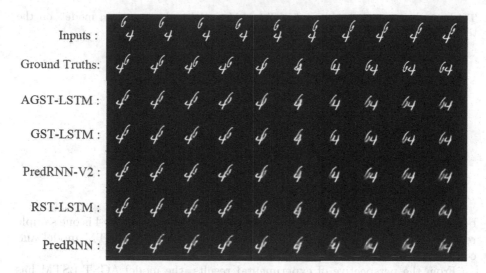

Fig. 3. Qualitative results of AGST-LSTM and comparison models on the Moving-MNIST dataset.

article, AGST-LSTM, generates relatively clearer images compared to comparative models. The effectiveness of the AGST-LSTM model has been demonstrated through experiments.

4.2 AirTemperature

Weather prediction is closely related to people's lives, and the accuracy of meteorological prediction has strong practical significance. Weather prediction is another challenge in the application of spatiotemporal series forecasting, which involves feeding continuous multi-frame meteorological images to the spatiotemporal series prediction model, extrapolating the next meteorological situation, and outputting meteorological pictures. AirTemperature [13] is based on NCEP Global Data Assimilation System GDAS, which is provided by the NOAA PSL, Boulder, Colorado, USA. The data obtained from global daily mean air temperature data available on their website at https://psl.noaa.gov (accessed April 1, 2023).

In this paper, the daily temperature average line experiment with a surface pressure value of 200 is selected as the data for the experiment. And the data from 2011 to 2019 years are used as the training set; the validation set is the data from 2020; and the data from 2021 to 2022 is the test set.

To conveniently input data to the model, this article crops all experimental data sizes to 72×144 (clipping out the last row of pixels), recomposes the data into sequence data of length 8, and uses the first 4 frames as input to predict the data of the last 4 frames, τ is set to 3 for MAU, STAU, STRPM, and AGST-LSTM in this experiment. In this experiment, MSE and MAE are used as evaluation indices. Table 2 represents the comparison of evaluation results

Table 2. Assessment results of AGST-LSTM model and comparison models on the temperature prediction task (4 frames by 4 frames).

Method	MAE	MSE
RST-LSTM [12]*	87.3	1.19
PredRNN-V2 [19]*	91.5	1.25
MAU [3]*	82.5	1.33
STAU [1]*	81.7	1.31
STRPM [2]*	81.5	1.32
AGST-LSTM (ours)	**81.0**	**1.12**

between the AGST-LSTM model and competition models. Figure 4 is one sample result of the visualization of the test results for the AGST-LSTM model and comparison models.

From the perspective of experimental results, the model AGST-LSTM has a better grade in temperature prediction than the comparison models and has a lower MAE value and MSE value. This validates the effectiveness of AGST-LSTM on the temperature prediction task.

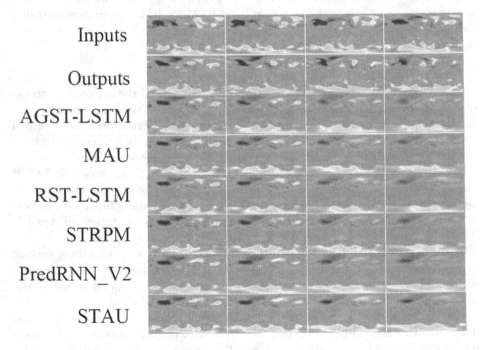

Fig. 4. Visual results of AGST-LSTM model and comparison models on the temperature prediction task using AirTemperature data.

4.3   Ablation Study

Looking at the data in Table 1, through the analysis of the effects of the ConvLSTM, PredRNN, and GST-LSTM models, it is found that GST-LSTM has improved its performance compared to ConvLSTM and PredRNN. This fully demonstrates the effectiveness of adding a memory unit to ConvLSTM that utilizes the original memory unit to update information and uses a spatiotemporal memory flow structure to focus on long-term and short-term features.

So as to illustrate the effectiveness of the attention module by aggregating long-term hidden state and short-term hidden state to increase the hidden state time-receptive field, this article compares the results of the AGST-LSTM and GST-LSTM models in Moving-MNIST dataset. For a more convenient comparison of experimental results, the ablation experimental results are placed in Table 1 and Fig. 3 in this paper. Through research on the experimental results, the SSIM value of the AGST-LSTM model proves some improvement in contrast with the GST-LSTM model, while The MSE value is reduced by 2.2%, fully demonstrating that the attention module is helpful in promoting the performance of the model.

5   Conclusion

On the basis of the ConvLSTM model unit, this paper studies the possibility of adding a spatiotemporal memory state with updated information from the original temporal memory state and designs a new network structure unit, GM-LSTM. The memory unit added by GM-LSTM can focus on long-term and short-term spatial features. Inspired by the network structure of MAU, the temporal receptive field of the hidden state is increased, and a model with attention module, AGST-LSTM, is designed.

In this article, two experiments are executed to examine the AGST-LSTM model. The experimental outcomes certify that the style of the spatiotemporal memory state M that uses the primitive memory cell C of the ConvLSTM network unit to update information is more excellent than that of the PredRNN network model. At the same time, the attention module is used to combine long-term and short-term hidden states, and it is helpful to promote the capability of the model by increasing the temporal perception domain of the hidden state. By comparing the model effects of GST-LSTM and AGST-LSTM, it is found that the effect of the attention module on the model is not obvious, but the increase in the computational cost is relatively large, so we will study how to increase the time-receptive field more effectively in the future, and we will experiment with AGST-LSTM on more types of datasets in the future.

References

1. Chang, Z., Zhang, X., Wang, S., Ma, S., Gao, W.: STAU: a spatiotemporal-aware unit for video prediction and beyond. arXiv preprint arXiv:2204.09456 (2022)
2. Chang, Z., Zhang, X., Wang, S., Ma, S., Gao, W.: STRPM: a spatiotemporal residual predictive model for high-resolution video prediction. In: Proceedings of the IEEE/CVF Conference on Computer Vision and Pattern Recognition, pp. 13946–13955 (2022)
3. Chang, Z., et al.: MAU: a motion-aware unit for video prediction and beyond. Adv. Neural. Inf. Process. Syst. **34**, 26950–26962 (2021)
4. Gao, Y., Wang, J., Zhang, X., Li, R.: Ensemble wind speed prediction system based on envelope decomposition method and fuzzy inference evaluation of predictability. Appl. Soft Comput. **124**, 109010 (2022)
5. Hochreiter, S., Schmidhuber, J.: Long short-term memory. Neural Comput. **9**(8), 1735–1780 (1997)
6. Kurani, A., Doshi, P., Vakharia, A., Shah, M.: A comprehensive comparative study of artificial neural network (Ann) and support vector machines (SVM) on stock forecasting. Ann. Data Sci. **10**(1), 183–208 (2023)
7. LeCun, Y.: The MNIST database of handwritten digits (1998). http://yann.lecun.com/exdb/mnist/
8. LeCun, Y., Bottou, L., Bengio, Y., Haffner, P.: Gradient-based learning applied to document recognition. Proc. IEEE **86**(11), 2278–2324 (1998)
9. Lee, S., Kim, H.G., Choi, D.H., Kim, H.I., Ro, Y.M.: Video prediction recalling long-term motion context via memory alignment learning. In: Proceedings of the IEEE/CVF Conference on Computer Vision and Pattern Recognition, pp. 3054–3063 (2021)
10. Li, M., Zhu, Z.: Spatial-temporal fusion graph neural networks for traffic flow forecasting. In: Proceedings of the AAAI Conference on Artificial Intelligence, vol. 35, pp. 4189–4196 (2021)
11. Ling, C., Zhong, J., Li, W.: Predictive coding based multiscale network with encoder-decoder LSTM for video prediction. arXiv preprint arXiv:2212.11642 (2022)
12. Luo, C., Xu, G., Li, X., Ye, Y.: The reconstitution predictive network for precipitation nowcasting. Neurocomputing **507**, 1–15 (2022)
13. NOAA PSL, Boulder, Colorado, USA: NCEP global data assimilation system GDAS. https://psl.noaa.gov/data/gridded/data.ncep.html. Accessed 1 Apr 2023
14. Rumelhart, D.E., Hinton, G.E., Williams, R.J.: Learning representations by back-propagating errors. Nature **323**(6088), 533–536 (1986)
15. Shi, X., Chen, Z., Wang, H., Yeung, D.Y., Wong, W.K., Woo, W.: Convolutional LSTM network: a machine learning approach for precipitation nowcasting. Adv. Neural. Inf. Process. Syst. **28**, 802–810 (2015)
16. Srivastava, N., Mansimov, E., Salakhudinov, R.: Unsupervised learning of video representations using LSTMS. In: International Conference on Machine Learning, pp. 843–852. PMLR (2015)
17. Wang, K., Wang, J., Zeng, B., Lu, H.: An integrated power load point-interval forecasting system based on information entropy and multi-objective optimization. Appl. Energy **314**, 118938 (2022)
18. Wang, Y., Long, M., Wang, J., Gao, Z., Yu, P.S.: PredRNN: recurrent neural networks for predictive learning using spatiotemporal LSTMS. Adv. Neural. Inf. Process. Syst. **30**, 879–888 (2017)

19. Wang, Y., et al.: PredRNN: a recurrent neural network for spatiotemporal predictive learning. IEEE Trans. Pattern Anal. Mach. Intell. **45**(2), 2208–2225 (2022)
20. Wang, Y., Zhang, J., Zhu, H., Long, M., Wang, J., Yu, P.S.: Memory in memory: a predictive neural network for learning higher-order non-stationarity from spatiotemporal dynamics. In: Proceedings of the IEEE/CVF Conference on Computer Vision and Pattern Recognition, pp. 9154–9162 (2019)
21. Werbos, P.J.: Backpropagation through time: what it does and how to do it. Proc. IEEE **78**(10), 1550–1560 (1990)
22. Wu, H., Yao, Z., Wang, J., Long, M.: MotionRNN: a flexible model for video prediction with spacetime-varying motions. In: Proceedings of the IEEE/CVF Conference on Computer Vision and Pattern Recognition, pp. 15435–15444 (2021)
23. Ye, X., Bilodeau, G.A.: VPTR: efficient transformers for video prediction. In: 2022 26th International Conference on Pattern Recognition (ICPR), pp. 3492–3499. IEEE (2022)
24. Ye, X., Bilodeau, G.A.: Video prediction by efficient transformers. Image Vis. Comput. **130**, 104612 (2023)

A Shape-Based Quadrangle Detector
for Aerial Images

Chaofan Rao, Wenbo Li, Xingxing Xie, and Gong Cheng(✉)

School of Automation, Northwestern Polytechnical University, Xi'an, China
chenggong1119@gmail.com

Abstract. The performance of oriented object detectors has been adversely impacted by the substantial variations in object orientation. In this paper, we propose a simple but efficient object detection framework for oriented objects in aerial images, termed QuadDet. Instead of adopting oriented bounding box to represent the object, we directly predict the four vertices of the object's quadrilateral. Specially, we introduce a fast sorting method for four vertexes of quadrangles, called the *Vertex Sorting Function*. The function confirms that the vertexes can compose a valid quadrangle by sorting tangents of the vertexes. Furthermore, we employ an efficient polygon IoU loss function, named the *PolyIoU Loss Function*, to progressively align the predicted quadrangle's shape with the ground truth. Under these strategies, our model achieves competitive performance. Without bells and whistles, our method with ResNet50 achieves 73.63% mAP on the DOTA-v1.0 dataset running at 23.4 FPS, which surpasses all recent one-stage oriented object detectors by a significant margin. Moreover, on the largest dataset DOTA-v2.0, our QuadDet with ResNet50 obtains 51.54% mAP. The code and models are available at https://github.com/DDGRCF/QuadDet.

Keywords: Oriented Object Detection · Quadrangle Representation · Vertex Sorting · Aerial Images

1 Introduction

Object detection [6,9,10,18,24–26] is fundamental yet challenging in computer vision. It aims to classify and locate objects in images. With significant advancements in deep learning, object detection has not only demonstrated its prowess in various domains such as face detection, autonomous driving, and optical text detection, but it has also laid the foundation for high-level image analysis and applications, including instance segmentation, scene understanding, object tracking, etc.

In generic object detection, the model commonly applies the horizontal bounding boxes to represent objects, which meets the requirements for most objects positioning in natural images. However, aerial images, being captured from bird's eye view, often contain objects with diverse orientations and large

© The Author(s), under exclusive license to Springer Nature Singapore Pte Ltd. 2024
Q. Liu et al. (Eds.): PRCV 2023, LNCS 14428, pp. 368–379, 2024.
https://doi.org/10.1007/978-981-99-8462-6_30

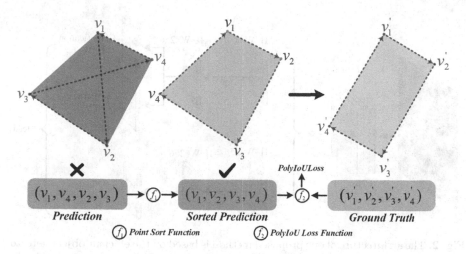

Fig. 1. The pipeline of the proposed method. The red quadrangle represents the original prediction, the blue quadrangle signifies the sorted prediction, and the green quadrangle represents the ground truth. (Color figure online)

aspect ratios, such as bridges and harbors. Representing objects with the oriented bounding box representation can potentially enable more precise localization. However, detectors with the oriented bounding box suffer from imprecise angle prediction [2–5,16,22,23,28,31,38]. To address this, many researchers have explored quadrangle representation for objects [19,21,30] in aerial images.

However, the issue of the boundary discontinuity caused by the vertexes exchange in edge cases hampers the performance of quadrangle-based detection. Most existing quadrangle-based approaches primarily focus on introducing novel quadrangle representations or improving loss function to mitigate the problem of boundary discontinuity. For instance, Gliding Vertex [30] proposes a unique quadrangle representation for the oriented object by sliding the vertexes of the horizontal bounding box (HBB) to predict the quadrangle. While it avoids complex vertex sorting, the boundary discontinuity problem still persists, especially for nearly horizontal objects. Moreover, constraining the vertexes to the HBB limits the learning capacity of the model. RSDet [21] alleviates the problem of boundary discontinuity by introducing a dynamic loss function, where the regression loss is determined by the distance between the sliding vertexes and the original vertexes of the horizontal box. However, the constraint of the learning capacity still exists. Additionally, the method increases the training time for calculating loss multiple times.

In this paper, we propose a simple yet effective framework for addressing the issue of quadrangle prediction, termed QuadDet. Unlike the aforementioned method, which regards the quadrangle regression as a one-to-one length regression between each predicted vertexes and the ground truth, our QuadDet regards the regression of all quadrangle vertexes as a whole. QuadDet employs a vertex sorting function to ensure that the vertexes can form a valid quadrangle without

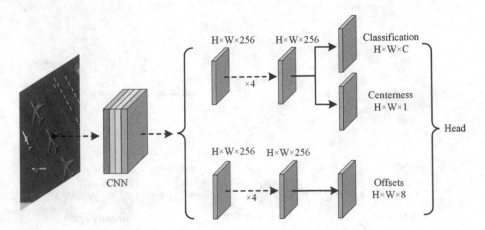

Fig. 2. The architecture of our proposed method is based on the efficient object detector ATSS [39]. To predict the quadrangle, we replace the original 4D convolutional layer with 8D convolutional layer in the regression branch.

crossed diagonal lines. Subsequently, we gradually align the predicted quadrangle's shape with the ground truth using a specialized polygon IoU loss function. The whole pipeline is illustrated in Fig. 1. The contributions of this paper are as follows:

- We propose an intuitive shape-based quadrangle detector called QuadDet that addresses the boundary discontinuity problem;
- Two shaped-base functions, *Vertex Sorting Function* and *PolyIoU Loss Function*, are introduced to achieve fast vertex sorting and better optimization, respectively;
- Extensive experiments have been implemented to demonstrate the effectiveness of our QuadDet. QuadDet with ResNet50 achieves 73.63% mAP on DOTA-v1.0 having 23.4 FPS, which surpasses all the recent one-stage oriented object detectors. Additionally, our QuadDet model with ResNet50 attains 51.54% mAP on the largest DOTA-v2.0 dataset.

2   Method

2.1   Overview

Our proposed method introduces modifications to enhance the accuracy of quadrangle prediction based on the detector ATSS [39]. As depicted in Fig. 2, our model comprises two main branches: a classification branch and a regression branch. In classification branch, we employ a convolutional layer with C dimensions (C representing the number of categories) and a 1D convolutional layer. These layers work together to generate separate predictions for category probabilities and centerness.

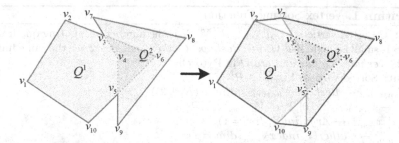

Fig. 3. The schematic diagram of the search of convex hull in *PolyIou Loss Function*. Here, $Q1$ and $Q2$ are the ground gruth and predicted bounding box separately. $v_1 \sim v_{10}$ is the key points in the intersection of $Q1$ and $Q2$.

Our primary contributions lies in regression branch. To predict the quadrangle, we replace the original 4D convolutional layer with an 8D convolutional layer, which predicts offsets for vertexes of the quadrangle. Subsequently, the predicted vertex offsets are added to the corresponding anchor points, resulting in the generation of unsorted vertex coordinates denoted as $V_{unsorted} \in R^{H \times W \times 8}$. To ensure that the vertexes form a valid quadrangle without intersecting diagonal lines, we employ the *Vertex Sorting Function*. The function sorts the vertexes based on a specific criterion, yielding sorted vertex coordinates denoted as $V_{sorted} \in R^{H \times W \times 8}$. For optimizing the model, we utilize a specialized IoU loss function named *PolyIoU Loss Function*. The loss function operates on the sorted vertex coordinates, progressively aligning the predicted quadrangle with the ground truth. By incorporating these modifications, our method achieves more accurate quadrangle prediction, and the whole piepline is shown in Fig. 1.

2.2 Vertex Sorting Function

Vertex Sorting Function is a fast shape-based vertex sorting method. Unlike Gliding Vertex, which constrains vertexes within the sides of HBBs, Vertex Sorting Function just makes sure that four vertexes can compose a valid-shaped quadrangle with non-crossing lines. Therefore, our method neither has the boundary discontinuity problem nor affects the learning capacity of the model. Vertex Sorting Function does not involve complex computation and only includes three steps. Firstly, it obtains the vertex with the smallest Y among the four vertexes of a quadrangle. If there are multiple vertexes with the same smallest Y, the one with the smallest X will be viewed as the minimum one. After that, it calculates the tangent value of the four vertexes relative to this minimum point. Finally, the tangent values are sorted in the descending order, getting four vertexes sorted in clockwise. These three steps are easily parallelized, so there is only a slight drop in inference speed. The pseudocode of *Vertex Sorting Function* can reference Algorithm 1.

Algorithm 1. Vertex Sorting Function

Input: Unsorted vertices $V_{unsorted} \in R^{N \times 8}$. N is the number of inputing quadrangles.
EPS is small enough, like $1E-6$. *Reshape*, *Gather* and *Atan2* are the same function
as the *reshape*, *gather* and *atan2* in Pytorch

Output: Sorted vertices $V_{sorted} \in R^{N \times 8}$

1: $V_{unsorted} \leftarrow Reshape(V_{unsorted}, dim = (N, 4, 2))$

2: $X \leftarrow V_{unsorted}[:, :, 0], Y \leftarrow V_{unsorted}[:, :, 1]$

3: $index_{X_{min}} \leftarrow Argmin(X, dim = 1)$

4: $Y_{x_{min}} \leftarrow Gather(Y, index_{X_{min}}, dim = 1)$
   $Y_{x_{min}} \leftarrow Y_{x_{min}} - EPS$

5: $index_{Y_{min}} \leftarrow Argmin(Y, dim = 1)$

6: $X_o \leftarrow Gather(X, index_{Y_{min}}, dim = 1)$
   $Y_o \leftarrow Gather(Y, index_{Y_{min}}, dim = 1)$

7: $X_{rel} \leftarrow X - X_o, Y_{rel} \leftarrow Y - Y_o$

8: $A_{Y/X} \leftarrow Atan2(Y, X), index_A \leftarrow Argsort(A)$

9: $V_{sorted} \leftarrow Gather(V, index_A, dim = 1)$
   $V_{sorted} \leftarrow Reshape(V, dim = (N, 8))$

2.3   PolyIoU Loss Function

PolyIoU Loss Function is a shape-based IoU loss function applied to the optimization of quadrangle regression. It directly enforces the maximal overlap between the predicted quadrangle and the ground truth boxes, and jointly regresses all the quadrangle vertexes as a whole unit. For two irregular quadrangles, the polygon formed by the intersection points between them is likely to be a concave polygon, and it is very complicated and time-consuming to determine the order of each point of the concave polygon. So, we adopt a polygon IoU calculation method based on the convex hull. As shown in Fig. 3, the polygon surrounded by green lines is the union of two quadrangles, $Q1$ and $Q2$, while the polygon surrounded by yellow dotted lines is the intersection of $Q1$ and $Q2$. We carry Graham's Scan [11] both on the union and the intersection, getting two convex hulls surrounded by the red line and the dotted blue line separately. Then, it calculates the areas of two convex polygons by the Shoelace Formula [1]. The IoU value range of this method is $[0, 1]$, and it hits 1 when $Q1$ and $Q2$ coincide, which is the same as the ideal IoU loss.

2.4   Implementation Details

Unless specified, we adopt ResNet50 [14] with FPN [17] as the backbone and use the same hyper-parameters as ATSS. All experiments are conducted on an NVIDIA GeForce RTX 3090 with the batch size of 2.

In the training, our loss L is defined as

$$L = \frac{1}{N_{pos}} \sum_{x,y} L_{cls} + \frac{\lambda_1}{N_{pos}} \sum_{x,y} \mathbb{I}_{c_{x,y}>0} L_{ctr}$$
$$+ \frac{\lambda_2}{N_{pos}} \sum_{x,y} \mathbb{I}_{c_{x,y}>0} L_{iou} \tag{1}$$

where L_{cls} is the focal loss [18], L_{ctr} is binary cross entropy (BCE) loss, and L_{iou} is calculated by *Vertex Sorting Function*. λ_1 and λ_2 are the balance weights for L_{ctr} and L_{iou}. Here, we set both λ_1 and λ_2 as 1. The summation is calculated over all locations on the feature maps. $\mathbb{I}_{c_{x,y}>0}$ is the indicator function, being 1 if the sample is positive and 0 otherwise. Then, we optimize the model by SGD algorithm with the initial learning rate of 0.0025, the momentum of 0.9 and the weight decay of 0.0001. Random horizontal and vertical flipping are applied during training to avoid over-fitting. All experiments run 12 epochs and reduce the learning rate by a factor of 10 at the end of epoch 9 and epoch 11. In the inference, we choose the location with $score_{x,y} > 0.05$ as positive samples and get the sorted vertices following the pipeline in Fig. 1.

3   Experiments

3.1   Datasets

DOTA-v1.0 [27] is a universal aerial image object detection dataset containing 2,806 images and 188,282 instances of 15 common object classes: Harbor (HA), Storage Tank (ST), Large Vehicle (LV), Swimming Pool (SP), Soccer Ball Field (SBF), Bridge (BR), Ship (SH), Plane (PL), Baseball Diamond (BD), Tennis Court (TC), Ground Track Field (GTF), Basketball Court (BC), Small Vehicle (SV), Roundabout (RA) and Helicopter (HC).

DOTAv-v2.0 [8] contains 11268 images and 1793658 instances, covering 18 common object classes, including Harbor (HA), Storage Tank (ST), Large Vehicle (LV), Swimming Pool (SP), Soccer Ball Field (SBF), Bridge (BR), Ship (SH), Plane (PL), Baseball Diamond (BD), Tennis Court (TC), Ground Track Field (GTF), Basketball Court (BC), Small Vehicle (SV), Roundabout (RA), Helicopter (HC), Container Crane (CC), Airport (AI), and Helipad (HP). The more images and more instances make it more challenging than DOTA-v1.0.

Since DOTA-v1.0 and DOTA-v2.0 have over 4000 × 4000 super-sized images, we adopt the same preprocessing strategy as some advanced methods (e.g. [7,12,

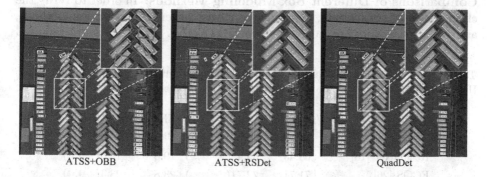

ATSS+OBB   ATSS+RSDet   QuadDet

Fig. 4. The visualization results of ATSS+OBB, ATSS+RSDet and QuadDet. To facilitate better comparison, we have enlarged a portion of the image and positioned it in the upper right corner.

13, 29]), which crop the images into 1024×1024 pathces with the stride of 824. For multi-scale experiments, we first resize original images into three scales (0.5, 1.0, 1.5) and then crop all the images of these three scales into 1024×1024 with a stride of 524. On the DOTA-v1.0 and DOTA-v2.0 datasets, the union of the training set and validation set is used for training, and the test set is used for model evaluation. All evaluation results are obtained from the official evaluation server.

Table 1. The results of QuadDet when using different detectors as baselines. Here, ATSS+OBB, ATSS+GV and ATSS+RSDet mean that implementing the representations of the oriented bounding box, Gliding Vertex and RSDet in ATSS separately. The red and blue fonts indicate the top two performances respectively. mAP_{50} and mAP_{75} are the results obtained at IoU thresholds of 0.5 and 0.75, respectively.

Method	GLOPS	#Param	$mAP_{50}(\%)$	$mAP_{75}(\%)$	FPS
ATSS+OBB	201.95	31.87M	72.28	40.20	23.7
ATSS+GV [30]	204.35	32.35M	72.45	42.25	22.9
ATSS+RSDet [21]	204.20	32.30M	**72.82**	**43.02**	23.1
QuadDet	**202.10**	**31.88M**	73.63	44.48	**23.4**

3.2 Ablation Experiments

Baseline: We employ ATSS as our baseline, where the horizontal bounding box are simply replaced with the oriented bounding box with the angle range within $[-\pi/4, 3\pi/4]$ [19]. For the model structure, we add a 1D convolutional layer on the last regression feature map of ATSS to predict the oriented bounding boxes. For the sample assignment, we only consider the points which fall in the oriented bounding box as positive sample points. Except for the above changes, other settings keep the same as the original ATSS.

Comparisons of Different Box Modeling Methods: In order to verify the effectiveness of our method, we compared it with other representation methods, and the experiments are carried out on DOTA-v1.0. The results are reported

Table 2. Comparisons of baseline and QuadDet implementation of different detectors. Here, baselines are based on oriented bounding box representation.

Method	mAP(%)		FPS	
	Baseline	QuadDet	Baseline	QuadDet
RetinaNet [18]	68.43	69.45(+1.02)	19.8	19.5(−0.3)
FCOS [26]	71.71	72.71(+1.00)	23.8	23.4(−0.4)
ATSS [39]	72.28	73.63(+1.35)	23.7	23.4(−0.3)
Oriented RCNN [29]	75.87	76.01(+0.14)	15.8	15.6(−0.2)

Table 3. Comparison with state-of-the-art methods on the DOTA-v1.0 dataset. R50-FPN stands for ResNet50 with FPN (likewise R101-FPN). ATSS-O means ATSS modified by adding angle prediction. All Results with † are reported at MMRotate [40]. * means multi-scale training and testing with rotation augmentation. The red and blue fonts indicate the top two performances respectively.

Method	Backbone	PL	BD	BR	GTF	SV	LV	SH	TC	BC	ST	SBF	RA	HA	SP	HC	mAP	FPS
Two-stage																		
RoI Transformer [7]	R50-FPN	88.34	77.07	51.63	69.62	77.45	77.15	87.11	90.75	84.90	83.14	52.95	63.75	74.45	74.45	59.24	73.76	12.8
SCRDet [34]	R101-FPN	89.98	80.65	52.09	68.36	68.36	60.32	72.41	90.85	87.94	86.86	65.02	66.68	66.25	68.24	65.21	72.61	-
Gliding Vertex* [30]	R50-FPN	89.64	85.00	52.26	77.34	73.01	73.14	86.82	90.74	79.02	86.81	59.55	70.91	72.94	70.86	57.32	75.02	13.2
Oriented R-CNN [29]	R50-FPN	89.46	82.12	54.78	70.86	78.93	83.00	88.20	90.90	87.50	84.68	63.97	67.69	74.94	68.84	52.28	75.87	15.8
Refine-stage																		
R³Det† [32]	R50-FPN	89.05	78.46	44.15	65.16	75.09	72.88	86.04	90.83	81.41	82.53	56.24	61.01	56.49	67.67	52.92	69.80	13.5
Oriented Reppoints† [37]	R50-FPN	87.78	77.68	49.54	66.46	78.52	73.12	86.59	90.87	83.75	84.35	53.14	65.63	63.70	68.71	45.91	71.72	16.1
SASM† [15]	R50-FPN	80.80	66.37	45.41	65.58	72.08	72.35	79.83	90.90	75.69	83.06	49.06	59.47	63.76	59.42	33.19	66.46	15.3
S²ANet [12]	R50-FPN	89.11	82.84	48.37	71.11	78.11	78.39	87.25	90.83	84.90	85.64	60.36	62.60	65.26	69.13	57.94	74.12	14.3
One-stage																		
DAL [20]	R50-FPN	88.68	76.55	45.08	66.80	67.00	76.76	79.74	90.84	79.54	78.45	57.71	62.27	69.05	73.14	60.11	71.44	-
CSL† [31]	R50-FPN	89.34	79.68	40.86	69.94	77.69	62.04	77.49	90.87	82.88	82.07	59.97	65.30	53.56	64.04	46.61	69.49	18.9
GWD† [33]	R50-FPN	89.40	79.64	40.27	70.37	77.90	63.84	80.68	90.90	82.41	82.19	56.27	65.57	56.82	65.06	44.18	69.70	20.3
KLD† [35]	R50-FPN	89.16	80.00	42.41	73.09	78.08	68.84	84.72	90.90	83.30	83.76	57.19	62.98	59.50	67.21	48.34	71.30	20.5
RSDet [21]	R50-FPN	89.30	82.70	47.70	63.90	66.80	62.00	67.30	90.90	85.30	82.40	62.30	65.40	65.70	68.60	64.60	70.80	-
RetinaNet-O*† [18]	R50-FPN	88.10	84.43	50.54	79.12	73.65	59.80	72.94	90.39	86.45	87.24	65.02	65.55	67.09	70.72	58.44	73.30	19.8
FCOS-O [26]	R50-FPN	89.01	78.32	47.99	58.23	78.92	76.38	86.04	90.91	81.62	84.62	56.99	63.24	64.19	71.90	47.39	71.71	23.8
ATSS-O [39]	R50-FPN	88.50	77.73	49.60	69.85	76.85	72.52	82.46	90.83	80.31	82.96	62.35	64.63	64.86	66.76	53.98	72.28	23.7
QuadDet	R50-FPN	88.93	78.57	50.30	70.65	80.12	78.87	87.00	90.90	81.94	83.97	58.35	66.56	64.89	68.51	54.90	73.63	23.4
QuadDet	R101-FPN	89.52	79.86	49.10	72.70	79.62	78.87	86.85	90.90	83.81	85.54	58.06	65.85	70.26	69.22	51.91	74.14	18.2
QuadDet*	R50-FPN	89.30	82.71	57.38	76.44	81.38	84.10	88.38	90.80	84.22	86.91	67.92	69.73	75.70	80.27	67.08	78.82	23.4

in Table 1. ATSS with the oriented bounding box, Gliding Vertex and RSDet representations receive 72.28%, 72.45%, 72.82% mAP_{50} and 40.20%, 42.25%, 43.02% mAP_{75}, respectively. In contrast, our method achieves the hightest 73.63% mAP_{50} and 44.48% mAP_{75} with only negligible parameter growth and a slight drop in speed. This demonstrates that our method is simple yet effective. The visualized results of different representation methods are compared in Fig. 4. As evident from the visualized results, QuadDet is capable of achieving a more precise regression compared to the oriented bounding box and RSDet representations.

The Performance of QuadDet When Using Different Detectors as Baselines: To verify the generalizability of QuadDet, we compare the results of QuadDet when using different detectors (RetinaNet, FCOS, and ATSS) as baselines. The results are reported on the Table 2. For one-stage detectors, compared to the baselines, the QuadDet method achieves an average improvement of more than 1% with a slight speed drop. And for the two-stage detector Oriented RCNN, QuadDet also exhibits some improvement in accuracy, but due to the already high precision of the two-stage regression, the increase in accuracy is quite limited. The result proves that our methods is not only suitable for ATSS, but also effective for common object detectors.

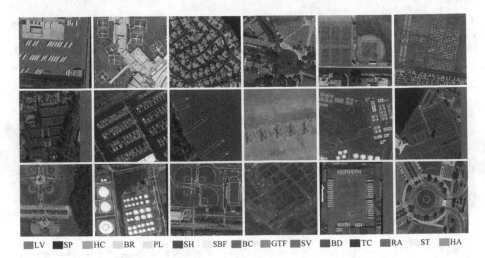

■LV ■SP ■HC ■BR ■PL ■SH ■SBF ■BC ■GTF ■SV ■BD ■TC ■RA ■ST ■HA

Fig. 5. The visualization results of QuadDet. Different categories of bounding boxes render with different colors.

Table 4. Comparison with state-of-the-art methods on the DOTA-v2.0 dataset. All Results with † are reported at MMRotate.

Method	Backbone	PL	BD	BR	GTF	SV	LV	SH	TC	BC	ST	SBF	RA	HA	SP	HC	CC	AI	HP	mAP	FPS
Two-stage																					
Faster R-CNN-O [25]	R50-FPN	71.61	47.20	39.28	58.70	35.55	48.88	51.51	78.97	58.36	58.55	36.11	51.73	43.57	55.33	57.07	3.51	52.94	2.79	47.31	15.5
Gliding Vertex [30]	R50-FPN	72.89	49.62	41.80	61.49	38.52	49.77	56.64	72.67	59.83	59.30	38.37	55.12	46.27	51.45	53.72	13.96	65.73	8.17	49.74	13.2
Refine-stage																					
R³Det† [32]	R50-FPN	71.72	50.71	39.95	62.54	42.73	47.33	53.80	78.41	55.00	60.32	38.23	51.85	46.81	48.30	33.22	0.01	57.84	0.11	46.60	13.5
Oriented Reppoints† [37]	R50-FPN	73.43	46.23	43.58	61.70	46.94	50.05	58.44	80.16	56.47	65.76	33.74	50.98	48.45	51.10	35.63	1.99	65.24	1.49	48.41	16.1
S²ANet [12]	R50-FPN	77.83	54.57	44.22	62.99	47.38	50.87	57.96	79.72	59.25	65.36	39.26	53.35	45.72	52.45	34.08	0.67	67.21	5.03	49.88	14.3
One-stage																					
GWD† [33]	R50-FPN	72.64	51.84	36.22	59.88	40.03	40.25	46.92	77.49	56.31	58.95	37.83	53.90	39.70	48.28	36.17	0.01	68.15	0.22	45.82	20.3
KLD† [35]	R50-FPN	73.50	47.22	38.58	60.96	39.57	43.46	47.24	76.31	56.60	57.72	38.96	52.64	40.74	47.61	41.19	0.41	68.54	10.59	46.77	20.5
KFIoU† [36]	R50-FPN	71.64	50.87	37.55	60.60	41.43	38.85	45.65	74.24	53.88	58.29	38.19	54.31	38.85	52.82	41.14	0.00	66.29	0.12	45.82	19.7
RetinaNet-O† [18]	R50-FPN	71.10	51.37	36.46	57.63	39.83	37.03	45.44	77.11	53.66	57.85	36.13	52.30	38.29	49.85	33.73	0.00	64.41	0.11	44.57	19.8
ATSS-O [26]	R50-FPN	72.90	48.03	43.01	60.05	46.25	51.67	58.29	79.40	57.74	64.41	41.68	52.79	46.79	55.02	45.26	14.63	65.37	10.53	50.77	23.8
QuadDet	R50-FPN	76.40	50.16	42.04	60.86	46.16	51.31	58.37	79.75	59.25	63.83	41.61	52.53	45.90	55.38	51.52	14.20	64.35	12.18	51.54	23.4

3.3   Comparison with State-of-the-Art Methods

DOTA-V1.0: We divide the advanced oriented object detection methods into three groups based on their architectures: one stage, two stages, and refined detectors. Here, refine stage denotes that there are cascade modules in the head of the model, which may bring a slight performance improvement with the huge speed loss. The comparison results between our method and state-of-the-arts on the DOTA-v1.0 are shown in Table 3. Without any bells and whistles, our method based on ResNet50-FPN achieves 73.63% mAP, which surpasses all one-stage detectors and is very close to the most advanced refine stage detector. Using ResNet101-FPN, our method obtains 74.14% mAP. With multi-scale training and testing and rotated augmentation, our model reaches 78.82% mAP with ResNet50-FPN as backbone, which is competitive on the DOTA-v1.0 dataset. We visualize some detection results in Fig. 5.

DOTA-V2.0: To verify the robustness for QuadDet, we also compare our model with other state-of-the-art methods on the much challenging dataset DOTA-v2.0. The results are reported in Table 4. As shown in Table 4, QuadDet achieves 51.54% mAP, which is the highest results among all one-stage and refine-stage detectors, and has only a tiny gap from the advanced two-stage methods.

4 Conclusions

In this paper, we introduced QuadDet, a simple yet effective framework for object detection in aerial images. QuadDet utilizes the innovative function, *Vertex Sorting Function*, to sort the four vertexes, ensuring the validity of the predicted quadrangle without constraining the learning capacity of the model. Subsequently, the *PolyIoU Loss Function* computes the IoU loss between the sorted quadrangle and the ground truth based on the search of the convex hull, and optimizes the model accordingly. Since QuadDet does not involve some rigid definitions that limit the capacity to learn, it can achieve competitive results. The results on DOTA-v1.0 and DOTA-v2.0 also verify the effectiveness of our method.

References

1. Braden, B.: The surveyor's area formula. Coll. Math. J. **17**(4), 326–337 (1986)
2. Chen, W., Miao, S., Wang, G., Cheng, G.: Recalibrating features and regression for oriented object detection. Remote Sens. **15**(8), 2134 (2023). https://doi.org/10.3390/rs15082134
3. Cheng, G., Li, Q., Wang, G., Xie, X., Min, L., Han, J.: SFRNet: fine-grained oriented object recognition via separate feature refinement. IEEE Trans. Geosci. Remote Sens. **61**, 1–10 (2023). https://doi.org/10.1109/TGRS.2023.3277626
4. Cheng, G., et al.: Anchor-free oriented proposal generator for object detection. IEEE Trans. Geosci. Remote Sens. **60**, 1–11 (2022). https://doi.org/10.1109/TGRS.2022.3183022
5. Cheng, G., et al.: Dual-aligned oriented detector. IEEE Trans. Geosci. Remote Sens. **60**, 1–11 (2022). https://doi.org/10.1109/TGRS.2022.3149780
6. Cheng, G., et al.: Towards large-scale small object detection: survey and benchmarks. IEEE Trans. Pattern Anal. Mach. Intell. (2023)
7. Ding, J., Xue, N., Long, Y., Xia, G.S., Lu, Q.: Learning RoI transformer for oriented object detection in aerial images. In: Proceedings of the IEEE International Conference on Computer Vision and Pattern Recognition, pp. 2849–2858 (2019)
8. Ding, J., et al.: Object detection in aerial images: a large-scale benchmark and challenges. IEEE Trans. Pattern Anal. Mach. Intell. **44**(11), 7778–7796 (2022). https://doi.org/10.1109/TPAMI.2021.3117983
9. Girshick, R.: Fast R-CNN. In: Proceedings of the IEEE International Conference on Computer Vision, pp. 1440–1448 (2015)
10. Girshick, R., Donahue, J., Darrell, T., Malik, J.: Rich feature hierarchies for accurate object detection and semantic segmentation. In: Proceedings of the IEEE Conference on Computer Vision and Pattern Recognition, pp. 580–587 (2014)

11. Graham, R.L.: An efficient algorithm for determining the convex hull of a finite planar set. Inf. Process. Lett. **1**, 132–133 (1972)
12. Han, J., Ding, J., Li, J., Xia, G.S.: Align deep features for oriented object detection. IEEE Trans. Geosci. Remote Sens. 1–11 (2021). https://doi.org/10.1109/TGRS.2021.3062048
13. Han, J., Ding, J., Xue, N., Xia, G.S.: ReDet: a rotation-equivariant detector for aerial object detection. In: Proceedings of the IEEE International Conference on Computer Vision and Pattern Recognition, pp. 2786–2795 (2021)
14. He, K., Zhang, X., Ren, S., Sun, J.: Deep residual learning for image recognition. In: Proceedings of the IEEE Conference on Computer Vision and Pattern Recognition, pp. 770–778 (2016)
15. Hou, L., Lu, K., Xue, J., Li, Y.: Shape-adaptive selection and measurement for oriented object detection. In: Proceedings of the AAAI Conference on Artificial Intelligence (2022)
16. Li, C., Cheng, G., Wang, G., Zhou, P., Han, J.: Instance-aware distillation for efficient object detection in remote sensing images. IEEE Trans. Geosci. Remote Sens. **61**, 1–11 (2023)
17. Lin, T.Y., Dollár, P., Girshick, R., He, K., Hariharan, B., Belongie, S.: Feature pyramid networks for object detection. In: Proceedings of the IEEE Conference on Computer Vision and Pattern Recognition, pp. 2117–2125 (2017)
18. Lin, T.Y., Goyal, P., Girshick, R., He, K., Dollár, P.: Focal loss for dense object detection. In: Proceedings of the IEEE International Conference on Computer Vision, pp. 2980–2988 (2017)
19. Ma, J., et al.: Arbitrary-oriented scene text detection via rotation proposals. IEEE Trans. Multimedia **20**, 3111–3122 (2018)
20. Ming, Q., Zhou, Z., Miao, L., Zhang, H., Li, L.: Dynamic anchor learning for arbitrary-oriented object detection. In: Proceedings of the AAAI Conference on Artificial Intelligence, pp. 2355–2363 (2021)
21. Qian, W., Yang, X., Peng, S., Yan, J., Guo, Y.: Learning modulated loss for rotated object detection. In: Proceedings of the AAAI Conference on Artificial Intelligence, pp. 2458–2466 (2021)
22. Qian, X., Wu, B., Cheng, G., Yao, X., Wang, W., Han, J.: Building a bridge of bounding box regression between oriented and horizontal object detection in remote sensing images. IEEE Trans. Geosci. Remote Sens. **61**, 1–9 (2023)
23. Rao, C., Wang, J., Cheng, G., Xie, X., Han, J.: Learning orientation-aware distances for oriented object detection. IEEE Trans. Geosci. Remote Sens. 1 (2023). https://doi.org/10.1109/TGRS.2023.3278933
24. Redmon, J., Divvala, S., Girshick, R., Farhadi, A.: You only look once: unified, real-time object detection. In: Proceedings of the IEEE Conference on Computer Vision and Pattern Recognition, pp. 779–788 (2016)
25. Ren, S., He, K., Girshick, R., Sun, J.: Faster R-CNN: towards real-time object detection with region proposal networks. In: Proceedings of Conference on Advances in Neural Information Processing Systems, pp. 91–99 (2015)
26. Tian, Z., Shen, C., Chen, H., He, T.: FCOS: fully convolutional one-stage object detection. In: Proceedings of the IEEE International Conference on Computer Vision, pp. 9627–9636 (2019)
27. Xia, G.S., et al.: DOTA: a large-scale dataset for object detection in aerial images. In: Proceedings of the IEEE Conference on Computer Vision and Pattern Recognition, pp. 3974–3983 (2018)
28. Xie, X., Cheng, G., Li, Q., Miao, S., Li, K., Han, J.: Fewer is more: efficient object detection in large aerial images. Sci. China Inf. Sci. (2023)

29. Xie, X., Cheng, G., Wang, J., Yao, X., Han, J.: Oriented R-CNN for object detection. In: Proceedings of the IEEE International Conference on Computer Vision, pp. 3520–3529 (2021)
30. Xu, Y., et al.: Gliding vertex on the horizontal bounding box for multi-oriented object detection. IEEE Trans. Pattern Anal. Mach. Intell. **43**(4), 1452–1459 (2020)
31. Yang, X., Yan, J.: Arbitrary-oriented object detection with circular smooth label. In: Vedaldi, A., Bischof, H., Brox, T., Frahm, J.-M. (eds.) ECCV 2020. LNCS, vol. 12353, pp. 677–694. Springer, Cham (2020). https://doi.org/10.1007/978-3-030-58598-3_40
32. Yang, X., Yan, J., Feng, Z., He, T.: R3Det: refined single-stage detector with feature refinement for rotating object. In: Proceedings of the AAAI Conference on Artificial Intelligence, pp. 3163–3171 (2021)
33. Yang, X., Yan, J., Ming, Q., Wang, W., Zhang, X., Tian, Q.: Rethinking rotated object detection with gaussian Wasserstein distance loss. In: Proceedings of IEEE International Conference on Machine Learning, pp. 11830–11841 (2021)
34. Yang, X., et al.: SCRDet: towards more robust detection for small, cluttered and rotated objects. In: Proceedings of the IEEE International Conference on Computer Vision and Pattern Recognition, pp. 8232–8241 (2019)
35. Yang, X., et al.: Learning high-precision bounding box for rotated object detection via kullback-leibler divergence. In: Proceedings of Conference on Advances in Neural Information Processing Systems, pp. 18381–18394 (2021)
36. Yang, X., et al.: The KFIoU loss for rotated object detection. In: Proceedings of International Conference on Learning Representations (2023)
37. Yang, Z., Liu, S., Hu, H., Wang, L., Lin, S.: Reppoints: point set representation for object detection. In: Proceedings of the IEEE International Conference on Computer Vision and Pattern Recognition (2022)
38. Yao, Y., et al.: On improving bounding box representations for oriented object detection. IEEE Trans. Geosci. Remote Sens. 1–11 (2022)
39. Zhang, S., Chi, C., Yao, Y., Lei, Z., Li, S.Z.: Bridging the gap between anchor-based and anchor-free detection via adaptive training sample selection. In: Proceedings of the IEEE International Conference on Computer Vision, pp. 9759–9768 (2020)
40. Zhou, Y., et al.: MMRotate: a rotated object detection benchmark using pytorch. In: Proceedings of ACM International Conference on Multimedia (2022)

End-to-End Unsupervised Style and Resolution Transfer Adaptation Segmentation Model for Remote Sensing Images

Zhengwei Li and Xili Wang[✉]

School of Computer Science, Shaanxi Normal University, Xi'an, China
wangxili@snnu.edu.cn

Abstract. For remote sensing image unsupervised domain adaptation, there are differences in resolution except for feature differences between source and target domains. An end-to-end unsupervised domain adaptation segmentation model for remote sensing images is proposed to reduce the image style and resolution differences between the source and target domains. First, a generative adversarial-based style transfer network with residual connection, scale consistency module, and perceptual loss with class balance weights is proposed. It reduces the image style and resolution differences between the two domains and maintains the original structural information while transferring. Second, the visual attention network (VAN) that considers both spatial and channel attention is used as the feature extraction backbone network to improve the feature extraction capability. Finally, the style transfer and segmentation tasks are unified in an end-to-end network. Experimental results show that the proposed model effectively alleviates the performance degradation caused by different features and resolutions. The segmentation performance is significantly improved compared to advanced domain adaptation segmentation methods.

Keywords: Image segmentation · Remote sensing image · Unsupervised domain adaptation · Feature and resolution differences

1 Introduction

Deep neural networks based on large amounts of labeled data have made significant progress in semantic segmentation, but labeling pixel-level labels is time-consuming and laborious [1]. Unsupervised domain adaptation leverages both labeled source domain data and unlabeled target domain data and aligns the two domains to make the segmentation network perform better in the target domain data set.

Most UDA (unsupervised domain adaptation) segmentation models are poorly applied in remote sensing images since there are significant feature and spatial resolution differences in remote sensing images (may come from different sensors, imaging regions, seasons, etc.). Recently, inspired by GAN [2] (Generative adversarial networks), GAN-based style transfer methods have been proposed. Some works combine them with UDA to make the images of two domains closer in appearance by style transfer. Toldo

Q. Liu et al. (Eds.): PRCV 2023, LNCS 14428, pp. 380–393, 2024.
https://doi.org/10.1007/978-981-99-8462-6_31

et al. [3] directly add CycleGAN before the UDA segmentation network for an inter-domain style transfer. Zhao et al. [4] add DualGAN with residual connection before the UDA segmentation network to retain detailed information while style transfer. These methods directly add the GAN-based generative network to transfer the source domain image into the target domain style image. They do not use the segmentation results to constrain the generation network, thus leading to the generated images being incomplete in semantic and detailed information. To address the problems, methods using the segmentation results to constrain the GAN-based style-generated images are proposed. Li et al. [5], Cheng et al. [6], and Yang et al. [7] constrain the generative network via a perceptual loss calculated by the semantic segmentation results before and after the source domain image and target domain image style transfer. The above methods retain more semantic and detailed information when style transfer and confirm perceptual loss effectively constrain the generative network. However, the perceptual loss does not pay enough attention to the classes with fewer samples, which makes the transfer results of the classes with fewer samples poor. Such an uneven number of samples in different classes is common in remote sensing images. In addition, the above methods do not take into account the spatial resolution difference of images coming from different sensors, while the resolution difference leads to significant discrepancy in the generated images structure and the original images, thus making a severe loss of semantic and detailed information of the generated images, further affects subsequent segmentation performance.

To this end, this paper proposes an end-to-end unsupervised domain adaptation model EUSRTA (end-to-end unsupervised style and resolution transfer adaptation model) for remote sensing image segmentation. This model reduces the style and resolution differences between the two domains, constrains the style transfer using segmentation results, fuses multilayer features based on spatial and channel attention, and unifies the style, resolution transfer, and segmentation tasks in an end-to-end model. In summary, our contributions are as follows: 1) we add residual connection and scale consistency modules to the generative adversarial network to retain semantic and detailed information of remote sensing images. 2) To improve the contribution of less-sampled classes on style transfer, we propose a perceptual loss with class-balanced weights to constrain the style transfer network. 3) The visual attention network [8] is used as the backbone network of the segmentation network to capture more details and semantic features in space and channels to improve the feature extraction capability of the network. The model has been tested on the Potsdam_Vaihingen and Tibetan Plateau 2022.2_2020.5 remote sensing image semantic segmentation datasets and compared with other unsupervised domain adaptation methods. The results show that the results of the proposed model are better than those of the comparison methods on the two sets of datasets.

2 EUSRTA Model

The model consists of three parts (shown in Fig. 1): style transfer network, segmentation network, and output-level discrimination network. X_S and X_T represent the source and target domain image. X_S^T represents a source domain image transferred to a target domain style, and conversely X_T^S. $X_S^{S'}$ represents the reconstructed image to source domain style from X_S^T, and similarly the meaning of $X_T^{T'}$. $RG_{S \to T}$ and $RG_{T \to S}$ are the generative networks in the style transfer network with residual connections and scale consistency modules, D_S and D_T are the discrimination networks in the style transfer network. VAN is the backbone network of the segmentation network, and ASPP is the atrous spatial pyramid pooling. D_F is the output-level discrimination network. P_S, P_T, P_S^T, P_T^S are the segmentation prediction results of X_S, X_T, X_S^T, X_T^S.

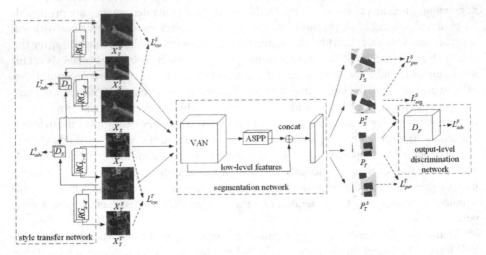

Fig. 1. The structure of EUSRTA

2.1 Style Transfer Network

Since remote sensing images contain more information and categories, features are more affluent and categories are unevenly distributed, the semantic and detailed information cannot be retained well by unrestrained image generation. In addition, the resolution is often different between the two domains in remote sensing images, which affects the segmentation performance of the UDA model. The traditional style transfer methods generally ignore the resolution difference, while eliminating the resolution discrepancies is conducive to the performance improvement of the UDA segmentation model.

Therefore, we propose modules based on DualGAN [9] to make the appearance and resolution of the two domains similar. The style transfer network adds residual connection after the generation network, which substantially preserves the semantic and detailed information of the image. To address the scale differences, a scale consistency module is added after the generated images to convert the source-domain scale images to the target-domain scale images by bilinear interpolation, which alleviates the difficulty of domain adaptation due to the scale differences between the source and target domains and further improves the quality of the generated images. The style transfer process can be expressed as follows.

$$X_S^T = RG_{S \to T}(X_S) = Resize_{S \to T}(G_{S \to T}(X_S) + X_S) \tag{1}$$

2.2 Segmentation Network

The segmentation network is based on deeplabV3+ [10], but the backbone network of the segmentation network, ResNet, is replaced by VAN. VAN imposes spatial and channel attention on features by decomposing large kernel convolution so that the segmentation network can effectively improve the feature extraction ability. VAN has four hierarchical structures, which are very similar. Each hierarchical structure is distinguished by the magnification R of the image down-sampling, the number of output channels C and the number of repetitions L of the module. The structure of one of the hierarchical structures is shown in Fig. 2. There are seven versions of VAN depending on the number of network parameters. In the experiments, deeplabV3+ with ResNet101 as the backbone network is selected as the comparison method. ResNet101 has 44.7M parameters. For a fair comparison, the B3 version with the same order of magnitude parameters as ResNet101 is chosen.

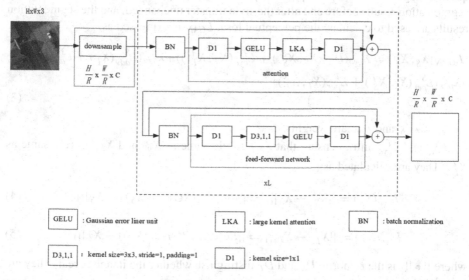

Fig. 2. The structure of VAN

2.3 Output-Level Discrimination Network

The output-level discrimination network aims to distinguish the segmentation results that come from which domain, thus forcing the internal representation of the model to be aligned between domains. The discrimination network consists of five convolution layers with a kernel size of 4×4, a step size of 2, and a padding of 1. Except for the last layer, each convolution layer is connected to a leakyReLU activation function with a negative slope of 0.2. The number of channels in each layer is set to 64, 128, 256, 512, and 1.

2.4 Loss Function

The overall loss function of EUSRTA contains five terms: style transfer loss L_{GAN}, segmentation loss L_{seg}, discrimination loss L_{DS}, L_{DT} and L_{DF} generated by the discrimination network D_S, D_T and D_F respectively. The overall loss function is:

$$L(X_S, X_T, X_S^T) = L_{GAN}(X_S, X_T) + L_{seg}(X_S^T, X_T) +$$
$$L_{DS}(X_S) + L_{DT}(X_S) + L_{DF}(P_T, P_S^T) \tag{2}$$

The style transfer network is designed to transfer the image styles of two domains to each other, but the structure of the image before and after the transfer is retained. L_{GAN} consists of three parts: the first part L_{cyc}^S and L_{cyc}^T are the cyclic consistency loss. The second part L_{adv}^S and L_{adv}^T are the adversarial loss of the two discrimination networks in the style transfer network. They constrain style transfer networks generate high-quality images. The third part L_{per}^S and L_{per}^T are the perception loss that ensure the image segmentation results before and after the style transfer unchanged, and the segmentation results are used to calculate the perceptual loss. L_{GAN} is expressed as:

$$L_{GAN}(X_S, X_T) = \lambda_{cyc}(L_{cyc}^S(X_S, X_S^{S'}) + L_{cyc}^T(X_T, X_T^{T'})) + \lambda_{adv}(L_{adv}^S(X_T) + L_{adv}^T(X_S)) +$$
$$\lambda_{per}(L_{per}^S(X_S, X_S^T) + L_{per}^T(X_T, X_T^S)) \tag{3}$$

λ_{cyc}, λ_{adv} and λ_{per} are weights.

L_{cyc}^S and L_{cyc}^T aim to ensure that X_S entirely same as $X_S^{S'}$ and X_T entirely same as $X_T^{T'}$. They are calculated as:

$$L_{cyc}^S(X_S) = ||X_S^{S'} - X_S||_1 = ||RG_{T \to S}(RG_{S \to T}(X_S)) - X_S||_1 \tag{4}$$

$$L_{cyc}^T(X_T) = ||X_T^{T'} - X_T||_1 = ||RG_{S \to T}(RG_{T \to S}(X_T)) - X_T||_1 \tag{5}$$

where $|| \bullet ||_1$ is the ℓ_1 norm. D_S and D_T distinguish whether the image is real. They are calculated as:

$$L_{adv}^S(X_T) = -D_S(RG_{T \to S}(X_T)) \tag{6}$$

$$L_{adv}^T(X_S) = -D_T(RG_{S \to T}(X_S)) \tag{7}$$

In order to avoid the gradient disappearance, Wasserstein GAN loss [11] is used to calculate the discrimination loss.

The perceptual loss L_{per}^S and L_{per}^T makes the style transfer network retain the semantic and detailed information better, and can be calculated by P_S, P_T, P_S^T and P_T^S. However, the perceptual loss is mainly affected by those classes with more samples, and the classes with fewer samples have little effect on it. The source domain has labels and the weights can be calculated. But the labels of the target domain are unknown. We segment the target domain images by the segmentation network pre-trained via the source domain and obtain the pseudo-labels, then calculate the weights. This newly defined perceptual loss balances the roles played by all classes:

$$L_{per}^S(X_S) = \frac{1}{HW} \sum_{i=1}^{c} \frac{\lambda_{max}}{\lambda_i} ||P_{S\_i} - P_{S\_i}^T||_2^2 \tag{8}$$

$$L_{per}^T(X_T) = \frac{1}{HW} \sum_{i=1}^{c} \frac{\lambda_{max}}{\lambda_i} ||P_{T\_i} - P_{T\_i}^S||_2^2 \tag{9}$$

where $|| \bullet ||_2^2$ is the square of the ℓ_2 norm, λ_i is the sample number of class i, and λ_{max} is the sample number of the class with the most sample.

L_{seg} consists of the segmentation loss of the source domain image L_{seg}^S and the adversarial loss L_{adv}^F that adapts the segmentation network to the changes in the target domain features:

$$L_{seg}(X_S^T, X_T) = L_{seg}^S(X_S^T) + \lambda_{adv}^F L_{adv}^F(X_T) = $$
$$-Y_S \log(P_S^T) - \lambda_{adv}^F \log(D_F(P_T)) \tag{10}$$

λ_{adv}^F is the loss weight. The segmentation loss of the source domain image is calculated using the cross-entropy loss function, while the segmentation loss of the target domain image is by the predicted result of the target domain image and the updated discriminator.

Discrimination loss L_{DS} and L_{DT} are calculated as:

$$L_{DS}(X_T) = D_S(X_S) - D_S(X_T^S) \tag{11}$$

$$L_{DT}(X_S) = D_T(X_T) - D_T(X_S^T) \tag{12}$$

where D_S treats X_S as a positive sample and X_T^S as a negative sample. D_T treats X_T as a positive sample and X_S^T as a negative sample. L_{DF} is calculated as:

$$L_{DF}(P_T, P_S^T) = \log(D_F(P_S^T)) - \log(D_F(P_T)) \tag{13}$$

D_F treats P_S^T as a positive sample and P_T as a negative sample. Each of the three discrimination network updates using its discrimination loss functions respectively.

3 Experiment Results

3.1 Dataset

EUSRTA is evaluated using two set of remote sensing image semantic segmentation datasets Potsdam_Vaihingen [12] and Tibetan Plateau 2022.2_2020.5. Potsdam_Vaihingen have a spatial resolution of 5 cm and 9 cm, thus, Potsdam is cropped to 896 × 896 pixels and Vaihingen is cropped to 512 × 512 pixels. Tibetan Plateau 2022.2_2020.5 have the same resolution. They are all cropped to 512 × 512 pixels. The training set contains all source domain images and target domain images, and the validation set and test set occupy 30% and 70% of the target domain. The class distribution of two sets of datasets are shown in Table 1 and Table 2. It is obvious that the classes are unevenly distributed.

Table 1. Class distribution of labels in the Potsdam and Vaihingen dataset

Class	Clu	Imp	Car	Tree	Low veg	Building
Potsdam	4.7%	31.4%	1.7%	15.3%	22.0%	24.8%
Vaihingen	0.8%	22.7%	1.2%	23.1%	21.3%	25.9%

Table 2. Class distribution of labels in the Tibetan Plateau 2022.2 and 2020.5 dataset

Class	road	farmland	snow	construction	building	greenbelt	mountainous	water
Potsdam	4.33%	3.62%	7.43%	11.98%	5.47%	2.54%	62.57%	2.05%
Vaihingen	4.33%	3.33%	1.46%	12.30%	5.39%	3.93%	66.32%	2.93%

3.2 Implementation Details

To train the style transfer network, the generative network is optimized using Adam with an initial learning rate set to 5×10^{-4}. The discrimination network is optimized using RMSProp with an initial learning rate set to 5×10^{-4}. To train the segmentation network, the optimization is performed using SDG with an initial learning rate set to 2×10^{-4} and a weight decay of 2×10^{-4}. When training the output-level discrimination network, the optimization is performed using Adam, with the initial learning rate set to 2×10^{-4}, the weight decay to 2×10^{-4}. For hyper-parameters, λ_{cyc} set to 10, λ_{adv} set to 5, λ_{per} set to 0.1, and λ_{adv}^{F} set to 0.02. Two images are read randomly for each training batch, and the number of iterations is set to 100.

3.3 State-of-the-Art Comparison

EUSRTA is compared with AdaptSegNet [13], SEANet [14], SAA [15], SDSCA-Net, MemoryAdaptNet [16], CCDA_LGFA [17], and ResiDualGAN [4]. The evaluation metrics are IoU, mIoU and F1-score. The data set partition and parameter setting of the above comparison methods are the same as those of EUSRTA.

As seen in Table 3, compared with other UDA methods in P_V dataset, the proposed EUSRTA achieves the highest values of mIoU for clutter, impervious surfaces, trees, low vegetation, buildings, and overall. Benefit from the improved style transfer networks and VAN, mIoU and F1-score of EUSRTA are improved by 19.35% and 19.67% compared with the traditional adversarial discrimination-based UDA segmentation network AdaptSegNet. In addition, both the mIoU and F1-score of EUSRTA are higher than other UDA methods, but only the results of car are lower than those of ResiDualGAN. Resnet-based ResiDualGAN downsamples the input images by 16 times in the encoding phase of the segmentation network to extract features, while VAN-based EUSRTA downsamples the input images by 32 times, this may result in an inferior feature presentation of EUSRTA for smaller objects (i.e., car). But ResiDualGAN is not an end-to-end model, thus the style transfer process cannot be constrained by the segmentation result, and the parameters of the style transfer network cannot be adjusted according to each iterated segmentation result, therefore the retained semantic and detailed information of the generated images are incomplete. In summary, EUSRTA shows better performance than other advanced methods. The segmentation comparison results are given in Fig. 3.

Table 3. Comparison of domain adaptation indices of different methods in P_V

Method	Clu		Imp		Car		Tree		Low veg		Building		Overall	
	IoU	F1	IoU	F1	IoU	F1	IoU	F1	IoU	F1	IoU	F1	mIoU	F1
AdaptSegNet	9.01	16.53	54.76	70.77	25.17	40.21	55.52	71.4	14.88	25.90	56.48	72.19	35.97	49.50
SEANet	16.91	28.93	49.69	66.39	34.02	50.77	55.28	71.20	16.90	28.91	66.13	79.61	39.82	54.30
SAA	14.83	25.83	58.38	73.72	38.43	55.53	58.37	73.72	29.03	44.99	59.99	74.99	43.17	58.13
SDSCA-Net	17.21	29.36	59.77	74.82	34.53	51.33	60.11	75.09	38.77	55.87	66.93	80.19	46.22	61.11
Memory-AdaptNet	21.26	35.07	65.97	79.49	41.66	58.51	53.55	69.75	44.43	61.52	68.99	81.65	49.31	64.38
CCDA_LGFA	17.83	30.26	67.55	80.63	39.72	56.85	58.56	73.86	43.30	60.53	76.00	86.37	50.51	64.75
ResiDualGAN	15.20	26.38	70.86	82.94	**46.99**	**63.94**	53.05	69.33	47.50	64.41	78.26	87.81	51.98	65.80
EUSRTA	**23.75**	**38.39**	**72.27**	**83.90**	41.88	59.03	**65.09**	**78.85**	**49.76**	**66.45**	**79.17**	**88.37**	**55.32**	**69.17**

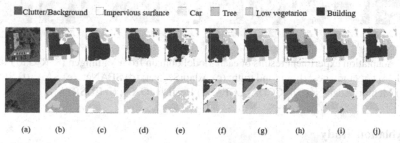

Fig. 3. Adaptation experiment results of each model domain from P_V. (a) target images; (b) ground truth; (c) AdaptSegNet; (d) SEANet; (e) SAA; (f) SDSCA-Net; (g) MemoryAdaptNet; (h) CCDA_LGFA; (i) ResiDualGAN; (j) EUSRTA

Table. 4 shows the results of V_P adaptation, and it can be seen that the model proposed in this paper still achieves the best overall mIoU and F1-score. The resolution of the Vaihingen dataset is lower than that of the Potsdam dataset because images with higher resolution will contain more detailed features, so comparing Table 3, the evaluation metrics are decreased, but in the overall mIoU and F1-score are better than the other models. The segmentation comparison results are given in Fig. 4.

Tables 5 and Tables 6 show the 22_20 and 20_22 domain adaptation results. Since SAA and SDSCA-Net can only be applied to datasets with different resolutions, there is no comparison between SAA and SDSCA-Net. Compared to other methods, EUSRTA still obtains the best overall mIoU and F1-score.

Table 4. Comparison of domain adaptation indices of different methods in V_P

Method	Clu		Imp		Car		Tree		Low veg		Building		Overall	
	IoU	F1	IoU	F1	IoU	F1	IoU	F1	IoU	F1	IoU	F1	mIoU	F1
AdaptSegNet	13.63	23.99	49.94	66.62	4.17	8.00	40.47	57.63	38.33	55.42	49.80	66.49	32.73	46.36
SEANet	11.77	21.07	52.16	68.56	35.48	52.38	8.42	15.53	21.89	35.92	38.15	55.23	27.98	41.45
SAA	17.72	30.10	52.07	68.49	43.89	61.01	33.72	50.44	43.44	60.57	50.93	67.49	40.30	56.35
SDSCA-Net	16.25	27.95	54.16	70.27	48.42	65.25	42.56	59.70	47.45	64.36	53.45	69.67	43.71	59.53
Memory-AdaptNet	7.28	13.57	65.29	79.00	55.90	71.72	42.34	59.50	38.01	55.09	72.30	83.93	46.86	60.47
CCDA_LGFA	13.67	24.05	**68.36**	**81.20**	48.72	65.52	26.33	41.68	42.55	59.70	**76.89**	**86.93**	46.08	59.85
ResiDualGAN	15.75	27.21	63.15	77.41	**62.47**	**76.89**	40.85	58.01	45.07	62.13	69.05	81.69	49.39	63.89
EUSRTA	**17.83**	**30.27**	65.36	79.05	51.48	67.97	**56.22**	**71.97**	**53.11**	**69.37**	73.04	84.42	**52.84**	**67.18**

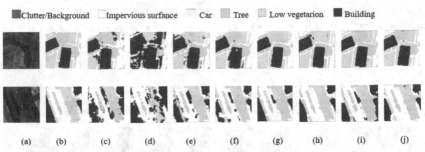

■Clutter/Background ▫Impervious surface ▫Car ■Tree ▫Low vegetarion ■Building

(a)   (b)   (c)   (d)   (e)   (f)   (g)   (h)   (i)   (j)

Fig. 4. Adaptation experiment results of each model domain from Vaihingen IR-R-G to Potsdam IR-R-G. (a) target images; (b) ground truth; (c) AdaptSegNet; (d) SEANet; (e) SAA; (f) SDSCA-Net; (g) MemoryAdaptNet; (h) CCDA_LGFA; (i) ResiDualGAN; (j) EUSRTA

3.4 Ablation Study

To demonstrate the effectiveness of the EUSRTA, we show an ablation analysis on VAN, residual connection, scale consistency module, perceptual loss, and class balance weights. The settings of the ablation experiments are shown in Table 7 and the results are shown in Table 8.

As can be seen from the experimental results, comparing Experiments 1 and 2, replacing the backbone network from ResNet101 to VAN, the mIoU and F1-score are improved by 5.5% and 6.47%. It indicates that the attention mechanism helps to select the same features in two domains and reduces the influence of irrelevant features. Comparing Experiments 2 and 3, using DualGAN as the base style transfer network. Because the remote sensing images contain more information and the two datasets have different scales, the ability to retain semantic and detailed information of the generative adversarial network is reduced, and the quality of the generated images is poor. Thus mIoU decreases by 4.66% and F1-Score decreases by 5.72%. Comparing Experiments 3 and 4, a scale-consistent module is added to the generative network of DualGAN. Because convolutional neural networks are sensitive to image scale, aligning the scale difference between two domains can effectively improve segmentation accuracy. The mIoU

Table 5. Comparison of domain adaptation indices of different methods in 22_20

Method	Road		farmland		snow		construction		building		greenbelt		mountainous		water		Overall	
	IoU	F1	IoU	F1	IoU	F1	IoU	F1	IoU	F1	IoU	F1	IoU	F1	IoU	F1	mIoU	F1
AdaptSegNet	13.50	23.79	31.12	47.47	46.59	63.56	23.75	38.39	38.76	55.86	0.89	1.77	77.81	87.52	60.91	75.71	36.67	49.26
SEANet	10.27	18.63	9.75	17.77	56.52	72.22	35.64	52.55	41.76	58.92	0.14	0.29	82.61	90.48	59.26	74.42	36.99	48.16
CCDA_LGFA	23.96	38.66	26.22	41.55	**75.39**	**86.62**	**35.86**	**52.78**	45.19	62.25	**24.84**	**39.80**	82.14	90.19	58.35	73.70	46.62	60.69
MemoryAdaptNet	**28.51**	**44.37**	**40.79**	**57.95**	62.23	76.72	34.60	51.41	**49.96**	**66.63**	14.75	25.70	82.65	90.50	61.08	75.83	46.82	61.14
ResiDualGAN	24.13	38.88	36.35	53.32	72.84	84.28	32.10	48.59	47.74	64.66	2.08	4.09	**83.12**	**90.78**	57.93	73.36	44.54	57.25
EUSRTA	24.22	39.00	40.56	57.72	68.85	81.55	31.16	47.52	47.24	64.16	20.22	33.64	83.01	90.72	**70.31**	**82.57**	**48.20**	**62.11**

Table 6. Comparison of domain adaptation indices of different methods in 22_20

Method	Road		farmland		Snow		construction		building		greenbelt		mountainous		water		Overall	
	IoU	F1	IoU	F1	IoU	F1	IoU	F1	IoU	F1	IoU	F1	IoU	F1	IoU	F1	mIoU	F1
AdaptSegNet	19.86	33.13	23.71	38.33	56.22	71.98	30.96	47.29	45.13	62.19	8.47	15.62	64.69	78.56	12.72	22.57	32.72	46.21
SEANet	2.90	5.63	5.76	10.90	61.45	76.12	26.05	41.33	30.32	46.53	16.14	27.78	69.37	81.92	13.09	23.15	28.14	39.17
CCDA_LGFA	20.08	33.45	24.25	39.03	19.38	32.47	25.71	40.91	48.76	65.55	3.41	6.60	77.24	87.16	70.92	82.99	36.22	48.52
MemoryAdaptNet	16.21	27.90	30.74	47.03	56.94	72.56	35.76	52.68	44.11	61.21	14.88	25.92	69.26	81.84	25.72	40.91	37.70	51.26
ResiDualGAN	29.96	46.11	37.66	54.72	19.12	32.10	35.80	52.73	55.08	71.03	4.87	9.29	77.12	87.08	54.71	70.72	39.29	52.97
EUSRTA	18.66	31.45	28.50	44.36	46.31	63.31	36.07	53.02	48.62	65.43	6.29	11.84	76.66	86.78	59.95	74.96	40.14	53.89

Table 7. Setup of ablation experiments from Potsdam to Vaihingen

number	ResNet101	VAN	DualGAN	resize model	residual connection	perceptual loss	class balance weight
1	√	×	×	×	×	×	×
2	×	√	×	×	×	×	×
3	×	√	√	×	×	×	×
4	×	√	√	√	×	×	×
5	×	√	√	√	√	×	×
6	×	√	√	√	√	√	×
7	×	√	√	√	√	√	√

Table 8. Results of ablation experiments from Potsdam to Vaihingen

Method	Clu		Imp		Car		Tree		Low veg		Building		Overall	
number	IoU	F1	IoU	F1	IoU	F1	IoU	F1	IoU	F1	IoU	F1	mIoU	F1
1	6.34	11.93	48.95	65.73	17.97	30.47	56.21	71.96	21.62	35.56	56.04	71.83	34.52	47.91
2	9.58	17.49	51.92	68.35	27.48	43.55	59.35	74.49	30.35	46.57	61.07	75.83	40.02	54.38
3	2.45	4.79	44.03	61.14	33.90	50.64	41.04	58.20	19.78	33.02	70.94	82.29	35.36	48.46
4	13.79	24.24	61.12	75.87	37.83	54.90	55.51	71.39	39.76	56.89	66.31	79.74	45.72	60.51
5	10.97	19.78	65.83	79.40	**45.20**	**62.26**	53.21	69.46	44.50	61.59	76.52	86.70	49.37	63.20
6	15.10	26.24	**72.34**	**83.95**	38.83	55.94	**65.55**	**79.19**	48.72	65.52	**80.85**	**89.41**	53.57	66.71
7	**23.75**	**38.39**	72.27	83.90	41.88	59.03	65.09	78.85	**49.76**	**66.45**	79.17	88.37	**55.32**	**69.17**

improves by 10.36%, and an F1-Score improves by 12.05%. Comparing Experiments 4 and 5, Experiment 5 adds residual connection to retain semantic and detail information in the generative network, mIoU improves by 3.65% and F1-Score improves by 2.69%. Comparing Experiments 5 and 6, Experiment 6 adds perceptual loss to 5, although the overall mIoU improves by 4.20% and F1-Score improves by 3.51%, the car class accuracy decreases. Because perceptual loss is biased to the correction of the category with more samples, it cannot give a better correction to the car class. Comparing Experiments 6 and 7, Experiment 7 adds class balance weights to improve the contribution of less-sampled categories to perceptual loss. The evaluation metrics of more-sampled categories such as impervious surface, trees, and buildings have a slight decrease, but the evaluation metrics of less-sampled categories such as background, cars, and low vegetation increase significantly, and the overall mIoU improves by 1.75% and F1-Score improves by 2.46%. The segmentation results are given in Fig. 5.

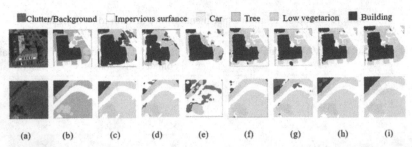

Fig. 5. Results of cross-domain semantic segmentation tasks from Potsdam to Vaihingen in ablation experiments. (a) The original image; (b) ground truth; (c) deeplabV3+ with backbone network of ResNet101; (d) deeplabV3+ with backbone network of VAN; (e) Add DualGAN to (d); (f) Add scale consistency modules to (e); (g) Add residual connection to (f); (h) Add perceptual loss to (g); (i) Add class balance weight to (h);

4   Conclusion

This paper proposes an end-to-end unsupervised style and resolution transfer adaptation segmentation model for remote sensing images. Firstly, the residual connections, scale consistency module, and perceptual loss with class balance weights are added to the generative adversarial-based style transfer network to improve the style transfer network's semantic information and detail information retention capability. Secondly, VAN is used as the backbone network to provide attention to feature extraction. The experiment results show that EUSRTA is better to cope with the transfer challenges brought by the different feature distributions and resolution of the source and target domain images, and the superiority of the proposed method over other state-of-the-art methods.

Acknowledgements. This work was supported by the Second Tibetan Plateau Scientific Expedition and Research (2019QZKK0405).

References

1. Liang, M., Wang, X.L.: Domain adaptation and super-resolution based bi-directional semantic segmentation method for remote sensing images. In: International Geoscience and Remote Sensing Symposium (IGARSS), pp. 3500–3503. IEEE (2022)
2. Ian, G., Jean, P., Mehdi, M.: Generative adversarial nets. In: Neural Information Processing Systems (NIPS), pp. 2672–2680. MIT Press (2014)
3. Toldo, M., Michieli, U., Agresti, G.: Unsupervised domain adaptation for mobile semantic segmentation based on cycle consistency and feature alignment. Image Vis. Comput. **95**, 103889 (2020)
4. Zhao, Y., Gao, H., Guo, P.: ResiDualGAN: Resize-Residual DualGAN for cross-domain remote sensing images semantic segmentation. Remote Sens. **15**(5), 1428 (2023)
5. Li, Y., Yuan, L., Vasconcelos, N.: Bidirectional learning for domain adaptation of semantic segmentation. In: International Conference on Computer Vision and Pattern Recognition (CVPR), pp. 6936–6945. IEEE (2019)

6. Cheng, Y., Wei, F., Bao, J.: Dual path learning for domain adaptation of semantic segmentation. In: International Conference on Computer Vision (CVPR), pp. 9082–9091. IEEE (2021)
7. Yang, J., An, W., Wang, S., Zhu, X., Yan, C., Huang, J.: Label-driven reconstruction for domain adaptation in semantic segmentation. In: Vedaldi, A., Bischof, H., Brox, T., Frahm, J.-M. (eds.) ECCV 2020. LNCS, vol. 12372, pp. 480–498. Springer, Cham (2020). https://doi.org/10.1007/978-3-030-58583-9_29
8. Guo, M.H., Lu, C.Z., Liu, Z.M.: Visual attention network. arXiv preprint arXiv:2202.09741 (2022)
9. Yi, Z., Zhang, H., Tai, P.: DualGAN: unsupervised dual learning for image-to-image translation. In: International Conference on Computer Vision (ICCV), pp.2849–2857. IEEE (2017)
10. Chen, L.-C., Zhu, Y., Papandreou, G., Schroff, F., Adam, H.: Encoder-decoder with atrous separable convolution for semantic image segmentation. In: Ferrari, V., Hebert, M., Sminchisescu, C., Weiss, Y. (eds.) ECCV 2018. LNCS, vol. 11211, pp. 833–851. Springer, Cham (2018). https://doi.org/10.1007/978-3-030-01234-2_49
11. Arjovsky, M., Chintala, S., Bottou, L.: Wasserstein generative adversarial networks. In: International Conference on Machine Learning (ICML), pp. 214–223. ACM (2017)
12. Gerke, M.: Use of the stair vision library within the ISPRS 2D semantic labeling benchmark (Vaihingen) (2015). https://doi.org/10.13140/2.1.5015.9683
13. Tsai, Y.H., Hung, W.C., Samuel, S.: Learning to adapt structured output space for semantic segmentation. In: International Conference on Computer Vision and Pattern Recognition (CVPR), pp. 7472–7481. IEEE (2018)
14. Xu, Y., Du, B., Zhang, L.: Self-ensembling attention networks: addressing domain shift for semantic segmentation. In: the AAAI Conference on Artificial Intelligence (AAAI), pp. 5581–5588. AAAI (2019)
15. Deng, X.Q., Zhu, Y., Tian, Y.X.: Scale aware adaptation for land-cover classification in remote sensing imagery. In: Winter Conference on Applications of Computer Vision (WACV), pp. 2160–2169. IEEE (2021)
16. Zhu, J., Guo, Y., Sun, J., Yang, L., Deng, M., Chen, J.: Unsupervised domain adaptation semantic segmentation of high-resolution remote sensing imagery with invariant domain-level prototype memory. IEEE Trans. Geosci. Remote Sens. **61**, 1–18 (2023)
17. Zhang, B., Chen, T., Wang, B.: Curriculum-style local-to-global adaptation for cross-domain remote sensing image segmentation. IEEE Trans. Geosci. Remote Sens. **60**, 1–12 (2021)

A Physically Feasible Counter-Attack Method for Remote Sensing Imaging Point Clouds

Bo Wei[1], Huanchun Wei[2]([✉]), Cong Cao[3], Teng Huang[3]([✉]), Huagang Xiong[1], Aobo Lang[4], Xiqiu Zhang[4], and Haiqing Zhang[4]([✉])

[1] School of Electronic and Information Engineering,
Beihang University, Beijing, China
[2] College of Beidou, Guangxi University of Information Engineering Guangxi,
Nanning, China
whc191021@163.com
[3] Institute of Artificial Intelligence and Blockchain,
Guangzhou University, Guangzhou, China
huangteng1220@gzhu.edu.cn
[4] School of Arts, Sun Yat-sen University, Guangzhou, China
zhanghq56@mail.sysu.edu.cn

Abstract. This research introduces an innovative approach to address the vulnerability of deep learning-based point cloud target recognition systems to adversarial sample attacks. Instead of directly tampering with the spatial location information of the point cloud data, the approach focuses on modifying specific attribute information. The modified point cloud data is then associated with a simulated physical environment for comprehensive attack testing. The experimental findings demonstrate the aggressive nature of the generated counter-samples achieved through signal amplitude modulation, effectively deceiving deep learning-based target recognition systems. The experimental results highlight the effectiveness of the adversarial object samples generated by modifying the signal amplitude of the point cloud data, showcasing their strong misguiding capabilities towards the deep learning-based target recognition algorithm. This approach ensures good stealthiness and practicality, making it a viable attack method applicable in physical scenarios.

Keywords: Point Cloud · Target Recognition · Counter-sample Attacks

1 Introduction

Point cloud data refers to a set of 3D point information that represents the surface of an object. It differs from traditional 3D data by having minimal

Supported by the National Natural Science Foundation of China under Grant 62002074.

redundancy and being less influenced by lighting conditions and image quality. This unique characteristic has led to its wide application in various domains, including 3D matching [1,2], target recognition [3–5], and semantic segmentation [6,7]. However, recent investigations have exposed the susceptibility of deep learning-based point cloud target recognition methods to adversarial attacks. In such attacks, adversaries can generate apparently innocuous point cloud samples designed to deceive deep learning models. Currently, some existing attack methods focus on perturbing point cloud data, such as modifying the coordinate information of some point clouds [8], adding interfering points [9,10], etc. However, these methods are mainly limited to attack tests in the digital world and are difficult to be migrated to real scenarios in the physical world. Therefore, it remains a challenging problem to map entities into point cloud data in the physical world and construct effective adversarial samples to interfere with the recognition of the model.

To tackle the aforementioned challenges, this study introduces a novel approach to combat adversarial attacks on point cloud data. Unlike directly modifying the coordinates of point cloud data or adding additional interference points, this method tests the attack by modifying specific attribute information and correlating the data to the simulated physical environment. As depicted in Fig. 1, the approach comprises two essential components: the RMD(Real Mapping to Digital) module and the DPR(Digital Project Back to Real) module, which is used to correlate the digital world with the physical world. The RMD module converts physical world entities into sensitive point clouds by modifying the signal amplitude. The DPR module transforms the sensitive point clouds into physical world target entities and constructs adversarial objects in the emulated physical environment. This paper makes the following significant contributions:

- An algorithm for adversarial sample generation of point cloud data is proposed, which is based on remote sensing imaging, extending the attack target from LiDAR to remote sensing radar.
- From a fresh angle, a novel approach is introduced to create impactful point clouds by adjusting the signal amplitude of the point cloud data.
- The utilization of techniques that modulate the diffuse reflectance of objects' surfaces for creating adversarial objects in the physical world presents a stealthy and manipulable strategy.

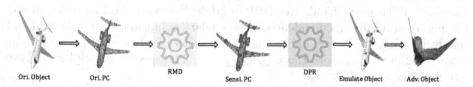

Ori. Object   Ori. PC   RMD   Sensi. PC   DPR   Emulate Object   Adv. Object

Fig. 1. Introduction diagram. (Ori.: Original. Gene.: Generate. PC.: Point Cloud. Disturb.: Disturbance. Sensi.: Sensitive. Adv.: Adversarial.)

2   Related Work

Point cloud target recognition employs neural networks [11–13] to extract features and enhance recognition efficiency. Voxelmorphology and point neural networks are the two main approaches currently used. Voxel-based methods classify point cloud data by dividing point cloud space into multiple regular subspaces (voxels) such as the Voxnet model [14] and Voxelnet model [15]. However, this method faces limitations such as low resolution leading to errors and performance restrictions, and high resolution leading to increased computation [16,17]. Point-based approaches employ models with powerful feature expression capabilities to directly learn features from point cloud data, achieving end-to-end target recognition. PointNet [18] and PointNet++ [19] are two popular methods in point cloud target recognition based on point-based approaches, and this paper will use these two classic models for experimentation. Specifically, the Point-Net model will be used to evaluate sensitive point clouds generation algorithms, while the PointNet++ model will be used to test the transferability of generated sensitive point clouds.

By manipulating point cloud data, attackers can carry out adversarial attacks on recognition models. These attacks involve generating adversarial samples that deceive the models. In physical attacks, adversaries can create adversarial objects using 3D printing techniques based on manipulated point cloud data. This approach allows them to evade detection by LiDAR systems, but it is prone to detection, leading to unsuccessful attacks. Another method involves injecting false point cloud data into the adversarial samples. This technique takes advantage of the LiDAR ranging principle and modifies signals during the scattering process to generate false points. However, it requires high-precision devices for real-time capture and transmission of LiDAR signals, making it impractical in complex real-world environments.

3   Methods

This article presents a novel method for generating adversarial object samples specifically targeting remote sensing point clouds. As shown in Fig. 1, two key modules (RMD and DPR) are employed to generate adversarial entities suitable for the physical world. The RMD module is based on the gradient concept and utilizes a binary search algorithm to adaptively adjust a balancing factor. By considering the characteristics of the target entity, it generates sensitive point clouds. On the other hand, the DPR module leverages the principles of ray tracing to establish correspondences between data points in the digital world and real-world target objects. This enables the transfer of adversarial perturbations from the digital domain to the physical world. The specific principles of these two modules are as follows:

3.1   Real Mapping to Digital Module

Point cloud data based on remote sensing imaging is a representation of how radar sensors perceive the physical world through electromagnetic waves. Each point cloud data corresponds to an echo signal received by the sensor, which contains information on the coordinates and amplitude of the signal. In the domain of remote sensing, the point cloud data, which comprises a set of n points, can be denoted as $X \subset R^{n \times 4}$. Each point x is depicted as a vector $[x_s, r_s, s_s, A_s]$, where x_s, r_s, and s_s represent the azimuth, distance, and elevation of the signal, respectively. A_s reflects the amplitude of the signal and captures the characteristics of the target being imaged. Building upon this characteristic, this research introduces an innovative approach in Fig. 2 to produce vulnerable point clouds by manipulating the signal amplitude of the point cloud. Specifically, an elaborate perturbation $p \in R^{n \times 1}$ is added to the fourth dimension of the original point clouds x to generate an sensitive point clouds x' such that $x' \leftarrow x \oplus p$.

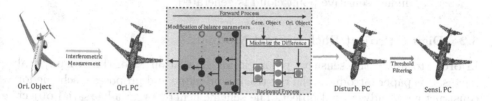

Fig. 2. The sensitive point clouds generation method diagram.

(i) *Point cloud data obtainment.* The retrieval of point cloud data related to the target object is accomplished through an interferometric technique [20].
(ii) *Gaining against perturbations.* In this study, Euclidean distance will be used as a metric for perturbations to control and generate small perturbations that are aggressive in nature. The size of this perturbation can be expressed as $\|p\|_2 = \sqrt{\sum_{i=1}^{n} p_i^2}$. The objective function set when performing an untargeted attack is Eq. 1.

$$min \|p\|_2 , s.t. F(x \oplus p) \neq t \tag{1}$$

Here, t represents the accurate labels assigned to clean samples, and $F(\bullet)$ represents the classifier that has been trained. Since the optimization problem for this objective function is difficult to solve directly, this paper converts it into a gradient-based optimization algorithm as shown in Eq. 2.

$$min \{L_{adv} (x \oplus p) + \lambda * \|p\|_2\} \tag{2}$$

where λ denotes the balancing factor, used to adjust the weights of the adversarial loss and perturbation size so that the perturbation can be optimized towards the targeted attack while being appropriately reduced. $L_{adv}(\bullet)$ quantifies the difference between the logit output of the classifier and the desired target outcome. This difference becomes smaller when the attack has a higher chance of succeeding. In the targetless attack condition, $L_{adv}(\bullet)$ denote Eq. 3.

$$L_{adv}(x) = Max \left\{ \mathbb{Z}(x)_t - \underset{i \neq t}{Max} \{\mathbb{Z}(x)_i\} , 0 \right\} \tag{3}$$

Here, $\mathbb{Z}(x)_i$ represents the ith component of the logit output produced by the classifier.

(iii) *Adaptive adjustment of the equilibrium coefficient.* The choice of the equilibrium coefficient λ directly impacts the effectiveness of the attack and the magnitude of the introduced perturbation. To generate an aggressive minimum perturbation, this paper uses a dichotomous search algorithm to set the balance factor adaptively on the input samples. Firstly, the balance factor λ and its upper and lower bounds (λ_{upper} and λ_{lower}) are initialized, secondly, a fixed number of iterations *eps* is defined, and the upper and lower limits of the equilibrium coefficient λ are adjusted based on the variation in perturbation magnitude at each iteration; finally, the complete remote sensing image sensitive point cloud is generated.

3.2   Digital Project Back to Real Module

In order to apply the sensitive point clouds generation algorithm to physical scenes, this paper introduces the ray-tracing principle and proposes a scheme for constructing an adversarial object. The scheme generates an adversarial object by modifying the diffuse reflectance of the target object surface, as shown in Fig. 3.

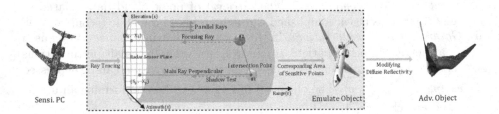

Fig. 3. The Adversarial Object Generation Method Diagram.

Ray tracing is used to obtain the reflected signal of a ray after it has interacted with an object by tracing the path of the ray through the scene. By tracing the path of the ray as it interacts with the object, we can obtain the correspondence between each sensitive point clouds and the object's surface position. Typically, electromagnetic rays emitted by radar are difficult to track and need to be completed in a modeled physical coordinate system. The specific procedure consists of the following steps:

(i) *Modelling of physical coordinate systems.* First, a 3D scene of the target is created (as shown in Fig. 3). The x, s, and r axes represent the azimuth, elevation, and distance directions, respectively; secondly, an orthogonal projection camera is defined to simulate the radar sensor; finally, the sensor emits multiple parallel rays simultaneously in the direction of the target through a static system to directly obtain the emulated object data.

(ii) *Tracking of scene objects.* Track the scene objects. Place two spherical objects in the 3D scene and perform the following operations. First, create a primary ray with coordinates (x_p, s_p) perpendicular to the plane; second, record the coordinates of the ray and the first reflected signal from object 1 $(x_s = x_p, r_s = r_1, s_s = s_p)$; next, perform a shadow test to determine if the intersection point is in shadow; finally, calculate the signal based on the type of reflection at the intersection point amplitude. The equations for diffuse and specular reflections are Eq. 4 and Eq. 5 respectively.

$$I_d = F_d \cdot I_{sig} \cdot (N \cdot L)^{F_b} \tag{4}$$

$$I_s = F_s \cdot (N \cdot H)^{\frac{1}{F_b}} \tag{5}$$

in this context, F_d and F_s represent the factors for diffuse and specular reflection, respectively. I_{sig} denotes the intensity of the incident signal, N represents the normal vector that is perpendicular to the surface, L is the normalized vector pointing towards the light source, F_b is the surface brightness factor (with a default value of 1), H is the vector pointing towards the light source L and the reciprocal component of the surface's normal vector N, and F_r is the roughness factor that defines the sharpness of the specular surface's high gloss.

If a secondary reflection is produced at point #1, the direction of the reflection is obtained based on the direction of the normal and primary ray at point #1. Next, a fresh ray is projected in this specific direction, assumed to intersect with object 2 at point #2. Following that, a shadow test is executed. To acquire the secondary reflection's coordinates, a ray that runs parallel to the primary ray (known as the focal ray) is generated, commencing from the intersection point #2. This ray will intersect the sensor plane at the point (x_0, s_0), and using this intersection point, the coordinates of the second reflection signal can be calculated by the following Eq. 6.

$$x_s = \frac{x_0 + x_p}{2} \quad r_s = \frac{r_1 + r_2 + r_3}{2} \quad s_s = \frac{s_0 + s_p}{2} \tag{6}$$

The amplitude of the second reflection signal can be obtained by weighting it according to the reflection coefficient of object 1. Continuing in a similar manner, the pursuit of the third reflection coordinates perseveres until the count of reflections attains the pre-defined upper threshold.

Moreover, each individual point within the point cloud can approximately depict a localized portion of data corresponding to a minute section of the object's surface, taking into account the gridded nature of the sensor plane. Based on the mapping relationship between signal amplitude and diffuse reflectance $Diffuse = f(Amplitude)$, the diffuse reflectance of the adversarial object surface corresponding to the sensitive point is calculated. By covering the surface of the object with a material of a specific material, the diffuse reflectance of the modified area is changed to the specified size, and the adversarial object is obtained.

4   Experimental Analysis and Discussion

4.1   Datasets and Classifiers

To generate the experimental dataset, first, three types of geometric models, denoted as *Step*1, *Step*2, and *Step*3, were generated using RaySAR. They are simulation entities composed of different numbers $(1, 2, 3)$ of cuboids. When rendering the model, the simulated radar sensor is located on a circle centered at the origin $(0, 0, 0)$ with a radius of 60 and a 45° angle to the horizontal plane; secondly, the sensor position is kept constant, and the model is rotated on the vertical axis where the origin is located, generating a new model with each 1° rotation; finally, the model data is collected using the RaySAR rendering operation and the first four dimensions are extracted Finally, the model data is collected using the RaySAR rendering operation and the first four dimensions are extracted to produce a point cloud sample.

The point cloud data samples obtained through the above steps contain information not only from multiple reflections of the radar signal on the target object but also from reflections from the ground. However, multiple reflections will result in less valid information carried by the electromagnetic waves, and in order to effectively generate counter samples, this thesis will only keep the point cloud data from one reflection. These point cloud data are pre-processed to give a dataset consisting of 360*3 samples. Experiments will use this dataset to train the PointNet classifier.

To train the PointNet classifier for target identification and attack, several steps need to be followed. Firstly, the dataset should be divided to create a training set and a test set. This division involves randomly selecting 70% of the samples for the training set, while the remaining 30% are allocated to the test set. Secondly, the PointNet classifier needs to be configured with specific training parameters. In this case, the sampling parameters are set to 256, utilizing both random sampling and farthest distance sampling techniques. These sampling methods assist in effectively handling the irregularity of the point cloud data. Lastly, after training the classifier using the defined parameters, it demonstrates a training accuracy of 96.2% and a test accuracy of 97.4%, indicating its proficiency in correctly identifying and classifying targets.

In Sect. 4.2 section of the experimental evaluation, each category of samples was sorted by increasing rotation angle, one sample was selected every 4°, and a total of 90 samples were selected to generate the adversarial samples, (The average number of points for each class of samples is 2564, 2188, and 1317, respectively.)

4.2   Attack Performance

To verify the feasibility of the algorithm, experimental results demonstrating the attack success rate and perturbation magnitude were presented. The data from the second column of Table 1 reveals that the attack success rate for the two adversarial samples belonging to the real classes Step1 and Step2 exceeds 96%.

Table 1. Targetless attack success rate and perturbation size assessment. (SR: Success Rate. Dist.: Distance.)

True Label	SR	Ave L2 Dist.	Min L2 Dist.	Max L2 Dist.
Step1	96.6%	0.1012	0.0824	0.1661
Step2	96.9%	0.9886	0.3962	3.024
Step3	55.4%	0.3246	0.0348	2.8241

Additionally, over half of the adversarial samples with the real class Step3 were successfully attacked. This indicates the practicality of the proposed adversarial point cloud generation algorithm in the context of target-less attacks. Furthermore, the data in the last three columns of Table 1 indicates that the generated adversarial samples exhibit minimal perturbation in target-less attack scenarios, further highlighting the algorithm's ability to generate covert perturbations.

4.3 Simulation of Physical Scene Attacks

To evaluate the workability of the technique put forward in Sect. 3.2, this experiment chooses a sample and executes an individualized assault without a predefined objective.

First, a clean sample of the real class Step1 and angle 0°, which comprises 1775 points, was selected for the experiment. The corresponding 3D object has two properties: (i) its faces are parallel to the coordinate plane, which simplifies the determination of the modification position; and (ii) the rays intersect the object's upper bottom and sides at a 45° angle, resulting in the received points exhibiting the same signal amplitude. This simplifies the relationship between diffuse reflectance and signal amplitude.

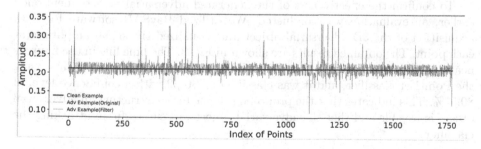

Fig. 4. The signal amplitude of the sample.

Next, the generated adversarial samples were subjected to a filtering process. The algorithm described in Sect. 3.1 was employed to generate adversarial samples specifically designed for non-targeted attacks, and this adversarial sample is identified by the PointNet classifier as Step2 with 96.84% confidence. In the filtering operation, the threshold ξ for perturbation is set to 0.125 and the adversarial sample is filtered. The filtered adversarial sample retains only 6 sensitive points of perturbation, and it is identified as Step2 with a confidence level of 88.68%, indicating that the sample still maintains a strong aggressiveness, and the adversarial object can be constructed based on this sample. As illustrated in Fig. 4, the signal amplitude in the clean samples is represented by the blue line, while the green line represents the signal amplitude in the adversarial samples, respectively. The red line represents the signal amplitude of the adversarial samples after filtering. It can be observed that the added perturbations are mostly small, with only a few points having larger perturbations. Furthermore, after the filtering process, the adversarial samples demonstrate a significantly reduced number of points with perturbation absolute values surpassing the predefined threshold.

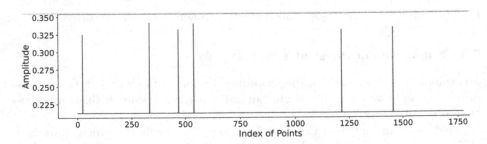

Fig. 5. The signal amplitude of physical adversarial samples.

Finally, the location of the modified area is determined and its diffuse reflectance is modified according to the resulting mapping relation $Diffuse = 1.41421369 * Intensity$ to generate the adversarial object.

To confirm the effectiveness of the generated adversarial objects mentioned earlier, an evaluation was conducted. We utilized RaySAR software for data acquisition of the 3D adversarial object and visualized the signal amplitude at each point. The obtained results are shown in Fig. 5. The blue line in the figure is nearly identical to the red line in Fig. 4. We identified the point cloud data using the PointNet classifier, and it was classified as Step2 with a confidence level of 89.57%. This indicates that the proposed scheme for constructing an adversarial object is feasible and that the adversarial object was successful in attacking the classifier.

5   Conclusion

Recent investigations have revealed the vulnerability of deep learning-based methods used for point cloud data target recognition to adversarial sample attacks. In this study, we propose a novel algorithm for generating adversarial samples in the digital domain by modifying the signal amplitude of point cloud data. We subsequently evaluate its effectiveness in compromising the performance of point cloud data target recognition methods. Our experimental results demonstrate the formidable attack potential of adversarial samples generated through signal amplitude modification in point cloud data. Additionally, we present a physically realizable attack strategy by combining ray-tracing techniques. In our experiments, we generate attack samples in the digital world and utilize RaySAR software to construct physical adversarial samples for targeting the recognition algorithm. The experimental findings substantiate the successful deception of the deep learning-based target recognition algorithm by the constructed physical adversarial samples.

References

1. Dai, A., Chang, A.X., Savva, M., Halber, M., Funkhouser, T., Nießner, M.: ScanNet: richly-annotated 3D reconstructions of indoor scenes. In: Proceedings of the IEEE Conference on Computer Vision and Pattern Recognition, pp. 5828–5839 (2017)
2. Lai, B., et al.: 2D3D-MVPNet: learning cross-domain feature descriptors for 2D-3D matching based on multi-view projections of point clouds. Appl. Intell. **52**(12), 14178–14193 (2022)
3. Yang, B., Luo, W., Urtasun, R.: PIXOR: real-time 3D object detection from point clouds. In: Proceedings of the IEEE Conference on Computer Vision and Pattern Recognition, pp. 7652–7660 (2018)
4. Shi, S., Wang, X., Li, H.: PointRCNN: 3D object proposal generation and detection from point cloud. In: Proceedings of the IEEE/CVF Conference on Computer Vision and Pattern Recognition, pp. 770–779 (2019)
5. Fernandes, D., et al.: Point-cloud based 3D object detection and classification methods for self-driving applications: a survey and taxonomy. Inf. Fusion **68**, 161–191 (2021)
6. Lawin, F.J., Danelljan, M., Tosteberg, P., Bhat, G., Khan, F.S., Felsberg, M.: Deep projective 3D semantic segmentation. In: Felsberg, M., Heyden, A., Krüger, N. (eds.) CAIP 2017. LNCS, vol. 10424, pp. 95–107. Springer, Cham (2017). https://doi.org/10.1007/978-3-319-64689-3_8
7. Wu, B., Wan, A., Yue, X., Keutzer, K.: SqueezeSeg: convolutional neural nets with recurrent CRF for real-time road-object segmentation from 3D lidar point cloud. In: 2018 IEEE International Conference on Robotics and Automation (ICRA), pp. 1887–1893. IEEE (2018)
8. Liu, D., Yu, R., Su, H.: Extending adversarial attacks and defenses to deep 3D point cloud classifiers. In: 2019 IEEE International Conference on Image Processing (ICIP) (2019)
9. Xiang, C., Qi, C.R., Li, B.: Generating 3D adversarial point clouds. In: 2019 IEEE/CVF Conference on Computer Vision and Pattern Recognition (CVPR) (2019)
10. Guo, Y., Wang, H., Hu, Q., Liu, H., Liu, L., Bennamoun, M.: Deep learning for 3D point clouds: a survey. IEEE Trans. Pattern Anal. Mach. Intell. **43**(12), 4338–4364 (2020)
11. Aloysius, N., Geetha, M.: A review on deep convolutional neural networks. In: 2017 International Conference on Communication and Signal Processing (ICCSP), pp. 0588–0592. IEEE (2017)

12. Pang, Y., et al.: Graph decipher: a transparent dual-attention graph neural network to understand the message-passing mechanism for the node classification. Int. J. Intell. Syst. **37**(11), 8747–8769 (2022)
13. Pang, Y., et al.: Sparse-Dyn: sparse dynamic graph multirepresentation learning via event-based sparse temporal attention network. Int. J. Intell. Syst. **37**(11), 8770–8789 (2022)
14. Maturana, D., Scherer, S.: VoxNet: a 3D convolutional neural network for real-time object recognition. In: 2015 IEEE/RSJ International Conference on Intelligent Robots and Systems (IROS), pp. 922–928. IEEE (2015)
15. Zhou, Y., Tuzel, O.: VoxelNet: end-to-end learning for point cloud based 3D object detection. In: Proceedings of the IEEE Conference on Computer Vision and Pattern Recognition, pp. 4490–4499 (2018)
16. Kang, Z., Yang, J., Zhong, R., Wu, Y., Shi, Z., Lindenbergh, R.: Voxel-based extraction and classification of 3-D pole-like objects from mobile lidar point cloud data. IEEE J. Sel. Top. Appl. Earth Obs. Remote Sens. **11**(11), 4287–4298 (2018)
17. Kuang, H., Wang, B., An, J., Zhang, M., Zhang, Z.: Voxel-FPN: multi-scale voxel feature aggregation for 3D object detection from lidar point clouds. Sensors **20**(3), 704 (2020)
18. Qi, C.R., Su, H., Mo, K., Guibas, L.J.: PointNet: deep learning on point sets for 3D classification and segmentation. In: Proceedings of the IEEE Conference on Computer Vision and Pattern Recognition, pp. 652–660 (2017)
19. Qi, C.R., Yi, L., Su, H., Guibas, L.J.: PointNet++: deep hierarchical feature learning on point sets in a metric space, arXiv preprint arXiv:1706.02413 (2017)
20. Dzurisin, D., Dzurisin, D., Lu, Z.: Interferometric synthetic-aperture radar (InSAR). In: Dzurisin, D. (ed.) Volcano Deformation: Geodetic Monitoring Techniques, pp. 153–194. Springer, Heidelberg (2007). https://doi.org/10.1007/978-3-540-49302-0_5

Adversarial Robustness via Multi-experts Framework for SAR Recognition with Class Imbalanced

Chuyang Lin, Senlin Cai, Hailiang Huang, Xinghao Ding⬥,
and Yue Huang(✉)⬥

School of Informatics, Xiamen University, Xiamen, China
yhuang2010@xmu.edu.cn

Abstract. With the rapid development of deep learning technology, significant progress has been made in the field of synthetic aperture radar (SAR) target recognition algorithms. However, deep neural networks are vulnerable to adversarial attacks in practical applications, inducing learning models to make wrong predictions. Existing works on adversarial robustness always assumed that the datasets are balanced. While in real-world applications, SAR datasets always suffer serious imbalanced distributions, which brings challenges to target recognition tasks and also affects the adversarial defense of models. So far few works have been reported on adversarial robustness for imbalanced SAR target recognition. Besides, single model is easily limited to a specific adversarial sample distribution. Motivated by this, a multi-expert collaborative diagnosis framework based on Contrastive Self-Supervised Aggregation (CS2AME) is proposed. The framework trains multiple personalized expert models precisely for dealing with specific SAR targets by formulating three expert guidance schemes, to better deal with different adversarial samples. In addition, a contrastive self-supervised aggregation strategy is designed to adaptively aggregate the professional expertise of each expert model. Extensive adversarial robustness recognition experiments on three publicly available imbalanced SAR datasets have demonstrated that the proposed CS2AME outperforms existing works in terms of standard performance and robust performance.

Keywords: SAR target recognition · Adversarial robustness · Class imbalanced distribution · Multi-expert

1 Introduction

Synthetic aperture radar (SAR) [14] is an active ground-based observation system capable of generating high-resolution radar images in low visibility and a

The work was supported in part by the National Natural Science Foundation of China under Grant 82172033, U19B2031, 61971369, 52105126, 82272071, 62271430, and the Fundamental Research Funds for the Central Universities 20720230104.

Q. Liu et al. (Eds.): PRCV 2023, LNCS 14428, pp. 405–417, 2024.
https://doi.org/10.1007/978-981-99-8462-6_33

variety of scenarios. It can overcome the influence of weather conditions and various scene factors, so it has important application value in many fields. The development of neural networks has enabled SAR target recognition technology based on deep learning to achieve notable results [20]. However, deep neural networks have been found to be vulnerable to slight adversarial interference [19], leading to erroneous predictions. The presence of adversarial samples exposes the vulnerability of neural networks and poses a serious security risk to radar target recognition systems. Improving adversarial robustness has become an essential requirement.

Existing research on adversarial robustness [22] typically assumes that the distribution of the dataset is balanced. However, in practical application scenarios, SAR target data can be easily affected by various factors during the complex imaging procedures, resulting in serious class imbalanced distributions. Categories with sufficient sample sizes are referred to as head classes, while those with insufficient sample sizes are referred to as tail classes. To investigate the challenges posed by adversarial robustness under imbalanced data distribution, we compared the recognition performance of models trained on the publicly available MSTAR [18] dataset under balanced and imbalanced conditions. Figure 1 shows the accuracy of the SAR target recognition model on original clean images and perturbed images after the PGD attack [12]. Standard accuracy (SA) is used to evaluate the recognition performance of clean images on normal and adversarial training (AT) models, i.e. Plain-SA and AT-SA, respectively. Robust accuracy (RA) refers to the performance of the AT model on perturbed images, i.e. AT-RA. Figure 1 reveals that: (1) On imbalanced datasets, both SA and RA exhibit a downward trend from head classes to tail classes. (2) When trained using the AT method, the standard and robust accuracy of the AT model declines more severely on the tail classes. The adversarial robustness training of the model will also be seriously affected under this imbalanced distribution of class. Learning the diversity features of SAR targets in the tail classes [7] is inherently challenging. When the features of tail classes are slightly disturbed, the model may misclassify them as features of other categories, leading to poor recognition performance on SAR targets in tail classes. However, there remains a lack of solutions for improving the adversarial robustness of SAR target recognition models under class imbalanced distribution.

To improve the adversarial robust recognition performance under class imbalance distribution, while considering the limitations of single models, we designed and proposed a multi-expert collaborative diagnosis framework based on contrastive self-supervised aggregation (CS2AME). The framework consists of personalized expert model training and a contrastive self-supervised aggregation strategy. The former trains different expert models to learn features of various categories in a more generalized manner, while the latter adaptively learns expert aggregation weights through contrast learning and integrates prediction information from different expert models to achieve more accurate recognition results. The proposed CS2AME is tested on the MSTAR dataset using various attack methods, including PGD, C&W, and AA. Adequate experiments are per-

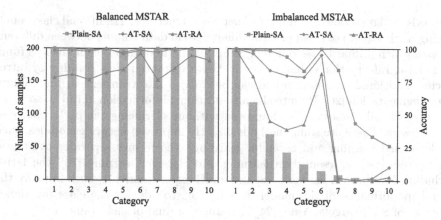

Fig. 1. Comparison chart of recognition performance of models under different data distributions.

formed to show the effectiveness of our approach, achieving 4–9% accuracy gains overall, and 2–12% gains in the tail classes. The main contributions of this work are summarized as follows:

- A multi-expert collaborative framework based on contrastive self-supervised aggregation is proposed for adversarial robustness in SAR recognition with imbalanced distribution.
- Considering imbalanced distribution, the three diverse expert guidance losses are developed for different adversarial sample distributions.
- A contrastive self-supervised aggregation strategy is designed to adaptively learn the aggregation weights of multiple expert models.

2   Related Work

2.1   Adversarial Robustness

Adversarial defense is a strategy that enables models to correctly identify adversarial samples and reduce the success rate of attacks. This includes model-level defenses such as gradient regularization [17], as well as data-level defenses such as PGD attack preprocessing to eliminate adversarial perturbations, hypersphere embedding [15], and vector image conversion [9]. In response to the adversarial robustness issue of SAR target recognition models, Xu [23] et al. introduced an unsupervised adversarial contrastive learning defense method for SAR target recognition.

2.2   Long-Tailed Recognition

To address the long-tailed recognition problem, some methods have made progress, e.g., traditional class rebalancing methods, information enhancement methods, and methods that improved network structures. Traditional class rebalancing methods aim to rebalance the differences in data volume between different classes, which primarily use different sampling methods and weight the loss function [3,16]. Information enhancement methods enhance model training by introducing additional information and can be divided into transfer learning [8] and data augmentation [26]. By introducing additional information in tail classes, the accuracy of tail classes can be improved without sacrificing the performance of head classes. There are some methods directly improved network modules, such as decoupled learning and ensemble learning. The former separates the training process into representation learning and classifier learning [10], the latter including dual-branch network [28] and multi-expert model [1] both improve the overall performance of the model. In addition, to address the class imbalance problem of SAR targets, Yang [24] introduced a dual-branch framework.

3   Methodology

Existing adversarial robustness methods are based on balanced datasets, and general long-tail methods do not take into account the challenges posed by adversarial attacks. At the same time, a single model is only good at dealing with one class-imbalanced training distribution, which often produces class bias and cannot adapt to a balanced test distribution. In addition, a specific training distribution cannot cover all possible distributions of adversarial samples generated, which makes it difficult for a single model to fitting the new adversarial sample of the tail class. The risk of misjudgment of tail-class SAR targets by a single model is thus greatly increasing, reducing the safety and reliability of radar target recognition systems.

For that, We propose an adversarial robustness solution for SAR target recognition models under class imbalanced distribution and design a contrastive self-supervised multi-expert collaborative diagnosis framework (namely CS2AME) to address the adversarial robustness recognition problem of imbalanced SAR targets. The structure of the framework is shown in Fig. 2, which can be divided into two parts: personalized expert model training and contrastive self-supervised aggregation strategy. The former trains different expert models to learn different categories of features in a more generalized way, while the latter learns expert aggregation weights adaptively through contrast learning and integrates prediction information from different expert models to obtain more accurate recognition results. Specifically, we formulate three expert training schemes by introducing different expert guidance losses. Multiple personalized expert models can be trained to deal with different adversarial sample distributions. In addition, we design the contrastive self-supervised aggregation strategy, which is expected to adaptively aggregate the professional advantages of each expert model, while making up for the lack of learning for SAR targets in the tail classes.

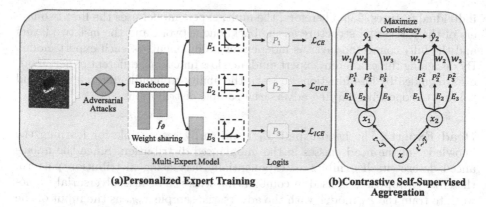

(a)Personalized Expert Training

(b)Contrastive Self-Supervised Aggregation

Fig. 2. The framework of the proposed CS2AME model: (a) extract features of adversarial samples through the backbone network, use three loss-guided expert models to learn knowledge of different distributions. (b) maximize the agreement between different augmented views.

3.1 Adversarial Robustness

The goal of adversarial attacks is to use adversarial samples to deceive the target model into searching for easily confused samples, thereby maximizing the adversarial loss. Suppose the training set is $D = (X, Y)$, where $x \in X$ is the training sample and $y \in Y$ is the class label. The deep neural network is defined as $f_\theta : x \rightarrow y$, where θ is the learnable parameter of the deep neural network. Adversarial training is one of the most effective defense methods against adversarial attacks and is usually described as solving a minimax problem. Its performance depends on the quality of adversarial samples generated by internal optimization [12] and is defined as shown in Eq. (1):

$$\arg \min_{\theta} E_{(x,y)\sim D}[\max_{\delta \in \Omega} \mathcal{L}_{CE}(x + \delta, y; \theta)] \tag{1}$$

where δ represents adversarial perturbation and $x + \delta$ represents the adversarial sample x_{adv} after the clean sample x is added with adversarial perturbation. $\Omega = \ell(x, \epsilon)$ represents an L_∞ norm ball with radius ϵ around x.

3.2 Personalized Expert Training

Ensemble learning with multiple expert models has been shown to be effective in addressing the class imbalanced problem [1]. To learn a multi-expert model with diverse skills from imbalanced SAR data, this chapter constructs a multi-expert model consisting of two main modules: (1) A shared backbone network f_θ, (2) Independent expert networks E_1, E_2, E_3.

In existing deep models, shallow networks are typically used to extract more general representations, while deep networks are used to extract more specific individual features [25]. Therefore, the multi-expert model uses the first two layers of the ResNet [5] structure as the backbone network and the last two layers and the fully connected layer as independent components of each expert model. By designing three different expert guidance loss functions, different expert models are guided by loss functions to learn the ability to handle different data and trained to cope with varying adversarial sample distributions.

Head Expert. The head expert E_1 is primarily responsible for learning the knowledge of the head classes in the imbalanced distribution. Since the imbalanced dataset itself exhibits an exponentially decreasing distribution pattern, the cross-entropy loss is used in combination with the attack adversarial framework to train the E_1 model, with the adversarial sample x_{adv} as the input of the model. The cross-entropy loss is defined as:

$$\mathcal{L}_{CE}(x_i, y_i) = -\log\left(p^1_{y_i}(x_i + \delta)\right) \tag{2}$$

where $p^1_{y_i}$ represents the logit output by the head expert E_1, that is, the predicted value of class y_i in P_1, and δ is the adversarial perturbation. Internal optimization is calculated in the same way as Eq. (1).

Middle Expert. The middle expert E_2 is used to balance the knowledge of the head and tail classes in the class imbalanced distribution. Inspired by [13], the goal of introducing prior distribution information is to compensate for the deviation of the prior distribution for logits and alleviate the inherent bias caused by category imbalance. The uniform cross-entropy loss is used for optimization and is defined as follows:

$$\mathcal{L}_{UCE}(x_i, y_i) = -\log\left(p^2_{y_i}(x_i + \delta) + \log(\rho_{y_i})\right) \tag{3}$$

where $p^2_{y_i}$ represents the logit output by the middle expert, that is, the predicted value of class y_i in P_2, ρ_{y_i} represents the frequency of class y under exponential distribution.

Tail Expert. The tail expert E_3 focuses on learning the knowledge of the tail class in the imbalanced distribution. To address the issue of a few training samples in the tail class, the idea of logit adjustment [13] is also adopted, and reverse prior distribution information is incorporated to improve the model's learning of the tail class. The loss function for the tail class expert is inverse cross-entropy loss, defined as:

$$\mathcal{L}_{ICE}(x_i, y_i) = -\log\left(p^3_{y_i}(x_i + \delta) + \log(\rho_{y_i}) - \log(\bar{\rho}_{y_i})\right) \tag{4}$$

where $p^3_{y_i}$ represents the Logit output by the tail expert, that is, the predicted value of class y_i in P_3, $\bar{\rho}_{y_i}$ is the frequency of class y under inverse exponential distribution.

3.3 Contrastive Self-supervised Aggregation Strategy

Different expert models have varying advantages. We designed a contrastive self-supervised aggregation strategy that assigns learnable weights to each expert and adaptively aggregates the strengths of each expert.

Assuming that the prediction logits of the multi-expert model are (P_1, P_2, P_3) and the corresponding aggregation weights are $\omega = (\omega_1, \omega_2, \omega_3)$, then the final prediction value $\hat{y}$ of the model can be defined as:

$$\hat{y} = \omega_1 \cdot P_1 + \omega_2 \cdot P_2 + \omega_3 \cdot P_3 \tag{5}$$

$$\text{s.t } \omega_1 + \omega_2 + \omega_3 = 1 \tag{6}$$

Using a contrastive learning strategy [11] and applying data augmentation twice $(t, t' \sim T)$, based on the cosine similarity between different views, we maximize the consistency of multi-expert model predictions $\hat{y}$ and learn higher aggregation weights ω for stronger expert models. Therefore, the contrastive self-supervised aggregation loss is defined as follows:

$$\arg\min_{\omega} \mathcal{L}_{CS2A} = \frac{1}{N} \sum_{x \in D} \cos(\hat{y}_1, \hat{y}_2) \tag{7}$$

where the dataset $D = \{x_i, y_i\}_{i=1}^{N}$, N is the total number of all samples, $\cos(\cdot)$ represents cosine similarity, and $\hat{y}_1$ and $\hat{y}_2$ refer to the model prediction values of two augmented views x_1 and x_2 of the same instance.

3.4 Loss Function

Inspired by some classic AT variants [27], additional regularization terms λ are introduced, and the mean square error (MSE) is used to measure the difference between the paired logits generated between clean samples and adversarial samples, further improving the model's robustness against adversarial attacks.

In summary, the overall loss of the CS2AME algorithm is defined as:

$$\mathcal{L}_{CS2AME} = \omega_1 \cdot \mathcal{L}_{CE} + \omega_2 \cdot \mathcal{L}_{UCE} + \omega_3 \cdot \mathcal{L}_{ICE}$$
$$+ \lambda \cdot \mathcal{L}_{MSE}\big(P(x+\delta), P(x)\big) \tag{8}$$

where λ is a regularization parameter, and P is the average prediction logits output by a multi-expert model.

4 Experiments

4.1 Setup

Dataset. The experimental data uses the open source data set MSTAR [18], which contains 10 categories, and the image pixels are 128×128. All samples with a pitch angle of $17°$ are used as a training set, and those with a pitch angle of 15° are used as a test set and are processed into class imbalanced datasets including Exp-MSTAR and Step-MSTAR. Exp-MSTAR is set by the $N_j^E = N_{max}^E \times (\frac{1}{\eta})^{\frac{j-1}{C-1}}$, where N_j^E is the number of samples of the j class, C is the total number of class and the imbalance rate $\eta = 100$. Step-MSTAR is set by $N_k^S = N_{max}^S \times (\frac{1}{\eta})^{k-1}$, where $\eta = N_{k-1}^S/N_k^S = 20$, only set two steps.

In addition, we used the natural class-imbalanced FUSAR-Ship [6] (CI-FUSAR). The dataset contains a total of 15 categories, and the image pixels are 512×512. Select 6 categories with sample sizes greater than 100 as the dataset, randomly select 100 samples from each category as the test set, and all the remaining samples as the training set.

To validate the effectiveness of our proposed framework on the Exp-MSTAR, Step-MSTAR, and CI-FUSAR, all classes are divided into the head and tail classes. The first half of classes with rich samples are selected as the majority classes and the rest with a few samples as the minority classes.

Implementation Details. We use ResNet-18 as the backbone network for the multi-expert model, where the first two layers are used as a shared backbone network and the last two layers and the fully connected layer are replicated three times to construct independent components for the multi-expert model. The aggregation weights ω are initialized to $(1/3, 1/3, 1/3)$ and updated every 50 epochs, with a regularization coefficient λ of 1. The Baseline method trains the model using the ordinary cross-entropy loss. To verify the defense performance of the model against adversarial attacks, We use three attack methods: PGD [12], C&W [2], and AA [4].

Evaluation Metrics. Accuracy, robust accuracy, and F1 score as evaluation metrics are used to evaluate the proposed method. Standard accuracy is used to measure the target recognition performance of the model, while robust accuracy is used to measure the adversarial robustness performance of the model. The F1 score is introduced as a comprehensive indicator to balance the impact of precision and recall. We define A_{all} as the overall precision, A_{maj} as the head class precision, and A_{min} as the tail class precision.

4.2   Experimental Results

To verify the robustness of the proposed framework in the imbalanced SAR target recognition model, several prominent methods were selected and compared. The relevant experimental results are shown in Table 1 to Table 2.

From tables we can see that the robust recognition accuracy of CS2AME under multiple attack indicators is better than other advanced methods, indicating that our model has better robustness performance and produces better quality model. At the same time, the recognition accuracy of the tail class under the attack is higher, which proves that our model has a stronger learning ability for the tail class.

Table 1. Experimental results on Exp-MSTAR and Step-MSTAR.

Method	Exp-MSTAR				Step-MSTAR			
	Clean	PGD	C&W	AA	Clean	PGD	C&W	AA
Baseline	74.00	0.00	0.00	0.00	89.58	0.00	0.00	0.00
PGD-AT [12]	66.33	44.87	44.92	44.08	79.17	57.46	57.96	56.75
TRADES [27]	66.54	45.96	46.29	45.29	80.17	58.83	59.21	58.13
PGD-HE [15]	67.13	47.50	46.25	45.00	80.75	59.29	58.71	57.37
Robal-N [21]	71.58	49.08	49.63	48.38	80.79	59.42	59.62	58.04
Robal-R [21]	**72.50**	48.71	48.71	47.42	81.13	59.29	60.00	57.67
CS2AME	71.42	**54.04**	**53.75**	**53.12**	**82.50**	**65.21**	**65.42**	**64.67**

Table 2. Experimental results on MSTAR and CI-FUSAR.

Method	AA						C&W		
Datasets	Exp-MSTAR			Step-MSTAR			CI-FUSAR		
	A_{maj}	A_{min}	F1	A_{maj}	A_{min}	F1	A_{maj}	A_{min}	F1
Baseline	0.00	0.00	0.00	0.00	0.00	0.00	0.33	0.00	0.00
PGD-AT [12]	66.78	23.72	37.69	83.35	32.89	54.66	87.33	44.57	63.45
TRADES [27]	69.16	23.87	38.61	84.23	34.70	56.48	87.33	43.12	62.78
PGD-HE [15]	66.34	25.85	38.63	83.17	34.23	55.66	**89.67**	44.20	64.53
Robal-N [21]	66.08	32.49	44.01	81.50	37.00	56.82	87.67	51.09	67.29
Robal-R [21]	63.26	33.20	43.74	80.62	37.08	56.57	86.67	50.72	66.65
CS2AME	**71.63**	**36.52**	**50.27**	**86.96**	**44.66**	**64.22**	88.33	**53.99**	**69.26**

4.3   Ablation Study

This chapter conducts a series of ablation experiments on the Exp-MSTAR dataset and provides a more detailed analysis of the proposed method.

To verify the contribution of the proposed multi-expert model to the overall performance of the framework, we primarily compare the performance of different expert combinations in Table 3. The results show that different expert models have unique learning capabilities and the overall recognition performance of the combination of three expert models is good. The three expert models complement each other to comprehensively improve the adversarial robustness recognition performance of the model.

Figure 3 explores the impact of different aggregation strategies on the learning of multi-expert models. EW represents the equal weight strategy, and the model trained with the CS2A strategy proposed in this paper performs more prominently in terms of robustness. By adaptively learning the aggregation weight coefficients to fully exploit the advantages of multi-expert collaborative diagnosis.

Table 3. Ablation experiments with different expert combinations.

Method			Clean				AA			
E_1	E_2	E_3	A_{all}	A_{maj}	A_{min}	F1	A_{all}	A_{maj}	A_{min}	F1
√			64.96	92.25	40.47	60.47	49.83	**74.63**	27.59	43.66
	√		69.37	**92.42**	48.70	66.88	51.62	71.10	34.15	47.87
		√	67.75	86.43	50.99	66.51	46.37	60.88	33.36	44.74
√	√		66.67	89.96	45.77	63.78	49.21	70.93	29.72	43.88
√		√	67.67	89.16	48.38	65.23	50.17	69.25	33.04	46.28
	√	√	70.00	87.75	**54.07**	69.19	49.46	62.82	**37.47**	48.18
√	√	√	**71.63**	91.89	53.44	**70.20**	**51.79**	70.84	34.70	**49.05**

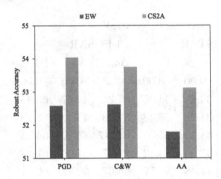

Fig. 3. Effect of aggregation weights

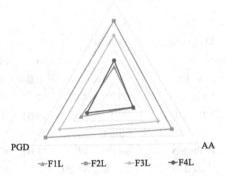

Fig. 4. Effect of backbone network.

To explore the impact of backbone network depth on the model's ability to extract general features, Fig. 4 shows the robust accuracy of the four different depths of backbone network structure under different attack metrics, where

Table 4. Effect of imbalance rate.

Imbalanced Rate η	Method	Clean	PGD	C&W	AA
100	Robal-N [21]	71.58	49.08	49.63	48.38
	Robal-R [21]	**72.50**	48.71	48.71	47.42
	CS2AME	71.42	**54.04**	**53.75**	**53.12**
50	Robal-N [21]	80.75	56.25	56.87	55.25
	Robal-R [21]	80.33	56.50	56.37	55.46
	CS2AME	**80.79**	**59.92**	**59.42**	**59.04**
20	Robal-N [21]	89.08	64.71	65.29	63.62
	Robal-R [21]	90.08	65.21	66.13	63.92
	CS2AME	**91.37**	**71.54**	**71.21**	**70.79**

F1L-F4L respectively represent using the first 1–4 Layers of the ResNet network structure as a shared backbone network. The curve area formed by F2L is the largest and the adversarial robustness recognition performance of the obtained model is better, which is more conducive to extracting more general representations by the backbone network.

To evaluate the generalization and promotion performance of the proposed framework, we tested it on datasets with different imbalanced rates. The results are shown in Table 4. CS2AME consistently maintains the best robust recognition performance.

5   Conclusion

We explores the problem of class imbalanced distribution and adversarial robustness in SAR recognition and proposes a multi-expert collaborative diagnosis framework based on contrastive self-supervised aggregation (CS2AME). Designed a multi-expert learning module for adversarial samples of different distributions, and adaptively aggregated the expertise of each expert through a self-supervised aggregation strategy. The proposed CS2AME achieves significant performance improvement on MSTAR and CI-FUSAR, with the key tail class recognition rate for the class imbalanced problem exceeding other methods and achieving the best performance in multiple attack modes.

References

1. Cai, J., Wang, Y., Hwang, J.N.: ACE: ally complementary experts for solving long-tailed recognition in one-shot. In: Proceedings of the IEEE/CVF International Conference on Computer Vision, pp. 112–121 (2021)
2. Carlini, N., Wagner, D.: Towards evaluating the robustness of neural networks. In: 2017 IEEE Symposium on Security and Privacy (SP), pp. 39–57. IEEE (2017)
3. Chawla, N.V., Bowyer, K.W., Hall, L.O., Kegelmeyer, W.P.: Smote: synthetic minority over-sampling technique. J. Artif. Intell. Res. **16**, 321–357 (2002)
4. Croce, F., Hein, M.: Reliable evaluation of adversarial robustness with an ensemble of diverse parameter-free attacks. In: International Conference on Machine Learning, pp. 2206–2216. PMLR (2020)
5. He, K., Zhang, X., Ren, S., Sun, J.: Deep residual learning for image recognition. In: Proceedings of the IEEE Conference on Computer Vision and Pattern Recognition, pp. 770–778 (2016)
6. Hou, X., Ao, W., Song, Q., Lai, J., Wang, H., Xu, F.: FUSAR-ship: building a high-resolution SAR-AIS matchup dataset of Gaofen-3 for ship detection and recognition. SCIENCE CHINA Inf. Sci. **63**, 1–19 (2020)
7. Jahan, C.S., Savakis, A., Blasch, E.: SAR image classification with knowledge distillation and class balancing for long-tailed distributions. In: 2022 IEEE 14th Image, Video, and Multidimensional Signal Processing Workshop (IVMSP), pp. 1–5. IEEE (2022)
8. Jing, C., et al.: Interclass similarity transfer for imbalanced aerial scene classification. IEEE Geosci. Remote Sens. Lett. **20**, 1–5 (2023)

9. Kabilan, V.M., Morris, B., Nguyen, H.P., Nguyen, A.: Vectordefense: vectorization as a defense to adversarial examples. In: Soft Computing for Biomedical Applications and Related Topics, pp. 19–35 (2021)

10. Kang, B., Li, Y., Xie, S., Yuan, Z., Feng, J.: Exploring balanced feature spaces for representation learning. In: International Conference on Learning Representations (2021)

11. Khosla, P., et al.: Supervised contrastive learning. Adv. Neural. Inf. Process. Syst. **33**, 18661–18673 (2020)

12. Madry, A., Makelov, A., Schmidt, L., Tsipras, D., Vladu, A.: Towards deep learning models resistant to adversarial attacks. arXiv preprint arXiv:1706.06083 (2017)

13. Menon, A.K., Jayasumana, S., Rawat, A.S., Jain, H., Veit, A., Kumar, S.: Long-tail learning via logit adjustment. arXiv preprint arXiv:2007.07314 (2020)

14. Moreira, A., Prats-Iraola, P., Younis, M., Krieger, G., Hajnsek, I., Papathanassiou, K.P.: A tutorial on synthetic aperture radar. IEEE Geosci. Remote Sens. Mag. **1**(1), 6–43 (2013)

15. Pang, T., Yang, X., Dong, Y., Xu, K., Zhu, J., Su, H.: Boosting adversarial training with hypersphere embedding. Adv. Neural. Inf. Process. Syst. **33**, 7779–7792 (2020)

16. Ren, J., Yu, C., Ma, X., Zhao, H., Yi, S., et al.: Balanced meta-softmax for long-tailed visual recognition. Adv. Neural. Inf. Process. Syst. **33**, 4175–4186 (2020)

17. Ross, A., Doshi-Velez, F.: Improving the adversarial robustness and interpretability of deep neural networks by regularizing their input gradients. In: Proceedings of the AAAI Conference on Artificial Intelligence, vol. 32 (2018)

18. Ross, T.D., Worrell, S.W., Velten, V.J., Mossing, J.C., Bryant, M.L.: Standard SAR ATR evaluation experiments using the MSTAR public release data set. In: Algorithms for Synthetic Aperture Radar Imagery V, vol. 3370, pp. 566–573. SPIE (1998)

19. Szegedy, C., et al.: Intriguing properties of neural networks. arXiv preprint arXiv:1312.6199 (2013)

20. Wang, J., Virtue, P., Yu, S.X.: Successive embedding and classification loss for aerial image classification. arXiv preprint arXiv:1712.01511 (2017)

21. Wu, T., Liu, Z., Huang, Q., Wang, Y., Lin, D.: Adversarial robustness under long-tailed distribution. In: Proceedings of the IEEE/CVF Conference on Computer Vision and Pattern Recognition, pp. 8659–8668 (2021)

22. Xia, W., Liu, Z., Li, Y.: SAR-PeGA: a generation method of adversarial examples for SAR image target recognition network. IEEE Trans. Aerosp. Electron. Syst. **59**(2), 1910–1920 (2022)

23. Xu, Y., Sun, H., Chen, J., Lei, L., Ji, K., Kuang, G.: Adversarial self-supervised learning for robust SAR target recognition. Remote Sens. **13**(20), 4158 (2021)

24. Yang, C.Y., Hsu, H.M., Cai, J., Hwang, J.N.: Long-tailed recognition of SAR aerial view objects by cascading and paralleling experts. In: Proceedings of the IEEE/CVF Conference on Computer Vision and Pattern Recognition, pp. 142–148 (2021)

25. Yosinski, J., Clune, J., Bengio, Y., Lipson, H.: How transferable are features in deep neural networks? In: Advances in Neural Information Processing Systems, vol. 27 (2014)

26. Zang, Y., Huang, C., Loy, C.C.: FASA: feature augmentation and sampling adaptation for long-tailed instance segmentation. In: Proceedings of the IEEE/CVF International Conference on Computer Vision, pp. 3457–3466 (2021)

27. Zhang, H., Yu, Y., Jiao, J., Xing, E., El Ghaoui, L., Jordan, M.: Theoretically principled trade-off between robustness and accuracy. In: International Conference on Machine Learning, pp. 7472–7482. PMLR (2019)
28. Zhou, B., Cui, Q., Wei, X.S., Chen, Z.M.: BBN: bilateral-branch network with cumulative learning for long-tailed visual recognition. In: Proceedings of the IEEE/CVF Conference on Computer Vision and Pattern Recognition, pp. 9719–9728 (2020)

Recognizer Embedding Diffusion Generation for Few-Shot SAR Recognization

Ying Xu, Chuyang Lin, Yijin Zhong, Yue Huang(✉)(iD), and Xinghao Ding(iD)

School of Informatics, Xiamen University, Xiamen, China
yhuang2010@xmu.edu.cn

Abstract. Synthetic Aperture Radar (SAR) has become a research hotspot due to its ability to identify targets in all weather conditions and at all times. To achieve satisfactory recognition performance in most existing automatic target recognition (ATR) algorithms, sufficient training samples is essentially owned. However, only few data for real-world SAR applications are generally available, arriving at efficient identification with few SAR training samples remains a formidable challenge that needs to be addressed. The success of denoising diffusion probabilistic model (DDPM) has provided a new perspective for few-shot SAR ATR. Consequently, this paper proposes the Recognizer Embedding Diffusion Generation (REDG), a novel approach for few-shot SAR image generation and recognition. REDG mainly consists of a generation module and a recognition module. The former is responsible for generating additional data from a limited SAR dataset by means of DDPM, to enhance the training of the recognizer. The latter is designed specifically to optimize data generation, which further enhances the recognition performance using the generated data. Extensive experiments conducted on three different datasets have demonstrated the effectiveness of proposed REDG, which represents a significant advancement in improving the reliability of SAR recognition systems, especially in scenarios lacking sufficient data.

Keywords: Synthetic Aperture Radar · Automatic Target Recognition · Few-shot Learning · Data Generation

1 Introduction

Synthetic Aperture Radar (SAR) is a crucial microwave remote sensing system that excels in target identification under diverse weather conditions. With the advancements in deep learning, SAR automatic target recognition (ATR)

The work was supported in part by the National Natural Science Foundation of China under Grant 82172033, U19B2031, 61971369, 52105126, 82272071, 62271430, and the Fundamental Research Funds for the Central Universities 20720230104.

Q. Liu et al. (Eds.): PRCV 2023, LNCS 14428, pp. 418–429, 2024.
https://doi.org/10.1007/978-981-99-8462-6_34

has witnessed significant improvements in recent years, with various Deep Neural Networks (DNNs) achieving remarkable results [2,22]. However, deploying DNNs in real-world scenarios poses challenges due to their high data requirements, often demanding hundreds to thousands of annotated training samples per class for optimal accuracy. Obtaining a sufficient number of annotated SAR images presents difficulties in practical applications, attributed to factors such as cost, labor-intensive labeling, and challenges in collecting observational objects of interest [13]. Therefore, it is important to identify SAR targets with only a few images, which is referred to as few-shot SAR target recognition.

SAR image generation offers a potential solution to address the challenge of insufficient training data. While the performance of generative models also affected by the scarcity of training data, resulting in lower quality of the generated images. For instance, in situations where only few samples are available for GAN-based model training, the generated data may not effectively capture the data distribution. Therefore, various GAN-based models [9,15,17] have been proposed for SAR generation, their application in the few-shot scenario still remains limited. Few-shot image generation for SAR is currently a sparsely researched area, and their generated data yielded limited quality.

Recently, denoising diffusion probabilistic model (DDPM) [5] have gained attention due to its superior performance in various image processing tasks. DDPM is an unconditional generative model that utilizes a latent variable model to reverse a diffusion process, gradually adding Gaussian noise to perturb the data distribution to the noise distribution. It has been successfully applied to tasks such as image generation, audio processing, as well as conditional generation tasks [7,12]. However, to date, there has been no application of DDPM in the remote sensing field, specifically for generating solutions for few-shot SAR ATR.

Motivated by above mentioned, this paper introduces a novel approach called Recognizer Embedding Diffusion Generation (REDG) to address the challenge of SAR ATR with extremely few training samples. REDG serves a dual purpose, Firstly, it includes a generation module based on a simplified version of DDPM, responsible for generating synthetic samples from few of available SAR images. Its aim is to augment the training data and enhance overall data diversity. Secondly, the recognition module of REDG performs SAR target recognition tasks. There introduces a feature layering and reuse algorithm in the recognition module to make better use of limited real samples. Through the integration of the generation and recognition modules, REDG is capable of generating synthetic SAR samples in a few-shot setting and leveraging them to improve SAR ATR. Overall, REDG aims to enhance the recognition capability of models by generating synthetic samples and combining them with real samples, thereby overcoming the limitations of few-shot SAR image recognition.

The contributions can be summarized as follows:

- A Recognizer Embedding Diffusion Generation (REDG) is proposed to address the few-shot SAR ATR.
- This paper design a generation module based on simplified DDPM to generate supplementary from few SAR data for the training of the recognizer.

- In this paper, a feature layering and reuse algorithm is designed for the recognition module, which enables a better balance between real and generated data and achieves improved recognition performance.
- The generation and recognition performance of proposed REDG is evaluated through extensive experiments.

2 Related Work

2.1 SAR Few-Shot Learning

SAR Few-Shot Learning methods encompass transfer learning and meta-learning approaches. Transfer learning leverages prior knowledge from a source domain to address the target domain problem. For example, Liu et al. [16] applied specialized transfer learning based on electromagnetic properties to improve performance. Tai et al. [21] introduced a novel few-shot transfer learning method using the connection-free attention module for selective feature transfer. Meta-learning accumulates experience by learning from numerous tasks in the source domain. For instance, Fu et al. [4] proposed the meta-learning framework MSAR and employed a hard task mining method to enhance effectiveness. Sun et al. [19] introduced the few-shot learning framework SCAN, which incorporates explicit supervision to learn the number and distribution of scattering points for each target type.

2.2 SAR Generative Models

Several studies have investigated SAR image generation using different techniques. Oh et al. [17] proposed PeaceGAN, a multi-task learning framework that combines pose and class information. Pu et al. [18] introduced a shuffle GAN with an autoencoder separation method to distinguish moving and stationary targets in SAR imagery.

Limited research has focused on SAR generation in the few-shot scenario. Sun et al. [20] introduced an attribute-guided GAN with an enhanced episode training strategy for few-shot SAR image generation. The AGGAN's attribute labels include category and aspect angle information, and spectral normalization was used to stabilize training and overcome few-shot learning challenges. Cao et al. [1] presented LDGANs, a novel image generation method that provides labeled samples for training recognition models. It incorporates a unique loss function based on Wasserstein distance to address the collapse mode issue and includes label information to prevent the generation of unlabeled target images.

3 Method

3.1 Overall

The overall framework of REDG is illustrated in Fig. 1, consisting of a generation module (highlighted in orange) and a recognition module (highlighted in blue).

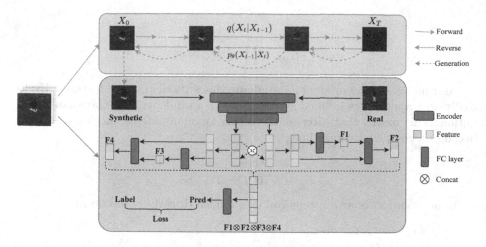

Fig. 1. Overall of REDG (Color figure online)

The generation module generates synthetic samples from random noise, and augmented samples are used as training data for the recognition module. To effectively utilize both real and synthetic samples in the recognition module, we designed a feature-layering and reuse algorithm. Feature layering can capture both low-level and high-level feature of SAR images, feature reuse enhances the guidance of real samples on the synthetic samples.

3.2   Generation Module

DDPM [5] is a mathematical model based on stochastic differential equations used to train a noise estimation model and iteratively generate images from random noise. Compared to GANs, the training of diffusion models is more stable and capable of generating more diverse samples [3] and it has been successfully applied to many tasks [7,12]. The diffusion model consists of two processes: the forward diffusion process and the reverse generation process. The forward diffusion process gradually adds Gaussian noise to an image until it becomes random noise, while the reverse generation process involves the denoising procedure. Based on the sparsity of SAR image features, we simplify the diffusion process in DDPM.

Forward Diffusion Process. Given an initial data distribution x_0 $q(x_0)$, define the forward process as gradually adding Gaussian noise to the data distribution. The noise is added T times, resulting in a series of noisy images $x_1, ..., x_T$. During the noise addition from x_{t-1} to x_t, the standard deviation of the noise is determined by β_t, which lies in the interval (0,1). The mean of the noise is determined by β_t and the current image data x_{t-1}. Specifically, the whole diffusion process is a Malkafe chain:

$$q(x_{1:T}|x_0) = \prod_{t=1}^{T} q(x_t|x(t-1))$$

As t increases, the original x_0 gradually loses its distinctive features. Eventually, as t approaches infinity, x_t converges to an independent Gaussian distribution. This means that after adding noise for many steps, the image almost becomes a completely noisy image:

$$\bar{\alpha}_T \approx 0$$
$$x_T \sim \mathcal{N}(0, I) \quad T \to \infty$$

In addition, x_t of any t steps can be sampled directly based on x_0:

$$\begin{aligned}
x_t &= \alpha_t x_{t-1} + \beta_t \varepsilon_t \\
&= \alpha_t(\alpha_{t-1} x_{t-2} + \beta_{t-1}\varepsilon_{t-1}) + \beta_t\varepsilon_t \\
&= \alpha_t(\alpha_{t-1}\cdots + \beta_{t-1}\varepsilon_{t-1}) + \beta_t\varepsilon_t \\
&= (\alpha_t\cdots\alpha_1)x_0 + (\alpha_t\cdots\alpha_2)\beta_1\varepsilon_1 + \cdots + \alpha_t\beta_{t-1}\varepsilon_{t-1} + \beta_t\varepsilon_t \\
&= \underbrace{(\alpha_t\cdots\alpha_1)}_{\bar{\alpha}_t} x_0 + \underbrace{\sqrt{1-(\alpha_t\cdots\alpha_1)^2}}_{\bar{\beta}_t}\bar{\varepsilon}_t, \quad \bar{\varepsilon}_t \sim \mathcal{N}(0, I) \\
&= \bar{\alpha}_t x_0 + \sqrt{1 - \bar{\alpha}_t^2}\bar{\varepsilon}_t, \bar{\varepsilon}_t \sim \mathcal{N}(0, I)
\end{aligned}$$

During the forward process, sampling the data at time t, input x_t into the model, and calculate the loss:

$$L_g = L_T + L_{t-1}^{\text{simple}} + L_0$$

$$L_t = \frac{1}{2\sigma_t^2}||\hat{\mu}_t(x_t, x_0) - \mu_\theta(x_t, t)||^2$$

$$L_{t-1}^{\text{simple}} = \mathbb{E}_{x_0, \varepsilon \sim \mathcal{N}(0, I)}\left[||\varepsilon - \varepsilon_\theta\left(\bar{\alpha}_t x_0 + \bar{\beta}_t\varepsilon, t\right)||^2\right]$$

Reverse Generation Process. On the contrary, continuously denoising the images from x_T, and after time t, obtained the denoised samples:

$$\varepsilon_t \sim \mathcal{N}(0, I)$$

$$x_{t-1} = \frac{1}{\alpha_t}\left(x_t - \beta_t\mu_\theta(x_t, t)\right) + \sigma_t\varepsilon_t$$

In order to generate the desired data based on the labels need, incorporating classifier-guidance during the sampling process:

$$\begin{aligned}
p(x_{t-1}|x_t, y) &= \frac{p(x_{t-1}|x_t)p(y|x_{t-1}, x_t)}{p(y|x_t)} \\
&= \frac{p(x_{t-1}|x_t)p(y|x_{t-1})}{p(y|x_t)} \\
&= p(x_{t-1}|x_t)e^{\log p(y|x_{t-1}) - \log p(y|x_t)}
\end{aligned}$$

3.3   Recognition Module

The recognition module utilizes a simple CNN network, and introduce the feature layering and reuse to enhance its generalization ability. Firstly, using a simple three-layer CNN as the encoder to extract features from both real and synthetic images. After obtaining the feature, we introduce a novel strategy by concatenating the first half of real feature and the second half of the synthetic feature, forming a new feature representation. This concatenation scheme aims to enhance the discriminative capabilities of the model by combining complementary information.

Additionally, we introduce feature layering in the recognition module, which involves feeding the obtained features into a linear network with multiple layers. This strategy enables the model to learn hierarchical representations of the features, capturing both low-level and high-level characteristics of the SAR images. Moreover, we innovatively leverage the reuse of features from real images. This means that the features obtained from real images are not only used for training the recognition module but also utilized to refine the features of synthetic samples. The main process is shown in Algorithm 1.

Algorithm 1. Recognition Module

Input: input:x_{real} ,x_{syn}, label:y

1: $f_{real} = Encoder(x_{real})$
2: $f_{syn} = Encoder(x_{syn})$
    # Feature Concatenate
3: $f_1 = f_{real}[: half] \otimes f_{syn}[half :]$
4: $f_2 = f_{syn}[: half] \otimes f_{real}[half :]$
    # Feature Layering and reuse
5: $F_1 = FC_1(f_1[: half])$
6: $F_2 = FC_2(F1 \otimes f_1[half :])$
7: $F_3 = FC_3(f_2[half :])$
8: $F_4 = FC_4(F3 \otimes f_2[: half])$
9: $Pred = FC(F_1 \otimes F_2 \otimes F_3 \otimes F_4)$
10: $L_r = Loss(Pred, y)$

---

4   Experiment

4.1   Dataset

The MSTAR dataset [11] is a well-known benchmark for evaluating SAR ATR algorithms. It contains a diverse collection of $0.3\,m \times 0.3\,m$ SAR images representing ten different classes of ground targets. These targets include tanks, trucks, jeeps, and various military vehicles, exhibiting different orientations, articulations, and operating conditions. The OpenSARShip dataset [8] is widely used for ship detection and classification tasks using SAR imagery. It consists of ship chips extracted from 41 Sentinel-1 SAR images, providing a comprehensive coverage of environmental conditions and various ship types. The dataset includes 11,346 ship chips, representing 17 ship classes across five typical scenes. The FuSARShip dataset [6] is a comprehensive SAR dataset specifically designed for marine target analysis. It contains 126 slices extracted from high-resolution Gaofen-3 SAR images, with a resolution of 1.124 m × 1.728 m. The dataset encompasses 15 major ship categories and includes diverse non-ship targets. It incorporates polarimetric modes DH and DV and covers a range of marine, land, coastal, river, and island scenes.

4.2 Implementation

In the generation module, the U-Net architecture is utilized as the backbone, with a timestep set to 500. The Adam optimizer is employed with the learning rate of 4e−4. In the recognition module, a simple three-layer convolutional network is used as the encoder. The network is trained for 200 epochs with a learning rate of 8e−4.

To comprehensively evaluate the effectiveness of the generation module and recognition module, conducted experiments from two aspects: similarity and recognition accuracy. In the similarity experiments, measured the image quality using three image quality metrics: Normalized Cross Correlation (NCC), Structural Similarity Index Measure (SSIM), and Peak Signal-to-Noise Ratio (PSNR). Additionally, using two distribution metrics: Earth Mover's Distance (EMD) and Maximum Mean Discrepancy (MMD). In the recognition accuracy experiments, testing the ATR accuracy of the REDG on test set and compared it with other methods.

4.3 Similarity Results

The evaluation of the generated synthetic data in terms of similarity to the training and testing datasets is crucial to assess the performance of the generation module. Table 1 presents the results of experiments conducted under 1-shot, 5-shot, and 10-shot. Higher values of NCC and SSIM indicate a closer resemblance between the generated and real images, while a higher PSNR value signifies better image quality and fidelity. On the other hand, lower values of EMD and MMD indicate a higher similarity between the generated image distribution and the real image distribution. Despite training on few data, the generated samples exhibit remarkable similarity to the entire training dataset, demonstrating the effectiveness of the generation module in capturing SAR image features. As the number of available samples increases, the generated images closely resemble the real dataset in terms of quality and distribution. Furthermore, the generated data maintains similarity with the testing dataset, indicating the generalization

Table 1. Generation Quality of REDG

| | Fake vs Real Training | | | | | | | | |
| | MSTAR | | | OpenSAR | | | FUSAR | | |
	1-shot	5-shot	10-shot	1-shot	5-shot	10-shot	1-shot	5-shot	10-shot
NCC	0.86	0.95	0.95	0.91	0.89	0.94	0.44	0.80	0.80
SSIM	0.71	0.87	0.86	0.80	0.78	0.88	0.71	0.78	0.84
PSNR	22.69	27.60	28.12	24.69	24.02	27.52	30.65	33.13	34.46
EMD	0.03	0.02	0.02	0.03	0.03	0.01	0.01	0.01	0.01
MMD	0.42	0.20	0.11	0.57	0.53	0.27	1.46	1.12	0.69
	Fake vs Real Test								
	MSTAR			OpenSAR			FUSAR		
	1-shot	5-shot	10-shot	1-shot	5-shot	10-shot	1-shot	5-shot	10-shot
NCC	0.83	0.93	0.94	0.90	0.88	0.93	0.34	0.61	0.63
SSIM	0.67	0.84	0.83	0.77	0.73	0.86	0.64	0.68	0.73
PSNR	21.98	26.89	27.66	24.01	22.89	26.63	28.57	29.76	30.68
EMD	0.03	0.02	0.02	0.03	0.03	0.02	0.01	0.01	0.01
MMD	0.46	0.20	0.10	0.61	0.55	0.32	1.55	1.25	0.84

capability of the generation module. These results strongly validate the effectiveness and reliability of the generation module in synthesizing SAR images that closely resemble real data. Partial generation samples are shown in Fig. 2.

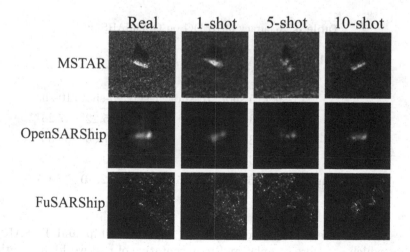

Fig. 2. REDG Generation Samples

4.4   Recognition Results

To validate the usability of the generated data and the effectiveness of the recognition module, we conducted experiments to evaluate the accuracy of ATR, as shown in Tables 2, 3, and 4.

Table 2. Accuracy of REDG on MSTAR SOC

MSTAR SOC						
Method	10way			5way		
	1-shot	5-shot	10-shot	1-shot	5-shot	10-shot
Scratch-Aconv [10]	20.90	47.40	65.00	31.60	55.90	72.30
SimCLROE [10]	28.80	60.50	75.60	39.90	69.70	83.30
DN4 [14]	33.25	53.48	64.88	**47.71**	62.78	70.64
DeepEMD [23]	36.19	53.14	59.64	40.99	58.04	62.72
PPSML [24]	36.56	59.56	73.36	38.43	77.89	83.57
REDG (Ours)	**38.79**	**62.66**	**88.22**	45.73	**79.45**	**89.97**

Table 2 presents the recognition accuracy on the MSTAR Standard Operating Conditions (SOC) dataset. From the results, it is evident that REDG achieves impressive recognition accuracy even with a simple CNN encoder. This demonstrates the effectiveness of the generated data from the generation module in complementing the training set and the efficient utilization of augmented data by the recognition module for training purposes.

Table 3. Accuracy of REDG on OpenSARShip

OpenSARShip						
Method	10way			5way		
	1-shot	5-shot	10-shot	1-shot	5-shot	10-shot
DN4	17.45	35.43	57.58	27.93	45.22	67.26
DeepEMD	18.53	34.08	55.00	29.81	47.60	66.33
PPSML	20.33	38.11	59.19	29.97	50.06	69.36
REDG (Ours)	**58.55**	**67.78**	**72.65**	**64.94**	**72.50**	**78.65**

The limited research conducted on the OpenSARShip and FUSARShip datasets, which are more complex and representative of real-world applications, caught our attention. To address this gap, extensive experiments were conducted on these datasets, and the results are presented in Tables 3 and 4. These results

Table 4. Accuracy of REDG on FuSARShip

FuSARShip						
Method	10way			5way		
	1-shot	5-shot	10-shot	1-shot	5-shot	10-shot
DN4	29.29	52.80	75.82	37.21	63.52	83.15
DeepEMD	35.13	54.32	74.79	48.21	61.84	79.85
PPSML	32.48	54.46	76.58	46.07	69.58	85.56
REDG (Ours)	**55.03**	**60.78**	**80.23**	**55.73**	**76.34**	**86.73**

serve as a valuable reference for future research, demonstrating the effectiveness of the approach even in challenging and realistic scenarios.

Overall, the experimental results show that REDG performs well when facing the simple dataset, and it gains a clear advantage when dealing with more complex datasets. This confirms the usefulness of generated data and the ability of the recognition module to enhance performance through the utilization of augmented data. It also validates that the proposed method is more suitable for complex real-world applications.

Table 5. Ablation Results, G: Generation Module; R: Recognition Module

Setting		10way			5way		
G	R	1-shot	5-shot	10-shot	1-shot	5-shot	10-shot
MSTAR SOC							
		21.08	44.75	60.25	24.83	48.03	61.31
+		31.28	60.25	79.21	37.70	69.30	76.98
+	+	38.79	62.66	89.22	45.73	79.45	89.97
OpenSARShip							
		10.31	11.42	13.67	10.46	20.65	53.35
+		35.16	49.36	59.54	42.99	51.24	61.55
+	+	58.55	67.78	72.65	64.94	72.50	78.65
FuSARShip							
		10.06	24.86	26.15	29.77	37.40	47.33
+		32.06	51.75	65.43	45.80	65.57	76.34
+	+	55.03	60.78	80.23	55.73	76.34	86.73

4.5   Ablation Studies

To validate the effectiveness of REDG framework's different modules, additional experiments were conducted. A baseline CNN model, similar to the recognition module but without the feature layering and reuse algorithm, was used for comparison. The results of these experiments are summarized in Table 5. The G represents the generation module, and the R represents the recognition module. These results clearly demonstrate the effectiveness of the modules in REDG.

5   Conclusion

The paper proposes the REDG for few-shot SAR image recognition. It consists of a generation module and a recognition module. The generation module is based on a simplified DDPM and is used to generate synthetic samples. The recognition module utilizes feature layering and reuse algorithm to enhance the model's recognition ability. Experimental results demonstrate the effectiveness of REDG in addressing few-shot SAR image recognition challenges and achieving superior performance only with a simple CNN model. This work contributes to advancing SAR image recognition with limited training data, offering insights for future research and practical applications.

References

1. Cao, C., Cao, Z., Cui, Z.: LDGAN: a synthetic aperture radar image generation method for automatic target recognition. IEEE Trans. Geosci. Remote Sens. **58**(5), 3495–3508 (2019)
2. Deng, J., Bi, H., Zhang, J., Liu, Z., Yu, L.: Amplitude-phase CNN-based SAR target classification via complex-valued sparse image. IEEE J. Sel. Top. Appl. Earth Obs. Remote Sens. **15**, 5214–5221 (2022)
3. Dhariwal, P., Nichol, A.: Diffusion models beat GANs on image synthesis. Adv. Neural. Inf. Process. Syst. **34**, 8780–8794 (2021)
4. Fu, K., Zhang, T., Zhang, Y., Wang, Z., Sun, X.: Few-shot SAR target classification via metalearning. IEEE Trans. Geosci. Remote Sens. **60**, 1–14 (2021)
5. Ho, J., Jain, A., Abbeel, P.: Denoising diffusion probabilistic models. Adv. Neural. Inf. Process. Syst. **33**, 6840–6851 (2020)
6. Hou, X., Ao, W., Song, Q., Lai, J., Wang, H., Xu, F.: FUSAR-ship: building a high-resolution SAR-AIS matchup dataset of Gaofen-3 for ship detection and recognition. SCIENCE CHINA Inf. Sci. **63**, 1–19 (2020)
7. Hu, M., Wang, Y., Cham, T.J., Yang, J., Suganthan, P.N.: Global context with discrete diffusion in vector quantised modelling for image generation. In: Proceedings of the IEEE/CVF Conference on Computer Vision and Pattern Recognition, pp. 11502–11511 (2022)
8. Huang, L., et al.: OpenSARShip: a dataset dedicated to Sentinel-1 ship interpretation. IEEE J. Sel. Top. Appl. Earth Obs. Remote Sens. **11**(1), 195–208 (2017)
9. Huang, Y., Mei, W., Liu, S., Li, T.: Asymmetric training of generative adversarial network for high fidelity SAR image generation. In: IGARSS 2022-2022 IEEE International Geoscience and Remote Sensing Symposium, pp. 1576–1579. IEEE (2022)

10. Inkawhich, N.: A global model approach to robust few-shot SAR automatic target recognition. IEEE Geosci. Remote Sens. Lett. (2023)
11. Keydel, E.R., Lee, S.W., Moore, J.T.: MSTAR extended operating conditions: a tutorial. In: Algorithms for Synthetic Aperture Radar Imagery III, vol. 2757, pp. 228–242 (1996)
12. Leng, Y., et al.: Binauralgrad: a two-stage conditional diffusion probabilistic model for binaural audio synthesis. Adv. Neural. Inf. Process. Syst. **35**, 23689–23700 (2022)
13. Li, H., Wang, T., Wang, S.: Few-shot SAR target classification combining both spatial and frequency information. In: GLOBECOM 2022-2022 IEEE Global Communications Conference, pp. 480–485. IEEE (2022)
14. Li, W., Wang, L., Xu, J., Huo, J., Gao, Y., Luo, J.: Revisiting local descriptor based image-to-class measure for few-shot learning. In: Proceedings of the IEEE/CVF Conference on Computer Vision and Pattern Recognition, pp. 7260–7268 (2019)
15. Li, X., Zhang, R., Wang, Q., Duan, X., Sun, Y., Wang, J.: SAR-CGAN: improved generative adversarial network for EIT reconstruction of lung diseases. Biomed. Signal Process. Control **81**, 104421 (2023)
16. Liu, J., Xing, M., Yu, H., Sun, G.: EFTL: complex convolutional networks with electromagnetic feature transfer learning for SAR target recognition. IEEE Trans. Geosci. Remote Sens. **60**, 1–11 (2021)
17. Oh, J., Kim, M.: PeaceGAN: a GAN-based multi-task learning method for SAR target image generation with a pose estimator and an auxiliary classifier. Remote Sens. **13**(19), 3939 (2021)
18. Pu, W.: Shuffle GAN with autoencoder: a deep learning approach to separate moving and stationary targets in SAR imagery. IEEE Trans. Neural Netw. Learn. Syst. **33**(9), 4770–4784 (2021)
19. Sun, X., Lv, Y., Wang, Z., Fu, K.: SCAN: scattering characteristics analysis network for few-shot aircraft classification in high-resolution SAR images. IEEE Trans. Geosci. Remote Sens. **60**, 1–17 (2022)
20. Sun, Y., et al.: Attribute-guided generative adversarial network with improved episode training strategy for few-shot SAR image generation. IEEE J. Sel. Top. Appl. Earth Obs. Remote Sens. **16**, 1785–1801 (2023)
21. Tai, Y., Tan, Y., Xiong, S., Sun, Z., Tian, J.: Few-shot transfer learning for SAR image classification without extra SAR samples. IEEE J. Sel. Top. Appl. Earth Obs. Remote Sens. **15**, 2240–2253 (2022)
22. Wang, D., Song, Y., Huang, J., An, D., Chen, L.: Sar target classification based on multiscale attention super-class network. IEEE J. Sel. Top. Appl. Earth Obs. Remote Sens. **15**, 9004–9019 (2022)
23. Zhang, C., Cai, Y., Lin, G., Shen, C.: DeepEMD: few-shot image classification with differentiable earth mover's distance and structured classifiers. In: Proceedings of the IEEE/CVF Conference on Computer Vision and Pattern Recognition, pp. 12203–12213 (2020)
24. Zhu, P., Gu, M., Li, W., Zhang, C., Hu, Q.: Progressive point to set metric learning for semi-supervised few-shot classification. In: 2020 IEEE International Conference on Image Processing (ICIP), pp. 196–200. IEEE (2020)

A Two-Stage Federated Learning Framework for Class Imbalance in Aerial Scene Classification

Zhengpeng Lv, Yihong Zhuang, Gang Yang, Yue Huang$^{(\boxtimes)}$ ⓘ, and Xinghao Ding ⓘ

School of Informatics, Xiamen University, Xiamen, China
yhuang2010@xmu.edu.cn

Abstract. Centralized aerial imagery analysis techniques face two challenges. The first one is the data silos problem, where data is located at different organizations separately. The second challenge is the class imbalance in the overall distribution of aerial scene data, due to the various collecting procedures across organizations. Federated learning (FL) is a method that allows multiple organizations to learn collaboratively from their local data without sharing. This preserves users' privacy and tackles the data silos problem. However, traditional FL methods assume that the datasets are globally balanced, which is not realistic for aerial imagery applications. In this paper, we propose a Two-Stage FL framework (TS-FL), which mitigate the effect of the class imbalanced problem in aerial scene classification under FL. In particular, the framework introduces a feature representation method by combing supervised contrastive learning with knowledge distillation to enhance the model's feature representation ability and minimize the client drift. Experiments on two public aerial datasets demonstrate that the proposed method outperforms other FL methods and possesses good generalization ability.

Keywords: Aerial scene analysis · Federated learning · Class imbalance · Two-stage framework · Remote sensing

1 Introduction

The success of deep learning in the field of computer vision has greatly contributed to the advancement of aerial imagery analysis, which cannot be achieved without the driving force of large-scale, diverse, and high-quality data. However, in practical application scenarios, aerial scene data are usually collected among remote sensing organizations in various industries. Each organization utilizes its

The work was supported in part by the National Natural Science Foundation of China under Grant 82172033, U19B2031, 61971369, 52105126, 82272071, 62271430, and the Fundamental Research Funds for the Central Universities 20720230104.

Q. Liu et al. (Eds.): PRCV 2023, LNCS 14428, pp. 430–441, 2024.
https://doi.org/10.1007/978-981-99-8462-6_35

own devices to collect data within the industry, which leads to data silos. The limited quantity and diversity of aerial images in each data silo present challenges in effectively applying deep learning methods. There are some attempts, such as semi-supervised learning [15,18] and few-shot learning [7] using limited supervision information to exploit the underlying pattern in the dataset, however, these methods rely on data center to train. It's difficult for some organizations to communicate and collaborate directly due to privacy and security concerns [1] about data.

Federated learning (FL) [14] can train a shared prediction model using data from multiple devices while protecting data privacy, which can well solve the collaboration problem among remote sensing organizations.

Despite these benefits, current FL algorithms still face some challenges. Firstly, traditional FL methods only focus on distributed model training under globally balanced class distributions. This leads to the FL model having a bias towards certain samples, making accurate identification more challenging. However, current imbalance-oriented FL methods [17,20] cannot yield satisfactory performance on aerial scene data. Secondly, during the collaboration of organizations, the distribution of aerial scene data may differ, resulting in statistical heterogeneity of non-independent and identically distributed (Non-IID) data across organizations. This poses difficulties for FL in ensuring convergence and effectiveness, particularly in complex real-world scenarios.

From neural network perspective, deep features are easily biased by commonly used cross-entropy loss function and labels, which will be propagated through the entire model via gradients.

Consequently, the bias of local model training on class imbalance and Non-IID data will seriously affect the classifier and further affect the feature representation model. Thus, FL methods are highly susceptible to the effects of class imbalance and require targeted solutions.

To this end, we propose a two-stage federated learning framework for class imbalance aerial scene classification. This framework decouples representation learning and classifier learning separately, mitigating client drift and model bias caused by Non-IID and class imbalance distributions. To further address the impact of class imbalance on the representation model, we propose a feature representation method based on federated knowledge distillation. By utilizing global data information, it learns an excellent feature representation model and effectively assists the training of the federated classifier, mitigating the impact of class imbalance on feature representation.

The contributions of this work can be summarized as follows:

- We propose a two-stage FL framework for aerial scene classification task with class imbalance distribution. This framework decouples representation learning and classifier learning separately to mitigate model bias caused by Non-IID and class imbalanced distributions during feature representation updates.
- To address the problem of performance loss due to traditional cross-entropy in imbalance distribution, we propose a federated feature representation method based on knowledge distillation, which utilizes supervised contrastive learning during client training process.

- To tackle client drift problem, we introduce global representation model applying global consistent distillation for the client representation training process.

2   Related Work

2.1   Class Imbalance Aerial Image Classification

In practical applications, aerial scene images are commonly imbalanced. To address this issue, some methods try to convert the distribution from an imbalanced distribution to a balanced one through rebalancing strategy. This includes oversampling the minority class [13,22] or reducing the weight of the majority class in the loss function and increasing the weight of the minority class [4,5]. However, these strategies can lead to information loss that impairs the learning of feature representations. Recent studies have shown that decoupling representation learning and classifier learning [8,23] can lead to less biased representations, effectively alleviating the impacts of class imbalance. However, we believe that further improvements are needed for task-specific federated representation learning.

2.2   Federated Learning

Federated learning provides a potential solution to the problems of privacy leaks and isolated data islands in aerial imagery analysis. The FedAvg algorithm [14], which serves as the basic framework of FL, cannot address heterogeneity well. There are FL solutions solve data heterogeneity through optimization strategies [12] or using mechanisms [2] on the server, however, they perform poorly on imbalanced data. Currently, there are some FL methods [17,20] for the imbalanced problem. Wang *et al.* [20] use balanced auxiliary data on the server to estimate the global class distribution for better local optimization. However, retrieving data from the server could lead to privacy issues. The current methods demonstrate limited effectiveness on imbalanced data and need further exploration.

3   Method

In this section, we first describe the overall framework of the two-stage FL, which combines federated representation learning and federated classifier learning, and then detail each stage of the two-stage FL process.

3.1   A Two-Stage Federated Learning Framework (TS-FL)

Our framework mainly focuses on addressing FL problems under class imbalance and Non-IID settings, which includes cross-client federated representation learning stage and federated classifier learning stage based on global representation model. As shown in Fig. 1, the left side represents the representation learning

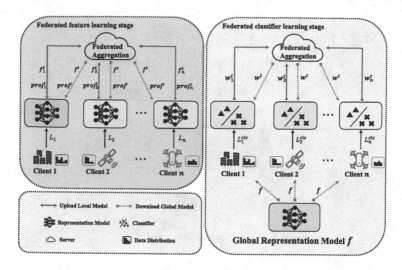

Fig. 1. The overall framework of TS-FL. (Color figure online)

stage, where blue, yellow and green backgrounds represent different local representation models, and gray one represents the global representation model. In each communication round t, each client trains their local representation models by minimizing the objective function L_n, and uploads them to the server after local training. At the server, these local representation models are aggregated into a new global representation model f^t. The optimal global representation model is finally obtained through multiple rounds of communication iterations. On the right side is the federated classifier learning stage. In this stage, clients create local classifier models using the global representation model and update the global classifier model using the FedAvg [14] method.

3.2 Feature Representation Method Based on Federated Knowledge Distillation

In the TS-FL, the performance of the global representation model learned in the first stage directly affects the effectiveness and the traditional methods using cross-entropy loss function are vulnerable to class imbalance. For this reason, we utilize a contrastive learning loss function that emphasizes inter-class and intra-class distance to reduce the impact of class imbalance. In addition, the Non-IID problem of FL can also lead to client bias during local training, which affects convergence. To address this issue, we use the idea of knowledge distillation and propose global consistency distillation, applying the knowledge of the global representation model in the local model training by aligning the feature space. The two parts are explained in detail below.

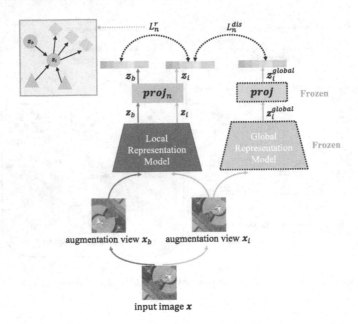

Fig. 2. Feature representation method based on federated knowledge distillation.

Local Supervised Contrastive Learning Loss. We use supervised contrastive learning [10] as the loss function to learn meaningful features from local data by introducing labels implicitly. As shown in Fig. 2, in contrastive representation learning, a local projection head $proj_n$ and a local representation model f_n are introduced to form the model F_n, and the output feature of the representation model for the sample x is projected onto a specific hypersphere, and its projection vector is represented as $z = F_n(x) = proj_n(f_n(x))$. The local representation model is learned by measuring the distance in the projection space. The projection head is generally defined as two non-linear layers, and it is removed after representation learning. F represents the global model of the local model F_n. For client p_n, the training set $(x, y) \in D_n$ has a sample size of d, and $2d$ augmented images are randomly generated for each set. the local supervised contrastive learning loss is:

$$L_n^r = \sum_{i \in I} \frac{-1}{|B(i)|} \sum_{b \in B(i)} \log \frac{\exp(z_i \cdot z_b / \tau)}{\sum_{k=1, k \neq i}^{2d} \exp(z_i \cdot z_k / \tau)} \tag{1}$$

The set $I \equiv \{1, \cdots, 2d\}$ represents the index set of randomly augmented samples. The set $B(i) \equiv \{b \in I | y_b = y_i, b \neq i\}$ represents the index set of all positive samples relative to x_i, where $|B(i)|$ is the number of indices in the set. z is the normalized projection vector, and $z_i \cdot z_b$ represents the inner product of the vectors z_i and z_b. τ is the temperature coefficient.

Global Consistent Distillation. Considering the impact of Non-IID, the inconsistent data distribution among the clients may lead to bias in the client models. In this case, the local representation model f_n may be suboptimal because local updates only minimize the loss function on the local datasets. The global representation model has powerful global data representation capabilities, which can be used to perform consistency distillation on the local training process to align their feature spaces and alleviate the client drift problem. The specific method is to use consistency regularization loss to reduce the distance between the projection vectors of the local representation model and the global representation model, as shown in Fig. 2. The loss can be described as

$$L_n^{\text{dis}} = \left\| z_i - z_i^{\text{global}} \right\|_2 \tag{2}$$

Here, $z_i^{\text{global}} = F(x) = proj(f(x_i))$, where $\|\cdot\|_2$ represents the L_2 loss function.

Therefore, the overall loss function for feature representation based on federated knowledge distillation, which includes the local supervised contrastive learning loss and the global consistency distillation loss, can be described as

$$L_n = L_n^r + \lambda L_n^{dis} \tag{3}$$

We use FedAvg [14] to obtain a global representation model during aggregation, where λ represents the weighting coefficient to balance the two terms.

3.3 Federated Classifier Learning

After the federated representation learning stage, learned representation model can be applied to our classification task. In order to evaluate the performance of the federated representation model, we use FedAvg [14] to train the federated classifier and the empirical loss of client P_n is defined as

$$L_n^{cls} = -\sum_{i=1}^{|D_n|} (p_i * \log q_i) \tag{4}$$

where p_i represents the true label of the sample in one-hot encoding form, and q_i represents the prediction vector of the sample after passing through the representation model and the local classifier, finally being normalized by $softmax$.

4 Experiment

4.1 Dataset

In this work, we use NWPU-RESISC45 [3] and AID [21] as the datasets to evaluate the proposed method. NWPU-RESISC45 dataset contains 31,500 remote sensing images, with 45 classes, each class having 700 images. The original size of the images is 256×256. AID contains 10,000 remote sensing images, with 30 classes, with 300 to 400 aerial images per category. The original size of the images is 600×600. Figure 3 illustrate some example images of the two datasets.

(a)NWPU-RESISC45   (b)AID

Fig. 3. Scene images of NWPU-RESISC45 [3] and AID [21] datasets.

NWPU-IM10 and NWPU-IM50. We defines the first 30 classes of the NWPU-RESISC45 dataset as majority classes and the last 15 classes as minority classes. For the minority classes, a NWPU-IM10 imbalanced dataset is constructed by taking 1/10 of each class's samples and a NWPU-IM50 dataset is constructed by taking 1/50 of each class's samples.

AID-IM10 and AID-IM50. For the AID dataset, the first 20 classes are defined as majority classes and the last 10 classes as minority classes. AID-IM10 and AID-IM50 datasets are constructed in the same way as in NWPU-RESISC45.

Dataset Non-IID Partition. The training and testing sets are partitioned according to the original number of samples for each class, taking 80% of each class's samples as the training set and the remaining 20% as the testing set, which is roughly balanced in terms of class distribution. This paper utilizes the widely used Non-IID sampling method in previous FL to sample class-imbalanced data [19]. The Dirichlet distribution is used to generate Non-IID datasets among different clients. Specifically, samples belonging to class k are sampled from $p_k \sim \mathrm{Dir}_N(\beta)$ and are distributed to client P_n based on the proportion of $p_{k,n}$, where the Dirichlet distribution with parameter β, represented by $Dir(\beta)$, is used, and $\beta = 100$ approximates the IID sampling method. As β decreases, the degree of Non-IID sampling increases. The results of dividing $\beta = \{100, 1, 0.5\}$ on NWPU-IM10 are shown in Fig. 4.

4.2   Implementation Details

The proposed method is implemented on PyTorch and FedML [6]. Each dataset corresponds to three Non-IID settings of $\beta = \{100, 1, 0.5\}$. We crop images of all the datasets into patches of 224×224 as the input model. Then we apply regular data augmentation to the input patches such as random crop, random flip and image normalization.

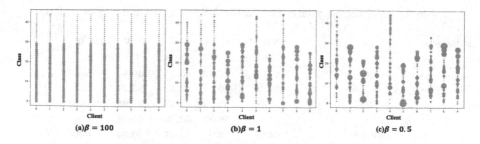

Fig. 4. Non-IID partition of NWPU-IM10

In FL training, this paper uses the Resnet34 as the backbone network and defines the projection head as a 2-layer fully connected layer, using ReLU as the activation function. The local training rounds are uniformly set to 10 rounds, and the total communication rounds are set to 100 rounds. The client participation rate for each round is set to 60%. The temperature coefficient τ for supervised contrastive learning is set to 0.1, the batch size is set to 128, and the optimizer is Adam, with a learning rate of 0.0001 and a weight decay of 0.0001. The number of communication rounds is set to 80 rounds. In the federated classifier training stage, we use FedAvg [14] to evaluate the performance of TS-FL. The representation model is also fine-tuned during the classifier learning stage with a batch size of 256 and the same optimizer settings as the federated feature learning stage. The communication round number is 20 rounds. For all other comparative methods, the communication round number is set to 100. All other settings keep consistent with the federated classifier training setup.

4.3   Experimental Results

We compare our proposed TS-FL with some classic FL methods, including FedAvg [14], FedProx [12], FedNova [19], SCAFFOLD [9], FedOpt [16], MOON [11] and two major class imbalance FL methods including Fed-Focal [17], FedAvg with τ-norm [8] and Ratio Loss [20]. The results are demonstrated in Table 1 and Table 2.

As seen, FedNova [19] utilizes the gradient information of the client training process, MOON [11] conducts comparative learning on the global model and local model, both have achieved good performance in traditional FL methods. Ratio Loss [20] requires auxiliary data on the server, achieves comparable performance with our method in some circumstances, but there are privacy concerns. Our method almost all achieves optimal accuracy. When facing Non-IID, our method achieves high accuracy, which also suggests that the method has good robustness.

Table 1. Top-1 accuracy rates (%) on the NWPU-IM10 and NWPU-IM50.

Family	Method	NWPU-IM10			NWPU-IM50		
		$\beta = 100$	$\beta = 1$	$\beta = 0.5$	$\beta = 100$	$\beta = 1$	$\beta = 0.5$
Traditional FL methods	FedAvg	54.63	57.57	50.24	48.86	46.38	45.56
	FedProx	59.98	56.95	52.21	53.27	52.44	49.63
	FedNova	65.78	60.83	55.67	57.95	55.84	55.57
	SCAFFOLD	64.49	59.92	50.46	55.59	51.92	49.75
	FedOpt	54.97	52.97	45.92	50.30	46.38	45.63
	MOON	68.33	63.89	53.05	59.29	56.52	53.43
Class imbalance FL methods	Fed-Focal	53.83	51.92	46.86	48.89	46.44	45.79
	FedAvg with τ-norm	54.29	51.87	45.68	48.84	46.68	44.71
	Ratio Loss	72.59	69.05	**66.11**	63.08	62.08	59.29
	TS-FL(Ours)	**81.19**	**75.05**	61.51	**67.46**	**62.41**	**60.41**

Table 2. Top-1 accuracy rates (%) on the AID-IM10 and AID-IM50.

Family	Method	AID-IM10			AID-IM50		
		$\beta = 100$	$\beta = 1$	$\beta = 0.5$	$\beta = 100$	$\beta = 1$	$\beta = 0.5$
Traditional FL methods	FedAvg	56.90	52.15	50.10	52.55	48.60	47.60
	FedProx	61.15	58.00	54.50	55.75	53.15	51.30
	FedNova	66.10	65.65	64.65	58.35	57.60	56.00
	SCAFFOLD	64.55	59.55	52.65	54.90	53.95	50.20
	FedOpt	56.15	50.35	50.15	52.70	49.35	48.50
	MOON	68.00	57.85	51.80	57.40	54.45	50.70
Class imbalance FL methods	Fed-Focal	57.60	51.65	50.60	52.35	48.80	48.45
	FedAvg with τ-norm	56.60	52.55	48.30	53.20	50.45	48.85
	Ratio Loss	71.00	68.29	64.50	61.85	59.20	55.95
	TS-FL(Ours)	**78.20**	**72.35**	**66.65**	**66.50**	**63.65**	**58.85**

4.4 Feature Visualization

To further illustrate the strengths and weaknesses of the representation model in our paper, we utilizes t-SNE to visualize the features learned by the global models. Figure 5 displays the visualizations of feature maps on the server under $\beta = 1$ of AID-IM10. Compared with other methods, our method demonstrates a more compact representation within each class under the global representation model, as well as a greater differentiation between classes. Other methods suffer from data bias and the minority classes are scattered and embedded in the feature space, which is also the reason for the poor recognition accuracy of these methods on the minority classes.

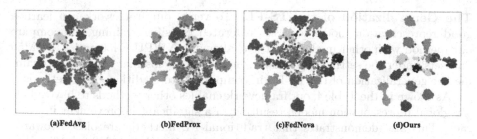

(a)FedAvg   (b)FedProx   (c)FedNova   (d)Ours

Fig. 5. Visualization feature maps of different methods on server. The large clusters belong to the majority class, and the point-like small clusters belong to the minority class.

4.5 Ablation Studies

Ablation Studies on Global Consistent Distillation. We conducted ablation experiments on the global consistency parameter under $\beta = 1$ in NWPU-IM10 and AID-IM10 to verify the effectiveness in addressing client drift. As shown in Table 3, the global consistent distillation loss can improve the accuracy of the two-stage federated learning framework.

Table 3. The Global consistent distillation impact on results.

	w/o Global Consistent Distillation	TS-FL
NWPU-IM10	73.43	**75.05**
AID-IM10	71.40	**72.35**

Table 4. Generalization improvement of TS-FL under different methods.

	NWPU-IM50	AID-IM50
FedAvg	46.38	48.60
+TS-FL(Ours)	62.41(+16.03)	63.65(+15.05)
FedProx	52.44	53.15
+TS-FL(Ours)	69.32(+16.88)	64.45(+11.3)
FedNova	55.84	57.60
+TS-FL(Ours)	70.3(+14.46)	65.25(+7.65)
SCAFFOLD	51.92	53.95
+TS-FL(Ours)	63.60(+11.68)	61.75(+7.80)
FedOpt	46.38	63.65
+TS-FL(Ours)	63.81(+17.43)	63.65(+14.30)
MOON	56.52	54.45
+TS-FL(Ours)	63.05(+6.53)	61.45(+7.00)

The Generalization of the TS-FL. To verify our framework can learn a good representation model to assist federated classifier training, we compare our method with other methods in the challenging NWPU-IM50 and AID-IM50 datasets under the setting of $\beta = 1$ for the federated classifier training, to examine the generalization of the TS-FL in comparison to traditional FL methods.

As shown in the Table 4, our framework enables other methods to have a better global representation model, resulting in a significant improvement in accuracy. This also demonstrates that traditional FL methods are biased towards both the classifier and feature representation due to class imbalance distributions, whereas our method mitigates this influence on the representation model more effectively.

5   Conclusion

In this study, we propose a two-stage FL framework, called TS-FL, to address the issues of class imbalance and statistical heterogeneity in aerial scene image classification under FL. The negative effects of classifier cross-entropy loss on feature representation are effectively addressed by dividing the FL process into federated representation learning and federated classifier learning. Specifically, during the federated representation learning stage, a feature representation method is introduced that leverages federated knowledge distillation and global consistency distillation loss to mitigate class imbalance and Non-IID problems. In the federated classifier learning stage, we utilize the learned feature representation model to assist in training the federated classifier. Experimental analysis on different aerial scene datasets provides further support for the significant improvement in recognition accuracy achieved by the proposed method, demonstrating its effectiveness and flexibility.

References

1. Alkhelaiwi, M., Boulila, W., Ahmad, J., Koubaa, A., Driss, M.: An efficient approach based on privacy-preserving deep learning for satellite image classification. Remote Sens. **13**(11), 2221 (2021)
2. Chen, H.Y., Chao, W.L.: FedBE: making Bayesian model ensemble applicable to federated learning. arXiv preprint arXiv:2009.01974 (2020)
3. Cheng, G., Han, J., Lu, X.: Remote sensing image scene classification: benchmark and state of the art. Proc. IEEE **105**(10), 1865–1883 (2017)
4. Cui, Y., Jia, M., Lin, T.Y., Song, Y., Belongie, S.: Class-balanced loss based on effective number of samples. In: Proceedings of the IEEE/CVF Conference on Computer Vision and Pattern Recognition, pp. 9268–9277 (2019)
5. Deng, Z., Liu, H., Wang, Y., Wang, C., Yu, Z., Sun, X.: PML: progressive margin loss for long-tailed age classification. In: Proceedings of the IEEE/CVF Conference on Computer Vision and Pattern Recognition, pp. 10503–10512 (2021)
6. He, C., et al.: FedML: a research library and benchmark for federated machine learning. arXiv preprint arXiv:2007.13518 (2020)

7. Ji, Z., Hou, L., Wang, X., Wang, G., Pang, Y.: Dual contrastive network for few-shot remote sensing image scene classification. IEEE Trans. Geosci. Remote Sens. **61**, 1–12 (2023)

8. Kang, B., et al.: Decoupling representation and classifier for long-tailed recognition. arXiv preprint arXiv:1910.09217 (2019)

9. Karimireddy, S.P., Kale, S., Mohri, M., Reddi, S., Stich, S., Suresh, A.T.: SCAFFOLD: stochastic controlled averaging for federated learning. In: International Conference on Machine Learning, pp. 5132–5143. PMLR (2020)

10. Khosla, P., et al.: Supervised contrastive learning. Adv. Neural. Inf. Process. Syst. **33**, 18661–18673 (2020)

11. Li, Q., He, B., Song, D.: Model-contrastive federated learning. In: Proceedings of the IEEE/CVF Conference on Computer Vision and Pattern Recognition, pp. 10713–10722 (2021)

12. Li, T., Sahu, A.K., Zaheer, M., Sanjabi, M., Talwalkar, A., Smith, V.: Federated optimization in heterogeneous networks. Proc. Mach. Learn. Syst. **2**, 429–450 (2020)

13. Li, Y., Lai, X., Wang, M., Zhang, X.: C-SASO: a clustering-based size-adaptive safer oversampling technique for imbalanced SAR ship classification. IEEE Trans. Geosci. Remote Sens. **60**, 1–12 (2022)

14. McMahan, B., Moore, E., Ramage, D., Hampson, S., Arcas, B.A.: Communication-efficient learning of deep networks from decentralized data. In: Artificial Intelligence and Statistics, pp. 1273–1282. PMLR (2017)

15. Miao, W., Geng, J., Jiang, W.: Semi-supervised remote-sensing image scene classification using representation consistency siamese network. IEEE Trans. Geosci. Remote Sens. **60**, 1–14 (2022)

16. Reddi, S., et al.: Adaptive federated optimization. arXiv preprint arXiv:2003.00295 (2020)

17. Sarkar, D., Narang, A., Rai, S.: Fed-focal loss for imbalanced data classification in federated learning. arXiv preprint arXiv:2011.06283 (2020)

18. Shi, J., Wu, T., Yu, H., Qin, A., Jeon, G., Lei, Y.: Multi-layer composite autoencoders for semi-supervised change detection in heterogeneous remote sensing images. SCIENCE CHINA Inf. Sci. **66**(4), 140308 (2023)

19. Wang, J., Liu, Q., Liang, H., Joshi, G., Poor, H.V.: Tackling the objective inconsistency problem in heterogeneous federated optimization. Adv. Neural. Inf. Process. Syst. **33**, 7611–7623 (2020)

20. Wang, L., Xu, S., Wang, X., Zhu, Q.: Addressing class imbalance in federated learning. In: Proceedings of the AAAI Conference on Artificial Intelligence, vol. 35, pp. 10165–10173 (2021)

21. Xia, G.S., et al.: AID: a benchmark data set for performance evaluation of aerial scene classification. IEEE Trans. Geosci. Remote Sens. **55**(7), 3965–3981 (2017)

22. Zhang, Y., Lei, Z., Yu, H., Zhuang, L.: Imbalanced high-resolution SAR ship recognition method based on a lightweight CNN. IEEE Geosci. Remote Sens. Lett. **19**, 1–5 (2021)

23. Zhuang, Y., et al.: A hybrid framework based on classifier calibration for imbalanced aerial scene recognition. In: Tanveer, M., Agarwal, S., Ozawa, S., Ekbal, A., Jatowt, A. (eds.) ICONIP 2022. LNCS, pp. 110–121. Springer, Cham (2022). https://doi.org/10.1007/978-3-031-30111-7_10

SAR Image Authentic Assessment
with Bayesian Deep Learning
and Counterfactual Explanations

Yihan Zhuang and Zhongling Huang[✉]

School of Automation, Northwestern Polytechnical University, Xi'an, China
huangzhongling@nwpu.edu.cn

Abstract. Synthetic Aperture Radar (SAR) targets vary significantly with observation angles, which leads to the deep neural network being over-confident and unreliable under the condition of limited measurements. This issue can be addressed by means of data augmentation such as simulation or exemplar synthesis for unknown observations, however, the authenticity of simulated/synthesized data is critical. In this study, we propose a Bayesian convolutional neural network (BayesCNN) for SAR target recognition to attain better generalization and assess predictive uncertainty. Based on the measured uncertainty of BayesCNN's prediction, we further propose to obtain the counterfactual explanation of an unrealistic SAR target generated by the unreliable simulation method. The experiments preliminarily demonstrates the proposed BayesCNN outperforms the counterpart frequentist neural network in terms of accuracy and confidence calibration when observation angles are constrained. In addition, the counterfactual explanation can reveal the non-authenticity of the augmented SAR target, which inspires us to filter the high-quality data, as well as to understand and improve the fidelity of generated SAR image in the future study.

Keywords: Synthetic Aperture Radar · SAR Target recognition · Explainable artificial intelligence · Bayesian deep learning · Counterfactual analysis

1 Introduction

SAR automatic target recognition with artificial intelligence has attracted great attention in the past years. We notice that recent deep learning algorithms can achieve high performance on SAR target recognition [7], especially on MSTAR dataset [8]. However, we believe the strong performance is a result of the dataset's cleanliness, as the MSTAR dataset contains SAR targets of almost full azimuth

This work was supported by the National Natural Science Foundation of China under Grant 62101459, and the China Postdoctoral Science Foundation under Grant BX2021248.

angles [8]. As is well-known, scattering properties of SAR targets vary considerably with observation angle. In practical application scenarios where the training data are limited to a narrow range of azimuth angles, the performance of deep learning model will decline significantly, and the prediction would be less confident particularly for test data with very different imaging conditions. The current deep learning model faces challenge of overfitting as well as over-confident to prediction, that makes users unable to trust the model's decision [11].

To overcome the limited training data problem in SAR target recognition, many data augmentation approaches have been explored, such as image simulation using electromagnetic physical model, image generation with generative deep model (e.g., GAN, VAE, etc.) [4,9] . However, the characteristics of SAR targets are considerably distinguished with those in optical images. For example, in the optical remote sensing image, the airplanes placed with different orientations can be generated with a simple rotation of the original image. Whereas, the SAR targets illuminated from different angles present diverse scattering centers and shadows, as shown in Fig. 1. SAR image generation is challenging due to the specific electromagnetic scattering properties, and thus the authenticity of the generated SAR image should be carefully concerned.

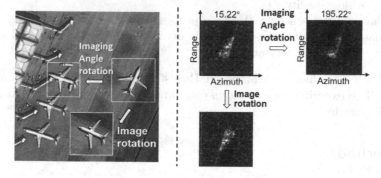

Fig. 1. The target characteristics change considerably in SAR image, instead of the rotation in-variant property in optical remote sensing image.

In this study, we firstly propose to introduce the Bayesian deep learning [12] and counterfactual analysis [1] to evaluate the authenticity of the generated SAR images with respect to model performance. Different from the current assessment approaches that compare the generated samples with ground truth in terms of some metrics (e.g., structural similarity index measure), our method aims to filter out the pseudo data that may prevent from training reliable deep model and to explain the details of low-quality data.

Bayesian deep learning introduces Bayes rules into deep neural network, which assigns the model weight a prior distribution and learns the posterior distribution with observed training data [12]. In this way, inference during the test stage is an integration of model predictions that considers all possible values of model weights according to the posterior distributions. It has been proved that Bayesian deep neural network (BNN) has advantages in improving the generalization ability, calibrating the confidence, and assessing the prediction uncertainty [3,6,13]. In addition, some related work apply BNN to discover the out-of-distribution data which would lead to a high degree of uncertainty in BNN's prediction [12]. If a BNN model for SAR target recognition is trained with real SAR images, it will probably predict the highly unrealistic test data incorrectly with a large uncertainty.

Besides filtering the inauthentic SAR image, it is also important to explain the inauthentic details of the pseudo sample, which would reveal the deficiency of the simulation approach and motivate future researches. Counterfactual explanation for uncertainty aims to understand which input patterns are responsible for predictive uncertainty [1]. Consequently, we propose to generate the counterfactual explanation for the inauthentic SAR target, which can not only point out the inauthentic details in the image, but also supplement missing information.

The main contributions are listed as follows:

- We propose a novel pipeline for SAR image authentic assessment based on Bayesian deep neural networks and counterfactual analysis.
- A Bayesian convolutional neural network (BayesCNN) is constructed and optimized with Bayes by Backprop for SAR target recognition, discovering pseudo samples of low quality.
- A variational auto-encoder (VAE) is proposed to generate the counterfactual explanations, revealing the inauthentic details in the SAR image and supplementing the missing scatterings.

Section 2 introduces the proposed BayesCNN and counterfactual analysis method. The experiments and discussions are presented in Sect. 3, and Sect. 4 draws the conclusion.

2   Method

2.1   Overview

The proposed SAR image authentic assessment pipeline based on Bayesian deep neural networks and counterfactual analysis is presented in Fig. 2. Firstly, the BayesCNN is trained with real SAR targets which is able to output trustworthy prediction for authentic images with correct label and low uncertainty. The inauthentic simulated data would result in a high degree of uncertainty and may be predicted as the wrong label. Secondly, a variational auto-encoder (VAE) is trained as a generator. In the third step, the latent counterfactual z is optimized given the expected prediction of BayesCNN, and then the counterfactual explanation can be generated from the decoder of VAE. The inauthentic details of the data can be explained finally.

2.2   Bayesian Convolutional Neural Network

BayesCNN Architecture. We construct a Bayesian convolutional neural network (BayesCNN) based on A-ConvNets [14], a simple all convolutional neural

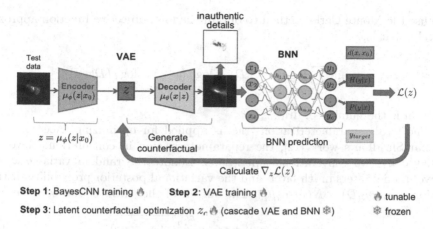

Fig. 2. The proposed SAR image authentic assessment pipeline based on Bayesian deep neural networks and counterfactual analysis.

network designed for SAR target recognition specifically. It comprises five convolutional layers and a Softmax activation layer at the end to obtain the prediction. The Dropout layer in A-ConvNets is removed to construct the BayesCNN. All parameters are replaced with random variables following Gaussian distributions $N(\theta|\mu, \sigma^2)$.

Optimization with Bayes by Backprop. Backpropagation and gradient descent are widely applied for neural network training. We apply one of the most popular BayesCNN optimization approaches, Bayes by Backprop (BBB) [2] to learn the posterior distribution of BayesCNN parameters. BBB is a variational inference-based method to learn the posterior distribution on the weights of a neural network from which weights $\omega \sim q_\theta(\omega|D)$ can be sampled in back propagation. It regularizes the weights by maximizing Evidence Lower Bound (ELBO).

To learn the Gaussian distribution parameters that weight follows, we need to find a way to approximate the intractable true posterior distribution $p(\omega|D)$. We use the method of variational inference. This is done by minimizing the Kullback-Leibler difference between the simple variational distribution $q_\theta(\omega|D)$ and the true posterior distribution $p(\omega|D)$, that is,

$$\arg \min_\theta KL[q_\theta(\omega)||p(\omega|D)] = \log(p(D)) + \arg \min_\theta(-ELBO). \qquad (1)$$

It can be achieved by maximizing the ELBO as follows:

$$- ELBO = KL[q_\theta(\omega)||p(\omega)] - E_{q_\theta}[\log(p(D|\omega))]. \qquad (2)$$

We use the Monte Carlo method to obtain the final objective function approximately, that is,

$$\arg\min_{\theta} \sum_{i=1}^{n} \log q_{\theta}(\omega^{(i)}|D) - \log p(\omega^{(i)}) - \log p(D|\omega^{(i)}) \tag{3}$$

where n is the number of draws.

The local reparameterization trick is applied for optimizing. According to Kumar Shridhar's work [12], the reparameterization in convolutional layer is achieved by two convolutional operations. Denote the random variable ω_i as convolutional filters in ith layer, and the variational posterior probability distribution as $q_{\theta}(\omega_i|D) = N(\mu_i, \alpha_i \mu_i^2)$. The output of the ith layer can be reparameterized as:

$$A_{i+1} = A_i * \mu_i + \epsilon \odot \sqrt{A_i^2 * (\alpha_i \odot \mu_i^2)} \tag{4}$$

where $\epsilon \sim N(0,1)$. A_i is the input feature map of the ith layer. $*$ and $\odot$ denote the convolution operation and component-wise multiplication, respectively.

Thus, the output of a Bayesian convolution layer can be realized by two steps of convolutions, that is, a convolution with the input feature map A_i and the mean values of kernels and another convolution with the square of A_i and the variances of kernels. In this way, the parameters μ_i and α_i can be updated separately in the two steps of convolution. We apply Adam for optimization.

Epistemic and Aleatoric Uncertainty. The uncertainty of BNN prediction is given by

$$H(y^*|x^*) = E_q[yy^T] - E_q[y]E_q[y]^T \tag{5}$$

It can be decomposed into the aleatoric and epistemic uncertainty [10,12], that is,

$$H(y^*|x^*) = \underbrace{\frac{1}{T}\sum_{t=1}^{T} diag(\hat{p}_t) - \hat{p}_t\hat{p}_t^T}_{aleatoric} + \underbrace{\frac{1}{T}\sum_{t=1}^{T}(\hat{p}_t - \overline{p})(\hat{p}_t - \overline{p})^T}_{epistemic}, \tag{6}$$

where $\hat{p}_t = Softmax(f_{\omega_t}(x^*))$ denotes the frequentist inference with parameter w_t sampled from the obtained posterior distribution, and $\overline{p} = \frac{1}{T}\sum_{t=1}^{T}\hat{p}_t$ denotes the Bayesian average of predictions, and T is the sampling number.

2.3   Counterfactual Explanation Generation

Definition of Counterfactual. If the simulated SAR image x_0 is inauthentic, the BayesCNN would output a high uncertainty or even the wrong label y_0. It is expected to find the counterfactuals that the BayesCNN can output accurate and trustworthy prediction. In a word, we aim to generate an x_c keeping the smallest change of the original data x_0 that makes the BayesCNN output the desired prediction y_c.

Denote the BayesCNN predictor as P_{BNN}, the counterfactuals x_c can be generated by solving the following optimization problem:

$$x_c = \arg\max_x P_{\text{BNN}}(y_c|x) - d(x, x_0) \quad s.t. \quad y_0 \neq y_c, \tag{7}$$

where $d(\cdot)$ denotes the L_1-norm distance metric.

The desired output of BNN $P_{\text{BNN}}(y_c|x)$ should satisfy the following conditions, that is, the expected label prediction of y_c and low entropy of $H(y_c|x)$ referring to a small uncertainty. As a result, $P_{\text{BNN}}(y_c|x)$ can be decomposed into:

$$\arg\max_x P_{\text{BNN}}(y_c|x) = \arg\min_x - \sum_{i=0}^c y_c log(p(y_i|x)) + H(y_c|x) \tag{8}$$

Counterfactual Generation Based on VAE. Counterfactual Latent Uncertainty Explanations (CLUE) was the first method proposed to explain the uncertainty [1]. In this work, we follow the basic idea of CLUE to generate the counterfactuals and inauthentic details of simulated data based on VAE.

Since the naive optimization of Eq. (7) in the high-dimensional input space is difficult, we aim to generate the counterfactuals in the lower-dimensinal latent space of VAE. As shown in Fig. 2, the input SAR image x_0 is embedded into the latent space of random variable z via an encoder $q_\phi(z|x)$ of a VAE. The generative output of VAE can be obtained from the decoder $p_\theta(x) = \int p_\theta(x|z)p(z)dz$. ϕ and θ are the parameters in the encoder and decoder of the VAE, respectively. The expectations of the latent space and the output are denoted as $E_{q_\phi(z|x)}[z] = \mu_\phi(z|x)$ and $E_{p_\theta(x|z)}[x] = \mu_\theta(x|z)$, respectively.

The VAE model is firstly trained with abundant data to obtain the representative encoder $q_\phi(z|x)$ and decoder $p_\theta(x|z)$ for counterfactual explanation generation. We expect the counterfactuals of an input x_0 can be generated from the decoder, i.e., $\mu_\theta(x|z)$. Therefore, the latent counterfactual z_c is optimized from the following objective function based on Eq. (7):

$$\mathcal{L}(z) = - \sum_{i=0}^c y_c log(p(y_i|\mu_\theta(x|z))) + H(y_c|\mu_\theta(x|z)) + d(\mu_\theta(x|z), x_0). \tag{9}$$

Then, the latent counterfactual z_c can be obtained from $z_c = \arg\min_z \mathcal{L}(z)$, and the counterfactual explanation in the image domain can be generated from the decoder of VAE:

$$x_c = \mu_\theta(x|z_c) \tag{10}$$

The inauthentic details of the test SAR image can be reported as the pixel changes of the counterfactual explanations and the original image. For a better visualization, the difference $\Delta x = x_c - x_0$ is multiplied with its absolute value $|\Delta x|$ while keeping the sign. The positive and negative values are colored with red and blue, respectively.

3   Experiments

3.1   Dataset and Settings

In the experiment, we use MSTAR [8] dataset under the standard operation condition (SOC), where the samples with depression angle 17° are for training and 15° for testing. It contains ten categories and each category in the MSTAR dataset has images with azimuths ranging from 0° to 360°.

To demonstrate the efficiency of BayesCNN under the limited observation conditions, we set up three different training sets, i.e., data with azimuth angles ranging from 0°–30° (trainset 0–30), 0°–90° (trainset 0–90), and 0°–180° (trainset 0–180). In the test stage, we calculate the performance also by azimuths, i.e., 12 groups evenly divided from 0°–360°. We conduct the comparative study of the proposed BayesCNN with frequentist CNN and Monte Carlo Dropout (MCD) approximation. MCD is considered as a straightforward approximation of Bayesian neural network, where the dropout is enabled in both training and testing stage to approximate the posterior distribution. The hyperparameter setting is given in Table 1.

Table 1. The hyperparameter setting of BayesCNN, CNN, and MCD training.

epochs	300	Start Learning Rate	0.001
Batch Size	25	Dropout rate for CNN	0.5
BNN priors	$N(0, 0.1)$	Dropout rate for MCD	0.1
BNN posterior initial value	$\mu \sim N(0, 0.1), \rho \sim N(-5, 0.1), \sigma = log(1 + e^{\rho})$		

The Expected Calibration Error (ECE) [5] metric is applied to assess the reliability of model predictions, defined as the expected difference between confidence and accuracy. To estimate the expected accuracy of finite samples, we group predictions into M interval bins (each of size 1/M) and calculate the accuracy and confidence of each bin.

$$
E_{\hat{P}}[|P(\hat{Y} = Y | \hat{P} = p) - p|] \approx \sum_{m=1}^{M} \frac{|B_m|}{n} |acc(B_m) - conf(B_m)|
$$

$$
= \sum_{m=1}^{M} \frac{|B_m|}{n} | \frac{1}{|B_m|} \sum_{i \in B_m} 1(\hat{y}_i = y_i) - \frac{1}{|B_m|} \sum_{i \in B_m} \hat{p}_i |
$$

(11)

where $\hat{y}_i$ and y_i are the predicted and true class labels for sample i. $\hat{p}_i$ is the confidence for sample i. The closer ECE is to zero, the better the model prediction reliability. In the experiment, we choose $M = 10$.

We crop the 128×128 image from the center to 88×88 size. The architecture of VAE are shown in the Table 2, where the latent dimension is 40, m is the batch size and we choose m = 25. In the optimization process of CLUE, the specific

Table 2. The hyperparameter setting of VAE

Encoder		Decoder	
Architecture	Output shape	Architecture	Output shape
Conv2d		Linear	
Resblock*2	(m,32,88,88)	LeakyReLU	(m,latent dimension)
Conv2d		BatchNorm	
Resblock*2	(m,64,44,44)	Linear	
Conv2d		LeakyReLU	(m,128*11*11)
Resblock*2	(m,128,22,22)	BatchNorm	
Conv2d		unFlatten	(m,128,11,11)
LeakyReLU	(m,128,11,11)	ConvTranspose2d	
Flatten		Resblock*2	(m,128,22,22)
BatchNorm	(m,128*11*11)	ConvTranspose2d	
Linear		ResBlock*2	(m,64,44,44)
LeakReLU	(m,2*latent dimension)	ConvTranspose2d	
BatchNorm		Resblock	(m,32,88,88)
Linear	(m,2*latent dimension)	Conv2d	(m,1,88,88)

Table 3. The recognition evaluation metrics of CNN, BayesCNN, and MCD.

	trainset 0–30		trainset 0–90		trainset 0–180	
	OA(%)	ECE	OA(%)	ECE	OA(%)	ECE
CNN	26.05	0.65	40.58	0.52	64.43	0.31
MCD	28.55	0.59	41.83	0.51	65.25	0.30
BayesCNN	**31.63**	**0.56**	**44.13**	**0.49**	**66.41**	**0.28**

parameters of BNN and VAE remain unchanged, and only the parameter z of the latent space is optimized, the initial value of z is chosen to be $z_0 = \mu_\phi(z|x_0)$. Optimization runs for 150 iterations, with a learning rate of 0.5. $d(x, x_0) = \lambda_x ||x - x_0||_1$ and $\lambda_x = 0.005$.

3.2 Performance of BayesCNN

Table 3 and Fig. 3 present the overall performance and the model performance of CNN, BayesCNN, and MCD under three different training settings are given. Figure 3 demonstrates the overall accuracy and ECE of frequentist A-ConvNets and Bayesian A-ConvNets under different training conditions.

When the training data is limited to a certain azimuth range, the model performs well on test data at the same observation angles and nearby (the test accuracy is higher, and ECE is closer to 0). If the test SAR targets are viewed at a different angle, e.g. 30° from the known azimuths, the model's performance will decline dramatically. It is due to the large variation of SAR targets with

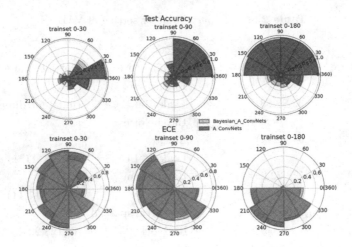

Fig. 3. The overall accuracy and the expected calibration error of frequentist A-ConvNets and Bayesain A-ConvNets under different training conditions.

changing azimuth angles. CNN performs as well as BayesCNN on the test data of which the azimuths are in the same range of training data. However, when it comes to un-observed angles, BayesCNN achieves higher accuracy and smaller ECE than CNN, which indicates BayesCNN has a better generalization ability and can well calibrate the confidence.

Table 3 shows that the overall performance of MCD has been improved compared with that of CNN, and MCD can also assess the prediction uncertainty. Among them, BayesCNN achieves the best performance, which indicates that the MCD approximation is less capable to estimate the posterior distribution of weights than optimizing a Bayesian neural network directly. Due to the shallow architecture of BayesCNN, the Bayes by Backprop can achieve a good optimization result and the complexity can be comparable with MCD.

3.3   Inauthentic Assessment Based on BayesCNN's Prediction

In order to evaluate the proposed inauthentic assessment method, we simulate some SAR images by image rotation, as shown in Fig. 1. As we know, the image rotation of SAR target cannot simulate the real imaging angle rotation like optical data. Consequently, it may result in a high degree of inauthenticity.

We take the data with an azimuth angle of 0°–30° and rotate it 90° counter-clockwise to simulate the data with azimuth angle of 270°–300°, and mix them with the real ones. The corresponding BayesCNN is trained using real data with azimuth angle of 270°–300°. The uncertainty of simulated and real SAR images, as well as the generated counterfactual explanation x_c, predicted by BayesCNN is recorded in Table 4. It can be found that the uncertainty of simulated data is significantly higher than that of real data, and the uncertainty of counterfactuals is significantly decreased. Figure 4 also shows that these simulated data have higher uncertainty. In other words, BayesCNN can well evaluate the authentic-

Table 4. Inauthentic Assessment Based on BayesCNN's Prediction (Simulated data: rotate 90°)

Category	Uncertainty of Simulated data	Uncertainty ofReal data	Uncertainty of Counterfactuals	ΔH
2S1	0.4382	0.1468	0.0307	−0.4075
BMP-2	0.4215	0.2413	0.0625	−0.3590
BRDM-2	0.3847	0.2316	0.0621	−0.3226
BTR-60	0.7237	0.1954	0.0491	−0.6746
BTR-70	0.4805	0.2599	0.0333	−0.4472
D7	0.3765	0.0226	0.0181	−0.3584
T-62	0.2895	0.1907	0.0622	−0.2273
T-72	0.5625	0.1761	0.0284	−0.5341
ZIL-131	0.2574	0.1479	0.0802	−0.1772
ZSU-234	0.3940	0.1883	0.0217	−0.3723
Average	**0.43285**	**0.18006**	**0.04483**	**−0.38802**

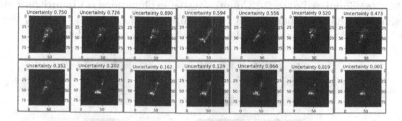

Fig. 4. The uncertainty of simulated data (red frame) and real data (blue frame) in class 2S1 is ranked from greatest to smallest (Color figure online)

ity of images by predicting the uncertainty of images. Different from traditional structural and pixel statistical similarity, BayesCNN can also evaluate the result of simulated data from the effect of image recognition. What's more, counterfactual can produce more realistic images.

3.4 Counterfactual Explanation

To further analyze the validity of the counterfactual results, we calculate the averaged images of simulated data, real data, and counterfactual explanations, respectively, within 30°, to demonstrate the statistic information of the data. As shown in Fig. 5 and Fig. 6, column (a), (b), and (d) demonstrate the simulated data, counterfactual explanation, and the real data, while column (c) and (e) present the inauthentic details of the simulated data compared with the obtained counterfactuals and the real data, respectively. It can be found that the obtained counterfactuals are close to the real image, and the inauthentic details of our method can successfully approximate the reality.

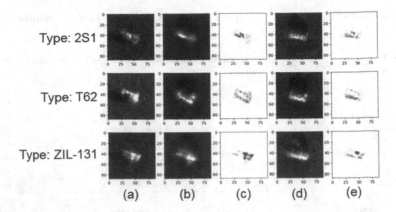

Type: 2S1

Type: T62

Type: ZIL-131

(a)   (b)   (c)   (d)   (e)

Fig. 5. The counterfactual explanation discussion of the simulated data by rotating image 90°. (a) Simulated data (b) Counterfactual explanations (c) Inauthentic details of the simulated data compared with the obtained counterfactuals (d) Real data (e) Inauthentic details of the simulated data compared with the real data.

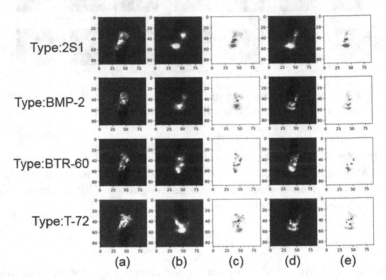

Type:2S1

Type:BMP-2

Type:BTR-60

Type:T-72

(a)   (b)   (c)   (d)   (e)

Fig. 6. The counterfactual explanation discussion of the simulated data by rotating image 180°. (a) Simulated data (b) Counterfactual explanations (c) Inauthentic details of the simulated data compared with the obtained counterfactuals (d) Real data (e) Inauthentic details of the simulated data compared with the real data.

4   Conclusion

In this paper, we propose a novel inauthentic assessment pipeline for simulated SAR images, which is composed of a BayesCNN and a VAE-based counterfactual explanation module. Apart from the current methods, our approach assess the simulated data from the perspective of recognition model. The trained

BayesCNN outputs the estimated uncertainty that can filter the inauthentic data with a high uncertainty. Based on this, we further propose to optimize the latent counterfactual embedding and apply the decoder of a pre-trained VAE model to generate the counterfactual explanation image as well as the inauthentic details of simulated data. The experiments demonstrate the effectiveness of the proposed BayesCNN in attaining the uncertainty and improving the generalization ability for target recognition, so that the inauthentic simulated data can be filtered out. Additionally, the proposed counterfactual explanation method can successfully assess the inauthentic details consistent with the fact.

References

1. Antorán, J., Bhatt, U., Adel, T., Weller, A., Hernández-Lobato, J.M.: Getting a clue: a method for explaining uncertainty estimates. arXiv preprint arXiv:2006.06848 (2020)
2. Blundell, C., Cornebise, J., Kavukcuoglu, K., Wierstra, D.: Weight uncertainty in neural network. In: International Conference on Machine Learning, pp. 1613–1622. PMLR (2015)
3. Datcu, M., Huang, Z., Anghel, A., Zhao, J., Cacoveanu, R.: Explainable, physics-aware, trustworthy artificial intelligence: a paradigm shift for synthetic aperture radar. IEEE Geosci. Remote Sens. Mag. **11**(1), 8–25 (2023). https://doi.org/10.1109/MGRS.2023.3237465
4. Goodfellow, I., et al.: Generative adversarial networks. Commun. ACM **63**(11), 139–144 (2020)
5. Guo, C., Pleiss, G., Sun, Y., Weinberger, K.Q.: On calibration of modern neural networks. In: International Conference on Machine Learning, pp. 1321–1330. PMLR (2017)
6. Huang, Z., Liu, Y., Yao, X., Ren, J., Han, J.: Uncertainty exploration: toward explainable SAR target detection. IEEE Trans. Geosci. Remote Sens. **61**, 1–14 (2023). https://doi.org/10.1109/TGRS.2023.3247898
7. Zhongling, H., Xiwen, Y.: Progress and perspective on physically explainable deep learning for synthetic aperture radar image interpretation. J. Radar **11**(1), 107–125 (2021)
8. Keydel, E.R., Lee, S.W., Moore, J.T.: MSTAR extended operating conditions: a tutorial. Algorithms Synth. Aperture Radar Imagery **III**(2757), 228–242 (1996)
9. Kingma, D.P., Welling, M., et al.: An introduction to variational autoencoders. Found. Trends® Mach. Learn. **12**(4), 307–392 (2019)
10. Kwon, Y., Won, J.H., Kim, B.J., Paik, M.C.: Uncertainty quantification using Bayesian neural networks in classification: application to biomedical image segmentation. Comput. Stat. Data Anal. **142**, 106816 (2020)
11. Liu, B., Ben Ayed, I., Galdran, A., Dolz, J.: The devil is in the margin: margin-based label smoothing for network calibration. In: Proceedings of the IEEE/CVF Conference on Computer Vision and Pattern Recognition (CVPR), pp. 80–88 (2022)

12. Shridhar, K., Laumann, F., Liwicki, M.: A comprehensive guide to Bayesian convolutional neural network with variational inference. arXiv preprint arXiv:1901.02731 (2019)
13. Wilson, A.G., Izmailov, P.: Bayesian deep learning and a probabilistic perspective of generalization. Adv. Neural. Inf. Process. Syst. **33**, 4697–4708 (2020)
14. Xu Feng, W.H., Yaqiu, J.: Deep learning as applied in SAR target recognition and terrain classification. J. Radar **6**(2), 136–148 (2017)

Circle Representation Network for Specific Target Detection in Remote Sensing Images

Xiaoyu Yang[1], Yun Ge[1,2(✉)], Haokang Peng[1], and Lu Leng[1]

[1] School of Software, Nanchang Hangkong University, Nanchang 330063, China
geyun@nchu.edu.cn
[2] Jiangxi Huihang Engineering Consulting Co., Ltd., Nanchang 330038, China

Abstract. Compared with natural images, remote sensing images have complex backgrounds as well as a variety of targets. The circular and square-like targets are very common in remote sensing images. For such specific targets, it is easy to bring background information when using the traditional bounding box. To address this issue, we propose a Circle Representation Network (CRNet) to detect the circular or square-like targets. We design a special network head to regression radius and it has smaller regression degrees of freedom. Then the bounding circle is proposed to represent the specific targets. Compared to the bounding box, the bounding circle has natural rotational invariance. The CRNet can accurately locate the object while carrying less background information. In order to reasonably evaluate the detection performance, we further propose the circle-IOU to calculate the mAP. The experiments evaluated on NWPU VHR-10 and RSOD datasets show that the proposed method has excellent performance when detecting circular and square-like objects, in which the detection accuracy of storage tanks is improved from 92.1% to 94.4%. Therefore, the CRNet is a simple and efficient detection method for the circular and square-like targets.

Keywords: specific target detection · remote sensing image · bounding circles · circle-IOU

1 Introduction

Remote sensing image target detection is a technique to automatically identify and locate the target of interest from remote sensing images. It has wide applications in remote sensing image analysis, resource investigation, urban planning and other fields. In recent years, with the development of deep learning technology and the rapid growth of remote sensing data, the remote sensing image target detection has developed rapidly [1]. Simultaneously, the requirements for specific object detection have gradually increased due to the wide range of target types in remote sensing images [2]. However, it is difficult to represent specific object which is circular or square-like well only using bounding box. Thus a more detailed model is urgently needed to handle this problem.

The method of remote sensing object detection tasks by using deep learning was usually based on regional feature extraction methods, such as R-CNN [3].

This method requires a lot of repeated operations for each region, resulting in a large amount of computation and low efficiency. It has gradually been replaced by the methods based on fully convolutional neural networks, such as faster R-CNN [4]. In 2016, with the introduction of YOLO (You Only Look Once) [5], the object detection model also transitioned from the two-stage model to the one-stage. For the characteristics of remote sensing images, J Deng et al. [6] proposed a classification method with non-maximum suppression threshold to improve the detection accuracy of YOLOV4 for remote sensing images without affecting the speed. Feng et al. [7] used the transformation of the CSP bottleneck to optimize the feature extraction capability, combined YOLO headers with adaptive spatial feature fusion blocks to enhance the feature fusion capability, and used label smoothing to reduce the classification loss. For the specific target detection task in remote sensing images, Z Liu et al. [8] propose a new feature extractor with a stronger feature extraction ability for the network to detect aircraft. S Wang et al. [9] propose a specific ship detection based on Gaussian Wasserstcin Distance achieving great performance.

The above anchor-base target detection methods have high model complexity and low flexibility. It needs to lay the prior boxes on the original image, which is not conducive to expanding the object detection model towards different shape of detection boxes [10]. Therefore, the focus has shifted to anchor-free detection methods (i.e., without prior anchors) with simpler network design, more flexible representation, fewer hyperparameters, and even better performance [11–13]. Among the anchor-free detection methods, the keypoint-based detection algorithm represented by CenterNet [12] detects the keypoints in the image, and then predicts the position and pose of the oriented target. Recently, Pan et al. [14] designed a dynamic refinement network for oriented target detection of densely aligned targets based on CenterNet [12], a target detection method without anchor. Cui et al. [15] proposed a directional ship keypoint detection network for specific target ships by combining orthogonal pooling, soft directional non-maximum suppression, and CenterNet's keypoint estimation detection method.

In both anchor-based and anchor-free models for remote sensing object detection, bounding box limits the detection performance of certain specific shaped objects, such as storage tanks, chimneys, and baseball fields in remote sensing images. It often brings more background information when detecting objects, and the detection box of the same object will also change with the angle [16]. Although using annotated boxes with rotation angles to train object detection models can address the above problems, it requires a lot of material resources to annotate datasets with rotation angles [17]. The model may encounter a series of problems such as difficult rotation angle regression and large number of parameters during training. Therefore, for objects with specific shapes (e.g., circle, square-like), changing the shape of the detection box in the object detection model can often achieve better results. Compared to bounding box, bounding circle has good performance in detecting objects of specific shapes as shown in Fig. 1. Since there are many circular and square-like objects in remote sensing

Bounding boxes carry more background information and inconsistent manual bounding boxes post-rotation.

Bounding circles have less background information and consistent manual bounding circles post-rotation.

Fig. 1. Comparison of bounding boxes and bounding circles. For targets with specific shapes, bounding circles have less background information and effectively solve the rotation variability of traditional horizontal bounding boxes.

images, such as storage tanks, chimneys, which are common sources of pollution. This paper proposes an anchor-free object detection method based on circle representation for detecting this object. The circular shape is introduced as the detection box, and after the model detects the corresponding object, it does not use the rectangular detection box format but instead surrounds the object with a circle. Unlike traditional bounding boxes, the circular-based object detection algorithm only needs to regress a radius parameter to achieve object positioning after detecting the center point of the object. In summary, the contributions of this paper are mainly reflected in three aspects:

- We propose a circle representation method named CRNet for specific target in remote sensing images, which achieves superior detection performance while having fewer degrees of freedom, and the training process is compatible with the normal labels.
- For circular and square-like targets in remote sensing images, CRNet achieves wonderful rotational consistency while having less background information.
- Due to the change in shape, the traditional IOU calculation is not suitable for bounding circle. In order to scientifically evaluate the performance of CRNet, we propose Circle-IOU based on circle detectors.

2 Methods

2.1 Anchor-Free Structure

The design of circle representation network follows the CenterNet [12] which achieves high performance while maintaining a simple and clear structure. CRNet follows the principle of a simple and efficient model structure. The overall architecture of the model is shown in Fig. 2. The feature extraction network can

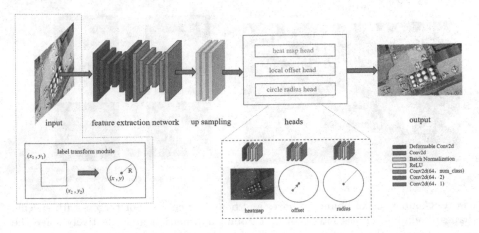

Fig. 2. The structure of CRNet. It consists of three main components: the feature extraction network, the upsampling module, and three network heads.

utilize classic networks such as ResNet50 [18] or 104-Hourglass [19]. The upsampling module employs bilinear interpolation. Finally, three different detection heads are used to obtain heatmap, offset and radius.

The annotation information for CRNet training is different from traditional label. To avoid additional labeling work, CRNet uses a label transformation module to automatically convert traditional bounding box labels $(x_1, y_1, x_2, y_2, class\_name)$ into bounding circle labels $(x_{center}, y_{center}, radius, class\_name)$, which are required for training. Subsequently, for an input image $I \in \Omega^{(W*H*3)}$, the output takes the form of a heatmap denoted as $\tilde{H} \in [0, 1]^{\frac{W}{R}*\frac{H}{R}*C}$, where each element represents the confidence of the center point position and its corresponding class membership. C represents the number of candidate classes, and R denotes the downsampling rate. Similarly, the ground truth is modeled in the same form using a Gaussian kernel function and is represented as $H_{xyc} \in [0, 1]^{\frac{W}{R}*\frac{H}{R}*C}$.

$$H_{xyc} = exp(\frac{(x - \tilde{p}_x)^2 + (y - \tilde{p}_y)^2}{2\sigma_p^2}) \tag{1}$$

$\tilde{p}_x$ and $\tilde{p}_y$ represent the downsampled center point coordinates of the target, and σ_p is the standard deviation obtained adaptively based on the size of the current target p. The classification loss L_p in this paper is formulated as a focal loss [20] to alleviate the problem of class imbalance:

$$L_p = -\frac{1}{N} \sum_{xyc} \begin{cases} (1 - \tilde{H}_{xyc})^\alpha \log(\tilde{H}_{xyc}), & \text{if } H_{xyc}=1 \\ (1 - H_{xyc})^\beta (\tilde{H}_{xyc})^\alpha \log(1 - \tilde{H}_{xyc}) & \text{otherwise} \end{cases} \tag{2}$$

where α and β are adjustable hyperparameters, which are set to 2 and 4 in this paper. N represents the number of objects in the image. Due to the downsampling modules, there is scale differences between the final size of feature map and the original image, which can cause a certain deviation in the predicted center point positions. To solve this problem, the model incorporates a trainable branch for the offset of the center points. The loss function is as follow:

$$L_{off} = \frac{1}{N} \sum_p \left| \tilde{O}_{\tilde{P}} - (\frac{P}{R} - \tilde{P}) \right| \tag{3}$$

where $\tilde{O}_{\tilde{P}}$ represents the learned center point offset, $\tilde{O} \in \Omega^{\frac{W}{R}*\frac{H}{R}*2}$, P denotes the ground truth center point, and $\tilde{P}$ represents the predicted value of the current center point.

2.2   Radius and Adaptive Network Heads

By using the results of the heatmap and offset, the center points of the targets can be accurately determined. Once the center points are determined, the proposed radius network head in this paper regresses the radius of the targets based on these center points.

$$L_{radius} = \frac{1}{N} \sum_{k=1} \left| \tilde{R}_p - r \right| \tag{4}$$

where $\tilde{R}_p$ represents the ground truth radius of the target, and r is the predicted radius by the network. Finally, the overall objective is

$$L_{det} = L_p + \lambda_{radius} L_{radius} + \lambda_{off} L_{off} \tag{5}$$

In object detection, the shape information of objects plays a crucial role in determining the classification scores obtained through the network head [21]. To better capture the shape information of objects, this paper proposes an adaptive perception layer within the network head. This layer allows the receptive field to continuously change and adapt, rather than being limited to a specific region. The offset of the special convolution kernel in the adaptive perception layer allows the classification detection head to achieve an adaptive target shape. This form of convolution allows the sampling grid to be freely deformed by adding 2D shifts to the mesh sampling locations of the standard convolution rules [22], and the offsets can be learned from the previous layer features by an additional convolution layer.

2.3   Circle-IOU

The mean Average Precision (mAP) is one of the commonly used evaluation metrics in object detection. The computation of mAP is based on the Intersection over Union (IOU) between predicted boxes and ground truth (GT) boxes. For traditional object detection models, the calculation of IOU is typically performed between two rectangles but is not suitable for the circular object detection boxes proposed in this paper. To address this issue, this paper proposes the Circle-IOU to compute the IOU between two circular boxes. The representation of Circle-IOU is as follows:

$$cIOU = \frac{Area(A \cap B)}{Area(A \cup B)} \tag{6}$$

where A and B represent two circles, and the coordinates of the center points of A and B are defined as (A_x, A_y) and (B_x, B_y) respectively. The distance between the center points of the two circles is defined as:

$$d = \sqrt{(B_x - A_x)^2 + (B_y - A_y)^2} \tag{7}$$

The union area of the two circles can be calculated using formula (8):

$$Area(A \cap B) = r_A^2 \sin^{-1}(\frac{L_y}{r_A}) + r_B^2 \sin^{-1}(\frac{L_y}{r_B}) - L_y(L_x + \sqrt{r_A^2 - r_B^2 + l_x^2}) \tag{8}$$

The parameters are shown in Fig. 3 where L_x and L_y are defined as:

$$L_x = \frac{r_A^2 - r_B^2 + d^2}{2d}, \quad L_y = \sqrt{r_A^2 - l_x^2} \tag{9}$$

The area where two circles intersect is shown in formula (10):

$$Area(A \cup B) = \pi r_A^2 + \pi r_B^2 - Area(A \cap B) \tag{10}$$

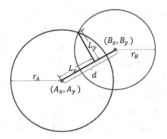

Fig. 3. Two circles with intersection area

3   Experiment and Analysis

3.1   Data

This paper conducted experiments on multiple remote sensing datasets. The NWPU VHR-10 [23] dataset is a collection of high-resolution remote sensing images developed and released by Northwestern Polytechnical University (NWPU). The dataset consists of 800 RGB remote sensing images, with 650 images containing objects and 150 background images. The images have a resolution of 1 m/pixel, and each image is 650 × 650 pixels in size. The dataset includes 10 object categories, named airplanes, ships, storage tanks, baseball fields, tennis courts, basketball courts, athletic fields, harbors, bridges, and vehicles. The RSOD dataset [24], developed and released by researchers from Wuhan University, is designed for object detection in remote sensing images. The dataset comprises four common categories of remote sensing images.

3.2   Experimental Environment and Evaluation Indicators

The experiments in this paper were conducted on an Ubuntu 18.04 environment, using an Intel Xeon(R) Gold 5218R@2.10 GHz processor and an NVIDIA GeForce RTX 3090 graphics card with 24 GB of memory. The deep learning framework used was PyTorch 3.8, accelerated with CUDA 11.2. We randomly divided the NWPU VHR-10 and RSOD into training, validation, and test set in the ratio of 7:2:1

The evaluation metrics used in this paper is Average Precision (AP) for individual classes, Mean Average Precision (mAP), and Frame Per Second (FPS). AP represents the average precision for a single class at an IoU threshold of 0.5, mAP50 represents the average precision for all classes at an IoU threshold of 0.5, and FPS represents the model's inference speed.

3.3   Results

To validate the effectiveness of the proposed method, four representative methods were selected for comparison: Faster R-CNN, RetinaNet, CenterNet, and CornerNet. The performance comparison between the proposed method and the other four methods on the NWPU VHR-10 dataset and RSOD dataset are shown in Table 1.

From Table 1, it can be observed that the proposed method achieves an mAP_{50} of 89.7% on NWPU VHR-10 dataset and 92.8% on RSOD dataset, demonstrating varying degrees of improvement in accuracy compared to the other methods. Moreover, the proposed method outperforms the other methods in terms of inference speed.

Table 1. Accuracy comparison (%) of different methods on the NWPU VHR-10 and RSOD dataset

methods	NWPU		RSOD	
	mAP_{50}	FPS	mAP_{50}	FPS
Faster R-CNN	87.6	47	92.0	54
CornerNet	88.9	91	90.7	92
RetinaNet	89.2	80	91.8	87
CenterNet	89.6	110	92.6	104
CRNet	89.7	119	92.8	121

To further demonstrate the excellent performance of our proposed method in detecting specific objects, experiments were conducted on the NWPU VHR-10 dataset to calculate the Average Precision (AP) for individual classes. The selected classes include Airplane, Storage tank, Baseball diamond, Tennis court, Basketball court, and Vehicle. The results are presented in Table 2. It can be observed that our circular-based object detection method shows significant improvement in accuracy when detecting specific classes of objects. Particularly, for circular objects such as storage tanks and Baseball diamonds, the detection performance is significantly enhanced. Moreover, for square-like objects, the detection accuracy shows a slight improvement while maintaining rotation invariance in the predicted boxes.

Table 2. Comparison of Average Precision (AP) (%) for specific classes using different methods on the NWPU VHR-10 dataset.

	methods				
	Faster R-CNN	RetinaNet	CornerNet	CenterNet	CRNet
Airplane	86.5	89.5	81.6	98.5	98.6
Storage tank	91.0	85.7	70.3	92.2	94.4
Baseball diamond	94.9	99.0	95.2	96.8	99.8
Tennis court	83.2	78.1	78.4	80.1	81.5
Basketball court	86.3	92.8	99.5	87.2	86.7
Vehicle	57.9	66.1	75.2	86.3	90.5
mAP_{50}	83.3	85.2	83.4	90.2	91.9

Figure 4 shows the visual results of our proposed method on the NWPU VHR-10 dataset for classes such as airplanes, vehicles, baseball diamonds, and storage tanks. From Fig. 4, it can be observed that our method accurately detects circular objects while capturing fewer background details. Additionally, for square-shaped objects, the bounding circle effectively mitigate the issue of horizontal bounding box variations caused by changes in object angles.

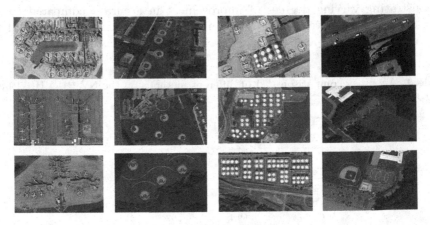

Fig. 4. Example of detection results of this method on NWPU VHR-10 dataset

4   Conclusion

The circular and square-like objects are very common in remote sensing images, such as storage tanks and chimneys, which are common sources of pollution. Therefore this paper proposes the CRNet to detect the circular and square-like targets. It achieves good performance in overall object detection as well as specific object categories. Firstly, we proposed the bounding circle efficiently locate objects by regressing their center points and radius, while also possessing rotational invariance that horizontal bounding boxes lack. Secondly, this paper proposes Circle-IOU to calculate the intersection over union between circular predicted boxes and ground truth, measuring the performance of the model. Finally, the proposed method is applicable to horizontal rectangular annotation datasets without the need for additional annotation information for circular boxes prior to training. Experimental results demonstrate that the proposed method achieves an mAP_{50} of 89.7% and 92.8% on the remote sensing datasets NWPU VHR-10 and on the RSOD datasets. Specifically, for certain categories such as storage tanks and baseball diamonds, it achieves an AP of 94.5% and 99.8%, respectively, showing significant improvements compared to other methods. In conclusion, the CRNet achieves good performance in overall object detection as well as specific object categories.

Acknowledgements. This work was supported by the National Natural Science Foundation of China under Grant NO. 42261070, and Grant NO. 41801288.

References

1. Qin, D.D., Wan, L., He, P.E., Zhang, Y., Guo, Y., Chen, J.: Multi-scale object detection in remote sensing image by combining data fusion and feature selection. Nat. Remote Sens. Bull. **26**(8), 1662–1673 (2022)
2. Kathiravan, M., Reddy, N.A., Prakash, V., et al.: Ship detection from satellite images using deep learning. In: 2022 7th International Conference on Communication and Electronics Systems (ICCES), pp. 1044–1050. IEEE (2022)
3. Girshick, R., Donahue, J., Darrell, T., et al.: Rich feature hierarchies for accurate object detection and semantic segmentation. In: Proceedings of the IEEE Conference on Computer Vision and Pattern Recognition, pp. 580–587 (2014)
4. Ren, S., He, K., Girshick, R., et al.: Faster R-CNN: towards real-time object detection with region proposal networks. In: Advances in Neural Information Processing Systems, vol. 28 (2015)
5. Redmon, J., Divvala, S., Girshick, R., et al.: You only look once: Unified, real-time object detection. In: Proceedings of the IEEE Conference on Computer Vision and Pattern Recognition, pp. 779–788 (2016)
6. Zakria, Z., Deng, J., Kumar, R., et al.: Multiscale and direction target detecting in remote sensing images via modified YOLO-v4. IEEE J. Select. Top. Appl. Earth Observations Remote Sens. **15**, 1039–1048 (2022)
7. Feng, J., Yi, C.: Lightweight detection network for arbitrary-oriented vehicles in UAV imagery via global attentive relation and multi-path fusion. Drones **6**(5), 108 (2022)

8. Liu, Z., Gao, Y., Du, Q., et al.: YOLO-extract: improved YOLOv5 for aircraft object detection in remote sensing images. IEEE Access **11**, 1742–1751 (2023)
9. Wang, S.: Gaussian Wasserstein distance based ship target detection algorithm. In: 2023 IEEE 2nd International Conference on Electrical Engineering, Big Data and Algorithms (EEBDA), pp. 286–291. IEEE (2023)
10. Kawazoe, Y., Shimamoto, K., Yamaguchi, R., et al.: Faster R-CNN-based glomerular detection in multistained human whole slide images. J. Imaging **4**(7), 91 (2018)
11. Law, H., Deng, J.: CornerNet: detecting objects as paired keypoints. In: Proceedings of the European Conference on Computer Vision (ECCV), pp. 734–750 (2018)
12. Zhou, X., Wang, D., Krähenbühl, P.: Objects as points. arXiv preprint arXiv:1904.07850 (2019)
13. Zhou, X., Zhuo, J., Krahenbuhl, P.: Bottom-up object detection by grouping extreme and center points. In: Proceedings of the IEEE Conference on Computer Vision and Pattern Recognition, pp. 850–859 (2019)
14. Cui, Z., Leng, J., Liu, Y., et al.: SKNet: detecting rotated ships as keypoints in optical remote sensing images. IEEE Trans. Geosci. Remote Sens. **59**(10), 8826–8840 (2021)
15. Cui, Z., Liu, Y., Zhao, W., et al.: Learning to transfer attention in multi-level features for rotated ship detection. Neural Comput. Appl. **34**(22), 19831–19844 (2022)
16. Yang, H., Deng, R., Lu, Y., et al.: CircleNet: anchor-free detection with circle representation. arXiv preprint arXiv:2006.02474 (2020)
17. Bai, M., Urtasun, R.: Deep watershed transform for instance segmentation. In: Proceedings of the IEEE Conference on Computer Vision and Pattern Recognition, pp. 5221–5229 (2017)
18. He, K., Zhang, X., Ren, S., et al.: Deep residual learning for image recognition. In: Proceedings of the IEEE Conference on Computer Vision and Pattern Recognition, pp. 770–778 (2016)
19. Newell, A., Yang, K., Deng, J.: Stacked hourglass networks for human pose estimation. In: Leibe, B., Matas, J., Sebe, N., Welling, M. (eds.) ECCV 2016. LNCS, vol. 9912, pp. 483–499. Springer, Cham (2016). https://doi.org/10.1007/978-3-319-46484-8_29
20. Lin, T. Y., Goyal, P., Girshick, R., et al.: Focal loss for dense object detection. In: Proceedings of the IEEE International Conference on Computer Vision, pp. 2980–2988 (2017)
21. Ding, J., Xue, N., Long, Y., et al.: Learning RoI transformer for oriented object detection in aerial images. In: Proceedings of the IEEE/CVF Conference on Computer Vision and Pattern Recognition, pp. 2849–2858 (2019)
22. Dai, J., Qi, H., Xiong, Y., et al.: Deformable convolutional networks. In: Proceedings of the IEEE International Conference on Computer Vision, pp. 764–773 (2017)
23. Cheng, G., Zhou, P., Han, J.: Learning rotation-invariant convolutional neural networks for object detection in VHR optical remote sensing images. IEEE Trans. Geo-sci. Remote Sens. **54**(12), 7405–7415 (2016)
24. Long, Y., Gong, Y.P., Xiao, Z.F., Liu, Q.: Accurate object localization in remote sensing images based on convolutional neural networks. IEEE Trans. Geosci. Remote Sens. **55**(5), 2486–2498 (2017)

A Transformer-Based Adaptive Semantic Aggregation Method for UAV Visual Geo-Localization

Shishen Li, Cuiwei Liu$^{(\boxtimes)}$, Huaijun Qiu, and Zhaokui Li

School of Computer Science, Shenyang Aerospace University, Shenyang, China
liucuiwei@sau.edu.cn

Abstract. This paper addresses the task of Unmanned Aerial Vehicles (UAV) visual geo-localization, which aims to match images of the same geographic target taken by different platforms, i.e., UAVs and satellites. In general, the key to achieving accurate UAV-satellite image matching lies in extracting visual features that are robust against viewpoint changes, scale variations, and rotations. Current works have shown that part matching is crucial for UAV visual geo-localization since part-level representations can capture image details and help to understand the semantic information of scenes. However, the importance of preserving semantic characteristics in part-level representations is not well discussed. In this paper, we introduce a transformer-based adaptive semantic aggregation method that regards parts as the most representative semantics in an image. Correlations of image patches to different parts are learned in terms of the transformer's feature map. Then our method decomposes part-level features into an adaptive sum of all patch features. By doing this, the learned parts are encouraged to focus on patches with typical semantics. Extensive experiments on the University-1652 dataset have shown the superiority of our method over the current works.

Keywords: UAV visual geo-localization · transformer · part matching

1 Introduction

Unmanned Aerial Vehicle (UAV) visual geo-localization refers to cross-view image retrieval between UAV-view images and geo-tagged satellite-view images. Recently, this technology has been applied in many fields, such as precision agriculture [1], rescue system [2], and environmental monitoring [3]. Using a UAV-view image as the query, the retrieval system searches for the most relevant

This work was supported in part by the National Natural Science Foundation of China (NSFC) under Grant No.62171295, and in part by the Liaoning Provincial Natural Science Foundation of China under Grant No.2021-MS-266, and in part by the Applied Basic Research Project of Liaoning Province under Grant 2023JH2/101300204, and in part by the Shenyang Science and Technology Innovation Program for Young and Middle-aged Scientists under Grant No. RC210427, and in part by the High Level Talent Research Start-up Fund of Shenyang Aerospace University under Grant No.23YB03.

© The Author(s), under exclusive license to Springer Nature Singapore Pte Ltd. 2024
Q. Liu et al. (Eds.): PRCV 2023, LNCS 14428, pp. 465–477, 2024.
https://doi.org/10.1007/978-981-99-8462-6_38

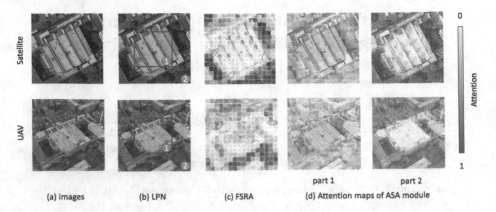

Fig. 1. An UAV-satellite image pair is shown in column(a). The square-ring partition strategy of LPN [5] is depicted in column(b). Column(c) illustrates the heat map and the part partition of image patches generated by FSRA [6]. Red ellipses mark patches that are similar in features but divided into different parts. Attention maps corresponding to two parts produced by the proposed ASA module are given in column(d).

satellite-view candidate to determine the geographic location of the UAV-view target. On the other hand, if a satellite-view image is used as the query, the corresponding UAV-view images can be retrieved from the UAV flight records, enabling UAV navigation. Compared to traditional geo-localization methods that rely on GPS or radar, UAV visual geo-localization does not require the UAV to receive external radio information or emit detection signals, making it possible to achieve UAV positioning and navigation in radio silence.

The key to UAV visual geo-localization is to extract discriminative features. Specifically, satellites acquire fixed-scale images from a vertical view, while UAVs take images from various distances and orientations at an oblique view, resulting in large visual and scale variations in UAV-satellite image pairs. Moreover, images of different locations share some local patterns, such as vegetation and similarly styled buildings, which also pose challenges to cross-view image retrieval. These issues make hand-crafted descriptors (e.g., SIFT [4]) perform poorly.

Early UAV visual geo-localization methods employ two-branch CNN models [7,8] to achieve cross-view matching between UAV-view and satellite-view images. Benefiting from the availability of multiple UAV-view images of the same location, the models are optimized in a location classification framework to learn view-invariant yet location-dependent features. By doing this, two-branch CNN models are desired to generate similar features for satellite-view and UAV-view images taken at a new location that is unseen during training. However, the learned features focus on the entire image, while neglecting fine-grained details that are crucial for distinguishing images of different locations. Upon two-branch CNN models, Local Pattern Network (LPN) [5] divides the feature map into several regions with square-rings to extract part-level representations as shown in Fig. 1 (b). Then part matching is performed to roughly align the geographic targets as well as their surroundings. Another iconic work is a two-branch transformer-based model called FSRA (Feature Segmentation and Region Alignment) [6], which clusters image patches into semantic parts according to the heat distribution of feature maps as shown in Fig. 1(c). Each part is desired to indicate certain semantics such as target or surroundings. Compared to LPN that adopts fixed-scale spatial partition, FSRA is more flexible in extracting parts and thus more robust against image shift and scale variations. However, FSRA employs a hard partition strategy where an image patch belongs to only one part and the part-level representations are calculated as mean of image patches. We argue that there are two issues pertaining to such strategy. First, image patches similar in features may be divided into different parts as shown in Fig. 1(c). Secondly, such strategy cannot extract the most representative semantics since it neglects the associations between image patches and parts.

To cope with these limitations, this paper proposes an Adaptive Semantic Aggregation (ASA) module. Unlike the hard partition strategy utilized in FSRA [6], the ASA module employs a soft partition strategy that considers correlations between parts and all image patches to generate global-aware part-level representations. Specifically, each part is regarded as one semantic and has a distribution of attention over all image patches as shown in Fig. 1(d). First, the most representative patch is selected as anchor of a part. Attentions of image patches are allocated by calculating similarities between patches and the anchor. A high attention expresses strong correlation of an image patch to the semantic part, while a low attention indicates weak correlation. Then all patch-level features are adaptively aggregated into global-aware part-level representations according to the learned attentions. Finally, the ASA module is integrated into the two-branch transformer-based framework [6] and explicitly enables the learned parts focus on distinctive patches, such as gray roof and circular road.

The remainder of this paper is organized as follows. Section 2 briefly describes the current work related to cross-view geo-localization. Section 3 introduces the overall framework of our method and describes the proposed ASA module in detail. Section 4 presents and analyzes the experimental results on the University-1652 dataset. Section 5 summarizes this paper.

2   Related Work

In 2020, Zheng et al. [7] formulated the UAV visual geo-localization problem as bidirectional cross-view image retrieval between UAV-view and satellite-view images. They employed a two-branch CNN model to extract global features from different domains and released the University-1652 dataset for model evaluation. Upon this work, Ding et al. [8] simplified the cross-view image retrieval task as a location classification problem during training, aiming to learn a common location-dependent feature space that is well scalable for unseen images. They noticed the imbalance of UAV-view and satellite-view images, and performed

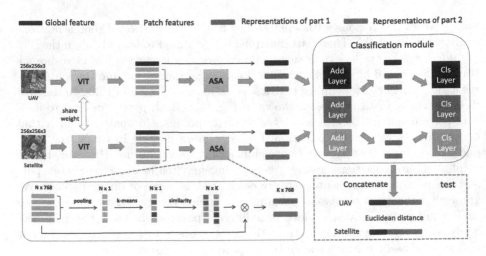

Fig. 2. Overall framework of our method.

data augmentation to expand training satellite-view images. Based on the above location classification framework, recent works [5,6,9–12] have made various attempts to improve the discriminative power of the learned feature space.

One typical solution is LPN [5] which achieves fine-grained part matching between UAV-view and satellite-view images. LPN applies a square-ring partition strategy to separate global feature maps into multiple parts according to their spatial position. Then rotation-invariant part-level features are obtained by performing average polling over points within each part. Tian et al. [9] produces synthesized UAV-view images by a generative model to reduce the gap between two views. Then they employed LPN [5] to achieve cross-view image retrieval. Zhuang et al. [10] improved LPN [5] by incorporating global features and adding KL loss to further close the distance between paired UAV-view and satellite-view images. Lin et al. [11] introduced a Unit Subtraction Attention Module (USAM) that forces the geo-localization model (e.g., LPN [5]) to focus on salient regions by detecting representative key points.

Considering that the square-ring partition strategy is not robust to scale variations, Dai et al. [6] aimed to extract semantic parts composed of patches scattered throughout the image. Small patches are ranked according to the feature heat map and uniformly divided into multiple parts, regardless of their spatial position. Then a part is represented by the average feature of patches within it. Zhuang et al. [12] improved this part partition strategy by searching for the optimal split based on the gradient between adjacent positions in the ranking results. However, the above methods [6,12] aggregate image patches equally, thus weakening the semantic characteristics of the learned parts. In this paper, we propose an Adaptive Semantic Aggregation module that regards parts as the most representative semantics in images and obtains part-level features by aggregating all patch features based on their correlations to parts.

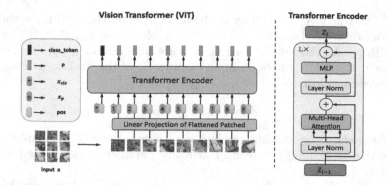

Fig. 3. Architecture of Vision Transformer (ViT).

3   Method

Figure 2 depicts the overall framework of our method. Inspired by siamese networks [13,14], two branches are designed to handle images captured by UAVs and satellites respectively. Following FSRA [6], we adopt ViT (Vision Transformer) [15] as the backbone to extract features from both UAV-view images and satellite-view images. Backbones of the UAV branch and the satellite branch share weights to learn a mapping function that is able to project images from both views to one common embedding space. ViT produces global features of the entire image as well as local features of image patches that are fed into the proposed ASA module to generate multiple part-level representations. Finally, global features and part-level representations are sent to the classification module, which regards geographic locations of training images as semantic categories. The classification module contains additive layers for representation transformation and classification layers for geo-location prediction.

In the test stage, our goal is to achieve cross-view retrieval between UAV-view and satellite-view images captured at new geographic locations. That is to say, the classifier cannot infer locations of query or gallery images during test, since the test data have their own location label space disjoint with the training data. Therefore, we concatenate the transformed representations before classification layers in the classifier module as the final descriptor of a test image. In order to measure the correlation between a query image from one view and gallery images from another view, we calculate the euclidean distance between them. Finally, we sort gallery images in terms of their distance to the query image and return the most similar one to achieve cross-view image retrieval.

3.1   Transformer-Based Backbone

Architecture of our backbone is illustrated in Fig. 3. Given an input image $x \in R^{H \times W \times C}$, it is first divided into fixed-size patches using pre-defined parameters. Then patches are linearly transformed to obtain embedding vectors

$x_p \in R^{N \times D}$, where N and D refer to the number of patches and the embedding dimension, respectively. Additionally, a learnable vector $x_{cls} \in R^{1 \times D}$ forwards along with x_p. All these embedding vectors are integrated with position embeddings $pos \in R^{(N+1) \times D}$ to obtain input vectors Z_0 of the Transformer Encoder. This procedure is formulated by

$$Z_0 = [x_{cls}; x_p] + pos. \tag{1}$$

To accommodate the varying resolution of input images, we employ learnable position embeddings instead of utilizing parameters pre-trained on ImageNet [16].

As shown in Fig. 3, the Transformer Encoder consists of alternating Multiple Head Self-Attention layers and Layer Norm (LN) operations, where residual connections are applied. Multi-Layer Perceptron (MLP) is a two-layer non-linearity block, each layer of which ends with a GELU activation function. The process of Transformer Encoder is formulated by

$$Z_l' = MHSA(LN(Z_{l-1})) + Z_{l-1}, \tag{2}$$

$$Z_l = MLP(LN(Z_l')) + Z_l', \tag{3}$$

where Z_l' and Z_l denote output vectors of the lth attention layer and the lth MLP layer, respectively. Finally, the output vector Z_L consists of a global feature $class\_token \in R^{1 \times D}$ generated upon x_{cls} and patch-level features $\{P_i \in R^{1 \times D}\}_{i=1:N}$ generated from x_p.

3.2   Adaptive Semantic Aggregation Module

Previous works have shown the effectiveness of part matching between UAV-view and satellite-view images in the UAV-view geo-localization task. In this paper, we propose a soft partition strategy to adaptively aggregate image patches into part-level representations. Different from hard partition strategies [5,6], the proposed soft partition strategy regards a part as semantic aggregation of image patches according to correlations between them. Details of the ASA module are illustrated in Fig. 2.

An input image x is first fed into ViT to extract global features as well as patch-level features $\{P_i \in R^{1 \times D}\}_{i=1:N}$. Assuming equal importance of feature dimensions, we apply average pooling on features of each patch to obtain 1-dimensional representations, denoted as $\{Q_i \in R^{1 \times 1}\}_{i=1:N}$.

$$Q_i = \frac{1}{D} \sum_{d=1}^{D} P_i^d, \tag{4}$$

where i and d represent the indices of patch and feature dimension, respectively. Patches with similar semantic information exhibit similar representations in $\{Q_i\}_{i=1:N}$, as demonstrated in FSRA [6] (see Fig. 1 (c)). Accordingly, we perform k-means algorithm upon $\{Q_i\}_{i=1:N}$ to extract representative semantics.

Specifically, the 1-dimensional representations are first sorted in descending order to obtain a sequence S, each item in which indicates a patch index. We introduce a hyper-parameter K, which denotes the number of semantic categories in the image. Center of the kth category is initialized as 1-dimensional representations of a patch S_{IC_k}, and the index IC_k is denoted by

$$IC_k = \frac{(2k-1)N}{2K}. \qquad k = 1, 2, ...K \qquad (5)$$

The k-means algorithm updates semantic categories iteratively and outputs the final category centers. Center of the kth category corresponds to a patch P_{C_k}, which is regarded as the anchor of a part. Then we utilize the original patch features produced by ViT to calculate Euclidean distance dis_k^i between anchor of a part and all image patches. This procedure is formulated by

$$dis_k^i = ||P_i - P_{C_k}||_2. \qquad (6)$$

Given the negative correlation between distance and similarity, we employ the cosine function to obtain the attention matrix by

$$A_k^i = \alpha \cdot cos(\frac{dis_k^i - dis_k^{min}}{dis_k^{max} - dis_k^{min}} \cdot \frac{\pi}{2}) + \beta, \qquad (7)$$

where α and β are factors for enhancing the robustness of the weight matrix and are respectively set to 1 and 0 in our specific experiments. Finally, we aggregate patch features $\{P_i\}_{i=1:N}$ into global-aware part-level features $\{\rho_k\}_{k=1:K}$ according to the customized attention matrix A.

$$\rho_k = \frac{\sum_{i=1}^{N} P_i \cdot A_k^i}{\sum_{i=1}^{N} A_k^i}. \qquad (8)$$

It should be noted that 1-dimensional semantic representations $\{Q_i\}_{i=1:N}$ are utilized to find anchors of parts at low clustering cost, while original D-dimensional features $\{P_i\}_{i=1:N}$ are employed for feature aggregation. In fact, clustering results of the k-means algorithm can also be used to achieve a hard partition strategy, which is compared to the proposed soft partition strategy on the University-1652 dataset in Sect. 4.

3.3 Classification Module

The classification module takes global features as well as part-level representations as input, aiming at classifying them into different geographic locations. As shown in Fig. 4, the classification module consists of additive layers and classification layers. The former achieve transformation of the input while the latter produce prediction vectors of geographic locations. Separate layers are constructed for global features and representations of each part, considering that they indicate different semantic characteristics. Take representations of the kth part as an example, operations of the classification module can be formulated as

$$f_k = F_{add}^k(\rho_k), \qquad (9)$$

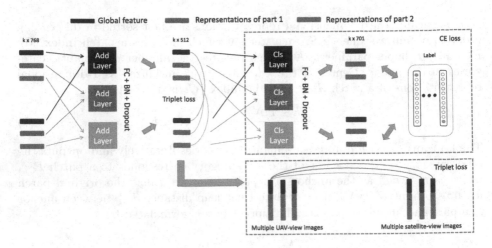

Fig. 4. Architecture of the classification module. In training, global and part-level features are fed into additive layers followed by classification layers. Suppose that the training data come from 701 locations, so a classification layer predicts a 701-dimensional vector. The model is optimized by CE loss and Triplet loss. Green and purple lines point at positive and negative samples for calculating the Triplet loss, respectively. (Color figure online)

$$z_k = F_{cls}^k(f_k), \tag{10}$$

where F_{add}^k and F_{cls}^k represents the additive layer and the classification layer for the kth part, respectively. f_k denotes output features of the additive layer and z_k indicates predicted logits for all geographic locations.

Following [6], the model is optimized by both CE loss L_{CE} and Triplet Loss $L_{Triplet}$. The total objective is defined as

$$L_{total} = L_{CE} + L_{Triplet}, \tag{11}$$

where L_{CE} is the cross entropy between the predicted logits and the ground-truth geo-tags, formulated by

$$L_{CE} = -\frac{1}{K} \sum_{i=1}^{K} \log \frac{exp(z_k(y))}{\sum_{c=1}^{C} exp(z_k(c))}, \tag{12}$$

where C indicates the number of geographic locations in the training data and $z_k(y)$ is the logit score of the ground-truth geo-tag y. The Triplet Loss is performed on the output features of additive layers to pull paired UAV-view and satellite-view images together, while pushing away mismatched image pairs. The Triplet Loss is formulated by

$$L_{TripletLoss} = \frac{1}{K} \sum_{i=1}^{K} max(d(f_k, p) - d(f_k, n) + M; 0), \tag{13}$$

where M is a hyper-parameter and empirically set to 0.3. $d(.)$ represents the distance function. p and n indicate positive sample and negative samples, respectively. As depicted in Fig. 4, both the positive (green line) and negative samples (purple lines) come from another view. For example, if we take part-level feature f_k of a UAV-view image as the anchor, then f_k is compared to the corresponding part features of the true-matched and mismatched satellite-view images.

4 Experiments

4.1 Dataset

University-1652 dataset [7] contains images of 1652 buildings from 72 universities in three different camera views, i.e. UAV view, satellite view, and ground view. This dataset is employed to evaluate the proposed method on the UAV visual geo-localization task, so only UAV-view and satellite-view images are utilized to achieve bidirectional cross-view image retrieval. For each building, one image was obtained by satellites and 54 images were captured by UAVs at different heights and perspectives. Thus, there exist large variations in image scale and viewpoints, which poses great challenges to the UAV visual geo-localization task.

The dataset is split into a training set including 38,555 images of 701 buildings from 33 universities and a test set containing 52,306 images of 951 buildings from the reminder 39 universities. During training, we number 701 buildings into 701 different categories of the classification module and learn an embedding space for UAV-view and satellite-view images. In the UAV-to-satellite image retrieval task, the query set consists of 37,855 UAV-view images of 701 buildings and the gallery includes 701 true-matched satellite-view images and 250 distractors. In the satellite-to-UAV image retrieval task, 701 satellite-view images constitute the query set and there are 51,355 gallery images, including 37,855 true-matched UAV-view images and 13,500 distractors from the rest 250 buildings.

4.2 Evaluation Protocol

Following the previous works [5,7–10], we employ two types of evaluation metrics, namely Recall@K and average precision (AP). Recall@K and AP are widely applied in image retrieval tasks. Recall@K refers to the ratio of query images with at least one true-matched image appearing in the top-K ranking list. AP computes the area under the Precision-Recall curve and considers all true-matched images in the gallery. Recall@K focuses on the position of the first true-matched image in the matching results and thus is suitable for evaluation on the UAV-to-satellite image retrieval task where the gallery only contains one true-matched satellite-view image for each UAV-view query. In the satellite-to-UAV image retrieval task, there are multiple true-matched UAV-view images for one satellite-view query and AP is able to comprehensively measure the matching results. In this paper, we report the mean Recall@K and mean AP of all queries.

Table 1. Cross-view image retrieval performance (%) of different methods.

Method	Year	Backbone	UAV-to-Satellite		Satellite-to-UAV	
			Recall@1	AP	Recall@1	AP
Zheng et al. [7]	2020	ResNet-50	58.49	63.31	71.18	58.74
LCM [8]	2020	ResNet-50	66.65	70.82	79.89	65.38
LPN [5]	2021	ResNet-50	75.93	79.14	86.45	74.79
PCL [9]	2021	ResNet-50	79.47	83.63	87.69	78.51
RK-Net (USAM) [11]	2022	ResNet-50	77.60	80.55	86.59	75.96
SGM [12]	2022	Swin-T	82.14	84.72	88.16	81.81
FSRA [6]	2022	Vit-S	84.51	86.71	88.45	83.37
ASA (Ours)	–	Vit-S	85.12	87.21	89.30	84.17

Table 2. Cross-view image retrieval performance (%) with different strategies.

Strategy	UAV-to-Satellite		Satellite-to-UAV	
	Recall@1	AP	Recall@1	AP
Uniform hard partition strategy	83.98	86.27	88.59	83.91
k-means hard partition strategy	84.97	87.12	88.16	83.85
k-means soft partition strategy	85.12	87.21	89.30	84.17

4.3   Implementation Details

All the input images are resized to 256×256 and the number of parts (K) is set to 2. Due to imbalance in the number of UAV-view and satellite-view images, we perform image augmentation on satellite-view images during training. A small Vision Transformer (ViT-S) pre-trained on the ImageNet [16] is employed as the backbone for feature extraction. The ASA module and the classification module are trained from scratch. Our model is learned by an SGD optimizer with a momentum of 0.9 and weight decay of 0.0005. The mini-batch size is set to 8. The learning rate is initialized to 0.003 and 0.01 for ViT-S and the rest layers, respectively. The model is trained for a total of 120 epochs and the learning rate is decayed by 0.1 after executing 70 epochs and 110 epochs.

4.4   Experimental Results

Comparison with Existing Methods. Table 1 presents our model's performance for bidirectional cross-view retrieval between UAV-view and satellite-view images, compared to the previous methods [5–9,11,12]. LCM [8] resizes input images into 384×384, and the image scale utilized in the other methods is 256×256. Methods in [5,7–9,11] all use ResNet-50 [17] as backbone to extract CNN features. Among them, methods in [7,8] learn global features, while LPN [5], PCL [9], and RK-Net [11] achieve part matching by using the

Table 3. Effect of the number of parts on cross-view image retrieval performance (%).

Number of parts (K)	UAV-to-Satellite		Satellite-to-UAV	
	Recall@1	AP	Recall@1	AP
1	72.11	75.59	79.46	71.83
2	85.12	87.21	89.30	84.17
3	84.73	86.88	88.45	83.55
4	84.48	86.74	87.59	83.22

Table 4. Cross-view image retrieval performance (%) with different image sizes.

Image Size	UAV-to-Satellite		Satellite-to-UAV	
	Recall@1	AP	Recall@1	AP
224×224	82.28	84.83	86.31	81.93
256×256	85.12	87.21	89.30	84.17
384×384	86.88	88.74	89.44	85.95
512×512	88.67	90.29	89.44	87.14

square-ring partition strategy, which is not robust to scale variations. The experimental results demonstrate the inferior performance of these methods [5,7–9,11]. FSRA [6] and SGM [12] adopt transformers as backbone and cluster patches into semantic parts with hard partition strategies. Our method performs better than FSRA [6] and SGM [12] on both tasks, even though SGM [12] employs a stronger backbone. In comparison to FSRA [6], both methods extract features with ViT-S backbone and utilize 3× sampling strategy to expand satellite-view images during training. The superior results achieved by our method demonstrate that the proposed ASA module reasonably aggregate patches into part-level representations, thus improving the performance of part matching.

Ablation Study. In this section, we investigate the effect of several key factors. First, we explore the effectiveness of the proposed soft partition strategy by comparing it with two baselines in Table 2. The baseline "Uniform hard partition strategy" uniformly groups patches into K parts according to the 1-dimensional semantic representation obtained by Eq. 4. The baseline "k-means hard partition strategy" utilizes the k-means clustering results for part partition. Both baselines take the average of patches within each part to generate part-level representations. As shown in Table 2, the proposed "k-means soft partition strategy" is more effective in aggregating part-level features and achieves optimal results for UAV visual geo-localization.

Next, we analyze the impact of the number of parts K on the cross-view image retrieval performance. As shown in Table 3, the performance is poor if K is set to 1, since the learned part cannot express representative semantics of the image. Our model achieves the optimal generalization performance when K is set to 2. We believe that the learned two parts indicate typical semantics of foreground and background.

Finally, we discuss the effect of different image sizes. As shown in Table 4, high-resolution images generally yield better results since they retain more fine-grained details at the cost of more memory resources and inference time. The performance of our method does not degrade significantly when the image size is reduced from 512 to 256. Therefore, we recommend using the image size of 256 if hardware resources are limited.

5   Conclusion

In this paper, we have presented an adaptive semantic aggregation method to learn global-aware part-level features that can express typical semantics of scenes. Unlike the existing hard partition strategies, we have developed a soft partition strategy that searches for the most representative semantics as parts and evaluates the significance of patches to different parts. Our method adaptively aggregates features of all patches into part-level representations. Compared to current works, the proposed method has achieved superior performance on the University-1652 dataset. Ablation studies also verified effectiveness of the proposed soft partition strategy and investigated several key factors of our methods.

References

1. Chivasa, W., Mutanga, O., Biradar, C.: Uav-based multispectral phenotyping for disease resistance to accelerate crop improvement under changing climate conditions. Remote Sensing **12**(15), 2445 (2020)
2. Rizk, M., Slim, F., Charara, J.: Toward ai-assisted uav for human detection in search and rescue missions. In: DASA, pp. 781–786. IEEE (2021)
3. Ecke, S., Dempewolf, J., Frey, J., Schwaller, A., Endres, E., Klemmt, H.J., Tiede, D., Seifert, T.: Uav-based forest health monitoring: a systematic review. Remote Sensing **14**(13), 3205–3249 (2022)
4. Chiu, L.C., Chang, T.S., Chen, J.Y., Chang, N.Y.C.: Fast sift design for real-time visual feature extraction. TIP **22**(8), 3158–3167 (2013)
5. Wang, T., Zheng, Z., Yan, C., Zhang, J., Sun, Y., Zheng, B., Yang, Y.: Each part matters: local patterns facilitate cross-view geo-localization. TCSVT **32**(2), 867–879 (2021)
6. Dai, M., Hu, J., Zhuang, J., Zheng, E.: A transformer-based feature segmentation and region alignment method for uav-view geo-localization. TCSVT **32**(7), 4376–4389 (2022)
7. Zheng, Z., Wei, Y., Yang, Y.: University-1652: a multi-view multi-source benchmark for drone-based geo-localization. In: ACM MM, pp. 1395–1403 (2020)
8. Ding, L., Zhou, J., Meng, L., Long, Z.: A practical cross-view image matching method between uav and satellite for uav-based geo-localization. Remote Sensing **13**(1), 47 (2020)
9. Tian, X., Shao, J., Ouyang, D., Shen, H.T.: Uav-satellite view synthesis for cross-view geo-localization. TCSVT **32**(7), 4804–4815 (2021)

10. Zhuang, J., Dai, M., Chen, X., Zheng, E.: A faster and more effective cross-view matching method of uav and satellite images for uav geolocalization. Remote Sensing **13**(19), 3979 (2021)
11. Lin, J., Zheng, Z., Zhong, Z., Luo, Z., Li, S., Yang, Y., Sebe, N.: Joint representation learning and keypoint detection for cross-view geo-localization. TIP **31**, 3780–3792 (2022)
12. Zhuang, J., Chen, X., Dai, M., Lan, W., Cai, Y., Zheng, E.: A semantic guidance and transformer-based matching method for uavs and satellite images for uav geo-localization. IEEE Access **10**, 34277–34287 (2022)
13. Bromley, J., Guyon, I., LeCun, Y., Säckinger, E., Shah, R.: Signature verification using a "siamese" time delay neural network. Neurips 6 (1993)
14. Koch, G., Zemel, R., Salakhutdinov, R., et al.: Siamese neural networks for one-shot image recognition. In: ICML Deep Learning Workshop, vol. 2. Lille (2015)
15. Dosovitskiy, A., et al.: An image is worth 16x16 words: Transformers for image recognition at scale. arXiv preprint arXiv:2010.11929 (2020)
16. Deng, J., Dong, W., Socher, R., Li, L.J., Li, K., Fei-Fei, L.: Imagenet: a large-scale hierarchical image database. In: CVPR, pp. 248–255. IEEE (2009)
17. He, K., Zhang, X., Ren, S., Sun, J.: Deep residual learning for image recognition. In: CVPR, pp. 770–778 (2016)

Lightweight Multiview Mask Contrastive Network for Small-Sample Hyperspectral Image Classification

Minghao Zhu[1]([⊠])([iD]), Heng Wang[1,2]([iD]), Yuebo Meng[1,2]([⊠])([iD]), Zhe Shan[1,2]([iD]), and Zongfang Ma[1]([iD])

[1] College of Information and Control Engineering, Xi'an University of Architecture and Technology, Xi'an 710055, China
mmhzhu@163.com, wanghengolwh@163.com, mengyuebo@163.com, zongfangma@xauat.edu.cn
[2] Xi'an Key Laboratory of Intelligent Technology for Building Manufacturing, Xi'an 710055, China

Abstract. Deep learning methods have made significant progress in the field of hyperspectral image (HSI) classification. However, these methods often rely on a large number of labeled samples, parameters, and computational resources to achieve state-of-the-art performance, which limits their applicability. To address these issues, this paper proposes a lightweight multiview mask contrastive network (LMCN) for HSI classification under small-sample conditions. Considering the influence of irrelevant bands, we construct two views in an HSI scene using band selection and principal component analysis (PCA). To enhance instance discriminability, we propose a combination of self-supervised mask learning and contrastive learning in the design of LMCN. Specifically, we train corresponding masked autoencoders using the obtained views and utilize the feature extraction part of the autoencoder as an augmentation function, conducting unsupervised training through contrastive learning. To reduce the number of parameters, we employ lightweight Transformer modules to construct the autoencoder. Experimental results demonstrate the superiority of this approach over several advanced supervised learning methods and few-shot learning methods under small-sample conditions. Furthermore, this method exhibits lower computational costs. Our code is available at https://github.com/Winkness/LMCN.git.

Keywords: Hyperspectral image (HSI) classification · Contrastive learning · Mask autoencoder · Lightweight network · Small samples

---

This work was supported in part by the Key Research and Development Project of Shanxi Province (No. 2021SF-429).

Supplementary Information The online version contains supplementary material available at https://doi.org/10.1007/978-981-99-8462-6_39.

1   Introduction

Hyperspectral image (HSI) contains rich and detailed spectral information as well as spatial information, making it advantageous in object classification and spectral analysis [1]. Currently, HSI has been applied in various fields such as earth monitoring, agricultural management, and medical analysis [2]. While HSI provides abundant information, it also poses several challenges for hyperspectral image classification (HSIC), such as high-dimensional data, sample imbalance, and diversity of objects [3].

To address these challenges, researchers have proposed numerous methods. Supervised deep learning methods typically employ deep learning models consisting of multiple layers to automatically learn features from a large amount of labeled data, eliminating the need for manual feature engineering. Representative methods include convolutional neural networks (CNN) and recurrent neural networks (RNN) [4]. Zhu et al. drew inspiration from human attention and proposed a residual spectral-spatial attention network (RSSAN) [5] to extract more effective features. These methods have achieved excellent performance in HSIC, but they suffer from the disadvantages of requiring a large amount of labeled samples and having overly complex models.

To overcome the dependency on annotated samples, many researchers propose using contrastive learning (CL) for unsupervised pretraining of deep networks [6]. Contrastive learning learns by comparing the similarity between samples without relying on label information and has recently been widely applied in HSIC [7]. For example, Guan et al. proposed a cross-domain contrastive learning framework (XDCL) [6] for unsupervised learning of HSI representations. Hou et al. [8] used residual networks (ResNet) as encoders for both the spatial and spectral domains instead of using two separate encoders. Recently, multimodal approaches and graph convolutional neural networks have also been employed in HSIC and achieved promising results [9].

The CL-based HSIC method has partially alleviated the reliance on labeled samples, but there are still several unresolved issues: (1) Due to equal treatment of all spectral bands, most methods are susceptible to the influence of irrelevant bands, leading to a decrease in accuracy. (2) Many existing CL models have high hardware requirements, limiting their practical applications. (3) The features extracted by commonly used encoders are often insufficient.

To address these issues, we draw inspiration from the multiview contrastive learning concept [7] and the BERT language model (masked language model (MLM) task) [10]. We propose a lightweight multiview mask contrastive learning network (LMCN) for HSI classification under small-sample conditions. Specifically, we believe that each band within the HSI reflects distinct characteristics of the terrain. However, fundamentally, they collectively represent the same semantic information. Therefore, images from different bands can be regarded as views of the terrain from various perspectives.

LMCN consists of two modules: band selection view autoencoder (SVE) and principal component view autoencoder (PVE). Each module consists of view construction and lightweight masked autoencoder components. The difference lies in the view construction strategy, where SVE adopts a band selection-based view construction strategy, while PVE employs a PCA-based band selection strategy. LMCN takes the original HSI as input, and the two modules construct two views of the same HSI in different ways. The autoencoder takes the views as input and, after training, the encoding part of the autoencoder serves as an effective feature extractor. After extracting features from the views, unsupervised training is performed through contrastive learning.

The main contributions of this paper are as follows: (1) We introduce the self-supervised masked learning method into the remote sensing field, which enhances the model's learning of important features in the data. (2) To enhance instance discriminability, we combine self-supervised masked learning with contrastive learning, demonstrating that the combination of these two methods yields superior results. (3) We develop a lightweight masked autoencoder that allows LMCN to maintain high accuracy while having lower computational costs. Additionally, we incorporate graph convolution-based band selection into view construction, mitigating the impact of irrelevant bands.

The remaining parts of this paper are organized as follows. Section 2 provides a detailed description of the proposed model. Section 3 presents the experiments conducted and analyzes the results. Section 4 concludes the paper and discusses future research directions.

2   Proposed Method

In this section, we will provide separate descriptions for the two proposed modules (SVE and PVE), and finally, present an overview of the network's overall framework.

2.1   SVE Module

The combination of graph convolution and self-representation methods enables the utilization of structural information from spectral bands [11]. Motivated by this, we consider an HSI cube $X = R^{N \times B}$, where N represents the number of pixels and B represents the number of bands, our objective is to obtain a representative subset of bands, denoted as $n(n < B)$, to construct views of the HSI. These views are then inputted into the masked autoencoder for self-supervised training. The structure of SVE is illustrated in Fig. 1.

First, we transform X into a K-nearest neighbor (KNN) graph $G = (\nu, \varepsilon, A)$. G is used to represent the structural information of spectral data in HSI, where each node ν_i represents a band x_i. By selecting a parameter k, we compute the k nearest nodes to each node ν_i to determine ε. A is the adjacency matrix of graph G, which can be defined as follows:

$$A^{(i,j)} = \begin{cases} 1 & x^i \in N_k(x^{(j)}) \ or \ x^j \in N_k(x^{(i)}) \\ 0 & otherwise \end{cases} \tag{1}$$

In this context, $N_k(x^{(i)})$ denotes the set of nearest neighboring nodes to $x^{(i)}$. Through this transformation, we convert the task of finding a subset of

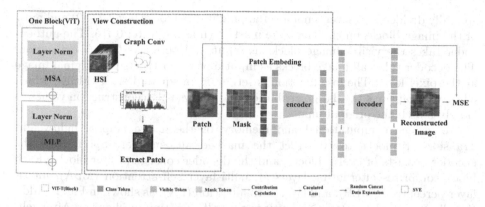

Fig. 1. Structure of SVE module. SVE takes HSI as input and constructs a spectral graph using graph convolution. The graph embedding is then used in self-expression to solve the self-expression matrix and obtain a subset of bands, which are used to construct a view. The autoencoder is trained with the masked view as input.

bands into the task of distinguishing a subset of significant nodes. Typically, a self-representation model can be defined as: $X = XC$, where C is the coefficient matrix for self-representation. The $L1$ norm of each row of C can serve as a metric to measure the contribution of the corresponding band. Usually, C can be obtained by minimizing the $L1$ norm or applying $L0$ norm regularization.

By employing graph convolution (GCN) [11] to compute the graph embedding of G and combining it with self-representation, the modified self-representation model can be defined as follows:

$$X = X\tilde{D}^{-1/2}\tilde{A}\tilde{D}^{-1/2}C, \ s.t. \ diag(C) = 0 \tag{2}$$

In the above equation, $X\tilde{D}^{-1/2}\tilde{A}\tilde{D}^{-1/2}$ represents the graph embedding matrix of the band graph. $\tilde{A}$ is a self-loop adjacency matrix, $\tilde{D}$ is the degree matrix of the nodes. Next, we utilize the frobenius norm to solve for the self-representation matrix C. The above equation can be transformed as follows:

$$C = (W^TW + \lambda I_b)^{-1}W^TX \tag{3}$$

The graph embedding matrix is denoted as W, and λ is a scaling factor. I_b is the identity matrix. After obtaining the solution for C, we normalize each column of C. Then, we calculate the $L1$ norm of each row in C to measure the contribution of each band to the reconstruction. Subsequently, we select the top-n bands with the highest contribution values as the chosen subset of bands.

We select a well-chosen subset of bands and randomly permute them, then concatenate them along the channel dimension to form a new HSI. Next, we perform boundary expansion on the image anddivide it into small patches with a neighborhood size of 28×28, centered around each land pixel. Each resulting patch image ($28 \times 28 \times n$) serves as a view of a land object instance and is inputted into the masked autoencoder after data augmentation. The views are spatially divided into several non-overlapping image blocks (size: 2 × 2) and most of the image blocks on the views are masked (masking rate: 0.75). The autoencoder takes the visible image blocks as input and performs feature extraction. The decoder takes all image blocks as input and reconstructs the original image at the pixel level. The loss function is set as mean squared error (MSE), which effectively captures the data distribution and allows the reconstruction values to closely match the original inputs.

To reduce computational and memory overhead, we employ lightweight Transformer blocks to construct the masked autoencoder. Specifically, the encoder consists of twelve blocks, and the decoder consists of four blocks. Each block comprises alternating layers of multi-layer self-attention (MSA), multi-layer perceptron (MLP), and layer normalization (LN), as shown in Fig. 1. Additionally, we use three attention heads in the self-attention mechanism. After self-supervised pretraining, the decoder is discarded, and the encoding part can serve as an effective image feature extractor, which is then applied to the complete view.

2.2 PVE Module

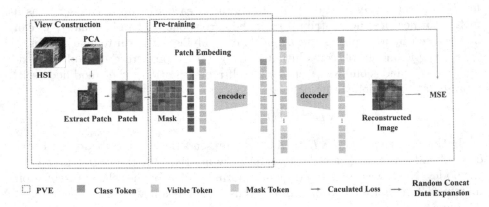

Fig. 2. Structure of PVE module. PVE takes HSI as input and constructs a view using PCA. Similar to SVE, the autoencoder takes the masked view as input and undergoes self-supervised training.

The structure of PVE is shown in Fig. 2. The first 100 bands of the HSI data are selected to generate views. Then, PCA is applied to extract the three principal components, forming a three-channel image $X_1^{h \times w \times 3}$.

Subsequently, similar to SVE, we perform boundary extension and image block segmentation on X_1, with a neighborhood size of 28 × 28. Each segmented image block has a size of 28 × 28 × 3 and is considered as a view for each land cover instance. Next, the generated views are randomly masked and used as inputs to the mask autoencoder. After self-supervised pretraining, similar to SVE, the decoder is discarded, and the encoding part serves as an effective spatial feature extractor that can be applied to the complete views.

2.3 Lightweight Multiview Mask Contrastive Network

In this study, to further enhance instance discriminability, after extracting view features separately in SVE and PVE, we utilize contrastive learning to further optimize the feature representation of the model. The loss function of LMCN is set as the Info NCE Loss. The structure of LMCN is shown in Fig. 3. First, LMCN takes the features extracted by SVE and PVE as input. Second, we utilize a projection head structure consisting of two fully connected layers (Linear), a batch normalization layer (BN), and an activation function (ReLU) after the class tokens of SVE and PVE respectively to enhance performance.

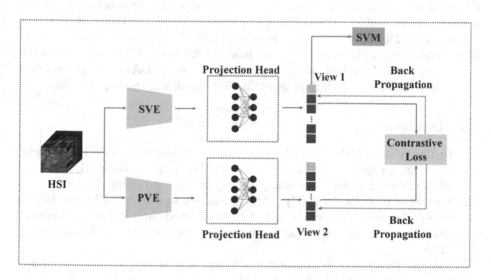

Fig. 3. Structure of LMCN. The feature extraction components of SVE and PVE serve as enhancement functions in LMCN. After obtaining features from different views, LMCN employs contrastive learning to enhance instance discriminability.

Let N be the number of samples used in each iteration during neural network training. In the unsupervised training process of LMCN, each iteration consists of $2 \times N$ views. Among these $2 \times N$ views, two views from the same land cover instance are defined as positive pairs, while two views from different land cover instances are considered negative pairs. The contrastive loss of LMCN is defined as follows:

$$\ell_{(i,j)} = -log \frac{\exp(sim(Z_i, Z_j))}{\sum_{k=l}^{2N} l[k \neq l]sim(Z_i, Z_k)} \tag{4}$$

Whereas, $sim(Z_i, Z_j)$ represents the cosine similarity between two feature vectors Z_i and Z_j. In the downstream classification stage, the encoded part of the trained autoencoder serves as a feature extractor. Subsequently, all samples of the HSI data are input into the encoder, producing corresponding feature vectors. Finally, we employ an SVM classifier to classify these feature vectors.

3   Experimental Results and Analysis

The proposed method in this paper is implemented using the PyTorch library. The experiments were conducted on a personal computer equipped with an Intel i5-13400F processor and an Nvidia GeForce RTX 3060 graphics card.

The optimizers used in this experiment are all Adam optimizers, with a batch size of 128. The image neighborhood size is set to 28 × 28, and the image masking rate is uniformly set to 0.75. Pre-training used 50% unlabled samples. The optimal values for k, λ, and the number of bands are 3, 0.0001, and 63, respectively. In order to achieve more stable convergence, a learning rate of 0.00015 and a weight decay coefficient of 0.05 are employed in the pre-training phase of the autoencoder. During the training process of LMCN, the learning rate is set to 0.001, and the weight decay coefficient is set to 0.000001. Furthermore, to evaluate the proposed method, we utilize overall accuracy (OA), average accuracy (AA), and the Kappa coefficient as evaluation metrics in the experiments.

3.1   Datasets

(1) **Indian Pines (IP)**: The IP dataset was acquired by the AVIRIS sensor, covering approximately 145 × 145 pixels. The original dataset consists of 10,249 labeled samples and includes 16 different land cover classes.
(2) **University of Pavia (UP)**: The UP dataset was acquired by the ROSIS sensor. The dataset consists of 103 spectral bands and includes 9 different land cover classes. The total number of labeled samples in this dataset is 42776.
(3) **Salinas (SA)**: The SA dataset is a hyperspectral image dataset collected from Salinas County, Indiana, USA. It consists of a grid of 512 × 217 pixels. The dataset contains 204 contiguous spectral bands covering the visible and infrared spectral ranges. It encompasses 16 distinct land cover classes.

3.2   Comparative Experiment

In this section, we compared our proposed LMCN model with five HSI classification methods: SVM [12], SSRN [13], DFSL [14], DCFSL [15], and CA-CFSL [16]. For these five methods and our proposed LMCN, we selected the same number of supervised samples, specifically, only five labeled samples were randomly selected for each class. The results were obtained from ten tests.

Quantitative Evaluation : Tables 1, 2, and 3 present the specific class classification accuracy, overall accuracy (OA), average accuracy (AA), and Kappa coefficient for the six different methods on three different HSI datasets. Table 4 presents the parameter count of six methods on the Indian Pines dataset. For the Indian Pines dataset, we achieve higher classification accuracy with a lower parameter count (Para: 7.35M), with classification OA and AA of 74.93% and 81.12%, respectively. Compared to the state-of-the-art few-shot classification method CA-CFSL [16], our LMCN shows an improvement of 4.30% and 4.92% in OA and k, respectively. On the Pavia University dataset, LMCN outperforms the best baseline CA-CFSL with an OA and k performance improvement of 4.67% and 6.00%, respectively. On the Salinas dataset, LMCN outperforms the best baseline method CA-CFSL with an increase of 3.67% in terms of overall accuracy and 2.94% in terms of kappa coefficient.

Table 1. Classification results of different methods for the Indian Pines dataset

Class	SVM	SSRN	DFSL	DCFSL	CA-CFSL	LMCN
1	70.73	**98.29**	90.24	97.07	96.59	91.30
2	37.24	36.91	36.73	46.42	59.95	**62.11**
3	42.99	52.78	37.78	53.47	58.95	**70.24**
4	47.63	80.56	56.38	79.91	80.52	**92.40**
5	62.64	74.96	64.27	73.08	**80.65**	74.53
6	67.49	80.54	71.82	85.93	**88.72**	76.71
7	81.30	100.0	98.70	100.0	100.0	**100.0**
8	58.16	81.90	74.97	81.54	88.03	**93.70**
9	89.33	99.33	99.33	**100.0**	98.67	90.00
10	40.66	60.89	54.95	**67.15**	65.78	63.58
11	37.46	54.30	52.23	60.16	60.03	**75.80**
12	26.05	51.33	28.61	48.04	52.13	**55.64**
13	90.45	96.90	95.80	**98.05**	97.15	89.75
14	66.09	80.19	81.70	85.21	**90.55**	87.50
15	28.82	66.35	51.92	69.37	78.19	**87.56**
16	89.20	98.64	95.91	**99.77**	99.20	87.09
OA(%)	47.07	62.15	56.10	66.38	70.63	**74.93**
AA(%)	58.52	75.87	68.21	77.82	80.94	**81.12**
k	40.90	57.62	50.65	62.16	66.91	**71.83**

Qualitative Evaluation : Figures 4, 5 and 6 respectively show the false-color images, ground truth annotations, and visual classification results of six different methods on three HSI datasets. On the whole, compared to other methods, LMCN generates more accurate classification maps and fewer large-scale misclassifications. This is because the masked autoencoder learns more important

Table 2. Classification results of different methods for the University of Pavia dataset

Class	SVM	SSRN	DFSL	DCFSL	CA-CFSL	LMCN
1	64.78	78.31	72.50	**80.28**	79.95	79.86
2	63.05	73.30	75.36	87.22	90.83	**97.23**
3	48.91	67.14	46.99	64.34	66.39	**90.51**
4	80.90	89.83	**92.32**	92.20	91.62	58.25
5	98.68	92.77	99.34	99.36	**99.73**	99.33
6	54.06	69.90	64.64	77.80	77.09	**96.48**
7	69.11	84.92	69.40	76.51	81.82	**96.76**
8	68.86	76.36	62.78	63.80	75.72	**97.41**
9	**99.73**	99.02	96.82	98.83	96.88	46.88
OA(%)	65.46	76.36	73.44	82.56	85.22	**89.89**
AA(%)	72.01	81.28	75.57	82.26	84.45	**84.75**
k	56.70	69.91	66.15	77.42	80.67	**86.67**

Table 3. Classification results of different methods for the Salinas dataset

Class	SVM	SSRN	DFSL	DCFSL	CA-CFSL	LMCN
1	98.19	**99.50**	95.40	98.86	99.35	94.67
2	94.77	98.02	95.15	99.36	**99.74**	94.92
3	82.14	81.37	83.15	91.17	90.78	**95.59**
4	98.25	**99.66**	99.55	99.36	99.52	87.01
5	94.96	**95.51**	88.81	91.99	92.64	89.76
6	98.81	97.76	99.02	**99.45**	99.23	94.54
7	99.28	**99.58**	99.08	98.94	98.98	96.20
8	54.28	71.21	73.54	72.67	73.93	**94.36**
9	97.98	97.87	97.18	**99.15**	99.61	98.33
10	76.80	77.11	71.97	86.03	89.36	**96.67**
11	87.80	94.46	94.64	**98.11**	98.02	89.41
12	93.77	98.66	93.07	99.01	**99.18**	91.38
13	95.85	98.92	99.37	99.24	**99.52**	84.06
14	90.93	96.09	**98.61**	98.15	98.56	81.30
15	52.33	72.93	65.51	75.38	79.47	**89.39**
16	82.97	87.02	87.27	92.35	93.74	**98.56**
OA(%)	79.64	86.79	85.20	88.74	89.93	**93.60**
AA(%)	87.45	91.61	90.08	93.70	**94.48**	92.25
k	77.43	85.35	83.57	87.51	88.82	**91.76**

Table 4. The number of parameters of different methods (Param(M))

Dataset	SVM	SSRN	DFSL	DCFSL	CA-CFSL	LMCN
UP	–	19.91	3.49	42.59	43.04	7.35
IP	–	34.67	3.49	42.69	43.04	7.35
SA	–	35.22	3.49	42.69	43.04	7.35

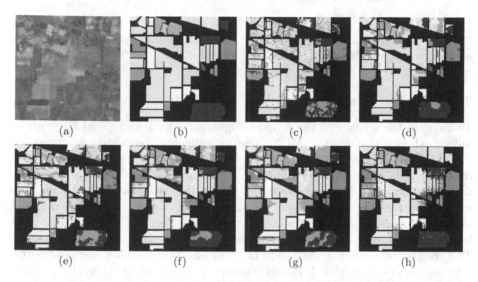

Fig. 4. Classification maps for IP dataset. (a) False color image, (b) Ground truth, (c) SVM, (d) SSRN, (e) DFSL, (f) DCFSL, (g) CA-CFSL, (h) Proposed LMCN

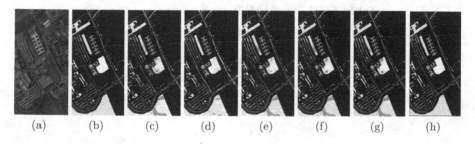

Fig. 5. Classification maps for UP dataset. (a) False color image, (b) Ground truth, (c) SVM, (d) SSRN, (e) DFSL, (f) DCFSL, (g) CA-CFSL, (h) Proposed LMCN

features through the task of predicting reconstruction, resulting in more comprehensive and representative extracted features. Additionally, LMCN exhibits fewer instances of large-scale misclassifications on each class and performs better in class boundaries. This is attributed to the ability of the autoencoder to extract more representative features, while the integration of contrastive learning further enhances instance discrimination.

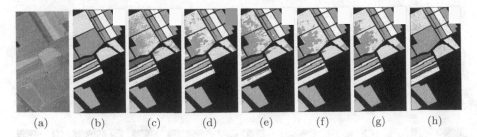

| | | | | | | | |
| (a) | (b) | (c) | (d) | (e) | (f) | (g) | (h) |

Fig. 6. Classification maps for SA dataset. (a) False color image, (b) Ground truth, (c) SVM, (d) SSRN, (e) DFSL, (f) DCFSL, (g) CA-CFSL, (h) Proposed LMCN

3.3   Ablation Study

To validate the efficacy of the proposed approach, we investigated the roles of individual modules within LMCN. The experimental outcomes, as presented in Table 5, reveal that the classification accuracy of LMCN with involvement in contrastive learning surpasses that of the SVE and PVE modules, which operate without contrastive learning, across all three HSI datasets. This phenomenon stems from the fact that the primary functions of the SVE and PVE modules are geared towards acquiring more representative features. As a result, their independent impact on classification performance is not notably pronounced. The results indicate that the fusion of representation learning and contrastive learning serves to further heighten instance discriminability, thereby yielding improved classification outcomes.

Table 5. OA(%) of LMCN with different modules

Module	UP	IP	SA
SVE	55.82±1.33	33.31±1.39	53.27±0.70
PVE	39.80±1.14	38.77±1.01	53.84±1.06
SVE+PVE+CL	89.89±1.74	74.93±2.15	93.60±1.32

4   Conclusion

The scarcity of training samples, the influence of irrelevant bands, and excessive memory consumption pose challenges to deep learning-based HSIC methods. To address the influence of irrelevant bands, we combine band selection based on graph convolution and self-representation with the construction of views. In order to alleviate the dependence on training samples and enhance instance discriminability, we integrate mask learning with contrastive learning, and demonstrate that this approach leads to improved classification results. The proposed LMCN method performs well under small-sample conditions and has lower computational costs, which widens its applicability.

In the future, we will further investigate the discriminability of features in contrastive learning and the classification of unknown categories. Additionally, most HSIC methods based on contrastive learning currently employ a two-stage strategy. Exploring the design of an end-to-end unsupervised classification network is another valuable research direction.

References

1. Wambugu, N., et al.: Hyperspectral image classification on insufficient-sample and feature learning using deep neural networks: a review. Int. J. Appl. Earth Obs. Geoinf. **105**, 102603 (2021)
2. Moharram, M.A., Sundaram, D.M.: Land use and land cover classification with hyperspectral data: a comprehensive review of methods, challenges and future directions. Neurocomputing (2023)
3. Su, Y., Li, X., Yao, J., Dong, C., Wang, Y.: A spectral-spatial feature rotation based ensemble method for imbalanced hyperspectral image classification. IEEE Trans. Geosci. Remote Sens. **61**, 1–18 (2023)
4. Mou, L., Ghamisi, P., Zhu, X.X.: Deep recurrent neural networks for hyperspectral image classification. IEEE Trans. Geosci. Remote Sens. **55**(7), 3639–3655 (2017)
5. Zhu, M., Jiao, L., Liu, F., Yang, S., Wang, J.: Residual spectral-spatial attention network for hyperspectral image classification. IEEE Trans. Geosci. Remote Sens. **59**(1), 449–462 (2020)
6. Guan, P., Lam, E.: Cross-domain contrastive learning for hyperspectral image classification. IEEE Trans. Geosci. Remote Sens. **60**, 1–13 (2022)
7. Liu, B., Yu, A., Yu, X., Wang, R., Gao, K., Guo, W.: Deep multiview learning for hyperspectral image classification. IEEE Trans. Geosci. Remote Sens. **59**(9), 7758–7772 (2020)
8. Hou, S., Shi, H., Cao, X., Zhang, X., Jiao, L.: Hyperspectral imagery classification based on contrastive learning. IEEE Trans. Geosci. Remote Sens. **60**, 1–13 (2022)
9. Wang, M., Gao, F., Dong, J., Li, H., Du, Q.: Nearest neighbor-based contrastive learning for hyperspectral and lidar data classification. IEEE Trans. Geosci. Remote Sens. **61**, 1–16 (2023)
10. He, K., Chen, X., Xie, S., Li, Y., Doll'ar, P., Girshick, R.B.: Masked autoencoders are scalable vision learners. In: 2022 IEEE/CVF Conference on Computer Vision and Pattern Recognition (CVPR), pp. 15979–15988 (2021)
11. Cai, Y., Zhang, Z., Liu, X., Cai, Z.: Efficient graph convolutional self-representation for band selection of hyperspectral image. IEEE J. Select. Top. Appl. Earth Observ. Remote Sens. **13**, 4869–4880 (2020)
12. Melgani, F., Bruzzone, L.: Classification of hyperspectral remote sensing images with support vector machines. IEEE Trans. Geosci. Remote Sens. **42**(8), 1778–1790 (2004)
13. Zhong, Z., Li, J., Luo, Z., Chapman, M.: Spectral-spatial residual network for hyperspectral image classification: a 3-D deep learning framework. IEEE Trans. Geosci. Remote Sens. **56**(2), 847–858 (2017)
14. Liu, B., Yu, X., Yu, A., Zhang, P., Wan, G., Wang, R.: Deep few-shot learning for hyperspectral image classification. IEEE Trans. Geosci. Remote Sens. **57**(4), 2290–2304 (2018)

15. Li, Z., Liu, M., Chen, Y., Xu, Y., Li, W., Du, Q.: Deep cross-domain few-shot learning for hyperspectral image classification. IEEE Trans. Geosci. Remote Sens. **60**, 1–18 (2022). https://doi.org/10.1109/TGRS.2021.3057066
16. Wang, W., Liu, F., Liu, J., Xiao, L.: Cross-domain few-shot hyperspectral image classification with class-wise attention. IEEE Trans. Geosci. Remote Sens. **61**, 1–18 (2023). https://doi.org/10.1109/TGRS.2023.3239411

Dim Moving Target Detection Based on Imaging Uncertainty Analysis and Hybrid Entropy

Erwei Zhao[1,2], Zixu Huang[1,2,3], and Wei Zheng[1(✉)]

[1] Key Laboratory of Electronics and Information Technology for Space System, National Space Science Center, Chinese Academy of Sciences, Beijing 100190, China
zhengwei@nssc.ac.cn
[2] School of Computer Science and Technology, University of Chinese Academy of Sciences, Beijing 100190, China
[3] School of Fundamental Physics and Mathematical Sciences, Hangzhou Institute for Advanced Study, University of Chinese Academy of Sciences, Hangzhou 310024, China

Abstract. The detection of weak moving targets is the fundamental work for target recognition and plays an important role in precise guidance systems, missile warning and defense systems, space target surveillance and detection tasks, satellite remote sensing systems, and other areas. In low signal-to-noise ratio conditions, the target signal is often submerged by strong background and noise signals, making it difficult to detect. Existing detection methods mainly focus on infrared bright target detection, which cannot solve the problem of darker targets compared to the background. Additionally, these methods do not sufficiently consider the temporal signal characteristics of moving targets in video sequences. To address these difficulties, this paper introduces an uncertainty analysis approach, establishes a target confidence model for weak moving target detection based on the principle of uncertainty in the time domain, and proposes a mixed entropy enhancement detection method based on single-pixel temporal signals. This method effectively suppresses background noise and enhances target signals in gaze scenes, enabling the detection of weak moving dark and bright targets. The proposed method demonstrates good performance in real-life scene experiments.

Keywords: Dim moving target detection · Uncertainty Analysis · Entropy

1 Introduction

Dim moving targets refer to small-sized and low-intensity moving targets detected under complex background conditions in long-range detection processes. The detection of faint and weak moving targets is a fundamental task in target recognition and is a core technology in fields such as precise guidance systems, surveillance and early-warning system, space target surveillance and detection tasks, and satellite remote sensing systems [13,19]. The methods for detecting moving targets can generally be divided into single-frame image-based methods and multi-frame image-based methods [15].

Single-frame image-based methods for detecting weak moving targets primarily utilize the spatial information of a single frame image to detect small targets in video sequences. With the assistance of filtering and learning classification algorithms in the field of signal processing, the single-frame image is preprocessed to obtain corresponding candidate target points. Currently, these methods are mostly applied in the detection of weak small targets in infrared imaging. Based on the principle of thermal infrared imaging, small targets often exhibit higher energy than the energy of the background and noise in infrared images. A series of spatial domain detection methods have been developed based on this characteristic. Chen [2] introduced a small target detection method based on local contrast measurement (LCM), which incorporates the theoretical principles of the human visual system into weak small target detection. Based on this, many scholars have conducted research and improvements on this type of method, such as improved local contrast measurement (ILCM) [7], novel local contrast measurement (NLCM) [14], weighted local difference measurement (WLDM) [4], multiscale tri-layer LCM (TLLCM) [8], and Weighted Strengthened Local Contrast Measure (WSLCM) [9]. However, most single-frame detection algorithms require high signal-to-noise ratio for the targets, which is not effective for detecting low-noise moving targets in complex backgrounds. Moreover, single-frame detection algorithms are mostly based on the assumption of target saliency, focusing on the detection of bright targets in dim backgrounds, and ignoring the situation where the environment is brighter than the target in real scenes.

Considering the limitations of single-frame detection algorithms, there has been extensive research on weak moving target detection using multi-frame spatial filtering or joint spatio-temporal filtering methods, such as frame differencing [10]. With the popularity of human visual mechanisms in single-frame detection, researchers have started to explore the fusion of temporal information. The spatial-temporal local contrast filter (STLCF) [5] integrates spatio-temporal local difference features to construct a simple and effective spatio-temporal filter. Furthermore, the spatial-temporal local difference measure (STLDM) [6] combines direction information on spatial local difference and the differences between adjacent three frames in the pixel's temporal domain. Although there are many weak moving target detection methods that combine temporal information from multiple frames, the exploration of temporal information in video sequences is still not fully sufficient, resulting in low detection rates and high false alarm rates for moving targets with very low signal-to-noise ratio.

During the process of detecting weak and dim moving targets at ultra-long distances, the target information is contaminated as it is transmitted through the signal acquisition channel due to factors such as atmospheric conditions and noise, as well as limitations and accuracy of the detection equipment. The accuracy of the observed information gradually decreases, making it difficult to accurately determine the position and time of the target from the detection data. The existence of the target becomes uncertain. Based on this, this paper introduces an analysis of uncertainty in weak moving target detection, discussing the pollution of the target signal by different stages in the observation and detection chain. An uncertainty model for the target signal is established, and a new hybrid entropy detection technique is proposed, which is applied to single-pixel temporal signals with the aim of fully exploring the temporal characteristics of the target signal. Through real-world scenarios verification, this method demonstrates improved detection performance in gaze scenes, enabling the simultaneous detection of both bright and dim targets.

2   Theoretical and Method

2.1   Uncertainty Model

For a specific target signal, the collected observation signal is contaminated by factors such as background noise, point spread effects, detector operating conditions, etc. The uncertainty of the target detection result increases as the number of intermediate links increases [3, 16]. Figure 1 illustrates the information transmission process of an optical imaging system.

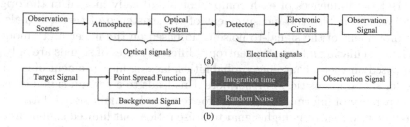

Fig. 1. The information transfer process of the optical imaging system.

In the temporal signal of target observation, the background signal refers to the grayscale value of the background environment. In gaze scenes, the grayscale value of the background signal remains constant. However, the background environment may experience slight fluctuations in practical observations. The fluctuations in the background signal itself result in an increase in information entropy. After the superposition of the target signal and the background signal, the combined signal passes through the atmospheric medium and the optical system to reach the focal plane of the detector. During this process, point spread effects occur, causing energy attenuation of the target signal. The sampling frequency is related to the observation parameters of the detector. A higher sampling frequency can allow the target to stay in the temporal domain for a longer time, increasing the total power of the target signal in the temporal domain. However, a shorter integration time can lead to a decrease in target energy and an increase in random noise. Figure 1 illustrates the transfer model of uncertainty. Each link in the imaging chain increases the uncertainty in target detection. Based on the principle of information entropy, we establish a target confidence model for weak moving target detection based on uncertainty.

$$\begin{cases} A_1 : Q = H_{Baakground} + H_{noise} + H_{target} \\ A_2 : Q = H_{Background} + H_{noise} \end{cases} \tag{1}$$

where the event A_1 indicates the presence of a target and A_2 indicates the absence of a target. Q is a measure of the likelihood of the existence of the target signal, $H_{Background}$ denotes the information entropy of background changes; H_{noise} denotes the information entropy caused by noise changes, and H_{signal} denotes the information entropy caused by changes in the target signal. Therefore, faint motion detection can be considered as a binary classification problem for the presence or absence of a moving target in the detection region.

2.2 Method Procedure

According to the uncertainty model, targets and non-targets can be theoretically separated. However, it is not possible to individually calculate the uncertainties introduced by targets, backgrounds, and noise signals because we cannot identify them. But the influences of each component are already hidden in the original signals, which enables the measurement of the existence of target signals and the improvement of the signal-to-noise ratio based on the uncertainty model.

When calculating information entropy, different states of signals are only processed based on their occurrence probabilities, lacking measurement of grayscale values for signal intensities. Furthermore, this leads to a certain signal enhancement capability of information entropy for temporal domain signal data of moving targets with relatively high signal-to-noise ratios, but limited signal enhancement capability for target signals with low signal-to-noise ratios. Here, we introduce the concept of anti-entropy and extend it to single-pixel temporal signals, proposing a new mixed variant entropy computation to enhance weak transient signals. The specific definition of variational entropy VE is as follow.

$$\text{VE} = -\sum_{i=1}^{I} M(i) * log(M^2(i)) \tag{2}$$

where I is the number of signal samples on the pixel; $M(i)$ is the target confidence function, which is used to assign the target confidence to each sampling point. Correspondingly, the variational Anti-entropy VAE is defined as:

$$\text{VAE} = -\sum_{i=1}^{I} M(i) * log((1 - M(i))^2) \tag{3}$$

Similar to information entropy, variant entropy and variational anti-entropy also satisfy the principles of maximum entropy and minimum entropy. They stretch the differences between the steady state and chaotic state at a larger scale, preserving and expanding the ability of information entropy to measure the degree of signal disorder. Furthermore, the hybrid entropy HE integrates the characteristics of both of the aforementioned measures and is defined as follows.

$$HE = VAE - VE$$

$$= -\sum_{i=1}^{I} M^2(i) * log((\frac{1 - M(i)}{M(i)})^2) \tag{4}$$

Figure 2 illustrates the functional distribution curves of variational entropy, variational anti-entropy, and hybrid entropy for a binary information source. The hybrid entropy is constructed based on the difference between the other two measures, while retaining the characteristic of having a minimal entropy from variational anti-entropy.

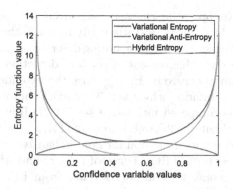

Fig. 2. Three proposed entropy function curves.

In practice, by constructing a sliding window and applying the concept of hybrid entropy to the sequential data of individual pixels, the mixed entropy curve can be obtained to enhance the temporal signal of the moving target. Here, the definition of the confidence allocation function $M(i)$ is provided, which is used to assign confidence to each element within a window of length I.

$$M(i) = \frac{f(i)}{\sum_{i=1}^{I} f(i)} \tag{5}$$

where $f(i)$ is the signal value sampled by the detector and i is the i-th sample in the window.

3   Experiment and Analysis

To evaluate the effectiveness of the theoretical framework and methods proposed in this study, we selected several sets of real remote sensing data from different scenarios for validation in this section. Six sets of representative baseline algorithms were used for comparison, including different types of single-frame detection algorithms as well as multi-frame detection algorithms."

3.1 Evaluation Metrics

Unlike the common visual target detection task, the core of weak target detection is to enhance the target signal to make it significant. Detection probability and false alarm rate are used in experiments as the most important performance metrics for measuring detectors. The detection probability indicates the probability that the detected target is a true target, while the false alarm rate indicates the probability that the detected target is a false target. They are defined as follows:

$$P_D = \frac{\text{number of detected true targets}}{\text{number of actual targets}} \tag{6}$$

$$P_F = \frac{\text{number of false pixels detected}}{\text{number of images pixels}} \tag{7}$$

Second, the receiver operating characteristic (ROC) curve and the area under the curves(AUC) are introduced to statistically measure the detection probability and false alarm rate of the detector under different segmentation thresholds τ to better evaluate the enhancement effect and detection performance of the proposed method comprehensively. Here we use the extended 3D-ROC to evaluate the detector performance, which is a 3D curve composed of P_D, P_F and τ, and can be further decomposed into three sets of 2D-ROC: (P_D, P_F), (P_D, τ) and (P_F, τ). Among them, the 2D ROC curve of (P_D, P_F) with its corresponding $\text{AUC}_{(D, F)}$ is a commonly used method for comprehensive evaluation of detector detection performance; the 2D ROC curve of (P_D, τ) and $\text{AUC}_{(D, \tau)}$ can be used for target detection alone, without interference from PF; while the 2D ROC curve of (P_F, τ) and $\text{AUC}_{(F, \tau)}$ play a key role in the evaluation of background suppression level. Based on the three AUCs, five evaluation metrics from [1] are presented to quantitatively evaluate the detector from five perspectives: target detection(TD), background suppression(BS), joint evaluation, overall detection probability (ODP) and signal-to-noise probability ratio (SNPR).

$$\text{AUC}_{\text{TD}} = \text{AUC}_{(D,F)} + \text{AUC}_{(D,\tau)} \tag{8}$$

$$\text{AUC}_{\text{BS}} = \text{AUC}_{(D,F)} - \text{AUC}_{(F,\tau)} \tag{9}$$

$$\text{AUC}_{\text{TDBS}} = \text{AUC}_{(D,\tau)} - \text{AUC}_{(F,\tau)} \tag{10}$$

$$\text{AUC}_{\text{ODP}} = \text{AUC}_{(D,\tau)} - (1 - \text{AUC}_{(F,\tau)}) \tag{11}$$

$$\text{AUC}_{\text{SNPR}} = \text{AUC}_{(D,\tau)}/\text{AUC}_{(F,\tau)} \tag{12}$$

3.2 Datasets

The experimental data used in this paper consists of two sets of sequence data from a publicly available dataset [11] and one set of sequence data taken by us. Where Seq. 1 and Seq. 2 are from the public dataset and Seq. 3 is private data. It is worth pointing out that the diverse scenarios with many different types of

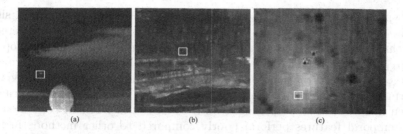

Fig. 3. Example image, target marked by red box. (a) Seq. 1; (b) Seq. 2; (c) Seq. 3. (Color figure online)

elements included in the three datasets verify the adaptability of the algorithm, and Fig. 3 shows the corresponding example plots.

From the perspective of target characteristics, all three data sets are faint motion targets with a size less than 7*7 at a low signal-to-noise ratio, among which Seq. 2 and Seq. 3 have signal-to-noise ratios lower than 3, which are typical of very faint targets. Among them, Seq. 1 and Seq. 3 is the sky background, and Seq. 2 is the ground background. From the target-scene interaction characteristics, Seq. 1 and Seq. 2 are infrared data, and the targets are bright targets with generally higher energy than the background scene brightness, while Seq. 3 is visible data, and the targets are dark targets with lower energy than the background scene brightness. Detailed information is shown in Table 1.

Table 1. The details of the sequence data.

Sequence	Frames	Image Resolution	SCR Average	Target Size	Scene
Seq.1	50	239*256	3.277	5*5	Cloudy sky
Seq.2	33	228*249	2.450	3*3	Ground with vegetation
Seq.3	319	119*149	1.588	7*7	Sky

3.3   Comparison of Detection Performance

Based on the characteristics of the test data, six groups of baseline methods were introduced in the experiments to evaluate the performance of the algorithms, taking into account the adaptability of the different methods. Among them, there are four groups of single-frame detection algorithms, absolute directional mean difference (ADMD) [12], WSLCM [9], local gradient and directional curvature (LGDC) [17], NRAM [18]; and the multi-frame detection algorithms include STLCF [5], and STLDM [6].

Figure 4 and Table 2 show the processing results regarding Seq. 1. Almost all algorithms perform well in target detection in the infrared weak target sequence in a typical gaze scenario like Seq. 1. Specifically, the relatively high signal-to-noise ratio of the targets in the test data makes all four airspace detection algorithms perform well, especially LGDC and our method obtains the optimal $AUC_{(D,F)}$, $AUC_{(F,\tau)}$ and AUC_{SNPR}. In contrast, although LGDC performs well in target detection, the lack of background rejection becomes the key to its performance, considering the performance of LGDC in $AUC_{(F,\tau)}$, which can also be seen in Fig. 4. Most surprisingly, both STLCF and STLDM combined with spatiotemporal features perform poorly compared to other methods in terms of detection and background suppression, with a large amount of unsuppressed spurious waves.

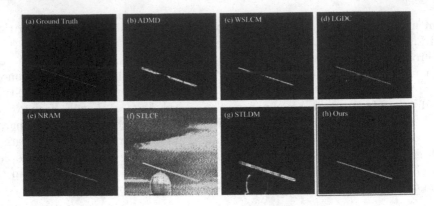

Fig. 4. Ground truth and corresponding seven methods experiment results of Seq. 1.

Table 2. AUC values calculated from Seq.1.

Methods	Single frame method				Multiframe method		
	ADMD	WSLCM	NRAM	LGDC	STLCF	STLDM	Ours
$AUC_{(D,F)}$	0.948	1.000	0.900	1.000	0.976	0.996	1.000
$AUC_{(D,\tau)}$	0.493	0.531	0.766	0.991	0.709	0.354	0.899
$AUC_{(F,\tau)}$	1.09E-03	5.01E-04	5.21E-04	9.03E-04	2.88E-01	6.13E-03	5.00E-04
AUC_{TD}	1.441	1.531	1.666	1.991	1.684	1.349	1.899
AUC_{BS}	0.947	0.999	0.899	0.999	0.688	0.989	1.000
AUC_{TDBS}	0.494	0.532	0.767	0.992	0.997	0.360	0.899
AUC_{ODP}	1.492	1.531	1.766	1.990	1.421	1.348	1.898
AUC_{SNPR}	453.7	1060.8	1471.1	1097.6	2.5	57.7	1797.2

The results of the seven groups of methods on Seq. 2 are shown in Fig. 5 and Table 3. Compared with Seq. 1, the target signal-to-noise ratio of Seq. 2

is significantly lower, which results in only three of the seven groups of algorithms whose performance maintains a degree of detection rate that is worth discussing, including NRAM, STLCF, and ours among them, STLCF's $AUC_{(F,\tau)}$ is the highest, which is consistent with the results of Seq. 1, where poorer background suppression constraints its performance. While NRAM achieves the best background suppression based on the principal component analysis algorithm, its target detection capability $AUC_{(D,F)}$ becomes a performance bottleneck. Collectively, our algorithm balances target detection and background suppression to achieve higher performance.

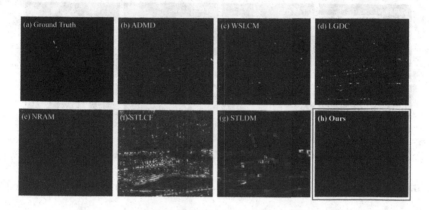

Fig. 5. Ground truth and corresponding seven methods experiment results of Seq. 2.

Table 3. AUC values calculated from Seq.2

Methods	Single frame method				Multiframe method		
	ADMD	WSLCM	NRAM	LGDC	STLCF	STLDM	Ours
$AUC_{(D,F)}$	0.49	0.77	0.74	0.98	0.97	0.95	1.00
$AUC_{(D,\tau)}$	0.01	0.02	0.33	0.07	0.49	0.10	0.23
$AUC_{(F,\tau)}$	6.1E-03	5.9E-03	5.5E-03	0.009	0.112	0.020	0.007
AUC_{TD}	0.49	0.79	1.07	1.05	1.46	1.05	1.22
AUC_{BS}	0.48	0.77	0.73	0.97	0.86	0.93	0.99
AUC_{TDBS}	0.01	0.02	0.34	0.08	0.60	0.11	0.23
AUC_{ODP}	1.00	1.01	1.33	1.06	1.38	1.08	1.22
AUC_{SNPR}	0.82	2.96	60.35	7.69	4.37	4.86	30.83

Finally, the tests were performed on Seq.3, whose saliency effect plot with AUC is shown in Fig. 6 and Table 4. First of all, the target in Seq.3 is an extremely low signal-to-noise ratio dark target, which makes many detection algorithms for bright infrared targets completely ineffective, especially single-frame detection methods like ADMD, WSLCM, LGDC, and NRAM. From Fig. 6, STLCF and Ours are better at enhancing the trajectory through which the target passes. Combined with Table 4, the performance of STLCF in terms of background suppression is not much different from other detectors due to the relatively simple background of Seq. 3, but still five times higher than our proposed method. Combining the results of Seq. 1 and Seq. 2, our proposed method consistently maintains high detection and good background suppression.

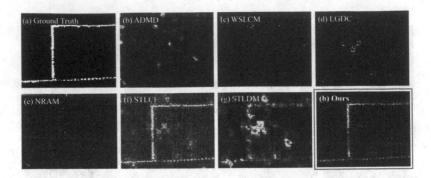

Fig. 6. Ground truth and corresponding seven methods experiment results of Seq. 3.

To better demonstrate the overall detection performance of the seven algorithms on the three sets of test data, the ROC curve of (P_D, P_F) are presented on Fig. 7. Intuitively, our methods almost always achieve higher detection rates in the ROC curves for the same false alarm rate.

Table 4. AUC values calculated from Seq.3

Methods	Single frame method				Multiframe method		
	ADMD	WSLCM	NRAM	LGDC	STLCF	STLDM	Ours
$AUC_{(D,F)}$	0.480	0.504	0.500	0.533	0.992	0.887	0.999
$AUC_{(D,\tau)}$	0.018	0.009	0.005	0.010	0.389	0.185	0.219
$AUC_{(F,\tau)}$	0.016	0.007	0.005	0.008	0.097	0.062	0.020
AUC_{TD}	0.498	0.513	0.505	0.543	1.380	1.072	1.217
AUC_{BS}	0.464	0.497	0.495	0.525	0.894	0.825	0.979
AUC_{TDBS}	0.034	0.016	0.010	0.018	0.486	0.247	0.238
AUC_{ODP}	1.002	1.002	1.000	1.002	1.291	1.123	1.199
AUC_{SNPR}	1.135	1.354	1.000	1.211	3.986	2.988	11.187

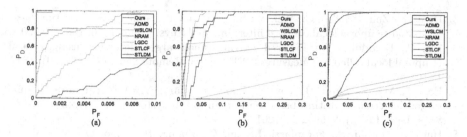

Fig. 7. (a)–(c) are 2D-ROC curves of (P_D, P_F) about Seq. 1–3.

4   Conclusion

During remote target monitoring, strong background and noise signals make the target signal extremely "weak" and "small" and difficult to be detected by traditional computer vision methods. Weak signal detection techniques are considered to detect target signals by deeply mining the features of target motion in the time-domain space of video sequences. In this paper, we introduce single-image time-domain analysis, which can solve the problem of low signal-to-noise ratio in the air domain and complex environmental interference in the air domain in a smooth background, and can solve the detection of both bright and dark targets. However, the default detector in this paper can receive the target signal, but the probability of receiving the photon signal during the target signal observation is uncertain in the case of low signal-to-noise ratio, and the specific situation needs further study and discussion. In addition, the in-depth analysis of the signal time domain performance also brings some limitations to the application of the proposed technique in this paper. The core technique of the proposed algorithm is done on a single image element, which requires the imaging process to be done in a relatively stable context to ensure its effectiveness. In the future, the limitations of the gaze scene can be considered to be compensated by background alignment, and can be combined with other spatial algorithms to repeatedly expand the significance of time-domain uncertainty analysis.

References

1. Chang, C.I.: An effective evaluation tool for hyperspectral target detection: 3d receiver operating characteristic curve analysis. IEEE Trans. Geosci. Remote Sens. **59**(6), 5131–5153 (2020)
2. Chen, C.P., Li, H., Wei, Y., Xia, T., Tang, Y.Y.: A local contrast method for small infrared target detection. IEEE Trans. Geosci. Remote Sens. **52**(1), 574–581 (2013)
3. Datla, R.V., Kessel, R., Smith, A.W., Kacker, R.N., Pollock, D.: Uncertainty analysis of remote sensing optical sensor data: guiding principles to achieve metrological consistency. Int. J. Remote Sens. **31**(4), 867–880 (2010)
4. Deng, H., Sun, X., Liu, M., Ye, C., Zhou, X.: Small infrared target detection based on weighted local difference measure. IEEE Trans. Geosci. Remote Sens. **54**(7), 4204–4214 (2016)

5. Deng, L., Zhu, H., Tao, C., Wei, Y.: Infrared moving point target detection based on spatial-temporal local contrast filter. Infrared Phys. Technol. **76**, 168–173 (2016)
6. Du, P., Hamdulla, A.: Infrared moving small-target detection using spatial-temporal local difference measure. IEEE Geosci. Remote Sens. Lett. **17**(10), 1817–1821 (2019)
7. Han, J., Ma, Y., Zhou, B., Fan, F., Liang, K., Fang, Y.: A robust infrared small target detection algorithm based on human visual system. IEEE Geosci. Remote Sens. Lett. **11**(12), 2168–2172 (2014)
8. Han, J., Moradi, S., Faramarzi, I., Liu, C., Zhang, H., Zhao, Q.: A local contrast method for infrared small-target detection utilizing a tri-layer window. IEEE Geosci. Remote Sens. Lett. **17**(10), 1822–1826 (2019)
9. Han, J., et al.: Infrared small target detection based on the weighted strengthened local contrast measure. IEEE Geosci. Remote Sens. Lett. **18**(9), 1670–1674 (2020)
10. He, L., Ge, L.: Camshift target tracking based on the combination of inter-frame difference and background difference. In: 2018 37th Chinese Control Conference (CCC), pp. 9461–9465. IEEE (2018)
11. Hui, B., et al.: A dataset for infrared detection and tracking of dim-small aircraft targets under ground/air background. China Sci. Data **5**(3), 291–302 (2020)
12. Moradi, S., Moallem, P., Sabahi, M.F.: Fast and robust small infrared target detection using absolute directional mean difference algorithm. Sig. Process. **177**, 107727 (2020)
13. Pan, Z., Liu, S., Fu, W.: A review of visual moving target tracking. Multimed. Tools Appl. **76**, 16989–17018 (2017)
14. Qin, Y., Li, B.: Effective infrared small target detection utilizing a novel local contrast method. IEEE Geosci. Remote Sens. Lett. **13**(12), 1890–1894 (2016)
15. Rawat, S.S., Verma, S.K., Kumar, Y.: Review on recent development in infrared small target detection algorithms. Procedia Comput. Sci. **167**, 2496–2505 (2020)
16. Viallefont-Robinet, F., Léger, D.: Improvement of the edge method for on-orbit mtf measurement. Opt. Express **18**(4), 3531–3545 (2010)
17. Wan, M., Xu, Y., Huang, Q., Qian, W., Gu, G., Chen, Q.: Single frame infrared small target detection based on local gradient and directional curvature. In: Opto-electronic Imaging and Multimedia Technology VIII, vol. 11897, pp. 99–107. SPIE (2021)
18. Zhang, L., Peng, L., Zhang, T., Cao, S., Peng, Z.: Infrared small target detection via non-convex rank approximation minimization joint l 2, 1 norm. Remote Sensing **10**(11), 1821 (2018)
19. Zhao, M., Li, W., Li, L., Hu, J., Ma, P., Tao, R.: Single-frame infrared small-target detection: a survey. IEEE Geosci. Remote Sens. Mag. **10**(2), 87–119 (2022)

Author Index

Q. Liu et al. (Eds.): PRCV 2023, LNCS 14428, pp. 503–504, 2024.
https://doi.org/10.1007/978-981-99-8462-6

Printed in the United States
by Baker & Taylor Publisher Services